Microbial Fuel Cell

Debabrata Das
Editor

Microbial Fuel Cell

A Bioelectrochemical System that Converts Waste to Watts

Springer

Editor
Debabrata Das
Department of Biotechnology
Indian Institute of Technology
Kharagpur, India

Co-published by Springer International Publishing, Cham, Switzerland, with Capital Publishing Company, New Delhi, India.

Sold and distributed in North, Central and South America by Springer, 233 Spring Street, New York 10013, USA.

In all other countries, except SAARC countries—Afghanistan, Bangladesh, Bhutan, India, Maldives, Nepal, Pakistan and Sri Lanka—sold and distributed by Springer, Tiergartenstr. 15, 69121 Heidelberg, Germany.

In SAARC countries—Afghanistan, Bangladesh, Bhutan, India, Maldives, Nepal, Pakistan and Sri Lanka—printed book sold and distributed by Capital Publishing Company, 7/28, Mahaveer Street, Ansari Road, Daryaganj, New Delhi 110 002, India.

ISBN 978-3-319-66792-8 ISBN 978-3-319-66793-5 (eBook)
https://doi.org/10.1007/978-3-319-66793-5

Library of Congress Control Number: 2017954952

Printed on acid-free paper

This Springer imprint is published by Springer Nature
The registered company is Springer International Publishing AG
The registered company address is: Gewerbestrasse 11, 6330 Cham, Switzerland

My beloved wife – Ms Sumana Das

Foreword

The increasing consumption of non-renewable fossil fuels has led to the global energy challenge with respect to climate change and local environment pollutions. The world is searching for alternative renewable sources of energy. Microbial fuel cell mechanism is an attractive renewable technology that can convert wastes into a directly usable form of energy, i.e. electricity. MFCs are unique in their ability to use bacteria as catalysts for the conversion of chemical energy to electrical energy without the need for combustion thereby making it a green technology.

In brief, microbial fuel cells comprise an anode and a cathode which may or may not be separated by a semipermeable membrane. The bacteria (known as exoelectrogens) are usually grown at the anode in an anaerobic environment such that they metabolize the substrate (waste stream) and donate the released electrons to the anode (which acts as a terminal electron acceptor). The fate of these electrons depends on the product of interest and type of oxidant present at the cathode. Usually, oxygen is the most commonly used oxidant for fuel cell applications due to its high electronegativity. MFCs as an area of research have observed a plethora of interest in the past 15 years, and significant advances have been made in the architectural design, electron transfer mechanisms, electrochemical reaction kinetics, membrane properties, etc. that have led to the development of a few pilot-scale MFCs. While the energy produced from MFCs is not yet practical for on-site applications, a thorough study is helpful to have understanding of bioenergy. It is also important to have knowledge of energy generation systems for applications.

Current research has extended the scope of MFC-based technologies to a wide variety of applications such as bioremediation (sediment MFC), desalination (microbial desalination cell), CO_2 sequestration (microbial carbon capture cells), hydrogen production (microbial electrolysis cell), synthesis of chemicals (microbial electrosynthesis), biosensors (microfluidic microbial fuel cell), etc. Though the basic underlying principles for all these applications remain the same, the reaction kinetics and the electron transfer mechanisms vary with respect to the final product to such an extent that a deeper understanding is required thus motivating studies relating to individual processes and applications.

The book provides a detailed description of the basic principles of MFC technology, the developments made till date, their applications and the associated challenges and the possible integration with the present energy generation systems to maximize energy recovery. Each aspect of MFC has been dealt with an interdisciplinary approach with easy to understand illustrations keeping in mind the diverse audience. I strongly recommend this book to energy researchers, scientists in industries, engineers and student enthusiasts who are interested in microbial fuel cell research and want to get abreast with the latest developments. This book serves as an ideal platform to discuss the wider scope of MFC-based applications beyond the laboratory conditions and thus shows promising aspects towards a green and sustainable future.

I appreciate the dedicated and sustained efforts of Prof. Debabrata Das who has compiled the book for interested students, researchers, industries and even policymakers.

National Institute of Advanced Studies
Bengaluru, India

Baldev Raj

Preface

Once we accept our limits, we go beyond them. (Albert Einstein)

Wastewater management and alternative energy are considered the most important fields of research in environmental biotechnology. The microbial fuel cell (MFC) is a sustainable process which is gaining popularity in the realm of increasing demand of energy sources and concern for global warming due to emission of greenhouse gases. The MFC utilizes waste to extract renewable biofuels and byproducts. This process can prove to be a boon to developed and developing countries. The widespread application of the process can help these countries to eradicate the major problems of remote power generation, waste management, sanitation, health, biofertilizer production and renewable energy. One of the major products of MFC is electricity, and bioelectricity is the greenest of all the biofuels. Thus, it can be reclassified as "super low-carbon fuel" from "low-carbon fuels". Bioelectricity produced in MFC can be utilized for decentralized power generation, and additional revenue can be obtained in the form of renewable energy credit, carbon dioxide credits and/or other greenhouse gas emissions credits.

The MFC is considered as a renewable natural power source derived from organic wastes and sewage treatment. Bioelectricity is usually produced through MFC in oxygen-deficient environment where a series of microorganisms convert the complex wastes to electrons via liquefaction through a cascade of enzymes in a bioelectrochemical process. Detailed description of MFC technology and their applications have been discussed in this book. Basic principles of MFCs would help to perceive new aspects of bioenergy conversions and how such system could be integrated with the present energy generation systems to maximize energy recovery. In brief, MFCs work by utilizing bacteria to drive oxidation of substrate (such as organic wastes from agriculture, industries or sewage) at anode and reduction of an oxidant (usually O_2) at cathode. The theories underlying the electron transfer mechanisms, the biochemistry and the microbiology involved and the material characteristics of anode, cathode and the separator have been described in detail. The physico-chemical parameters affecting the performance of MFC have been included. Technological advancements based on MFC designs as well as process economy are also discussed.

MFC technology represents a multidisciplinary approach to the quest for alternative sources of energy. It symbolizes the confluence of the chemical, physical and life sciences and is a meeting point of basic and applied research. The working principle of MFCs is based on microbial physiology coupled with electrochemistry. Several diagnostic tools that are used to assess the performance of MFCs have been discussed in detail ranging from electrochemical workstation to molecular biological tools. Mathematical modelling of the process has been discussed which would help in scaling up the MFC. A multidisciplinary approach such as electrochemistry, materials science, microbiology, engineering, etc. has been considered for the improvement of MFC.

MFC-based technology has a dual purpose: power generation and wastewater treatment. Several applications of MFC have been discussed such as bioremediation (sediment MFC), desalination (microbial desalination cell), CO_2 sequestration (microbial carbon capture cells), hydrogen production (microbial electrolysis cell), synthesis of chemicals (microbial electrosynthesis), biosensors (microfluidic MFC), etc.

This book is aimed at a wide audience, mainly undergraduates, postgraduates, energy researchers, scientists in industries and research organizations, energy specialists, policymakers and others interested in the latest developments concerning MFCs. Each chapter in the book begins with the fundamental explanation for general readers and ends with in-depth scientific details suitable for expert readers. One of the chapters highlights the concise summary of biochemistry for bioelectricity production as well as important major factors involved in the process towards the realization of a stable bioelectricity-based economy. A successful application of MFC was found in a pilot-scale wastewater treatment along with renewable energy production. The proper configuration of MFC and efficiently pretreated "feedstock" are the major factors to maximize the production of bioelectricity. The fundamentals of reactor design based on process economy have been discussed.

The contributions of all the authors in the book are gratefully acknowledged. Valuable suggestions for the book given by Dr. Soumya Pandit, Postdoctoral Research Fellow, Ben-Gurion University of the Negev, Israel, and Ms. Jhansi L. Varanasi, Senior Research Fellow, Indian Institute of Technology Kharagpur, are worth mentioning.

Kharagpur, India Debabrata Das

Contents

Acronyms

e^-	Electron
ΔE°	Standard electrode potential
ΔG°	Gibbs free energy
H^+	Proton
2D	Two dimensional
3D	Three dimensional
A	Ampere
Ag	Silver
AgCl	Silver chloride
AgNPs	Silver nanoparticles
AC	Alternate current
AD	Anaerobic digester
ADP	Adenosine diphosphate
AFMBR	Anaerobic fluidized bed membrane bioreactor
AEM	Anion exchange membrane
AFM	Atomic force microscopy
AL	Activation losses
APHA	American Public Health Association
ARB	Anode respiring bacteria
ARDRA	Amplified ribosomal DNA restriction analysis
AQDS	Anthraquinone 2-6-disulphonate
ATP	Adenosine triphosphate
BEAMR	Bioelectrochemically assisted microbial reactor
BES	Bioelectrochemical system
BET	Bioelectrochemical treatment
BFC	Biological fuel cell
BOD	Biochemical oxygen demand
BPM	Bipolar membrane
BPV	Biophotovoltaic system
ClO^{4-}	Perchlorate

CO_2	Carbon dioxide
CoTMPP	Cobalt tetramethoxyphenylporphyrin
$Cr_2O_7^{2-}$	Dichromate
CAPEX	Capital expense
CE	Columbic efficiency
CEA	Cathode electrode assembly
CEM	Cation exchange membranes
CFCs	Chemical fuel cells
COD	Chemical oxygen demand
CNT	Carbon nanotubes
CP	Concentration polarization
CR	Cathode resistance
CSLM	Confocal scanning laser microscopy
CTC	*c*-type cytochromes
CV	Cyclic voltammetry
CW-MFC	Constructed wetland microbial fuel cell
DC	Direct current
DET	Direct electron transfer
DGGE	Denaturing gradient gel electrophoresis
DMRB	Dissimilatory metal reducing bacteria
DNA	Deoxyribonucleic acid
DO	Dissolved oxygen
EAB	Electrochemically active bacteria
EC	Electrochemical cell
ECP	Electronically conducting polymers
EDS	Energy dispersive X-ray spectroscopy
EE	Energy efficiency
EES	Ecological engineered system
EET	Extracellular electron transfer
EFC	Enzymatic fuel cells
EIS	Electrochemical impedance spectroscopy
ELTs	Emergency locator transmitters
EO	Electrooxidation
EPS	Exo-polysaccharide
ETC	Electron transport chain
$Fe(CN)_6^{3-}$	Ferricyanide
FAME	Fatty acid methyl ester
FC	Fuel cell
FISH	Fluorescence in situ hybridization
FNB	Fluoronitrobenzene
FO	Forward osmosis
FTIR	Fourier transform infrared spectroscopy
GAC	Granular activate carbon
GC	Gas chromatography

GO	Graphene oxide
GR	Graphene
H_2	Hydrogen
H_2O_2	Hydrogen peroxide
HER	Hydrogen evolution reaction
HPLC	High-pressure liquid chromatography
HRT	Hydraulic retention time
IABF	Intermittently aerated biological filter
IEM	Ion exchange membrane
LC	Liquid chromatography
LCA	Life cycle assessment
LCC	Layered corrugated carbon
LED	Light-emitting diode
LHC	Light-harvesting complex
LSV	Linear sweep voltammetry
MnO^{4-}	Permanganate
MPc	Metal phthalocyanine
M+MMFC	Membraned microfluidic microbial fuel cell
MB	Methylene blue
MBR	Membrane bioreactor
MCA	Membrane cathode assembly
MCD	Maximum current density
MDC	Microbial desalination cell
MEC	Microbial electrolysis cell
MES	Microbial electrosynthesis
MET	Mediated electron transfer
MFC	Microbial fuel cell
MMFC	Microfluidic microbial fuel cell
M-MMFC	Membraneless microfluidic microbial fuel cell
MRECC	Microbial reverse-electrodialysis electrolysis
MPD	Maximum power density
MS	Mass spectrometry
MSC	Microbial solar cell
MPN	Most probable number
NO^{3-}	Nitrate
NA	Not available
NADH	Nicotinamide adenine dinucleotide
NADPH	Nicotinamide adenine dinucleotide phosphate
NER	Normalized energy recovery
NHE	Normal hydrogen electrode
NMRI	Nuclear magnetic resonance imaging
NQ	Naphthoquinone
NR	Neutral red
O_2	Oxygen

OCP	Open circuit potential
OCV	Open circuit voltage
OL	Ohmic losses
OLR	Organic loading rate
OM	Outer membrane
OPEX	Operational expense
ORR	Oxygen reduction reaction
P_i	Inorganic phosphate
Pt	Platinum
PANI	Polyaniline
PAHs	Polycyclic aromatic hydrocarbons
PBI	Polybenzimidazole
PBR	Photobioreactor
PD	Power density
PDMS	Polydimethylsiloxane
PEC	Photoelectrochemical cell
PEM	Proton exchange membrane
PhFC	Photosynthetic microbial fuel cell
PMFC	Plant microbial fuel cell
PMS	Power management system
PPy	Polypyrrole
PQ	Plastoquinone
PTFE	Polytetrafluoroethylene
PVA	Polyvinyl alcohol
q-PCR	Quantitative polymerase chain reaction
R_{int}	Internal resistance
RRDE	Rotating ring disk electrode
RDE	Rotating disk electrode
RNA	Ribonucleic acid
RO	Reverse osmosis
RVC	Reticulated vitreous carbon
$S_2O_8^{2-}$	Peroxydisulphate
SEM	Scanning electron microscope
SERS	Surface-enhanced Raman scattering
SHE	Standard hydrogen electrode
SIP	Stable isotope probing
SMFC	Sediment microbial fuel cell
SPEEK	Sulphonated poly(ether ether ketone)
SRT	Solid retention time
SS	Stainless steel
SSCP	Single-strand conformation polymorphism
STA	Silicotungstic acid
T-MEC	Two-chamber microbial electrolysis cell
T-MFC	Two-chamber microbial fuel cell

T-RFLP	Terminal restriction fragment length polymorphism
TCE	Trichloroethylene
TDS	Total dissolved solids
TEA	Terminal electron acceptor
TEM	Transmission electron microscopy
TGGE	Temperature gradient gel electrophoresis
TN	Total nitrogen
TOC	Total organic carbon
TPH	Total petroleum hydrocarbons
UASB	Upflow anaerobic sludge blanket
US-EPA	United States Environmental Protection Agency
V	Volt
VFA	Volatile fatty acid
W	Watt
WHO	World Health Organization
WWTP	Wastewater treatment plant
XRD	X-ray diffraction
ZVI	Zero valent iron

Chapter 1
Introduction

Debabrata Das

1.1 Background

Security for water and energy sources is gaining importance throughout the world. Increasing population and climate changes pose serious challenges that involve energy, water resources, land use and waste treatment issues. Throughout the world there is intense interest in evaluating and implementing alternative energy sources (Schröder 2008). Lots of research is going in quest of renewable energy sources (Chandrasekhar et al. 2015). The hydropower, biomass, wind, geothermal and solar radiation are among major sources for renewable energy generation. In recent years, Microbial Fuel Cell (MFC) technology has been emerging as one of the popular wastewater treatment-based technology to provide clean water and green energy (Pant et al. 2012). MFCs are bio-electrochemical devices where organic wastes degrade to smaller molecules, releasing electrons and protons, thereby generating electricity. MFCs can directly convert chemical energy into electrical energy through bioelectrochemical reactions utilizing microorganism or enzymatic catalysis. MFCs have several advantages as compared to the traditional fuel cells and enzymatic fuel cells. It is possible to utilize a wide range of organic or inorganic matter such as organic wastes, soil sediments as a source of fuel generation. High conversion efficiency can be achieved with such devices due to the direct or a single step conversion of substrate energy to electricity. Unlike a conventional fuel cell, MFCs can run at ambient temperature and atmospheric pressure (Du et al. 2007). In addition it can be useful for widespread application in locations lacking electrical

D. Das (✉)
Department of Biotechnology, Indian Institute of Technology, Kharagpur, West Bengal 721302, India
e-mail: ddas.iitkgp@gmail.com

D. Das (ed.), *Microbial Fuel Cell*,
https://doi.org/10.1007/978-3-319-66793-5_1

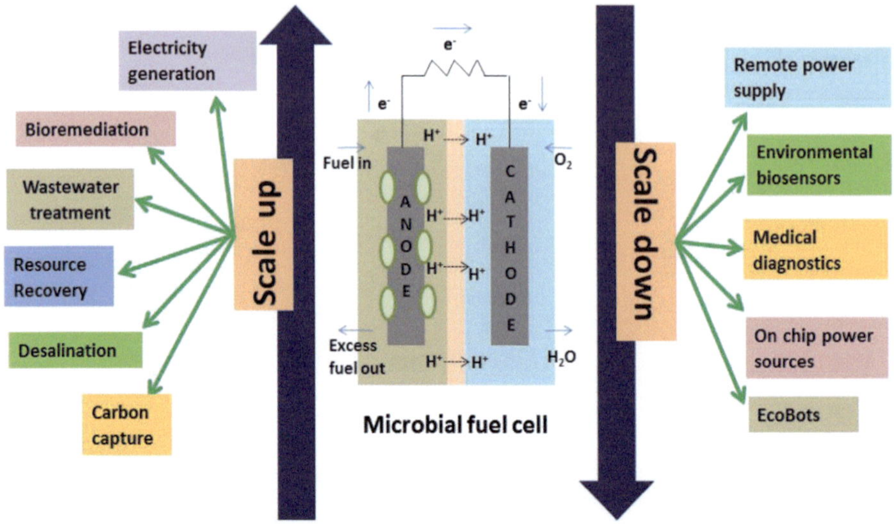

Fig. 1.1 An overview of different applications of MFC-based technologies

facilities (Stams et al. 2006). MFCs have outperformed other technologies like anaerobic digester, aerated lagoon etc. (Logan 2008) with a wide array of applications as depicted in Fig. 1.1. The objective of this chapter is to give an overview of the book.

1.2 Basic Principles of Microbial Fuel Cell (MFC)

MFC is a bioelectrochemical tool comprised of an anaerobic anode chamber and an aerobic cathode chamber physically separated by anion exchange membrane (AEM). In a typical MFC, microbes are utilized for oxidation of substrate in anode chamber; subsequently the electron released from the microorganism goes to cathode via external wire. The anode chamber consists of microorganism (catalyst) and an electrode (anode) and it can be fed with growth media or wastewater named as anolyte and redox mediator (not required in case of mediator-less MFC). The necessary protons and electrons extracted during bacterial substrate catabolism combine with oxygen to form water on cathode (Fig. 1.2). Usually, electrons flow to the cathode via a conductive material having an external resistance. The protons which migrate through membrane are reduced by accepting these electrons. Terminal electron acceptors (e.g. O_2 to water) at the cathode is similar to chemical fuel cell (Venkata Mohan et al. 2014) (Fig. 1.2). Chapter 2 deals with an overview of power generation in MFC including bioelectrochemical reactions, and major

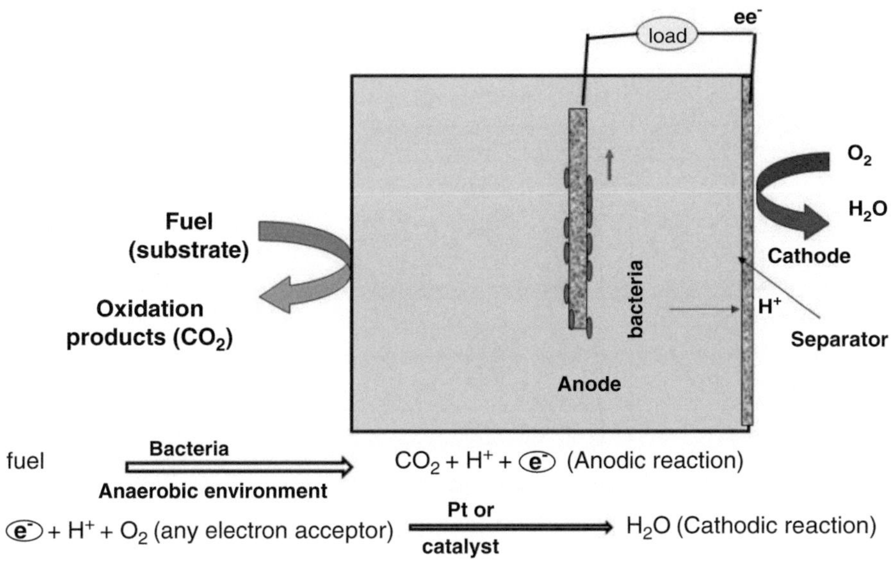

Fig. 1.2 Schematics of principle of a single chamber microbial fuel cell

considerations in an MFC system like benefits, limitations and calculations of the energy generation from various feedstocks.

The microbes in MFC play very important role. On the basis of their electron transfer mechanisms, they are basically classified into two types viz. indirect and direct (Fig. 1.3). In indirect MFCs, microbes require some form of mediator for electron transfer to anode which can be obtained from outside (e.g. *E. coli*, *Bacillus* sp. etc.) or can be produced endogenously (e.g. *Pseudomonas, Lactobacillus* etc.). On the other hand, in direct MFCs bacteria known as electroactive bacteria (e.g. *Shewanella putrefaciences* and *Geobacter sulferreducens*) donate electrons directly to anode. They are also known as anodophiles or exoelectrogens and are able to form conductive biofilm on anode. Apart from these, mixed cultures can also be used as biocatalyst in anode chamber of MFC that are dominated by different phylogenetic group such as *Firmicutes*, *Acidobacteria*, *Proteobacteria* etc. Under suitable acclimatization conditions, the MFC system is expected to select and enrich its own electrochemically active consortia from the mixed culture in course of time. Depending on the source of the inoculum, operational conditions, substrate availability and MFC system architecture, the microbial communities differ from each other. Many electroactive bacteria are also commonly present in syntrophy with fermentative organisms and methanogens that divert the degradation pattern of organic matter present in the wastewater.

Due to their utmost importance in MFCs, Chaps. 3, 4 and 5 have been dealt exclusively with the principle characteristics of electroactive bacteria including their microbiology, ecology, the effect of environmental factors (such as temperature, pH, nutrients etc.) on their growth, and their interaction with the electrodes in details.

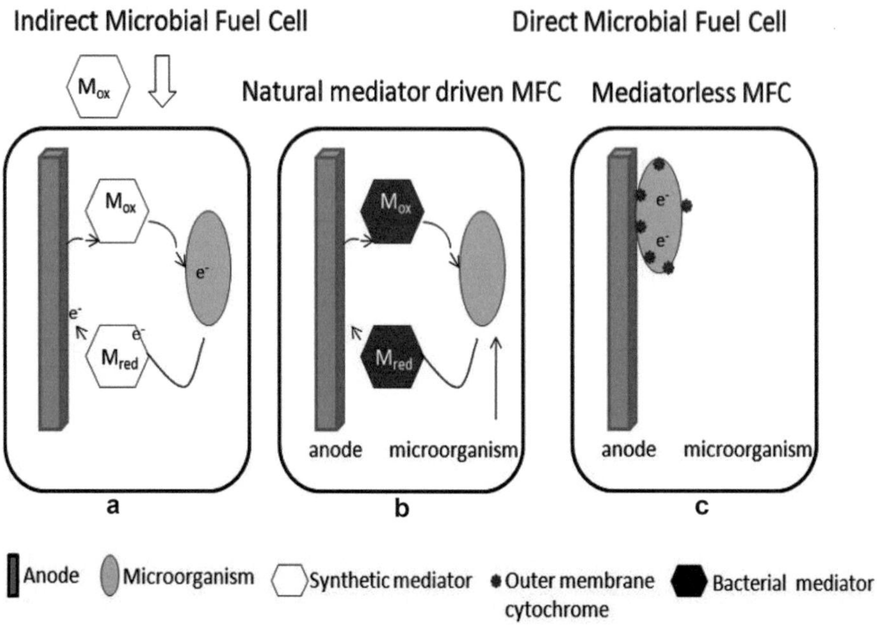

Fig. 1.3 Different mechanisms of electron transfer in MFC: (**a**) Indirect MFC; (**b**) Mediator driven MFC; and (**c**) Mediatorless MFC

1.3 Components of MFC

Like chemical fuel cell, MFC is composed of basic components such as the anode, the cathode, the separator and the external circuits (Fig. 1.2). Depending upon the application, different configurations of MFCs have been developed such as single chamber or double chambered (Fig. 1.4). However the basic principles for these configurations remain the same. Over the past few decades several advances have been made to improve the individual components of MFCs to increase the overall performance of MFCs. Table 1.1 shows some recent developments made on improving power generation in MFCs.

1.3.1 Anode Materials

Anode is an important component in MFC which allows EABs to form electroactive biofilm. Therefore, quality of anode material should have different attributes like superior conductivity and good biocompatibility. The other qualities

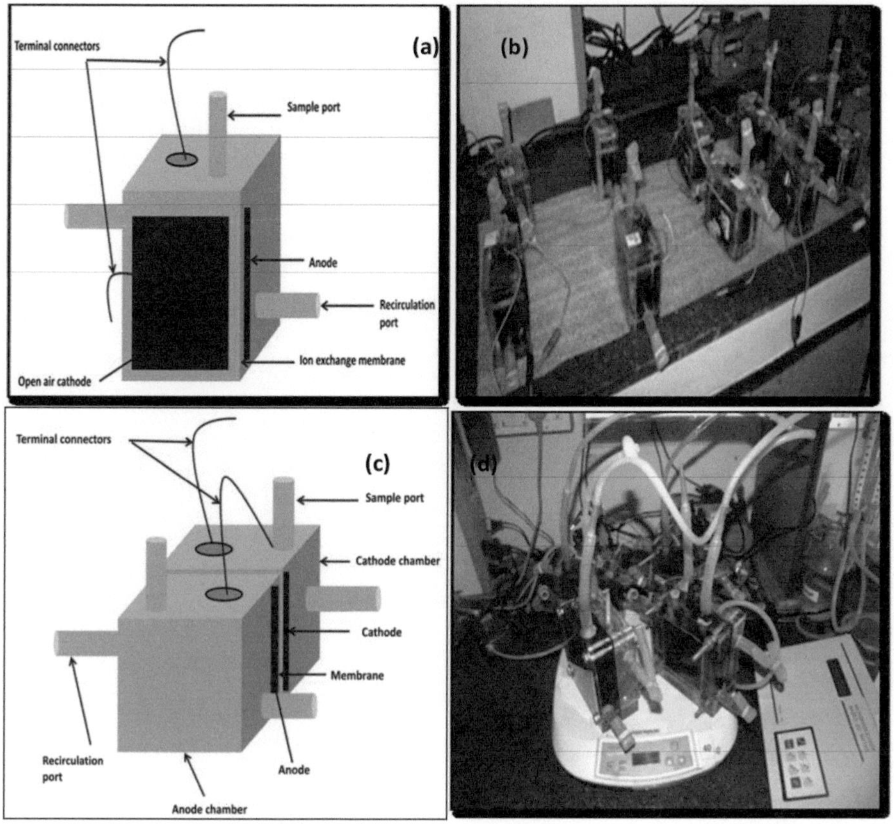

Fig. 1.4 (**a**) Schematic diagram of a single chambered MFC; (**b**) Experimental setup of single chambered MFC; (**c**) Schematic diagram of a double chambered MFC; and (**d**) Experimental setup of double chambered MFC

like chemical stability, resistance to corrosion, mechanical strength and toughness are also important parameters to choose anode materials. Anode should provide large specific surface area for adhesion and biofilm formation. Carbon materials are predominantly applied as anode materials which include graphite rod and fibre brush, carbon felt, reticulated vitreous carbon (RVC), carbon cloth etc. (Qiao et al. 2010). Chapter 6 deals extensively with the recent progress towards anode materials used for MFC applications.

1.3.2 Types of Separators/Membranes

Separator plays an important role in MFC. It physically divides anode chamber and cathode but ionically and electronically conjugated. In a dual chambered MFC, separator facilitates in developing both anodic and cathodic half-cell potential by

Table 1.1 Recent progress on power generation in MFCs

MFC configuration	Inoculum	e^- donor	e^- acceptor	Anode material	Cathode material	Membrane	Power density ($mW\ m^{-2}$)	Columbic efficiency (%)	References
Dual chambered	*Geobacter sulfurreducens*	Acetate	Oxygen	Graphite stick	Graphite stick	Nafion- 117	n.a.	9	Bond and Lovley (2003)
Single chambered	Mixed culture (Anaerobic sludge)	Petroleum sludge	Oxygen	Graphite plates	Graphite plates	Nafion-117	53.11	n.a.[a]	Chandrasekhar and Venkata Mohan (2012)
Dual chambered	Mixed culture (activated sludge)	Acetate	Oxygen	Toray carbon paper (20% Teflon)	Toray carbon paper (10% Teflon)/Pt coated	Sterion membrane	1.01	n.a.[a]	Gonzalez del Campo et al. (2013)
Single chambered	Mixed culture (domestic wastewater)	Acetate	Oxygen	Carbon cloth	Carbon cloth/Pt	Non-woven fabric	1200	35	Abourached et al. (2014)
Single chambered	Mixed culture (MFC effluent)	Fermented primary effluent	Oxygen	Graphite fiber brushes	Carbon cloth/Pt	PTFE[a]	320	57	Yang et al. (2013)
Dual chambered	Mixed culture (Anaerobic sludge)	Acetate	Oxygen	Graphite rod	Graphite rod	Nafion-117	13.59	29.5	Oh et al. (2014)
Single chambered	Mixed culture (Anaerobic sludge)	Acetate	Oxygen	Carbon cloth	Carbon cloth/Pt	PVA – STA[a]	58.8	2.94	Khilari et al. (2013)
Single chambered	*Shewanella putrefaciens*	Lactate	Oxygen	Carbon cloth	$MnCo_2O_4$-NRs[a]	PVA – PDDA[a]	n.a[a]	7.5	Pandit et al. (2014)

[a]*n.a* not available, *PTFE* Polytetrafluoroethylene, *PVA* Polyvinyl alcohol, *STA* Silicotungstic acid, *PDDA* Polydiallyldimethylammonium chloride

splitting anolyte at the anode chamber to cathode. The electrons producing through bacterial metabolic activity pass via an external circuit while the proton migrates via a separator, separating the anode chamber from the one in which the cathode is immersed (Leong et al. 2013). Use of separator has several advantages in MFC like transport of anolyte or substrate from anode to cathode.

Though separators have proven to be beneficial, they usually contaminate the cathode and generate mixed potential, consequently reducing MFC performance. The single chambered MFC without a separator has been recently reported to have increased current density. The absence of a separator results in increasing oxygen and substrate diffusivity and consequently reduces the CE. Chapter 7 briefs the benefits as well as challenges associated with the use of separator in MFCs.

1.3.3 Cathode Materials

Cathode is a place where oxygen reduction reaction (ORR) occurs. In cathode, air spurges to increase dissolved oxygen concentration which uptake electrons and proton generation during substrate metabolism in anode chamber. The performance of cathode is critical to high current generation in MFCs. The graphite, carbon felt, carbon cloth and carbon paper have been chosen by researchers as common cathode materials. Power generation is limited by the cathode due to sluggish reduction kinetics which limits these non-catalyzed materials for the commercial use as cathode (Rismani-Yazdi et al. 2008). Large over-potential develops (η_{act}) due to non-catalytic activity of common cathode materials which reduces the current generation (Rabaey and Keller 2008). The materials used for cathode and the catalyst used for ORR are discussed in details in Chap. 8. Basis of choosing materials, their synthesis and fabrication are highlighted in brief. Performances of these materials for the improvement of power generation have been analyzed critically.

Apart from biotic cathodes, recent advances in MFC research has focused on the development of biotic cathodes (biocathode) which have opened up a wide array of new and emerging applications for MFCs. Biocathodes can either be aerobic, anoxic or anaerobic in nature. Depending upon the nature and composition of biocathode, they can produce electricity or H^+ or methane in MECs with the help of small additional energy. Chapter 9 discusses the recent developments made in biocathode application in MFCs. Their mode of electron transfer to or from the electrodes and their composition and characteristics have been briefly described.

1.4 MFC Architecture

The reactor configuration of the MFC affects significantly the performance of MFC by controlling the internal resistance. Selection of a proper reactor configuration is of prime importance in MFC for anaerobic processes. The relatively low biosynthesis rate of methanogens in an anaerobic system demands special consideration for reactor design. The selection of reactor is based on the requirement of a high solid retention time/hydraulic retention time (SRT/HRT) ratio, mixing and shear control so as to prevent the washout of slow-growing exoelectrogens. Chapter 10 deals with different strategies used for the reactor design of bioelectrochemical systems.

For designing of scale down MFCs, a microfluidics approach is usually employed to achieve multiplexing, automation, and high-throughput screening. Apart from representing a miniature form of macro-scale MFCs, the micro-sized MFCs aid in on-chip powering and rapid screening of optimal operating conditions. The micro-sized MFCs possess a large surface area-to-volume ratio, short inter-electrode separation (hence low internal resistance), low response time and low Reynolds number. Their fabrication and assembly is precise and less expensive if done using micro-fabrication processes. They exhibit better performance than large-scale MFCs in terms of volumetric current and power densities. The laminar flow model to reduce internal resistance is also applicable to micro-sized MFCs only. The performance of a μMFC is limited by factors like heterophase catalytic reactions, transport/diffusion of reactants (substrate, electrons and electron-accepting species) from the bulk electrolyte to the electrode surfaces and the reaction kinetics on these surfaces. Details of microfluidics in MFC has been discussed in Chap. 11.

1.5 MFC Performance Indicators

Operating conditions in anode chamber-anaerobic wastewater treatment involves several groups of bacteria, having their own optimum working conditions. MFC operation in optimized anodic operating condition will facilitate proper growth of electrogenic biofilm during start up and enhance the subsequent performance in long term operation. Several process parameters such as pH, alkalinity, substrate type, substrate concentration and, organic loading rate (OLR) affect the activity of these anaerobic microorganisms (Gil et al. 2003). The anodic conditions affecting the performance of MFC include the inoculum, type of substrate, influent COD concentration, feed pH etc. Chapter 12 discusses in details the physicochemical parameters governing the performance of MFC.

MFC performance and efficiency is generally measured in terms of energetic parameters (volumetric power density, current, potential difference, cell internal resistance) and biological treatment efficiency (COD removal) (Logan 2012). Coulombic efficiency (CE) describes the efficiency with which charge (electrons) is transferred in a system facilitating an electrochemical reaction using organic substrate. The assessment of MFCs performance is usually measured through various techniques such as: polarization studies that analyze and characterize quality of fuel cells in terms of power generation; the current interruption method that is commonly used for evaluation of total internal resistance of MFCs;

electrochemical impedance spectroscopy (EIS) analysis that helps in differentiating the different components of internal resistance i.e. solution resistance, charge transfer resistances, and diffusion resistance in electrochemical systems; and cyclic voltammetry (CV) which is used to investigate the mechanisms of oxidation or reduction reactions on the electrode surface. Different types of diagnostic tools for the assessment of MFC have been included in Chap. 13. These include primarily electrochemical tools along with case studies with possible interpretation.

1.6 Modelling of Reaction and Transport Processes in MFCs

The electrochemically active microorganisms in MFC convert organic material to carbon dioxide, protons and electrons, and subsequently transfer the produced electrons to the anode surface. The conversion of the organic material into carbon dioxide, protons and electrons occurs inside the microorganism and the conversion rate is determined by enzyme kinetics. The transfer of electrons from the microorganism to the electrode occurs at the interface between the microorganisms and the electrode surface, and is described by various models such as Nernst-Monod model and Butler Volmer model. The Nernst-Monod model specifically aims at modelling the bioanode performance with respect to the effect of substrate and anode potential. The Nernst-Monod model is a modified version of the Monod model by considering the electrode as the final electron acceptor. Butler Volmer Monod model comprises kinetics of the electron transfer. Numerical modelling of an MFC system is a valuable tool for investigating system parameters with reduced time and money, as models can be easily modified to simulate various configurations and operating conditions. The details of modelling approach are highlighted in Chap. 14.

1.7 Applications of MFCs

Over the past few decades, the capacity of MFCs vary from μL to few litres which have been studied to evaluate their performance (Logan 2010; Schröder 2007). The advancement in MFC research domain increased power output from nW to kW per m^3 with simultaneously achieving effective wastewater treatment and other resources recovery. Though an appreciable improvement has been observed in the power outputs of MFCs, the practical utilization of these technologies has not been achieved so far due to certain inevitable limitations. With the improvement of design, cost effectiveness and performance efficiency of these systems, several progress have been made over the past few decades. Apart from electricity generation, MFCs have also been employed for various other applications such as

Table 1.2 Recent developments made in MFC based technologies

Type of MFC	e^- donor	e^- acceptor	Main products	References
Single chambered flat plate MFC	Acetate	Oxygen	Electricity	Varanasi et al. (2016)
Microbial carbon capture cell	LB media	CO_2	Algal biomass, CO_2 sequestration, Electricity	Pandit et al. (2012)
Biocathode MFC	Lactate	VFA rich fermentation effluent	Electricity	Sharma et al. (2013b)
Air cathode MFC	Molasses mixed sewage wastewater	Oxygen	Wastewater treatment, Electricity	Sevda et al. (2013)
Sediment MFC	Polyaromatic hydrocarbon	Nitrate/sulphate	Bioremediation, Electricity	Sherafatmand and Ng (2015)
Microbial electrolysis cell	Glycerol, milk and starch	Proton	Hydrogen	Montpart et al. (2014)
Microbial electrolysis cell	Sodium acetate	–	Methane	Ding et al. (2016)
Potentiostat poised half-cell	Electrode	Acetate and butyrate	Alcohols, acetone and caproate	Sharma et al. (2013a)
Potentiostat poised H-cell	Electrode	CO_2	Acetate, oxobutyrate and formate	Nevin et al. (2011)
Microbial desalination cell	Acetate	Potassium ferricyanide	Desalination, Electricity	Zhang et al. (2012)

bioremediation, resource recovery, biosensors etc. The present research is thus focused on targeted application to yield maximum efficiency of the system. Table 1.2 enlists the recent developments made in the different applications of MFCs.

1.7.1 Bioremediation and Wastewater Treatment

MFCs cannot be at par with conventional fuel cell as far as current generation is concerned. Therefore, efforts have been made to project MFCs as useful wastewater treatment devices. MFCs are bioelectrical devices that use microorganisms to oxidize the organic matter present in wastewaters from different sources (domestic, industrial, agricultural) generating electrical power that can be harvested for numerous purposes (Chandrasekhar and Venkata Mohan 2012). Sustainability of this process augurs in using existing wastewater treatment facility which would result in reduction in energy demands and operational costs (Chandrasekhar et al. 2015).

In addition, an MFC has great potential for treating recalcitrant wastewater with the help of microorganisms as biocatalysts. A large number of recalcitrant wastes which include dyes, pesticides, polyalcohol, and heterocyclic compounds are produced by industry. Currently, the discharge of wastewaters containing these pollutants is an important environmental hazard, owing to their mutagenicity, recalcitrance and tendency to accumulate in the environment. Also, their discharge in surface water leads to aesthetic problems and obstructs the light penetration and oxygen transfer into water, hence affecting aquatic life (Umbuzeiro et al. 2005). The Chap. 15 provides comprehensive insights on various aspects of MFC as a tool for efficient recalcitrant waste treatment.

1.7.2 Removal and Recovery of Heavy Metals

Heavy metals play an important role in the environmental pollution problem in several industrial, medical and domestic wastewater. Effluents from many industries, heavy metals pose a serious problem to the environment and public health due to their toxicity, bioaccumulation and non-biodegradability. Conventional physical, chemical and biological methodologies to treat wastewater containing heavy metals are energy-intensive and become ineffective if metals concentrations are below 1–100 mg L^{-1}. Currently, MFC technology is being investigated extensively for its state-of-the-art features and environmental welfare. Interestingly, treatment of heavy metal-containing wastewater can be attempted in both anode and cathode chambers of microbial fuel cells. Chapter 16 deals with the treatment of metal-containing effluents using MFCs.

1.7.3 Constructed Wasteland Management

The sediment MFC (SMFC) or marine MFC or benthic MFC is a special MFC which operates on ocean floor and/or river sediment. SMFCs are used as renewable power sources for sensors in fresh and ocean waters. It is an electrochemical device which generates electrical current from the natural redox gradients that commonly occur across the sediment-water interface. This interface separates the two half-cells of the SMFC. There is no need for a proton-permeable membrane because protons are exchanged over the oxic-anoxic interface at the water-sediment boundary. A key difference between MFCs with wastewater and SMFCs is the mass transport losses. Mass transport is often the main limiting factor in SMFCs due to the relatively low concentration of electron donors and the slow rate of mass transport in sediments. SMFCs are very promising power sources for low power marine sensors, water-based sensors, telemetry devices including acoustic receivers/transmitters, meteorological buoy and other oceanographic and environmental instruments. It can be helpful to operate pressure gauges deployed in the ocean bottoms to predict Tsunamis in oceans. A thorough discussion on SMFC has been provided in Chap. 17.

1.7.4 Water Desalination

Microbial desalination cells (MDCs) are a new, energy-sustainable method for using organic matter in wastewater as the energy source for desalination. In MDC, microbial electron harvesting ability is used to desalinate saline water. EABs on anode created a negative electric potential gradient which allows cation and anion from a concentrated saline chamber to migrate through ion-exchange membranes (IEMs) to balance the electro-neutrality. The electric potential gradient created by exoelectrogenic bacteria desalinates water by driving ion transport through a series of ion-exchange membranes (IEMs). The extent of desalination is affected by water transport through IEMs by both osmosis and electro osmosis. Membrane performance affects both anode and cathode activity. Thus water can be desalinated in MDC without any power consumption. Kim et al. demonstrated stackable MDCs with parallel desalination chambers and concentrated chambers which are separated by compartmental AEMs and CEMs (Kim and Logan 2013). The major factors which influence the performance of a MDC are membrane type, ion exchange ability, binder content for MCA fabrication, fouling on membrane/ cathode etc. MDC badly suffers from the membrane fouling during long time operation. It results in decreasing current density of the MFC due to the reducing ion transfer capabilities of the membrane. A detail description of operation of different MDCs has been documented in Chap. 18.

1.7.5 Biophotovoltaics

In a plant microbial fuel cells (PMFC), MFCs are associated with living higher plants (Strik et al. 2011). In PMFCs, electrochemically active bacteria on anode utilize excreted Rhizo deposits from plant roots to directly generate electricity. Therefore, it is possible to convert solar energy to electrical energy directly in PMFC. Rhizo deposits are rich in carbon sources like arabinose, ribose, fructose etc. Rhizo deposits consist of a different array of substances that generate from pilled off root cells, and soluble lysates and exudates. These compounds can be degraded by a wide array of EABs. Photosynthetic bacteria (e.g. *Synechocystis* pcc 6803) are able to donate electron directly to anode in PMFC.

Microbial carbon capture cells (MCC) are different type of PMFCs where microalgae can be used in cathode of a conventional dual chambered MFC to sequester atmospheric CO_2 and provide oxygen to cathode via biophotolysis. Wang et al. demonstrate an MCC where CO_2 from the anode chamber is redissolved in catholyte where a photosynthetic microorganism (*Chlorella vulgaris*) utilize the CO_2 in presence of light (Pandit et al. 2012). Microalgae biomass from the cathode chamber can be utilized for downstream processing of different value added products like lipid extraction or ethanol production. Chapter 19 deals with conversion of light energy to bioelectricity through photosynthetic MFC and associated technology.

1.7.6 Biosensors

BOD is a measure of biodegradable organic content and gives an indication on how much oxygen would be required for microbial degradation. High BOD of a sample indicates the extent of biodegradable substance present in it. BOD measures the oxygen uptake by bacteria in a water sample at a temperature of 20 °C over a period of 5 days in the dark. The sample is diluted with oxygen saturated de-ionized water, inoculating it with a fixed aliquot of microbial seed, measuring the dissolved oxygen (DO) and then sealing the sample to prevent further oxygen addition. The sample is kept at 20 °C for 5 days, in the dark to prevent addition of oxygen by photo-synthesis, and the dissolved oxygen is measured again. The American Public Health Association Standard Methods Committee adopted the 5-day biochemical oxygen demand (BOD_5) test. This method has been widely used as the standard method for determining the concentration of biodegradable organics in wastewater (APHA 1995). However, this conventional method is time-consuming (five days of incubation) and usually requires experience and skill to achieve reproducible results. As such, studies have already been conducted to develop alternative methods for real-time or on-line BOD monitoring. In MFC, wastewater could be used as fuel. Bacteria in anode of MFC can utilize biodegradable substrate easily and convert it in to electricity (DC) in a single step process named as electrogenesis. The coulomb generated from a mediator-less MFC is proportional to the concentration of fuel used. MFCs can be used as a BOD sensor (Kim et al. 2003). MFC as BOD biosensor has been discussed in Chap. 20.

1.7.7 MFC as Alternate Power Tool

Fuel cells convert chemical energy from various compounds into electrical energy. An MFC is almost similar to traditional fuel cells (FCs). MFCs use microorganisms as catalysts, while FCs use metal or oxide catalysts. MFCs use microbial metabolism to convert organic chemical energy to electrical energy. MFCs show great promise for providing small amounts of electrical energy to power systems like sensor nodes and sensor networks. Chapter 21 has included basic design elements, power density, prototyping methods and power output determination. In each of these aspects MFCs face challenges to produce useable electrical power to allow practical applications. Soil or sediment MFCs have some key comparative advantages particularly when deployed in situations such as under a dense tree canopy where sunlight or wind are not at sufficient levels to generate power via solar photovoltaic panels or traditional wind turbines (Chap. 21).

1.7.8 Biochemical Production via Microbial Electrosynthesis

Microbial electrosynthesis (MES) is an emerging theme of study involving microbes that are capable of taking up electrons from cathode surfaces and utilizing them for a series of electrochemical transformations. Various conventional biological processes can be integrated with bioelectrochemical systems (BES) for improved production, by retrofitting with the existing infrastructures or designing new and more efficient ones. Novel concepts in application as well as continuing research on developing efficient and state-of-the-art infrastructure designs, materials for electrodes, separators and catalysts along with genetically modified microorganisms have made BES a very promising technology. However, several factors need to be optimized for better performance. In Chap. 22, a general description about microbial electrosynthesis is provided along with information about the factors that influence the performance of BES. Few case studies where MES are used for improved production, the major challenges encountered in this research theme and some applied solutions for tackling these challenges have been also highlighted in Chap. 22.

1.8 Scaling Up of MFC

The development of MFC technology needs selection of suitable and cost effective electrode materials, membrane and bioreactor to find out its feasibility for the wastewater treatment under optimum operating conditions. The challenges for bringing MFC technologies out of the lab i.e. for practical applications, is to take into account the number of factors influencing performance of the MFC. The main challenge of MFC is to scale up the energy output with respect to volume of wastewater being treated. Total power from MFC can be improved either by increasing the capacity (volume) of MFC or by connecting number of MFCs in electrically stacked arrangement (series or parallel). Till date, only few pilot scale studies have been performed considering these factors as described in Table 1.3. Further details on the recent progress made towards scaling up of MFCs, the limitations and challenges associated with the scale up processes and the case studies have been discussed in Chaps. 23 and 24.

Table 1.3 Performance of pilot scale MFCs used for wastewater treatment

MFC configuration	Substrate	Working volume (L)	Power density (mW m^{-2})	Columbic efficiency (%)	Removal efficiency (%)	References
4-stack single chambered membraneless MFC	Municipal wastewater	45	73	24.8	24	Hiegemann et al. (2016)
12-stack multi anode/cathode MFC	Domestic wastewater	20	500	0.04	80	Jiang et al. (2011)
Stackable horizontal MFC	Domestic wastewater	250	116	7	79	Feng et al. (2014)
Modified two-chamber MFC	Cassava mill wastewater	30	1771	20	72	Kaewkannetra et al. (2011)
Tubular multi-cathode MFC	Synthetic wastewater	16	–	6	92	Tao et al. (2011)
Stackable baffled MFC	Brewery wastewater	90	171	–	84.7	Dong et al. (2015)
Column-type MFC	Diesel contaminated soil	50	3.4	–	89.7	Wang et al. (2015)

1.9 Challenges in MFC and Future Scope

Significant advances have been made in MFC architecture design, understanding the microbial exoelectrogenic consortia, and in describing electrochemical reactions kinetics. Economic feasibility is also becoming important consideration when assessing microbial electrochemical systems. However, there are many challenges that need to be addressed in order for MFCs to be utilized for various applications. Power output, shape and size of the reactors, electrodes and suitable microorganisms or a consortium of microorganisms pose serious questions on the ability of MFCs for various applications. Techno-economic analysis of MFC has been discussed by Jhadhav et al. in Chap. 23 of this book. An understanding of the microbiology of the current producing process for limiting growth of non-electrogenic microbes is required before further advances in power output are possible. Mixed culture when used as inoculum make the system cost effective but decreases the power output when methanogens grow over a period of time. The major bottleneck of MFC is low power generation. Series connection of MFC causes voltage reversal; therefore, parallel MFC with proper power management system is required to harvest worth amount of electricity. This also suggests that the cost of unit MFC cell should be minimized. Development of proper bacterial culture in the anode chamber of MFC requires optimization of operating condition in the anodic microenvironment. In addition, major challenges in MFC and future scope of this technology have been reviewed in Chap. 25 with perspective.

1.10 Conclusion

MFC technology is a promising field of research, which can solve the energy crisis to some extent, and reduce the amount of emission gases released into the atmosphere. The major breakthrough in MFC research are the development of air cathode MFC, sediment MFC, scaling up issues and bacterial ecology. In last few decades, power output from MFC boost in order of several magnitudes. However, it is important to increase it further for useful applications from scalable level with innovative design and cost effective materials. The combination of wastewater treatment along with electricity production helps in compensating the cost of wastewater treatment, making it sustainable. MFC is coming up as a promising technology to treat wastewater; still several challenges remain which need to overcome to commercialize this technology. Therefore, operating conditions such as influent COD concentration, HRT, feed pH and specific organic loading rate should be properly decided for the operation in order to obtain optimum COD removal efficiency and power production. This issue remains unsolved till date. Full-scale implementation of such bio-electrochemical wastewater treatment system is not straight forward because certain microbiological, technological and economic challenges need to be resolved that have not previously been encountered in any other wastewater treatment system which makes life cycle assessment complicated.

References

Abourached, C., Lesnik, K. L., & Liu, H. (2014). Enhanced power generation and energy conversion of sewage sludge by CEA–microbial fuel cells. *Bioresource Technology, 166*, 229–234.

American Public Health Association (APHA). (1995). *Standard methods for the examination of water and wastewater*. Washington, DC: American Public Health Association.

Bond, D. R., & Lovley, D. R. (2003). Electricity production by geobacter sulfurreducens attached to electrodes. *Applied and Environmental Microbiology, 69*, 1548–1555.

Chandrasekhar, K., & Venkata Mohan, S. (2012). Bio-electrochemical remediation of real field petroleum sludge as an electron donor with simultaneous power generation facilitates biotransformation of PAH: Effect of substrate concentration. *Bioresource Technology, 110*, 517–525.

Chandrasekhar, K., Amulya, K., & Venkata Mohan, S. (2015). Solid phase bio-electrofermentation of food waste to harvest value-added products associated with waste remediation. *Waste Management, 45*, 57–65.

Ding, A., Yang, Y., Sun, G., & Wu, D. (2016). Impact of applied voltage on methane generation and microbial activities in an anaerobic microbial electrolysis cell (MEC). *Chemical Engineering Journal, 283*, 260–265.

Dong, Y., Qu, Y., He, W., Du, Y., Liu, J., Han, X., & Feng, Y. (2015). A 90-litre stackable baffled microbial fuel cell for brewery wastewater treatment based on energy self-sufficient mode. *Bioresource Technology, 195*, 66–72.

Du, Z., Li, H., & Gu, T. (2007). A state of the art review on microbial fuel cells: A promising technology for wastewater treatment and bioenergy. *Biotechnology Advances, 25*, 464–482.

Feng, Y., He, W., Liu, J., Wang, X., Qu, Y., & Ren, N. (2014). A horizontal plug flow and stackable pilot microbial fuel cell for municipal wastewater treatment. *Bioresource Technology, 156*, 132–138.

Gil, G.-C., Chang, I.-S., et al. (2003). Operational parameters affecting the performance of a mediator-less microbial fuel cell. *Biosensors & Bioelectronics, 18*, 327–334.

Gonzalez del Campo, A., Lobato, J., Canizares, P., Rodrigo, M. A., & Fernandez Morales, F. J. (2013). Short-term effects of temperature and COD in a microbial fuel cell. *Applied Energy, 101*, 213–217.

Hiegemann, H., Herzer, D., et al. (2016). An integrated 45L pilot microbial fuel cell system at a full-scale wastewater treatment plant. *Bioresource Technology, 218*, 115–122.

Jiang, D., Curtis, M., Troop, E., et al. (2011). A pilot-scale study on utilizing multi-anode/cathode microbial fuel cells (MAC MFCs) to enhance the power production in wastewater treatment. *International Journal of Hydrogen Energy, 36*, 876–884.

Kaewkannetra, P., Chiwes, W., & Chiu, T. Y. (2011). Treatment of cassava mill wastewater and production of electricity through microbial fuel cell technology. *Fuel, 90*, 2746–2750.

Khilari, S., Pandit, S., Ghangrekar, M. M., Pradhan, D., & Das, D. (2013). Graphene oxide-impregnated PVA–STA composite polymer electrolyte membrane separator for power generation in a single-chambered microbial fuel cell. *Industrial and Engineering Chemistry Research, 52*, 11597–11606.

Kim, Y., & Logan, B. E. (2013). Microbial desalination cells for energy production and desalination. *Desalination, 308*, 122–130.

Kim, B. H., Chang, I. S., Gil, G. C., Park, H. S., & Kim, H. J. (2003). Novel BOD (biological oxygen demand) sensor using mediator-less microbial fuel cell. *Biotechnology Letters, 25*, 541–545.

Leong, J. X., Daud, W. R. W., Ghasemi, M., Liew, K. B., & Ismail, M. (2013). Ion exchange membranes as separators in microbial fuel cells for bioenergy conversion: A comprehensive review. *Renewable and Sustainable Energy Reviews, 28*, 575–587.

Logan, B. E. (2008). *Microbial fuel cells* (1st ed.). Hoboken: Wiley-Interscience. https://www.engr.psu.edu/ce/enve/logan/publications/2010-Logan-AMB.pdf

Logan, B. E. (2010). Scaling up microbial fuel cells and other bioelectrochemical systems. *Appllied Microbiology and Biotechnology, 85*(6), 1665–1671.

Logan, B. E. (2012). Essential data and techniques for conducting microbial fuel cell and other types of bioelectrochemical system experiments. *ChemSusChem, 5*, 988–994.

Montpart, N., Rago, L., Baeza, J., & Guisasola, A. (2014). Hydrogen production in single chamber microbial electrolysis cells with different complex substrates. *Water Research, 68C*, 601–615.

Nevin, K. P., Hensley, S. A., et al. (2011). Electrosynthesis of organic compounds from carbon dioxide is catalyzed by a diversity of acetogenic microorganisms. *Applied and Environmental Microbiology, 77*, 2882–2886.

Oh, S.-E., Yoon, J. Y., Gurung, A., & Kim, D.-J. (2014). Evaluation of electricity generation from ultrasonic and heat/alkaline pretreatment of different sludge types using microbial fuel cells. *Bioresource Technology, 165*, 21–26.

Pandit, S., Nayak, B. K., & Das, D. (2012). Microbial carbon capture cell using cyanobacteria for simultaneous power generation, carbon dioxide sequestration and wastewater treatment. *Bioresource Technology, 107*, 97–102.

Pandit, S., Khilari, S., Roy, S., Pradhan, D., & Das, D. (2014). Improvement of power generation using Shewanella putrefaciens mediated bioanode in a single chambered microbial fuel cell: Effect of different anodic operating conditions. *Bioresource Technology, 166*, 451–457.

Pant, D., Singh, A., Van Bogaert, G., et al. (2012). Bioelectrochemical systems (BES) for sustainable energy production and product recovery from organic wastes and industrial wastewaters. *RSC Advances, 2*, 1248–1263.

Qiao, Y., Bao, S.-J., & Li, C. M. (2010). Electrocatalysis in microbial fuel cells – From electrode material to direct electrochemistry. *Energy & Environmental Science, 3*, 544.

Rabaey, K., & Keller, J. (2008). Microbial fuel cell cathodes: From bottleneck to prime opportunity? *Water Science and Technology, 57*, 655.

Rismani-Yazdi, H., Carver, S. M., Christy, A. D., & Tuovinen, O. H. (2008). Cathodic limitations in microbial fuel cells: An overview. *Journal of Power Sources, 180*, 683–694.

Schröder, U. (2007). Anodic electron transfer mechanisms in microbial fuel cells and their energy efficiency. *Physical Chemistry Chemical Physics, 9*, 2619–2629.

Schröder, U. (2008). From wastewater to hydrogen: Biorefineries based on microbial fuel-cell technology. *ChemSusChem, 1*, 281–282.

Sevda, S., Dominguez-Benetton, X., Vanbroekhoven, K., et al. (2013). High strength wastewater treatment accompanied by power generation using air cathode microbial fuel cell. *Applied Energy, 105*, 194–206.

Sharma, M., Aryal, N., Sarma, P. M., et al. (2013a). Bioelectrocatalyzed reduction of acetic and butyric acids via direct electron transfer using a mixed culture of sulfate-reducers drives electrosynthesis of alcohols and acetone. *Chemical Communications (Cambridge), 49*, 6495–6497.

Sharma, M., Jain, P., Varanasi, J. L., Lal, B., et al. (2013b). Enhanced performance of sulfate reducing bacteria based biocathode using stainless steel mesh on activated carbon fabric electrode. *Bioresource Technology, 150*, 172–180.

Sherafatmand, M., & Ng, H. Y. (2015). Using sediment microbial fuel cells (SMFCs) for bioremediation of polycyclic aromatic hydrocarbons (PAHs). *Bioresource Technology, 195*, 122–130.

Stams, A. J. M., de Bok, F. A. M., et al. (2006). Exocellular electron transfer in anaerobic microbial communities. *Environmental Microbiology, 8*, 371–382.

Strik, D. P. B. T. B., Timmers, R. A., et al. (2011). Microbial solar cells: Applying photosynthetic and electrochemically active organisms. *Trends in Biotechnology, 29*, 41–49.

Tao, H.-C., Zhang, L.-J., Gao, Z.-Y., & Wu, W.-M. (2011). Copper reduction in a pilot-scale membrane-free bioelectrochemical reactor. *Bioresource Technology, 102*, 10334–10339.

Umbuzeiro, G. A., Freeman, H. S., Warren, S. H., Oliveira, D. P., Terao, Y., Watanabe, T., & Claxton, L. D. (2005). The contribution of azo dyes to the mutagenic activity of the Cristais River. *Chemosphere, 60*(1), 55–64.

Varanasi, J. L., Nayak, A. K., Sohn, Y., Pradhan, D., & Das, D. (2016). Improvement of power generation of microbial fuel cell by integrating tungsten oxide electrocatalyst with pure or mixed culture biocatalysts. *Electrochimica Acta, 199*, 154–163.

Venkata Mohan, S., Velvizhi, G., Annie Modestra, J., & Srikanth, S. (2014). Microbial fuel cell: Critical factors regulating bio-catalyzed electrochemical process and recent advancements. *Renewable and Sustainable Energy Reviews, 40*, 779–797.

Wang, H., Luo, H., Fallgren, P. H., Jin, S., & Ren, Z. J. (2015). Bioelectrochemical system platform for sustainable environmental remediation and energy generation. *Biotechnology Advances, 33*, 317–334.

Yang, F., Ren, L., Pu, Y., & Logan, B. E. (2013). Electricity generation from fermented primary sludge using single-chamber air-cathode microbial fuel cells. *Bioresource Technology, 128*, 784–787.

Zhang, F., Chen, M., Zhang, Y., & Zeng, R. J. (2012). Microbial desalination cells with ion exchange resin packed to enhance desalination at low salt concentration. *Journal of Membrane Science, 417–418*, 28–33.

Chapter 2
Principles of Microbial Fuel Cell for the Power Generation

Soumya Pandit and Debabrata Das

2.1 Introduction

Ever increasing energy demand has induced fossil fuel consumption, consequently pollution and global warming driving the world towards an unprecedented high and potentially devastating energy crisis. Therefore, water and energy securities are considered as major concerns in present scenario. Organic waste/wastewater signifies as a potential renewable feedstock to generate various forms of bioenergy aside from the remediation process by regulating the biological process. Bioenergy has gained significant attention as a sustainable and futuristic alternative to fossil fuels. Using waste for bioenergy through its remediation has instigated considerable interest and has further opened a new avenue for the use of renewable and inexhaustible energy sources. Therefore, the field of wastewater management and alternative energy are the most unexplored fields of Biotechnology and Science (Massoud et al. 2009). Microbial fuel cell (MFC) is gaining popularity as a promising tool for simultaneous waste treatment and current generation without polluting environment. The complete breakdown of a wide range of organic substrates to carbon dioxide and water is usually only possible with several enzymatic reaction steps which is easily achieved in MFCs. Though research on MFCs was initiated in the 1960s during NASA's space explorations, rapid gain in MFC research was observed during the last two decades.

S. Pandit • D. Das (✉)
Bioprocess Engineering Laboratory, Department of Biotechnology, Indian Institute of Technology, Kharagpur, West Bengal 721302, India
e-mail: ddas@hijli.iitkgp.ernet.in

D. Das (ed.), *Microbial Fuel Cell*,
https://doi.org/10.1007/978-3-319-66793-5_2

2.1.1 Fuel Cell and Brief Development of MFC

A fuel cell can convert the energy trapped in a fuel into electricity directly in a single step. A fuel cell can generate power constantly as long as it has the source of fuels. This makes it a better choice for electricity generation as compared to a battery which needs to be recharged. Hydrogen is mostly used as substrate utilized for oxidation at anode side where it encounters a catalyst. H_2 converts to electron and proton. The electron and proton are transported via the external circuit and an ion exchange membrane to the cathode chamber, respectively. The oxygen molecules combine with proton to form water via reduction on cathode. Water is the final product in this electrochemical reaction where current can be generated in single step. Due to its eco-friendly nature, high thermodynamic efficiency and sustainability fuel cell research gained popularity soon. Researchers paid significant attention on fuel cell research owing to an increase of oil price during 1960–1970.

The fuel cells are categorized in two families: abiotic and biotic fuel cell. Abiotic fuel cells are conventional one which is comprised of inorganic catalyst – for example solid oxide fuel cell, molten carbonate fuel cell, proton exchange membrane fuel cell etc. These usually contain precious catalyst for hydrogen oxidation and oxygen reduction at cathode. Abiotic fuel cell generates very high magnitude of power as compared to the biotic one. The biotic fuel cells are comprised of living organisms or enzymes or their derivatives as major catalyst. Contrary to abiotic fuel cell, the biological fuel cell (BFC) uses cheap substrate and inexpensive catalyst. These can be categorized into two main areas: (1) Microbial fuel cells (MFC) and (2) Enzymatic fuel cells (EFC). In an enzymatic fuel cell, enzymes are placed or adsorbed on anode materials and the oxidation occurs in presence of organic substrate. Alternatively, when the catalyst is a microbe, the cell is called an MFC.

The idea of using microbes to produce electricity was first noticed in the beginning of twentieth century. However, before that biologist Luigi Galvani (1780) already demonstrated relationship between electricity and biology. Michael Cresse Potter in 1911 generated current using *Escherichia coli*. However, it didn't come into limelight because work received little coverage. In 1931, Cohen prepared microbial half fuel cells and connected in series which produced around 35 V with 2 mA current. In the late 1970s, Allen and Bennetto explained the working principle of microbial fuel cells using bacteria that need supplemented electron mediator externally. In early 1990s, the fuel cells became far more appealing devices; consequently MFCs were considered as promising technology. However, the major breakthrough came in late nineties when MFC was run using bacteria which can directly transfer electron during oxidation of waste to anode without using mediator. MFC has few advantages over conventional fuel cells. It can utilize cheap renewable sources like wastewater and biomass to convert current unlike fuel cell where costly hydrogen/methanol is oxidized to generate current. In fuel cell chemical expensive substance like platinum is used as catalyst. In MFC, different microbes can be utilized for oxidation in anode which makes the process cost effective. Further, it can be operated at ambient temperature and atmospheric pressure unlike some fuel cells.

2.2 Basic Principle of MFC

Being a fuel cell, an MFC comprises an anode, a cathode and electrolyte(s). The anodic and cathodic chambers may or may not be separated by an ion exchange membrane (IEM). Live microbes in planktonic state or by forming biofilm in the anodic chamber oxidize substrates and as a result produce electrons, protons and other metabolites as end products. The electrons released by the microbes are collected by the anode and pass into the cathode through an external load. On the other hand, the protons percolate via the IEM or simply diffuse to the cathode to get reduced by the arriving electrons, thus completing the circuit. The flow of electrons through the external load generates electric current. As the protons get reduced in presence of oxygen at the anode, water is formed as a by-product which makes the process environment-friendly. MFCs can be either large (litre-scale) or small (microlitre-scale or millilitre-scale).

The major components of MFCs are anode, cathode and membrane or separators. Different carbon materials like graphite plates, rods, carbon paper, carbon cloth, carbon foam, glassy carbon, stainless steel mesh, carbon felt, Pt, Pt black and, reticulated vitreous carbon have been applied as anode materials. The anode material should allow bacteria to form biofilm and should be conductive and noncorrosive. On the other hand, the cathode usually contains catalyst materials like Pt, Pt black, MnO_2, and polyaniline for better oxygen reduction reaction (ORR) (Varanasi et al. 2016). Graphite, carbon felt, carbon paper, carbon cloth, carbon foam, glassy carbon, stainless steel, and titanium mesh are among the materials used as cathode. In dual chambered MFC, ion exchange membrane or separator is used. In single chambered MFC, IEM was hot-pressed or chemically fixed to cathode to prepare cathode membrane assembly for passive aeration to provide oxygen directly for ORR. Initially research started with salt-bridge as separator between anode and cathode chambers (Khilari et al. 2013). Gradually it was replaced by IEMs. Both CEM (Nafion, Ultrex) and anion exchange membrane (Ralex AEM, Ultrex) were reported to utilize as membrane. However, AEM was found effective in terms of power generation owing to its ability to generate stable voltage and reduced salt deposition.

2.2.1 Advantages of MFC over Other Bioenergy Processes

MFCs received much attention in recent years due to its advantages over the current technologies applied in biological wastewater treatment like anaerobic digester (AD), activated sludge process (ASP) etc. (Logan 2008). MFC is capable of direct conversion of waste to energy in terms of electricity in a single step and this ensures high conversion efficiency. Rabaey et al. (2007) showed that the conversion efficiency of bioethanol, biogas and biohydrogen is around 10–25%; 25–38% and 15–30%, respectively. In comparison to this conversion efficiency, MFC has the conversion efficiency as high as 80%. In anaerobic digestion, methane or hydrogen gas is produced along with carbon dioxide and little amount of hydrogen sulfide. Unlike these technologies, an MFC does not require gas treatment. The energy input is required for aeration in ASP where MFCs do not need the aeration as cathode is passively aerated. Compared to ASP and AD, MFC produces less amount of sludge. In addition, MFCs stack can be applied as decentralized source of power.

2.3 Power Generation and Evaluation of MFC Performance

In the respiratory metabolism, terminal electron acceptor is the electrode inside the anode chamber of the MFC. Anaerobic environment is required at the electro active biofilm containing anode chamber to convert biodegradable matter to electricity. The metals like Fe^{3+} or Mn^{4+} oxides, insoluble sulfate and nitrate are considered as nearest parallel substance to anodes in natural environments for anaerobic respiration (Qiao et al. 2010). The resemblance lies in the fact that all are basically insoluble exocellular electron acceptors (Fig. 2.1). The principle of microbial assisted current generation diverge from conventional biogeochemical processes,

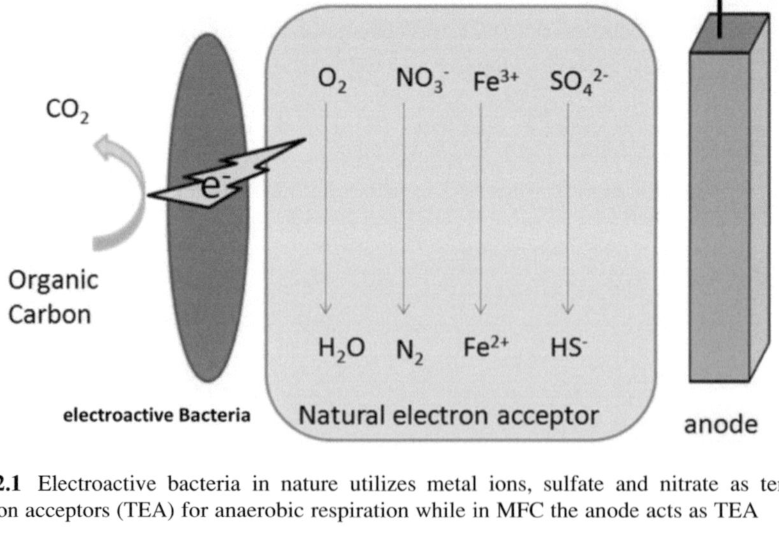

Fig. 2.1 Electroactive bacteria in nature utilizes metal ions, sulfate and nitrate as terminal electron acceptors (TEA) for anaerobic respiration while in MFC the anode acts as TEA

such as Mn^{4+} reduction, is that the electrons released during microbial respiration reduce anode instead of a natural electron acceptor (Erable et al. 2010). The EAB can reduce anode via substrate oxidation in absence of their natural electron acceptors like O_2 and SO_4^- in the anode chamber. These anodophilic bacteria generate electricity through this metabolism and generate ATP and thus grow. However, anodic potential limits the energy gain; hence the growth yields are very small as compared to those observed during aerobic metabolism. Smaller growth yield means less waste sludge and larger coulombic efficiency; this is the reason why it is attractive as wastewater treatment technology.

Electricity generation process in MFC is spontaneous since the overall reaction is thermodynamically favourable (Eqs. 2.1, 2.2, 2.3, 2.4 and 2.5). Considering the Nernst equation, when NADH acts as electron provider and oxygen acts as terminal electron receiver, the theoretical voltage generation is 1.1 V.

$$\text{Adode reaction : } NAD^+ + H^+ + 2e^- \rightarrow NADH \tag{2.1}$$

$$E^{01}_{\text{and}} = E_0 - \frac{RT}{2F} \ln \frac{[NADH]}{[NAD^+][H^+]^2} = -0.32\text{V} \tag{2.2}$$

$$\text{Cathode reaction : } O_2 + 4H^+ + 4e^- \rightarrow 2H_2O \tag{2.3}$$

$$E^{01}_{\text{cat}} = E_0 - \frac{RT}{nF} \ln \frac{1}{[O_2]^{0.5}[H^+]^2} = -0.84\text{V} \tag{2.4}$$

$$E^{01}_{\text{tot}} = E^{01}_{\text{cat}} - E^{01}_{\text{and}} = 0.84 - (-0.32)\text{V} = 1.16\text{V} \tag{2.5}$$

MFC utilizes acetate as fuel substrate in anode chamber. With a given concentration of carbonate (HCO_3^-) of 5 mM, acetate (CH_3COO^-) of 5 mM and an oxygen reducing cathode at pH 7 has a cell EMF of [0.805 – (−0.296)] of 1.101 V. Variation in electric potential (V) is created between anode and cathode chambers due to dissimilar liquid solutions and current (A) is generated due to flow of electron resulting in generation of electrical power (Watt). The electrode potential generated using glucose as anodic feed is given as follows:

Anode reaction:

$$C_6H_{12}O_6 + 6H_2O = 6CO_2 + 24H^+ + 24e^- \qquad E^0 = 0.014\text{ V} \tag{2.6}$$

Cathode reaction:

$$24H^+ + 24e^- + 6O_2 - 2H_2O \qquad E^0 = 1.23\text{ V} \tag{2.7}$$

Overall reaction:

$$C_6H_{12}O_6 + 6O_2 = 6CO_2 + 6H_2O + \text{electrical energy} \tag{2.8}$$

(Theoretically approaching 2840 kJ mol^{-1})

The energy output can be evaluated which depends on power generation and time required for bioelectrochemical reaction:

$$V = I \times R \tag{2.9}$$

$$E = P \times T \tag{2.10}$$

$$P = V \times I \tag{2.11}$$

where P, R and T denote the power (Watts), resistance (Ohm) and time (s) respectively. The output operating voltage (V) and corresponding current (I) influence power generation. The operating or actual voltage due to internal resistance of the cell can be calculated by Eq. 2.12:

$$V = E^{o} - \eta_a - \eta_b - I \times R \tag{2.12}$$

where E^{o} is theoretical cell voltage which can be calculated using Nernst equation. The η_a and η_b, indicate overpotential or voltage losses at the both anode and cathode. The $I \times R$ signifies Ohmic loss due to both anolyte and catholyte resistances. Owing to these losses the actual voltage output is always lower than theoretical voltage. The measured open circuit voltage is around 750–800 mV, which is lower than theoretical one of 1.1 V. Under close circuit condition, the operating voltage will reduce significantly.

2.3.1 Classifications of MFCs

MFC can be classified into two types based on the electron transfer mechanism of microbes to the anode. Two types of MFCs are available, namely, mediator MFC and mediator-less MFC depending on the requirement of external mediator (Fig. 2.2).

In mediator MFC, supplementation of synthetic mediators facilitates continuous electron transport to the anode via redox coupling. In both cases of direct and indirect mediator driven MFCs, soluble mediators are recycled in a process through simultaneous reduction (receiving electrons from cell) and oxidation (donating electrons to anode) (Fig. 2.2a, b). In indirect mediator MFC, microbes are unable to donate electrons directly to anode due to its non-conductive cell surface structures and incapability of producing self-mediators. Externally added suitable electron mediators can improve the electron transfer efficiencies in such MFCs. The value of oxidation reduction potential of electrochemical mediators is in between anode and NADH inside bacterial cell. Therefore, these have the ability to squeeze electrons from microbial cell and transfer it to anode. In this process, it can simultaneously reduce and oxidize itself and also regenerate itself. Various organometallic compounds were applied for this purpose which include phenothiazines

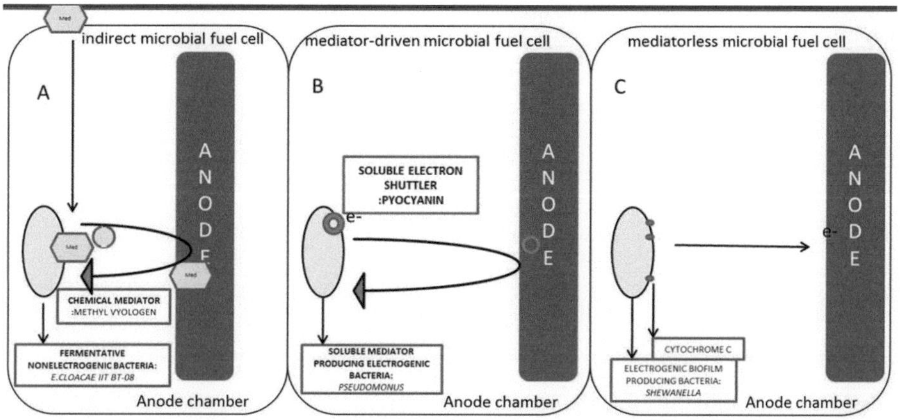

Fig. 2.2 (**a**) A mediator microbial fuel cell – usually non-electrogenic microorganisms were utilized which are incapable of directly donating electron to anode; these were used as catalyst and can be used along with external redox mediators in *indirect mediator microbial fuel cell*; (**b**) Microbes are capable of producing mediators for donating electron to anode directly in a *direct mediator microbial fuel cell*; and (**c**) Mediator-less MFC where bacteria form conductive biofilm

(alizarine brilliant blue, *N*, *N*dimethyl-disulfonated thionine, methylene blue, phenothiazine, toluidine blue) and phenoxazines (brilliant cresyl blue, gallocyanine, resorufin).

The mediator-less MFCs are further categorized based on the ability of the microbes to transfer electron either directly (Fig. 2.2c) or through self-synthesized electron shuttles from the cell to the electrode (direct mediator MFC; Fig. 2.2b). Direct transfer of electrons in mediator-less MFCs is done by organisms called as 'exoelectrogens' or electro active bacteria (EAB) whereas any other microorganism can be used in a mediator MFC. In mediatorless MFC, the anode half cell potential is developed as microorganism can directly reduce anode through the oxidation of substrate. Different ranges of bacteria like acidobacteria, firmicutes and α, β, γ and δ classes of proteobacteria were found effective as electron generator on anode to produce current. Further, it was reported that yeast strains *Saccharomyces cerevisiae* and *Hansenula anomala* were capable of producing electricity under anaerobic conditions. Different types of isolated pure cultures able to generate electricity in MFC are shown in Table 2.1. The dissimilatory metal reducing bacteria (DMRBs) and sulphate reducing bacteria (SRB) are major players contributing significantly on bioanode. In natural condition in absence of conductive anode they will utilize oxidized metal ions [Fe(III) or Mn(IV)] as terminal electron acceptor to continue respiration process. Probably, this is the reason why the mechanism of extracellular electron transfer (EET) activity for metal reduction and anode reduction is considered to be identical.

Table 2.1 Pure cultures used as inoculum in the anode of MFC

Reactor type	Biocatalysts	Fuel used	Power density ($mW\ m^{-2}$)	Power generated ($W\ m^{-3}$)	References
Single chamber	*Escherichia coli* acclimated	Complex medium	600		Erable et al. (2010)
Miniature	*Shewanella oneidensis*	Lactate	3000	500	Ringeisen et al. (2006)
H-type	*Geobactersul furreducens*	Acetate	13	0.356	Bond and Lovley (2003)
H-type	*Rhodoferax ferrireducens*	Glucose	33.4		Chaudhuri and Lovley (2003)
Two-chamber	*Clostridial* isolate	Complex medium		5.62	Prasad et al. (2006)

2.3.2 *Potential Losses in MFC*

MFC usually generates less operating voltage (V_{op}) with respect to the electromotive force (E_{thermo}). These potential losses or overpotentials are responsible for the lowering of performance of MFC. Overpotentials are of three different types and these are irreversible in nature. Activation overpotentials occur due to loss of electrons/energy utilized to initiate extracellular electron transfer to the anode and oxygen reduction by proton in cathode. A second type of voltage loss is found in the form of ohmic losses which is majorly owing to proton diffusion resistance and charge transfer resistance. The concentration overpotential is prominent at high current densities regime; cell operating voltage falls sharply with increase in current (Du et al. 2007). Unavailability of substrate to the biofilm or sluggish rate of substrate oxidation and reduction of terminal electron acceptor in cathode is due to proton and oxygen transfer limitation and is responsible for concentration overpotential (Qiao et al. 2010).

The operating voltage of MFC can be deduced by subtracting irreversible losses from theoretical maximum potential in MFC using Eq. 2.13 (Rismani-Yazdi et al. 2008):

$$V_{op} = E^{o} - \left[(\eta_{act} + \eta_{ohmic} + \eta_{conc})_{cathode} + (\eta_{act} + \eta_{ohmic} + \eta_{conc})_{anode}\right] \quad (2.13)$$

where η_{act}, η_{ohmic} and η_{conc} are the activation loss, ohmic loss and concentration loss respectively. These voltage losses occur due to slow electrochemical reaction kinetics, development of ionic and electronic resistances, and mass transport limitations. A typical polarization curve represents the current dependent voltage loss or overpotential of the electrodes at various phases (Fig. 2.3).

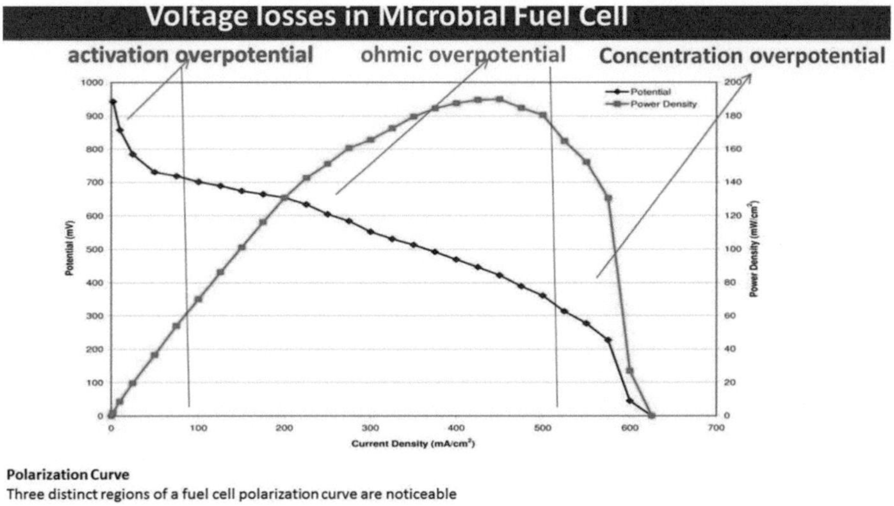

Fig. 2.3 Typical polarization curve of a fuel cell

2.3.3 *Factors Affecting the Performance of MFC*

The upper level of power generation that can be achieved in MFC is still unknown as the power output is limited by the high internal resistance of the MFC. The power generation in MFC is several order magnitude less than that obtained in chemical fuel cells. This is due to the various constraints associated with the MFC such as reactor configuration, electrode and separator materials, anode and cathode catalysts, electron acceptor, substrate type, substrate concentration, feed pH, HRT and temperature that increase the internal resistance of the MFC.

2.3.4 *Performance Evaluation for MFC*

The performance of MFC is generally assess based on power density (both volumetric and normalized to anode surface area, current density, electric potential difference, cell internal resistance) and biological treatment efficiency (COD removal) (Logan 2012). A polarization curve generator is a powerful experiment to analyze and characterize quality of fuel cells in terms of power generation. It is basically a voltage-current curve that helps to determine maximum power generation and internal resistance of the MFC (Zhao et al. 2009). Polarization study can be carried out by varying external resistances to obtain a voltage-current curve. The volumetric current and power density were calculated using Eqs. 2.14 and 2.15.

$$i_{\mathrm{d}} = \frac{V}{Rv_{\mathrm{and}}} \tag{2.14}$$

$$P_{\mathrm{d}} = \frac{V^2}{Rv_{\mathrm{and}}} \tag{2.15}$$

where volumetric current (i_{d}) and power density (P_{d}) were evaluated by normalizing to anolyte volume (v_{and}). Similarly, many researchers provided output of MFC in the form of the current density. Power density can be calculated by normalizing to electrode (mostly, anode surface area).

2.3.5 *Coulombic Efficiency and Energy Efficiency*

The Coulombic efficiency (CE) is basically the ratio of the total charge (in terms of coulombs) transported to the anode after bioelectro-oxidation of the substrate to maximum charge available or stored if complete substrate can be converted to current theoretically i.e. the total electron charge stored in the substrate (Logan et al. 2006). One major useful index to evaluate the performance of an MFC is to find out coulombic as well as energy efficiency. CE can be evaluated as per Eq. 2.16.

$$\mathrm{CE} = \left(M_{\mathrm{O2}} \int idt \right) / (n_e F V_{\mathrm{and}} \Delta \mathrm{COD}) \tag{2.16}$$

where V_{and} is the volume of the anode chamber; ΔCOD is change in chemical oxygen demand of the wastewater during MFC operation by the subtraction of the final to initial COD (g/L); M_{O2} is the molecular mass of O_2 and n_{e} is the required number of electrons to reduce oxygen to water. The CE thus indicates what fraction of the substrate was utilised by the bacteria for current production.

2.4 Microbes as Catalyst in MFC and Their Various Mode of Exo-cellular Electron Transfer to Electrode

Microbial biofilms present at electrodes are of a fundamental importance in the processes of biodegradation and current generation in MFC. These current producing bacteria are called electroactive bacteria (EAB) or exoelectrogens because of their ability of direct or indirect extracellular electron transfer (EET) to electrodes. Bacterial growth is supplemented by coupling of energy conservation with transfer of electrons from exoelectrogens leading to cathode/anode respiration (Mustakeem 2015). Due to the growing fame of MFC technology, the concepts of both electrode respiration and electromicrobiology have come in the limelight.

The microorganisms capable of carrying out extracellular electron transfer to electrodes in an MFC are either biofilm forming or planktonic ones (Logan 2009). A majority of these utilize exogenous or endogenous mediator molecules e.g. *E. coli* and *Pseudomonas* sp. (Pandit et al. 2015). The mediators are primarily responsible for EET in planktonic microorganisms. Biofilm-forming microorganisms can transfer the electron either directly using membrane bound cytochrome proteins in cell envelope to the anode or via nanowires, conductive matrices etc. *Shewanella* and *Geobacter* sp. can work by direct electron transfer from the cell membrane or with the help of nanowire appendages (Pham et al. 2009). Kinetic rates would be higher in case of electron transfer through the latter than the former. Hence the primary attribute in magnifying MFC power density lies on the capability of a EAB to adhere to anode and develop efficient monolayered or preferably multilayered electroactive biofilm (Yang et al. 2012).

Form of microbial growth i.e. either as planktonic cells or as biofilms depends on the metabolic status and environmental factors. For example, *Shewanella oneidensis* can form a biofilm on the anode electrode in lactate-fed MFCs. However, both planktonic cells and biofilms are able to exist together simultaneously and synergistically produce power. Contrary to it, *Geobacter sulfurreducens* sp. requires direct electrical contact with the electrode for electron transfer (Bond and Lovley 2003). These organisms can build an electroactive layered biofilm. The capability of *Shewanella* sp. of biosynthesis of redox active mediator molecules which enables planktonic cells to perform EET gives rise to the reported differences between the two most focused EABs. *Shewanella* are Gram negative, facultative bacteria for which aquatic and sedimentary environments provide optimum growth conditions (Kim et al. 2002). *Geobacters* are Gram-negative obligatory anaerobic bacteria, which thrive under aquatic and sedimentary environments. In the respiratory metabolism, terminal electron acceptor is the electrode, bacteria generate electricity through this metabolism and EABs can generate ATP and thus grow. Anodic potential limits the energy gain; hence the growth yields are very small as compared to those observed during aerobic metabolism. Smaller growth yields means less waste sludge and larger CE; this is the reason why it is attractive as wastewater treatment technology.

2.4.1 Electron Transfer by C-type Cytochromes

The C-type cytochromes (CTCs) play most useful role in electron transfer technique for electricity harvesting by EABs. C-type cytochrome is basically widespread heme-containing proteins available in most of archaea and eubacteria. *S. oneidensis* MR-1 contains 42 putative CTCs where 80% of these are located in the outer membrane and occupy 8–34% of the total cell surface (Fig. 2.4). CymA is an example of tetraheme-CTC with the N-terminal bound in the inner membrane and the C-terminal exposed to the periplasm. CymA is of a great importance because it participates in many *Shewanella* anaerobic respiration

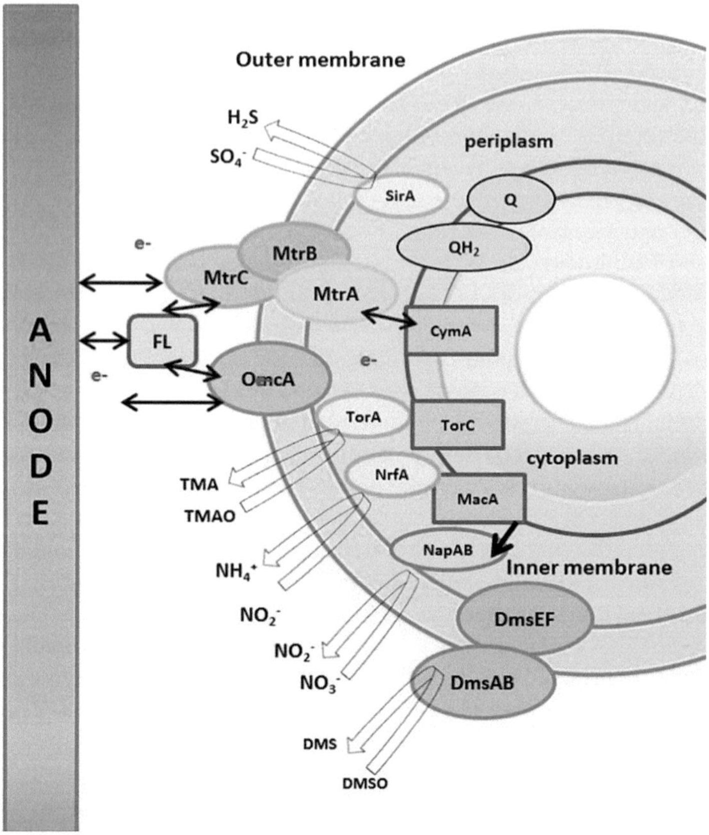

Fig. 2.4 Schematic image of the proposed exoelectron transfer of *Shewanella oneidensis* and its interactions with an anode. Mtr-pathway and terminal reductases of *Shewanella oneidensis* play major role in EET. (Quinones (Q) in the inner membrane liberate electrons which eventually transported to terminal reductases or an MtrCAB complex via CymA or TorC. Flavin molecules (FL) act as shuttle for transferring electron from MtrCAB complex to anode (Kracke et al. 2015)

processes. The deletion of a cymA gene caused >80% reduction in current generation in electrode reduction. CymA has an ability to directly interact with many terminal reductases located in periplasm, such as fumarate reductase (FR) and nitrate reductase.

CymA is the major electron channel to periplasmic space. Bacterial two-hybrid displayed pairwise interactions among CymA, MtrA and some other periplasmic redox proteins. CymA can also directly interact with periplasmic redox proteins by forming a transient protein complex. This hypothesis is further supported by an analysis in which electrode was used as an electron donor for fumarate reduction by *S. oneidensis*. MR-1 showed that 85% electrons were transferred to FR via CymA and 15% via MtrA. This indicates the presence of a CymA–FR–MtrA complex in periplasm. The MtrABC conduit can be understood as an extended branch of the periplasmic complex. An alternative electron transfer conduit can be made by replacement of IS-insertion activated SirCD expression partially by CymA. MtrABC complex is a tunnel type which facilitate in connecting intracellular electron flux to anode. The electron transfer rate of anode reduction by *S. oneidensis* MR-1 is more prominent than insoluble haematite reduction, but much slower as compared to the reductions of soluble electron acceptors in anaerobic environment. It was found that current generation could be still carried out if the primary gene is undergone single or multiple deletions. It suggested the presence of alternative electron transfer conduits. A serial of mtrABC homologues encoding MtrDEF has been identified by genome analysis in the same operon. MtrF, which is the homologue of MtrC and OmcA, displayed higher reducing tendency of flavins and soluble electron acceptors than OmcA and MtrC. In *S. oneidensis*, MtrB could be functionally replaced by MtrE. The complementation of mtrF/mtrA gene pair displayed a similar Fe(III)-reducing tendency as the complementation of mtrC/mtrA (Coursolle et al. 2010).

According to comparative genomic analysis, six *Geobacter* species displayed an average of 79 putative CTCs in each *Geobacter* genome. Only 14% of these are found to be conserved in all genomes. The MacA transported electrons from inner membrane to PpcA in periplasm, and PpcA further transfer electrons to the OMCs (Fig. 2.5).

The type of OMCs actively involved in *Geobacter* electrode reduction differs from those in dissimilatory metal reduction. For example, OmcB and OmcS are primarily responsible for Fe(III) reduction while the deletion of omcB or omcS gene did not stop producing current generation process where mutant lacking omcB or omcS were used. For *G. sulfurreducens*, OmcZ is the only OMC vital for current generation. However, omcZ is not conserved in all *Geobacter* species. A deletion of omcZ resulted in more than 90% of current reduction. However, no impact was observed on the reductions of other electron acceptors. The OmcZ were found predominant in biofilm-anode interface in MFC when OmcZ were tagged with immunogold labelling (Richter et al. 2012). The spatial orientation and the electrostatic reaction might be the reason for OmcZ to play pivotal role electron transfer in biofilms. In addition to it, there are greater significant impacts on the electron transfer in thick *G. sulfurreducens* biofilms than in the thinner biofilm when omcZ gene was removed. Thus, OmcZ is particularly essential for long-distance electron transfer in electrode biofilms.

Exoelectrogens are mostly the Gram-negative (G^-) bacteria since the Gram-positive (G^+) cell wall is unfavourable for EET. However, G^+ bacteria behave as the dominant DMRB in some environments. A few of Gram positive species such as *Thermincola potens* are capable of performing direct electron transfer mechanism. *T. potens* has 32 putative CTC genes in its genome and many CTCs may be anchored and have a close arrangement on outer membrane, thus establishing an electron passage for anode reduction.

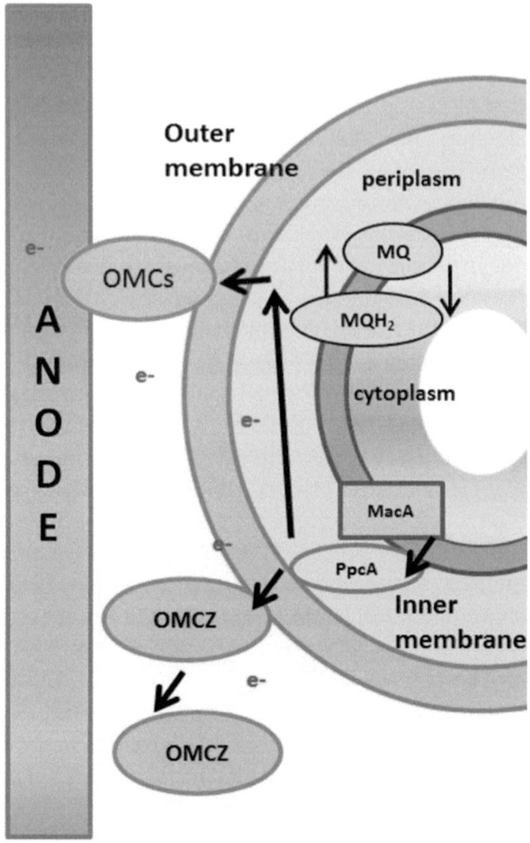

Fig. 2.5 Schematic picture of the exocellular electron transfer of *Geobacter sulferreducens* and its interactions with an anode (Kracke et al. 2015)

2.4.2 *Microbial Nanowire*

Microbial or bacterial nanowire technique is considered as new way of transferring electron to electrode. Bacterial nanowire is an electrically conductive pili which was discovered while reducing Fe(III) oxide by *G. sulfurreducens*. These were also found in *S. oneidensis* MR-1 and some other bacteria, and many other species signifying a wide environmental distribution of these bacterial appendages. The *S. oneidensis* MR-1 has bunches of pilus-like individual conductive outgrowth with a radius of 1.5–2.5 nm in the electron acceptor limiting regime and/or diffusion limiting or low agitating regime. The membrane pores divided nanowires. These nanowires helped in electron transfer due to its extended projection which can easily find contact with insoluble electron receivers or other bacteria partners (Steidl et al. 2016). The nanowire resistivity of *S. oneidensis* MR-1 is in the range of semiconductor. The k_o of *S. oneidensis* MR-1 nanowire and the entire cell is $1 \times 10^9\ s^{-1}$ and $2.6 \times 10^6\ s^{-1}$; both the values are several fold higher as compared to the k_o of OMCs. This indicates that nanowires are potentially most efficient part in EET (El-Naggar et al. 2010). The *S. oneidensis* MR-1 sp. produces various kinds of filamentous outgrowth or projections such as type IV pili, Msh-pili and flagella. The gene deletion analysis revealed that Msh-pili plays most important role in electron transport and, therefore, is primarily responsible for optimal electricity production. The deletions of other genes from the domain of type IV pili or flagella generation were responsible for loss of power production partially.

G. sulfurreducens nanowires are expressed when it utilizes metal oxides as electron receiver or while reducing fumarate at low temperatures. Two PilA isoforms with distinctive features were observed in *G. sulfurreducens* nanowires (Steidl et al. 2016). The electronic conductivity of *G. sulfurreducens* nanowires is 6 mS cm^{-1}, which is close to the synthetic metallic nanostructures. The electronic conductivity of *G. sulfurreducens* nanowires is significantly large compared to its proposed threshold value of 10^{-3} mS cm^{-1}, which might be the reason of high volumetric power generation. Contrary to the OMC-dependent conductivity of *Shewanella* nanowires, *Geobacter* nanowires possess typical metallic electronic property. The nanowires play significant roles in uranium reduction and development of electroactive biofilm for *G. sulfurreducens*.

2.4.3 Electron Shuttles or Mediators

The mediators or electron shuttles in MFC are secreted by mostly Gram negative (G^-) bacteria. The expected properties of these mediators or electron shuttle are: (1) dissolvability, (2) stability, (3) reusability, (4) environment-friendliness and (5) redox potential range between bacterial membrane protein and anode material. Power production in MFCs is stimulated via mediator secretion (Rabaey et al. 2007). The most documented electron shuttles in MFCs are endogenously produced flavins by *Shewanella* species, which majorly includes riboflavin (RF) and flavin mononucleotide (FMN). FMN has the ability to percolate through the outer membrane and then be transformed to RF in extracellular space while RF cannot be returned to *Shewanella* periplasm. In absence of mediator, the electron transfer rate of pure *S. oneidensis* MR-1 OMCs to insoluble iron ion (Fe^{+3}) is 100–1000 times lower than the whole cell. The addition of flavin can increase the reduction capacity to a state comparable with the entire cell. The electron transfer efficiency can be enhanced by secretion of microgram level flavin by *S. oneidensis* MR-1 by over 3.7-folds (Canstein et al. 2008). The ATP cost on flavin secretion was negligible when compared with the obtained energy benefit. The assumed k_o of flavin absorbed on an electrode is nearly two orders of magnitude lesser than that of the OMCs. *Shewanella* Mtr complex plays a vital part in flavin reduction. MtrC is responsible for 50% of the flavin reduction activity. Recent investigations suggest the possibility of flavin to be the by-products of cell lysis. Although flavin secretion is common in many microbes, the function of flavin may be limited in field-applied MFC since they are light-degradable compounds and can be utilized by the organism as carbon resource (Marsili et al. 2008).

Phenazines are intrinsic electron shuttles in MFCs generated by a diverse bacterial species, mainly *Pseudomonas*. The power producing ability of a non-electroactive Gram positive bacterium *Brevibacillus* sp. is optimized due to phenazines. This indicates that presence of soluble mediators in MFCs confers a synergic effect for other non-EABs to electricity harvesting. Apart from the bio-generated organic compounds, H_2 evolved during fermentation or other chemical response can also behave as electron shuttles or donors for *Shewanella* and *Geobacter* sp. in MFC (Stams et al. 2006). Apart from these humics, manganese species, polysulfide and other redox chemicals distributed in environments play role in electron transport. Nevertheless, the bacterial self-generated mediator shuttling has few bottlenecks as compared to electroactive biofilm based direct electron transport to anode. MFC experiences higher voltage loss owing to low diffusion of mediator. A part of bacterial energy is used to consume during mediator synthesis to continue anaerobic respiration. Further, these mediators are used to wash out in continuous system which causes serious power drop in continuous mode MFC operation. Some reports suggest that these mediators or shuttles are biodegraded or utilized by other types of bacteria.

2.5 MFC for Wastewater Treatment

As far as power generation is concerned, MFC cannot compete with conventional fuel cell which produces 10^3–10^5 fold higher current as compared to MFC. Henceforth, more attention was paid on MFC with wastewater treatment using different kind of organic waste materials. These involve wide range of electroactive bacteria which can enzymatically degrade a variety of wastewater from different sources like domestic, industrial, agricultural etc. Efforts have been made to utilize MFC stack as a decentralized source for power generation from waste resources. When the organic substrates to be treated are fermented, bacteria often play a crucial role in carbon conversion in anodic communities (Varanasi et al. 2015). A large number of recalcitrant wastes which includes dyes, pesticides, heavy metals, polyalcohols and heterocyclic compounds are produced by industry. These pollutants pose threat to environment and natural habitat when discharged without treatment. These substances are detrimental and known as potential recalcitrant and mutagenic. Therefore, these create serious trouble as far as health issue is concerned. MFC provides a solution to this problem since it can oxidize or reduce these materials bioelectrochemically in anode or cathode respectively to convert it into a harmless product.

2.6 Other Applications of MFC

MFCs have been largely used in wastewater treatment. Recently a number of MFC adapted technologies are drawing attraction, namely microbial solar cells (MSC), microbial desalination cells (MDCs), microbial electrolysis cell (MEC), microbial electrochemical snorkel (MES) etc.

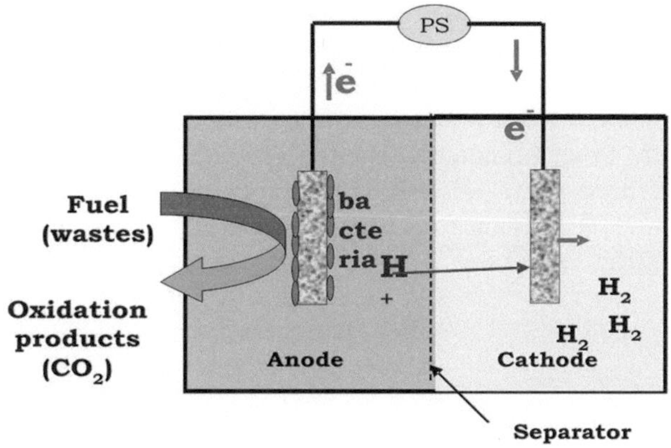

Fig. 2.6 Schematic representation of the microbial electrolysis cell (MEC)

MEC is a novel technology used for simultaneous biohydrogen generation and wastewater treatment. Like MFC, MEC has components like anode containing electroactive bacteria and cathode with catalysts facilitate in driving the reaction of proton and electron to form hydrogen in oxygen-free environment (Logan et al. 2008). The hydrogen production on cathode in this electrohydrogenesis is not a spontaneous reaction (Pierozynski 2011). Theoretically, the additional voltage required to get hydrogen from a MEC is [$E_{cathode} - E_{anode}$]. But practically more voltage ($\geq$0.114 V) is required due to overpotential at cathode to overcome the voltage loss or overpotentials associated with MEC reactors and its components. Nevertheless, this applied voltage is still lower as compared to the voltage required for water electrolysis (1.2 V). A traditional power supply or alternative source like solar or dye sensitized cell is required to provide 0.3 V or higher applied across the two electrodes. Sometimes CO_2 is getting reduced to hydrogen in cathode to produce methane gas. Incorporation of ion exchange membrane or separator is a big issue as it poses high internal resistance (Fig. 2.6). As a promising technology, lot of research is going on MECs in the last 10 years, with increasing numbers of reactor designs reported (Liu et al. 2005).

2.6.1 MFC as Toxic Sensor and BOD Biosensor

The biochemical oxygen demand (BOD) is an index of organic content and gives an indication on how much oxygen would be required for microbial degradation. High BOD of a sample indicates the extent of polluted substance present in it. It usually takes five days to get a result of the BOD of a biodegradable sample according to the American Public Health Association Standard Methods, which is time consuming and tedious. In MFC, wastewater could be used as fuel. Bacteria in anode of MFC can utilize biodegradable substrate easily and convert it in to electricity (DC) in a single step process named electrogenesis. The current produced in MFC is directly proportional to the concentration of substrate oxidized. So, MFCs can be used as a BOD sensor (Chang et al. 2004). Usually micro sized MFC of microlitre was used for that. The micro-sized MFCs possess a large surface area-to-volume ratio, short inter-electrode separation (hence low internal resistance), low response time and low Reynolds number. Their fabrication and assembly is precise and less expensive if done using micro-fabrication processes. The laminar flow model to reduce internal resistance is also applicable to micro-sized MFCs only (Elmekawy et al. 2013).

2.6.2 Preparation of Metal Nanoparticles

Metal nanoparticles such as gold nanoparticles (AuNPs), silver nanoparticles and cysteine capped silver nanoparticles (cys-AgNPs) were synthesized using the reducing ability of EABs in presence of organic substrate as electron provider (Khan et al. 2013). EAB used oxides of metal as electron acceptor for different metal nanoparticle synthesis where oxides were used as precursor. Sodium acetate was bioelectrochemically oxidized on a support (stainless steel) which generates electron to reduce precursor of nanoparticles to form different metal nanoparticles. Presence of conductive support like stainless steel provides electron-rich environment to convert metal ions to metal nanomaterial having zero valency. Further, these synthesized nanoparticles were utilized for different applications like biohydrogen production and antimicrobial activity (Kalathil et al. 2013).

2.6.3 Other Bioelectrochemical System Adapted from MFC

In microbial desalination cell (MDC), microbial electron harvesting ability is used to desalinate saline water. EABs on anode created a negative electric potential gradient which allows cation and anion from a concentrated saline chamber to migrate through ion-exchange membranes (IEMs) to balance the electro-neutrality. Thus water can be desalinated in MDC without any power consumption. Kim et al. demonstrated stackable MDCs with parallel dilute and concentrate chambers separated by IEMs (Kim and Logan 2013). In a plant microbial fuel cell (PMFC), MFCs are associated with living higher plants (Strik et al. 2011). In PMFCs, electrochemically active bacteria on anode utilize excreted rhizo-deposits from plant roots to directly generate electricity. Therefore, it is possible to convert solar energy to electrical energy directly in PMFC. Rhizo deposits are rich in carbon sources like arabinose, ribose, fructose etc. Rhizo deposits consist of a different array of substances that generate from pilled off root cells, and soluble lysates and exudates. These substrates can be utilized by a wide array of EABs.

Microbial carbon capture cells (MCC) are different type of PMFCs where microalgae can be used in cathode of a conventional dual chambered MFC to sequester atmospheric CO_2 and provide oxygen to cathode via biophotolysis. Wang et al. demonstrate an MCC where off gas CO_2 from the anode chamber is re-dissolved in catholyte where a photosynthetic algae (*Chlorella vulgaris*) was cultivated in presence of light and the CO_2 (Pandit et al. 2012). Microalgae biomass from the cathode chamber can be utilized for downstream processing of different value-added products like lipid extraction or ethanol production.

2.7 Conclusion

MFC technology is an upcoming popular research area which can be implemented to meet energy demand and water requirement for irrigation in agricultural sector. The major breakthroughs in MFC research like development of air cathode MFC, sediment MFC, scaling up issues and bacterial ecology have been reviewed. In last few decades, power output from MFC boost in order of several magnitudes. However, it is important to increase it further for useful applications from scalable level with innovative design and cost effective materials. MFC is coming up as a promising technology to treat wastewater, still several challenges remain which need to be overcome to commercialize this technology. Therefore, operating conditions such as influent COD concentration, HRT, feed pH and specific organic loading rate should be properly decided for operation in order to obtain optimum COD removal efficiency and power production. This issue remains unsolved till date.

References

Bond, D. R., & Lovley, D. R. (2003). Electricity production by *Geobacter sulfurreducens* attached to electrodes. *Applied and Environmental Microbiology, 69*, 1548–1555.

Canstein, H. v., Ogawa, J., Shimizu, S., & Lloyd, J. R. (2008). Secretion of flavins by *Shewanella* species and their role in extracellular electron transfer. *Applied and Environmental Microbiology, 74*, 615–623.

Chang, I. S., Jang, J. K., Gil, G. C., Kim, M., Kim, H. J., Cho, B. W., & Kim, B. H. (2004). Continuous determination of biochemical oxygen demand using microbial fuel cell type biosensor. *Biosensors & Bioelectronics, 19*, 607–613.

Chaudhuri, S. K., & Lovley, D. R. (2003). Electricity generation by direct oxidation of glucose in mediatorless microbial fuel cells. *Nature Biotechnology, 21*, 1229–1232.

Coursolle, D., Baron, D. B., Bond, D. R., & Gralnick, J. A. (2010). The Mtr respiratory pathway is essential for reducing flavins and electrodes in *Shewanella oneidensis*. *Journal of Bacteriology, 192*, 467–474.

Du, Z., Li, H., & Gu, T. (2007). A state of the art review on microbial fuel cells: A promising technology for wastewater treatment and bioenergy. *Biotechnology Advances, 25*, 464–482.

Elmekawy, A., Hegab, H. M., Dominguez-Benetton, X., & Pant, D. (2013). Internal resistance of microfluidic microbial fuel cell: Challenges and potential opportunities. *Bioresource Technology, 142*, 672–682.

El-Naggar, M. Y., Wanger, G., Leung, K. M., Yuzvinsky, T. D., Southam, G., Yang, J., Lau, W. M., Nealson, K. H., & Gorby, Y. A. (2010). Electrical transport along bacterial nanowires from Shewanella oneidensis MR-1. *Proceedings of the National Academy of Sciences, 107*, 18127–18131.

Erable, B., Duţeanu, N. M., Ghangrekar, M. M., Dumas, C., & Scott, K. (2010). Application of electro-active biofilms. *Biofouling, 26*, 57–71.

Kalathil, S., Khan, M. M., Ansari, S. A., Lee, J., & Cho, M. H. (2013). Band gap narrowing of titanium dioxide (TiO2) nanocrystals by electrochemically active biofilms and their visible light activity. *Nanoscale, 5*, 6323.

Khan, M. M., Ansari, S. A., Amal, M. I., Lee, J., & Cho, M. H. (2013). Highly visible light active $Ag@TiO_2$ nanocomposites synthesized using an electrochemically active biofilm: A novel biogenic approach. *Nanoscale, 5*, 4427.

Khilari, S., Pandit, S., Ghangrekar, M. M., Das, D., & Pradhan, D. (2013). Graphene supported α-MnO_2 nanotubes as a cathode catalyst for improved power generation and wastewater treatment in single-chambered microbial fuel cells. *RSC Advances, 3*, 7902.

Kim, Y., & Logan, B. E. (2013). Microbial desalination cells for energy production and desalination. *Desalination, 308*, 122–130.

Kim, H. J., Park, H. S., Hyun, M. S., Chang, I. S., Kim, M., & Kim, B. H. (2002). A mediator-less microbial fuel cell using a metal reducing bacterium, *Shewanella putrefaciens. Enzyme and Microbial Technology, 30*, 145–152.

Kracke, F., Vassilev, I., & Krömer, J. O. (2015). Microbial electron transport and energy conservation – The foundation for optimizing bioelectrochemical systems. *Frontiers in Microbiology, 6*, 575.

Liu, H., Grot, S., & Logan, B. E. (2005). Electrochemically assisted microbial production of hydrogen from acetate. *Environmental Science & Technology, 39*, 4317–4320.

Logan, B. E. (2009). Exoelectrogenic bacteria that power microbial fuel cells. *Nature Reviews. Microbiology, 7*, 375–381.

Logan, B. E. (2012). Essential data and techniques for conducting microbial fuel cell and other types of bioelectrochemical system experiments. *Chem Sus Chem, 5*, 988–994.

Logan, B. E., Hamelers, B., Rozendal, R., Schröder, U., Keller, J., Freguia, S., Aelterman, P., Verstraete, W., & Rabaey, K. (2006). Microbial fuel cells: Methodology and technology. *Environmental Science & Technology, 40*, 5181–5192.

Logan, B. E., Call, D., Cheng, S., Hamelers, H. V. M., Sleutels, T. H. J. A., Jeremiasse, A. W., & Rozendal, R. A. (2008). Microbial electrolysis cells for high yield hydrogen gas production from organic matter. *Environmental Science & Technology, 42*, 8630–8640.

Marsili, E., Baron, D. B., Shikhare, I. D., Coursolle, D., Gralnick, J. A., & Bond, D. R. (2008). *Shewanella* secretes flavins that mediate extracellular electron transfer. *Proceedings of the National Academy of Sciences, 105*, 3968–3973.

Massoud, M. A., Tarhini, A., & Nasr, J. A. (2009). Decentralized approaches to wastewater treatment and management: Applicability in developing countries. *Journal of Environmental Management, 90*, 652–659.

Mustakeem, F. (2015). Electrode materials for microbial fuel cells: Nanomaterial approach. *Materials for Renewable and Sustainable Energy, 4*, 22.

Pandit, S., Nayak, B. K., & Das, D. (2012). Microbial carbon capture cell using cyanobacteria for simultaneous power generation, carbon dioxide sequestration and wastewater treatment. *Bioresource Technology, 107*, 97–102.

Pandit, S., Khilari, S., Roy, S., Ghangrekar, M. M., Pradhan, D., & Das, D. (2015). Reduction of start-up time through bioaugmentation process in microbial fuel cells using an isolate from dark fermentative spent media fed anode. *Water Science and Technology: a Journal of the International Association on Water Pollution Research, 72*, 106–115.

Pham, T. H., Aelterman, P., & Verstraete, W. (2009). Bioanode performance in bioelectrochemical systems: Recent improvements and prospects. *Trends in Biotechnology, 27*, 168–178.

Pierozynski, B. (2011). On the hydrogen evolution reaction at nickel-coated carbon fibre in 30 wt.% KOH solution. *International Journal of Electrochemical Science, 6*, 63–77.
Prasad, D., Sivaram, T. K., Berchmans, S., & Yegnaraman, V. (2006). Microbial fuel cell constructed with a micro-organism isolated from sugar industry effluent. *Journal of Power Sources, 160*, 991–996.
Qiao, Y., Bao, S.-J., & Li, C. M. (2010). Electrocatalysis in microbial fuel cells – From electrode material to direct electrochemistry. *Energy & Environmental Science, 3*, 544–553.
Rabaey, K., Rodríguez, J., Blackall, L. L., Keller, J., Gross, P., Batstone, D., Verstraete, W., & Nealson, K. H. (2007). Microbial ecology meets electrochemistry: Electricity-driven and driving communities. *The ISME Journal, 1*, 9–18.
Richter, L. V., Sandler, S. J., & Weis, R. M. (2012). Two isoforms of geobacter sulfurreducens PilA have distinct roles in pilus biogenesis, cytochrome localization, extracellular electron transfer and biofilm formation. *Journal of Bacteriology, 194*, 2551–2563.
Ringeisen, B. R., Henderson, E., Wu, P. K., Pietron, J., Ray, R., Little, B., Biffinger, J. C., & Jones-Meehan, J. M. (2006). High power density from a miniature microbial fuel cell using Shewanella oneidensis DSP10. *Environmental Science & Technology, 40*, 2629–2634.
Rismani-Yazdi, H., Carver, S. M., Christy, A. D., & Tuovinen, O. H. (2008). Cathodic limitations in microbial fuel cells: An overview. *Journal of Power Sources, 180*, 683–694.
Stams, A. J. M., de Bok, F. A. M., Plugge, C. M., van Eekert, M. H. A., Dolfing, J., & Schraa, G. (2006). Exocellular electron transfer in anaerobic microbial communities. *Environmental Microbiology, 8*, 371–382.
Steidl, R. J., Lampa-Pastirk, S., & Reguera, G. (2016). Mechanistic stratification in electroactive biofilms of *Geobacter sulfurreducens* mediated by pilus nanowires. *Nature Communications, 7*, 12217.
Strik, D. P. B. T. B., Timmers, R. A., Helder, M., Steinbusch, K. J. J., Hamelers, H. V. M., & Buisman, C. J. N. (2011). Microbial solar cells: Applying photosynthetic and electrochemically active organisms. *Trends in Biotechnology, 29*, 41–49.
Varanasi, J. L., Roy, S., Pandit, S., & Das, D. (2015). Improvement of energy recovery from cellobiose by thermophillic dark fermentative hydrogen production followed by microbial fuel cell. *International Journal of Hydrogen Energy, 40*, 8311–8321.
Varanasi, J. L., Nayak, A. K., Sohn, Y., Pradhan, D., & Das, D. (2016). Improvement of power generation of microbial fuel cell by integrating tungsten oxide electrocatalyst with pure or mixed culture biocatalysts. *Electrochimica Acta, 199*, 154–163.
Yang, Y., Xu, M., Guo, J., & Sun, G. (2012). Bacterial extracellular electron transfer in bioelectrochemical systems. *Process Biochemistry, 47*, 1707–1714.
Zhao, F., Slade, R. C. T., & Varcoe, J. R. (2009). Techniques for the study and development of microbial fuel cells: An electrochemical perspective. *Chemical Society Reviews, 38*, 1926–1939.

Chapter 3
Characteristics of Microbes Involved in Microbial Fuel Cell

Jhansi L. Varanasi and Debabrata Das

3.1 Introduction

The ability of certain microorganisms to transfer electrons outside the cell has given rise to plethora of applications. These bacteria utilize different electron acceptors – usually metals like iron, manganese etc. In nature, these bacteria play a pivotal role in carbon cycle, metal oxidation or reduction, removal of waste organic matter, decomposition of aromatic compounds etc. Apart from this, certain bacteria known as exoelectrogens can use solid electrodes as terminal electron acceptors. These bacteria are utilized in microbial fuel cells (MFC) for bioenergy in the form of electricity. An MFC is different from the typical fuel cell. In an MFC, bacteria create electrical power by oxidising the organic matter present in wastewater which stabilizes the same at the same time.

MFCs are made of two compartments, an anode and a cathode, separated by a membrane that is usually selective for positively charged ions. In the anode, fuel is oxidised by microorganisms, generating electrons and protons. The electrons are transferred into the cathode by an external electronic circuit, while the protons can pass through the membrane owing their positive charge. The protons are consumed by combining with oxygen to form water. The reduction step at cathode, known as ORR (oxidation reduction reaction) usually requires a metal catalyst (e.g. platinum) for improving the reaction kinetics. Although, theoretically 1.1 V can be produced from MFCs, it is reported that an average voltage of 0.3–0.6 V is obtained with glucose or acetic acid as substrates (Logan et al. 2006). The unique organisms that degrade organics at anode are known by several names such as exoelectrogens, electroactive bacteria (EAB), anode respiring bacteria (ARB) or electricigens etc.

J.L. Varanasi • D. Das (✉)
Department of Biotechnology, Indian Institute of Technology, Kharagpur, West Bengal 721302, India
e-mail: ddas.iitkgp@gmail.com

D. Das (ed.), *Microbial Fuel Cell*,
https://doi.org/10.1007/978-3-319-66793-5_3

These EAB have the ability to direct the electrons obtained from the degradation of substrates to the solid electrode i.e. anode which acts as terminal electron acceptor. In MFCs, a wide variety of substrates ranging from simple sugars (e.g. glucose) to complex organic matter (e.g. agricultural or industrial wastewater etc.) can be utilized at anode for power generation. This flexibility of EABs for range of substrates makes MFCs ideal process for renewable energy generation.

Although, the knowledge on microbial interactions with solid electron acceptors exists from past ten decades, the detailed mechanisms of electron transfer have only been deduced for certain microorganisms like *Shewanella* sp. or *Geobacter* sp. This is due to the complexity of the nature of the EABs which makes it difficult to decipher the electrogenic route to electrodes *in vitro*. Earlier experiments suggest the use of artificial mediators with non-EAB to undergo the electron transfer process. Later on, electron transfer via self-secreted mediators (mediated electron transfer, MET) or physical attachment vial cell surface proteins or conductive nanowires (direct electron transfer, DET) are identified. The discovery of cathodic grown organisms further complicated the understanding of electron transfer mechanisms. These bacteria have a unique feature of accepting electrons from the solid electrodes, thereby, reversing the electron trajectory from electrode surface to bacterial cell surface. Detailed investigations are required to understand the properties of these cathodically grown bacteria.

Apart from diverse pure species of EABs, studies with enriched mixed consortium have been conducted to enhance power generation of MFCs (Jung and Regan 2007). A mixed consortium is advantageous as it can be directly utilized for wastewater degradation in non-aseptic conditions. To understand the diversity of microbial community in MFCs, different environmental sources have been explored such as sludge from anaerobic digester, cow dung, anaerobic sediment, soil etc. In such systems, species identification from the planktonic and biofilm community have revealed the occurrence of both DET-MET (direct electron transfer-mediated electron transfer) based bacteria (Beecroft et al. 2012). Being isolated from a similar environment, these species are used in MFCs as promising source for the treatment of wastewater treatment plants, brewery effluents etc. Although it is supposed that bacteria may belong to the rare biosphere, they may dominate when electrode are in contact with sample (Chabert et al. 2015).

The present chapter deals with different characteristics of EAB such as their nature, source, different growth conditions and the electron transfer mechanisms (Fig. 3.1). Table 3.1 enlists some of the known electrogenic microorganisms used for MFC applications.

3.2 Electrocigens – Nature and Source

EABs are widely spread in nature and have been isolated from various natural ecosystems. Majority of these bacteria are found in anaerobic environments. Initially it was suggested that most of these bacteria were Gram negative in nature

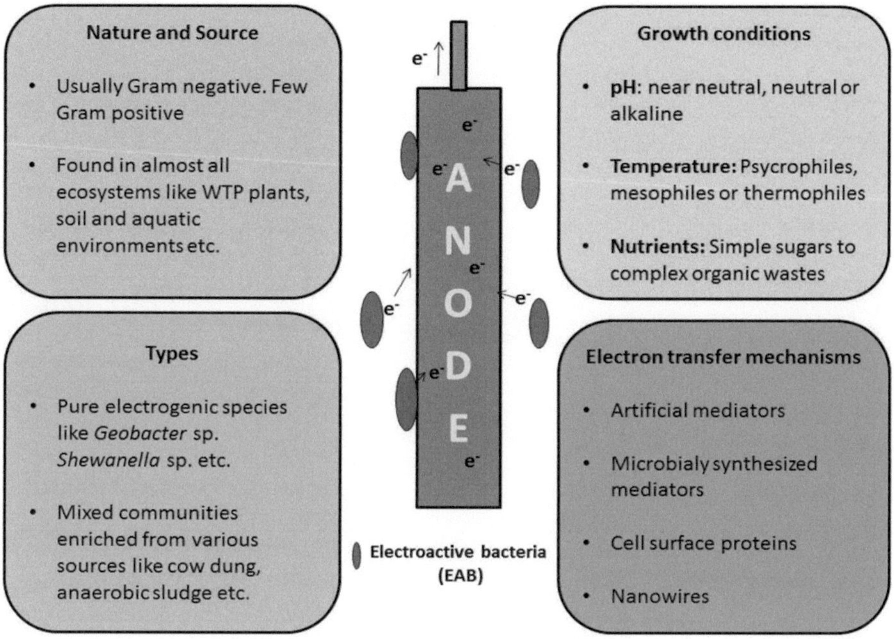

Fig. 3.1 Characteristics of exoelectrogens

belonging to the group Proteobacteria (*Shewanella* sp., *Geobacter* sp. etc.). However, recent advances have shown certain Gram positive bacteria which exhibited electrogenic properties e.g. *Thermincola potens* strain JR. This is a Gram-positive isolate obtained from the anode surface of a microbial fuel cell, using insoluble electron acceptors (Wrighton et al. 2011). EABs are observed in almost all ecosystems (Chabert et al. 2015). Wastewater treatment plants, marine and aquatic sediments, soil, activated sludge etc. are the rich sources for exoelectrogenic bacteria. Knowledge of these sources of EAB would help in identifying the suitable ecosystems to harbour the exoelectrogen of interest. The ability of EABs to generate electricity depends upon several factors such as alkalinity, temperature, nutrient availability, type of electron acceptor etc. Most studies conducted on MFCs use different environmental conditions for different bacteria which make it difficult to compare the power generating capability of these organisms. Since power generation in MFC is cathode limiting, it becomes unclear whether the poor performance observed in MFC is due to lower bacterial activity or other electrochemical losses. Thus it becomes essential to understand the EABs outside MFC environment to understand their actual electrogenic capabilities. The present section deals with the natural and man-made sources of EAB. A comprehensive description of the ecology of electrode specific EABs and their selection and interactions are provided in Chap. 4.

Table 3.1 Electroactive bacteria used in MFCs

Biocatalysts	Electron donor	MFC type	Power density (mW m^{-2})	References
Pure cultures				
Escherichia coli	Complex medium	Single chamber	600	Zhang et al. (2006)
Shewanella oneidensis	Lactate	Miniature	3000	Ringeisen et al. (2006)
Geobacter sulfurreducens	Acetate	H-type	13	Bond and Lovley (2003)
Rhodoferax ferrireducens	Glucose	H-type	33.4	Chaudhuri and Lovley (2003)
Clostridial isolate	Complex medium	Two-chamber	n.a.	Prasad et al. (2006)
Mixed cultures				
Activated sludge	Landfill leachate	H-type	n.a.	You et al. (2006b)
-do-	Glucose	H-type	115.6	You et al. (2006a)
Biofilm from domestic wastewater	Glucose	Single chamber	766	Cheng et al. (2006a)
-do-	Glucose	Single chamber	1540	Cheng et al. (2006b)
-do-	Domestic wastewater	Single chamber	464	-do-
Thermophilic effluent from anaerobic digestion of brewery wastewater	Acetate	Two-chamber	1030	Jong et al. (2006)
Activated sludge	glucose, glutamate	Two-chamber	560	Moon et al. (2006)
Granular anaerobic sludge	Sucrose	Up-flow	n.a.	He et al. (2006)
Fly ash leachate	Fermentation effluent	Two-chamber	85.07	Varanasi et al. (2015)

3.2.1 Natural Sources for EAB

Both terrestrial and aquatic habitats have been extensively exploited for the isolation of EABs. However, since most of these organisms are anaerobic in nature, aerial habitats rarely host them. A wide variety of organic and inorganic matter is present in the terrestrial environment which permits the growth of diverse group of bacteria. Organic soil is rich in nutrients like carbohydrates, vitamins, minerals, amino acid etc. which makes it suitable for growth of EAB. Many reports suggest compost as an ideal source for the isolation of electrogenic bacteria (Chabert et al. 2015). MFCs fed with soil is rich in bacterial communities belonging to the class *Proteobacteria*, *Firmicutes* and *Actinobacteria*, while MFCs fed with composted organic matter is majorly dominated by species belonging to γ-*Proteobacteria* (Mocali et al. 2013). These studies suggest that employing MFCs in soil-based environment could utilize their inherent EABs and can be further used as a tool for the mineralization and humification of the soil organic matter.

Aquatic habitats are observed as abundant of EABs. Most of the known electrogens have been isolated from marine or coastal environments (Chabert et al. 2015). The bacteria isolated from river or beach sediments show highest electrogenic activity. These environments provide proper anaerobic conditions that enhance EAB growth. Due to the abundance of these organisms in sediments, benthic or sediment MFCs have been developed which make use of EABs on site and provide low power applications like autonomous oceanographic or environmental sensors at marine or coastal regions. In fact, due to its success, the first practical application of MFC as a viable power supply was based on the principle of sediment MFC. In their study, they reported the development of a meteorological buoy which measures air temperature, pressure, relative humidity, and water temperature (Tender et al. 2008). These systems proved to be cost effective and durable allowing the average consumption of 100 mW power. Another major application of benthic MFCs is wasteland management which is described in details in Chap. 17. In principle, a benthic or sediment MFC comprises two electrodes that are placed *in situ* such that anode is kept below the sediment while the cathode is kept floating on the water (upper side serving as air cathode). The EABs present in the sediment create reducing power either via direct oxidation of organic compounds or through indirect redox reactions including inorganic compounds. Several researchers have investigated the typical EABs involved in different aquatic sediments like marine or freshwater and have found the dominance of bacteria belonging to class δ-*Proteobacteria*. The most often species among this class of bacteria are the ones belonging to the family *Geobacteraceae* of which *Geobacter sulfurreducens* is the most studied organism (De Schamphelaire et al. 2008).

3.2.2 Artificial Sources for EAB

The industrial or domestic effluents are rich in organic matter which makes them suitable sources for EAB. Many reports have suggested the isolation of EABs from effluents of wastewater treatment plants which include starchy wastewater, municipal wastewater, domestic wastewater etc. (Chabert et al. 2015). In fact majority of these studies utilize these sources directly as inoculum in MFCs. Apart from wastewaters and sewage treatment plants, EABs have also been isolated from other hazardous wastes such as petroleum-hydrocarbon contaminated soils (Zhou et al. 2016), landfill leachate (Iskander et al. 2016), fly ash from thermal power plant (Varanasi et al. 2015) etc. These studies suggest that MFCs have potential applications in bioremediation and recalcitrant waste treatment as described in Chap. 15.

The type of microbial community dominating a mixed EAB depends upon the source. β-*Proteobacteria* and δ-*Proteobacteria* (17%). They are observed to be dominant species when anaerobic sludge is used as inoculum in MFC. *Firmicutes* is the major class dominating fermentable sources while using starch processing wastewater (Oh et al. 2010). Apart from the source of inoculum, the different operating conditions of MFC also influence the bacterial community of the anode (Sect. 3.3). The wide range of diverse microbial communities in anode suggests the possibility of interplay of coexisting bacteria to facilitate electron transfer to the anode via different mechanisms. It is a known fact that in MFCs, mixed culture shows higher performance in terms of power outputs as compared to a single isolated specie (Rabaey et al. 2003). In addition, mixed cultures have the flexibility of adapting to robust environments making them suitable for recalcitrant waste treatment. Another strategy of utilizing EABs for the applications of MFCs in wastewater treatment is the development of synthetic mixed cultures (Curtis and Sloan 2005). These synthetic cultures have specific composition of EAB communities and thus can be used to deduce the electron transfer mechanisms occurring at the anode.

3.3 Growth Conditions of EAB

With the knowledge of diverse group of EAB communities and their ecology it is clear that the dominance of particular species over the other depend upon its surrounding environment, operating conditions, system architectures, electron donors and electron acceptors (at the cathode). Practical application of MFC either in lab-scale or *in situ* requires the need of maintaining stringent environmental conditions for the growth of particular EAB of interest. For the growth of mixed electroactive communities at laboratory conditions, the first fundamental process is to inoculate, acclimatize and enrich at MFC-suited conditions (Logan and Regan 2006). Nevertheless, most of the studies conducted on MFCs make use of different MFC configurations and operating conditions. In such cases which particular species or mixed EAB is capable of producing high power densities cannot be established. Thus, a thorough knowledge of different growth conditions and their effect on microbial community structure would help in determining which operating conditions could be used for maximizing MFCs performance.

3.3.1 pH

The pH in the microenvironment of the anode is the most crucial parameter which influences the growth and substrate metabolism of the EABs. This has a direct role in electron and proton generation from the substrate. Initially it is suggested that high MFC performance can be obtained at pH 6 (Raghavulu et al. 2009). However, later it is observed that influence of pH on performance of MFC is dependent on the type of microbial community (mixed or pure) and the feedstock used (Kaushik and

Chetal 2013). Thus, based on the type of organism, the pH microenvironment is varied in MFC systems. Several bacteria grow best at neutral or near-neutral pH environments i.e. in the range of 6.5–7.5. Most of the EABs fall under this category and thus several studies have been conducted in this optimum range (Patil et al. 2011). Bacterial metabolism usually leads to the production of weak acid compounds to maintain their intracellular pH (He et al. 2008). So, a slightly higher pH is usually recommended while operating MFCs to balance out the biological and electrochemical reactions. However, any slight change in pH can influence the microbial community. A favourable range of anodic pH in MFC is found to vary from 7 to 9 (Kaushik and Chetal 2013). It is a known fact that increased pH levels in the anolyte causes the anode potential to be more negative which in turn improves the MFC performance (Puig et al. 2010).

At extreme pH conditions (>8 or <5), the active sites of the enzymes involved in the biochemical reactions might get damaged that could severely affect the electrogenic activity of microbes. Nevertheless, there are certain exceptions. Many studies suggest the exploitation of acidophiles (pH < 5) for power generation in MFCs (Borole et al. 2008). These studies suggest that low pH conditions has certain advantage of reducing the diffusional limitations as compared to the neutral pH conditions. This indicates that the current generation at the acidic conditions is possible and more favourable. However, chances of electron diverting to alternate electron acceptors apart from anode is higher and it is the major drawbacks of these systems.

Acidophiles and alkaliphiles (pH range of 8.5–11) have also been exploited in MFCs. Operating the bioanode of MFCs at alkaline conditions has several advantages for electricity production such as reduced competition for substrate by methanogens (as methanogens cannot grow above pH 9), lower anode potentials etc. (Dopson et al. 2016). Nevertheless, sustaining such high pH levels in practical application of MFCs would require addition of high-price chemicals which would not be economical. Thus it is essential to obtain a suitable alternative to maintain alkalinity of the anolyte for the improvement in performance of MFCs. Yong et al. (2013) investigated the molecular mechanism behind the increased MFC performance by *S. oneidensis* MR-1 at alkaline pH and reported that in this particular organism the extracellular electron transfer is mediated by riboflavin. The riboflavin concentration increases with time which in turn increases the pH from 6 to 9 and thus leads to higher current generation. Similarly, most of the alkaliphiles used in MFCs are reported to produce exogenous mediators or utilize the mediators released by other coexisting bacteria to facilitate electron transfer to anode (Dopson et al. 2016). Nevertheless, though the alkaliphiles increase the pH of the medium, the anodic oxidation reaction which is strongly acidifying suggests that many of these systems actually operate under near-neutral pH conditions.

Other extremophiles that are reported to be used in MFCs are halophiles or haloalkaliphiles that are capable of growing in high salt conditions. As in alkaline conditions, a high salt concentration also has a positive effect on the performance of MFCs due to increased conductivity of the electrolyte. However, it is debatable whether the bacterial activity or the reduced internal resistance (due to increased conductivity of high salt medium) causes the improvement in MFC performance.

3.3.2 Temperature

Temperature is an important parameter affecting the power generation of MFCs as it impacts the activity of microorganisms and the undergoing electrochemical reactions. Most of these systems are operated in the ambient temperature range i.e. at mesophilic temperatures (20–45 °C) (Venkata Mohan et al. 2008). Any change in temperature of MFC affects its kinetic and thermodynamic properties as well as nature of the microbial communities (Larrosa-Guerrero et al. 2010). The enzymes involved in the biochemical reactions usually perform best at their optimum pH while the rate of electrochemical reactions is directly proportional with temperature. Nevertheless, it is observed that EABs are less sensitive to change in temperature as compared to other bacteria and can adapt themselves to psychrophilic ($<$20 °C) as well as thermophilic temperatures ($>$50 °C) (Michie et al. 2011).

Psychrophilic temperatures have an added advantage of utilizing low-temperature wastewaters directly as substrates (Dopson et al. 2016). Another major application is in the development of biosensors for low power consuming applications. Most of the psychrophiles used in MFCs have been isolated from marine sediments and belong to the group *Geobacteraceae* which are reported to achieve columbic efficiencies as high as 90% (Holmes et al. 2004). However, a major drawback of these systems is the decreasing current over time probably due to inactivation of certain electron transfer mediating enzymes at extreme low temperatures. A possible solution to this problem is the addition of trehalose, a compatible solute to the medium that can protect microorganisms against low temperatures (Dopson et al. 2016). However, such a system would not be economically favourable.

MFCs operated at thermophilic conditions have several benefits such as higher electrogenic activity, increased solubility of feed, lower mass transfer limitations and lowered risk of contamination (Dopson et al. 2016). However, one of the drawbacks is the higher operational and maintenance costs. A community analysis of the thermophilic MFC showed 80% of the population from the *Firmicutes* that were responsible for electricity generation (Wrighton et al. 2008). Further studies suggest the utilization of high-temperature anaerobic sludge as sources of thermophilic inocula in MFCs (Dopson et al. 2016). Potential applications of thermophilic MFCs include the treatment of high-temperature industrial wastewaters including distillery effluent, petroleum refineries, pulp and paper mills, chemical plants etc.

3.3.3 Substrate

In MFC, the type of substrate used as electron donor influences the predominant species of EAB as well as its performance in terms of power density and columbic efficiency (Chae et al. 2009). Different substrates have been explored as feed in MFC from simple carbohydrates (glucose, sucrose etc.) to complex organic compounds found in wastewaters and lignocellulosic biomass (Pant et al. 2010). However, as in case of other parameters like pH and temperature, most of the studies dealing with different substrates are conducted under different operating conditions making it difficult to compare which substrate would provide maximum performance. It is observed that lower molecular compounds like acetate provide much higher columbic efficiencies as compared to simple sugars like glucose probably because acetate cannot be diverted to alternative microbial conversions like fermentation at room temperature. Being the end product of several other metabolic pathways, utilizing acetate in MFCs has an added advantage of enhanced energy recovery by combining with fermentation process of simple sugars.

To achieve higher performance from MFCs, knowledge of the prevailing microbial community and the dominant species that contribute for electricity generation becomes necessary. The microbial community in the anode biofilms is found to vary widely with the type of substrate used. When MFCs are operated with five different substrates (acetic acid, formic acid, lactic acid, succinic acid or ethanol), a considerable phylogenetic diversity in the communities of the anode biofilms are observed that belonged to the groups *Proteobacteria* and *Firmicutes* (Chae et al. 2009). However, when MFCs are operated with three different substrates (glucose, acetate and lactate), not much difference is observed in the microbial composition and most of the community is dominated *Geobacter* species (Jung and Regan 2007). Nevertheless, both the studies are conducted in different operating conditions at different period of time. These results suggest that the community dynamics depends upon several other factors and long-term MFC operation should be considered for understanding the development of microbial communities that develop with different substrates.

3.3.4 Electrode Material and Membranes

Although electrode material and membrane majorly affect the electrochemical properties of MFCs, yet few studies are reported on the change in microbial community dynamics with the change in the electrode material or membrane used. Various carbon-based electrode materials have been used in MFCs such as graphite rod, carbon paper, carbon felt, carbon cloth etc. due to their biocompatible nature (Mashkour and Rahimnejad 2015). These electrode materials have different porosity and conductivity that can affect the bacterial adhesion on the surface (Baranitharan et al. 2015). The bacterial attachment to the anode surface is essential for biofilm formation which in turn leads to electron transfer from bacteria to anode. During biofilm formation, an extracellular polymeric substance (EPS) is released which forms a matrix on the electrode surface. The higher the electrode surface area, higher will be the possibility of uniform biofilm formation. Furthermore, the bacterial colonization increases with the increase in surface roughness. Similar to the anode material, cathode material plays significant role in deciding the type of microbial species dominating the biocathode (Sun et al. 2012). When granular activated carbon, carbon felt cube and granular semicoke are used as electrode materials, most of the dominant species are observed belonging to *Comamonas* of *β-Proteobacteria* while *Acidovorax* is dominant in granular graphite packed MFCs (Sun et al. 2012). Since similar bacteria are found in anodes of MFCs, it can be concluded that the type of electrode can influence both electron donating and electron consuming EABs.

With the introduction of nanoparticle coated electrodes for improving the conductivity in MFCs, new interactions between the nanoparticle-bacteria have emerged that can lead to distinct dominance of one species over the other (Varanasi et al. 2016). As compared to the conventional electrode, nano-modified electrodes have reduced internal resistances that enhance the electron transfer from bacteria to the electrode substratum. However, so far the effect of nanoparticle on the microbial community has not been discussed in details. Different densities of Au particles are decorated on to the carbon paper substratum and the change in microbial community is examined (Alatraktchi et al. 2014). It has been observed that the microbial community of the biofilm differs with the different density of Au nanoparticle loaded. This suggests that any change in electrocatalytic activity of the electrodes can also affect the microbial community structure in MFCs. Thus, these possibilities should be taken into consideration while modifying the electrodes with nano-catalysts.

Apart from electrodes, ion exchange membranes (IEM) have also shown to affect both the growth and community structure of EABs. Ion exchange membranes (like cation exchange membrane or proton exchange membrane) are typically used in MFCs as separators. They are placed in-between anode and cathode for the selective transport of protons from anode (low potential) to cathode (high potential). Three different kinds of membrane (Nafion N-117, Ultrex CMI-7000 and UltrexAMI-7000) are reported and examined on the microbial community of *Eubacteria* and *Archaea* (Sotres et al. 2015). Their experimental results indicate no effect of membranes on *Eubacterial* community but *Archaeal* community is observed to be highly dependent on type of membrane. However, the inter-relation of IEMs with the electrogenic microbial community is not yet clear and further exploration is needed to understand this behaviour.

3.4 Bioelectrogenesis and Mechanisms of Exocellular Electron Transfer (EET)

As explained in the previous sections, the bacteria that are able to transfer electrons outside the cell are exploited in MFCs for electricity generation from organic wastes. Numerous studies have been proposed on the electrons produced from the oxidation of substrates in a bioanode which are transferred to the electrode by means of several mechanisms. These are similar to those in extracellular anaerobic respiration. There are three possible reasons for microorganisms to perform EET (Logan 2009):

(a) for carrying out cell respiration using extracellular electron acceptors, usually metal oxides like iron or sulphate etc.
(b) for interspecies interactions through cellular appendages to fulfil mutual nutrient and energy requirements and
(c) for cell-cell communications which require exogeneous mediators to generate quorum sensing.

In MFCs, the EABs are exploited in such a way that the electrons are diverted from their natural electron acceptors to the solid electrode i.e. anode. In general, the EET mechanisms from EAB to anode have been broadly classified into two major types i.e. indirect or mediated electron transfer and direct electron transfer (Fig. 3.2).

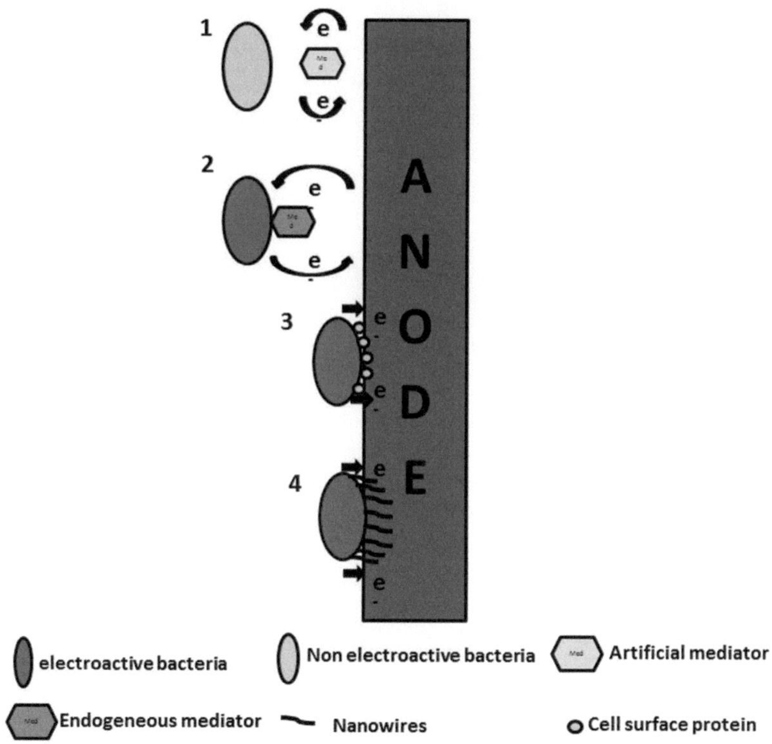

Fig. 3.2 Mechanisms of exocellular electron transfer: (*1*) Indirect/mediated EET in the presence of exogeneous mediator; (*2*) EET mediated by microbialy produced endogeneous mediator; (*3*) Direct EET through cell surface proteins; and (*4*) Bacterial attachment and EET through nanowires

3.4.1 Mediated Electron Transfer (MET)

MET mechanisms represent effective means by which any bacteria can be utilized for electricity generation in MFCs. Mediators or redox shuttles are essential for microorganisms that cannot transfer electrons because they enable electron shuttle from cell membrane to electrode even at larger distance (Logan 2009). On the basis of the type of organism and the redox shuttle utilized for EET, the MET mechanisms can be further classified into artificial MET or MET via endogeneous mediators. Artificial mediators are generally used for specific cases i.e. when non-EABs (like fermentative bacteria) are used in MFC for power generation e.g. introduction of inorganic or organic substances of the type potassium ferricyanide or benzoquinone can facilitate electron transfer from bacteria to the immersed electrodes (Schroder 2007). Other examples of artificial mediators include methyl viologen, anthraquinone 2–6-disulfonate (AQDS), phenazines etc. However, the major disadvantage of using artificial mediators is that it is not economically viable and the fate of the mediator is environmentally questionable. In the absence of soluble electron acceptors and externally available redox mediators, microorganisms evolve themselves to produce low-molecular endogeneous mediators during the secondary metabolic pathway. These secondary metabolites can be replaced with the artificial mediators and make the process more cost effective. Examples of endogeneous mediators include pyocyanine and phenazine-1-carboxamide, produced by *Pseudomonas aeruginosa*, quinone-type redox shuttles in *Shewanella oneidensis* etc.

3.4.2 Direct Electron Transfer (DET)

The DET mechanism occurs via direct physical attachment of bacterial cells to the electrode surface. The major requirement for DET by EAB is to possess membrane bound electron transport proteins or bacterial appendages that can relay electrons from the inside to the outside of the bacterial cell. Example of the cell surface proteins used for EET is c-type cytochromes that are found in *Geobacter* sp. and *Shewanella* sp. These cytochromes majorly comprise multi-heme proteins that are used to shuttle electrons to their natural electron acceptor (iron-oxides). In MFC conditions, anode takes the role of the solid electron acceptor. However, it is argued that only the first layer of organisms that are attached to the electrode surface are electrochemically active while the other bacteria (planktonic) do not play any role in EET (Schroder 2007). Thus further investigations are required to elucidate the electron transfer mechanisms occurring at anode.

Another form of EET by direct physical attachment is through bacterial appendages (pili) or nanowires that electrically conduct electrons to and from the electrode surface. The bacteria possessing nanowires are identified in some *Geobacter* sp. and *Shewanella* sp. These conducting pili offers microorganism to reach and utilize distant solid electron acceptors. It is observed that the bacterial attachment through nanowires leads to the formation of thick electroactive biofilms and thus can produce higher current densities. Nevertheless, though many reports suggest bacterial EET through nanowires, certain contradictory reports suggest that the *c*-type cytochromes indirectly transfer electrons through pili. Thus EET mechanisms of the biofilms have to be further elucidated.

3.5 Factors Affecting EAB Performance in MFC

The EAB performance in MFCs is usually evaluated in terms of bioanode potential and substrate utilization. Several factors contribute in decreasing the EAB performance in MFCs such as mass transfer limitations, bacterial metabolism losses, activation losses etc. which increase the overall overpotential at bioanode (Pham et al. 2009).

3.5.1 Mass Transfer Limitations

The mass transfer limitations at bioanode imply an insufficient supply of substrates to the EAB. During biofilm formation, various products such as intermediates or protons get accumulated which needs to be prevented as these intermediates disturb the activity of EABs. In addition, limitations in proton transfer can lead to pH gradients at anode that can severely affect the growth of bacteria. Limited mass transfer leads to concentration losses in MFC which limit the discharge of redox species from or to the electrode surface. This in turn increases the ratio between the oxidized and the reduced species at the electrode surface which can further lead to an increase in the electrode potential.

3.5.2 Bacterial Metabolism Losses

With the help of electron transport chain, bacteria gains energy by transporting electrons from a substrate of low potential to an electron acceptor at high potential (e.g. oxygen). In MFCs, the anode acts as the terminal electron acceptor and its potential directly relates to the energy gain by bacteria. The higher the difference between the redox potential of the substrate and the anode potential, the higher will be the metabolic energy gain for the bacteria. Thus, for attaining high performance from MFCs, the anode potentials must be as low as possible. However, if the anode potential is too low, the electron transfer gets inhibited by fermentative pathways due to the possibility of higher energy gain through fermentation route.

3.5.3 Activation Losses

To undergo a redox reaction, bacteria need to overcome an energy barrier which leads to activation losses in MFCs. These losses are due to the kinetic limitations in electron transfer from bacteria to the electrode surface. EABs play a vital role in reducing the activation losses. EABs optimize their electron-transferring strategies so as to increase their metabolic energy gain. This in turn would help in improving the electrode catalysis. The electron transfer from the bacteria to the anode can be further enhanced by improving the bacterial-electrode interactions with the help of biocompatible electrode surfaces.

3.5.4 Electron-Quenching Reactions

In mixed microbial communities, several bacteria are present apart from EABs that would compete for substrate in MFCs. These competing microbes can divert the electrons to non electrogenic pathways such as fermentation or methanogens. Thus this would lead to loss of electrons reaching the electrode surface lowering the substrate conversion efficiency to electrical current. Thus, the types of microbes, community composition, interactions between microbes etc. should be considered while operating MFCs for achieving high performances.

3.6 Strategies for Studying EAB

With the discovery of different species of EABs and other communities over the past decade, a proper understanding of the electricity generation capability of these organisms becomes prerequisite so as to obtain high power densities in MFCs. Several microbiological and electrochemical tools have been developed that help to choose the appropriate conditions for a particular species or a group of bacteria (Zhi et al. 2014). Knowledge of these tools would help in identifying the most appropriate method for studying EABs in MFCs.

3.6.1 Microbiological Methods

Microbiological techniques are the most primitive methods used for the characterization of bacterial communities. Isolation of pure cultures helps to understand the morphological, physiological and biochemical characteristics of particular EAB specie of interest. The most common strategy used for isolating pure EAB cultures is dilution-to-extinction method. Single colonies can be isolated using different plating techniques such as streaking, spreading or pour plating. However, as described in previous sections, interplay of mixed EAB communities play a major role in improving performance of MFCs reaching power densities much higher than single isolated specie. To study the population dynamics of a mixed microbial community, most probable number (MPN) method is usually employed for enumeration of viable bacteria of interest. The cultures are first incubated and then assessed by colony counting or microscopic counts.

3.6.2 *Molecular Methods*

The 16S rRNA-gene sequencing is the most widely used method for identification of pure bacterial species as well as for understanding the mixed community dynamics. Other methods used for analysing the community structures include fingerprinting methods like DGGE (denaturing gradient gel electrophoresis). In DGGE, DNA fragments are separated on the basis of their mobilites under increasingly denaturing conditions. Thus DGGE helps in rapid analysis of multiple samples making it a useful technique for studying EAB community dynamics in MFC. Other methods used for studying the microbial community dynamics in MFCs include formation of clone libraries, fluorescence *in situ* hybridization (FISH), pyrosequencing and DNA microarray. The details of these techniques can be found elsewhere.

3.6.3 *Electrochemical Methods*

Several electrochemical methods that are used for chemical fuel cells have been modified for application in MFCs like polarization techniques, cyclic voltammetry (CV), electrochemical impedance spectroscopy (EIS) etc. (Zhi et al. 2014). These methods are usually employed to study the kinetics of single electrode reactions. While studying MFCs using these techniques, many microbial reactions including the undergoing electron transfer mechanisms can be elucidated. Polarization techniques help in understanding the different overpotentials incurred by the MFC system. EIS analysis provides the detailed information on most influential resistance prevailing in the MFC system by the analysis of Nyquists and Bode plots. CV is a basic tool that has been regularly exploited to study the redox processes, electron discharge patterns with respect to time, the metabolic changes, and the carriers involved during electron transfer in MFC. These principles and methodology of these techniques are discussed in detail in Chap. 13.

3.7 Microbial Composition of Biocathode

Like exoelectrogens, there are other EABs known as exoelectrotrophs that have the ability to accept electrons from the electrode (Lovley 2011). These organisms are exploited in biocathode serving as an economic alternative to chemical or oxygen-based electron acceptors in cathodes. Moreover, by using biocathode a plethora of naturally available electron acceptors can be utilized in MFCs which can further be used for the generation of value-added products (Huang et al. 2011). Furthermore, cathode biofilm communities help in degradation of environmental contaminants like nitrates, radioactive waste, toxic heavy metals etc.

Biocathodes are of two major types: aerobic (which use oxygen as final electron acceptor) and anaerobic (which use inorganic compounds like nitrates, sulfates etc. as final electron acceptors). The electrotrophic properties are first identified in *Geobacter* sp. with nitrate and fumarate as electron acceptors and cathode poised at −500 mV (vs. Ag/AgCl) (Gregory et al. 2004). Other anaerobic electrotrophs identified thus far include some *Pseudomonas* sp. and *Desulfovibrio* sp. (Huang et al. 2011). In case of aerobic electrotrophs, several species have been identified which include *P. aeruginosa, Shewanella putrefaciens, Escherichia coli, Enterobacter cloacae, Bacillus subtilis* etc. (Rosenbaum et al. 2011). However, with the emergence of mixed microbial communities in biocathodes, several other electrotrophs and their electron transfer pathways remain undiscovered. Also, as in case of electron transfer to anode, the cathodic electron transfer mechanisms are poorly understood. However, possibilities of both indirect (through redox shuttles) and direct electron transfer (via cytochromes) have been observed in few species (Rosenbaum et al. 2011). These mediators/cytochromes function at different redox potentials than those of anode-based shuttles/cytochromes.

3.8 Challenges and Future Prospects

Though many EABs have been discovered over the past decade, the EET mechanisms have been elucidated only for few model organisms like *Shewanella* sp. and *Geobacter* sp. By analysing the different microbial diversity in the EAB biofilms under different environmental conditions, revelation of many new electrogenic species and the EET mechanisms can be expected. Apart from this, the secondary metabolites and the conductive pili (nanowires) from the EABs can be harvested in view of their electrochemical properties and possible economic value. The biofilms grown on the electrode surfaces are of special interest due to their role in biocatalytic applications. A deeper knowledge of the biofilm conductivity and growth kinetics is required to understand their role in natural systems. Based on this understanding, mutagenic strains can be developed that could decipher similar roles with enhanced activity. With the advancements in molecular techniques, the community dynamics could be coupled with metagenomic and transcriptomic analysis. Several efforts have been made to improve the EET to anode by improving the conductivity of the electrode material with the help of nanoparticles. However, the role of these particles and their interactions with the bacteria are still unknown and require further investigations. Since the power output of MFCs is far below than that of fuel cells, utilizing MFCs for integrated energy recovery as a polishing step in the wastewater treatment process could be the best possible near-future application. Moreover, with the recent advances in biocathode applications, cathodic EET mechanisms need to be further elucidated while the possibilities of generating wide range of value added products should be explored.

3.9 Conclusion

The electroactive bacteria are the heart of the MFC technology and a deeper understanding of their physiological and molecular characteristics as well as the electron transfer mechanisms would help in exploiting these organisms for plethora of applications. Understanding the optimum conditions for maximum bacterial activity would help in extracting maximum current densities in MFC systems. Since wide range of feedstock can be degraded by EABs, they have an important role in reducing environmental pollution and treating recalcitrant wastes. Novel and integrative strategies to enhance the microbial activity are warranted to further improve the performance of bioanodes. Feasible applications of EABs in the near future could be in wastewater treatment, biosensor applications and production of valuable products.

References

Alatraktchi, F. A., Zhang, Y., & Angelidaki, I. (2014). Nanomodification of the electrodes in microbial fuel cell: Impact of nanoparticle density on electricity production and microbial community. *Applied Energy, 116*, 216–222.

Baranitharan, E., Khan, M. R., et al. (2015). Effect of biofilm formation on the performance of microbial fuel cell for the treatment of palm oil mill effluent. *Bioprocess and Biosystems Engineering, 38*, 15–24.

Beecroft, N. J., Zhao, F., et al. (2012). Dynamic changes in the microbial community composition in microbial fuel cells fed with sucrose. *Applied Microbiology and Biotechnology, 93*, 423–437.

Bond, D. R., & Lovley, D. R. (2003). Electricity production by geobacter sulfurreducens attached to electrodes. *Applied and Environmental Microbiology, 69*, 1548–1555.

Borole, A. P., O'Neill, H., Tsouris, C., & Cesar, S. (2008). A microbial fuel cell operating at low pH using the acidophile *Acidiphilium cryptum*. *Biotechnology Letters, 30*, 1367–1372.

Chabert, N., Amin Ali, O., & Achouak, W. (2015). All ecosystems potentially host electrogenic bacteria. *Bioelectrochemistry, 106*, 88–96.

Chae, K.-J., Choi, M.-J., Lee, J.-W., Kim, K.-Y., & Kim, I. S. (2009). Effect of different substrates on the performance, bacterial diversity, and bacterial viability in microbial fuel cells. *Bioresource Technology, 100*, 3518–3525.

Chaudhuri, S. K., & Lovley, D. R. (2003). Electricity generation by direct oxidation of glucose in mediatorless microbial fuel cells. *Nature Biotechnology, 21*, 1229–1232.
Cheng, S., Liu, H., & Logan, B. E. (2006a). Increased performance of single-chamber microbial fuel cells using an improved cathode structure. *Electrochemistry Communications, 8*, 489–494.
Cheng, S., Liu, H., & Logan, B. E. (2006b). Increased power generation in a continuous flow MFC with advective flow through the porous anode and reduced electrode spacing. *Environmental Science & Technology, 40*, 2426–2432.
Curtis, T. P., & Sloan, W. T. (2005). Microbiology. Exploring microbial diversity – A vast below. *Science, 309*, 1331–1333.
De Schamphelaire, L., Rabaey, K., Boeckx, P., Boon, N., & Verstraete, W. (2008). Outlook for benefits of sediment microbial fuel cells with two bio-electrodes. *Microbial Biotechnology, 1*, 446–462.
Dopson, M., Ni, G., & Sleutels, T. H. (2016). Possibilities for extremophilic microorganisms in microbial electrochemical systems. *FEMS Microbiology Reviews, 40*, 164–181.
Gregory, K. B., Bond, D. R., & Lovley, D. R. (2004). Graphite electrodes as electron donors for anaerobic respiration. *Environmental Microbiology, 6*, 596–604.
He, Z., Wagner, N., Minteer, S. D., & Angenent, L. T. (2006). An upflow microbial fuel cell with an interior cathode: Assessment of the internal resistance by impedance spectroscopy. *Environmental Science & Technology, 40*, 5212–5217.
He, Z., Huang, Y., Manohar, A. K., & Mansfeld, F. (2008). Effect of electrolyte pH on the rate of the anodic and cathodic reactions in an air-cathode microbial fuel cell. *Bioelectrochemistry, 74*, 78–82.
Holmes, D. E., Nicoll, J. S., Bond, D. R., & Lovley, D. R. (2004). Potential role of a novel psychrotolerant member of the family Geobacteraceae, *Geopsychrobacter electrodiphilus* gen. nov., sp. nov., in electricity production by a marine sediment fuel cell. *Applied and Environmental Microbiology, 70*, 6023–6030.
Huang, L., Regan, J. M., & Quan, X. (2011). Electron transfer mechanisms, new applications, and performance of biocathode microbial fuel cells. *Bioresource Technology, 102*, 316–323.
Iskander, S. M., Brazil, B., Novak, J. T., & He, Z. (2016). Resource recovery from landfill leachate using bioelectrochemical systems: Opportunities, challenges, and perspectives. *Bioresource Technology, 201*, 347–354.
Jong, B. C., Kim, B. H., Chang, I. S., Liew, P. W. Y., Choo, Y. F., & Kang, G. S. (2006). Enrichment, performance, and microbial diversity of a thermophilic mediatorless microbial fuel cell. *Environmental Science & Technology, 40*, 6449–6454.
Jung, S., & Regan, J. M. (2007). Comparison of anode bacterial communities and performance in microbial fuel cells with different electron donors. *Applied Microbiology and Biotechnology, 77*, 393–402.
Kaushik, A., & Chetal, A. (2013). Power generation in microbial fuel cell fed with post methanation distillery effluent as a function of pH microenvironment. *Bioresource Technology, 147*, 77–83.
Larrosa-Guerrero, A., Scott, K., Head, I. M., Mateo, F., Ginesta, A., & Godinez, C. (2010). Effect of temperature on the performance of microbial fuel cells. *Fuel, 89*, 3985–3994.
Logan, B. E. (2009). Exoelectrogenic bacteria that power microbial fuel cells. *Nature Reviews. Microbiology, 7*, 375–381.
Logan, B. E., & Regan, J. M. (2006). Electricity-producing bacterial communities in microbial fuel cells. *Trends in Microbiology, 14*, 512–518.
Logan, B. E., Hamelers, B., et al. (2006). Microbial fuel cells: Methodology and technology. *Environmental Science & Technology, 40*, 5181–5192.
Lovley, D. R. (2011). Powering microbes with electricity: Direct electron transfer from electrodes to microbes. *Environmental Microbiology Reports, 3*, 27–35.
Mashkour, M., & Rahimnejad, M. (2015). Effect of various carbon-based cathode electrodes on the performance of microbial fuel cell. *Biofuel Research Journal, 2*, 296–300.

Michie, I. S., Kim, J. R., Dinsdale, R. M., Guwy, A. J., & Premier, G. C. (2011). The influence of psychrophilic and mesophilic start-up temperature on microbial fuel cell system performance. *Energy & Environmental Science, 4*, 1011–1019.

Mocali, S., Galeffi, C., Perrin, E., Florio, A., Migliore, M., Canganella, F., Bianconi, G., Di Mattia, E., Dell'Abate, M. T., Fani, R., & Benedetti, A. (2013). Alteration of bacterial communities and organic matter in microbial fuel cells (MFCs) supplied with soil and organic fertilizer. *Applied Microbiology and Biotechnology, 97*, 1299–1315.

Moon, H., Chang, I. S., & Kim, B. H. (2006). Continuous electricity production from artificial wastewater using a mediator-less microbial fuel cell. *Bioresource Technology, 97*, 621–627.

Oh, S. T., Kim, J. R., Premier, G. C., Lee, T. H., Kim, C., & Sloan, W. T. (2010). Sustainable wastewater treatment: How might microbial fuel cells contribute. *Biotechnology Advances, 28*, 871–881.

Pant, D., Van Bogaert, G., Diels, L., & Vanbroekhoven, K. (2010). A review of the substrates used in microbial fuel cells (MFCs) for sustainable energy production. *Bioresource Technology, 101*, 1533–1543.

Patil, S. A., Harnisch, F., Koch, C., Hubschmann, T., Felzer, I., Carmona-Martinez, A. A., Muller, S., & Schroder, U. (2011). Electroactive mixed culture derived biofilms in microbial bioelectrochemical systems: The role of pH on biofilm formation, performance and composition. *Bioresource Technology, 102*, 9683–9690.

Pham, T. H., Aelterman, P., & Verstraete, W. (2009). Bioanode performance in bioelectrochemical systems: Recent improvements and prospects. *Trends in Biotechnology, 27*, 168–178.

Prasad, D., Sivaram, T. K., Berchmans, S., & Yegnaraman, V. (2006). Microbial fuel cell constructed with a micro-organism isolated from sugar industry effluent. *Journal of Power Sources, Special issue, 160*, 991–996.

Puig, S., Serra, M., Coma, M., Cabré, M., Balaguer, M. D., & Colprim, J. (2010). Effect of pH on nutrient dynamics and electricity production using microbial fuel cells. *Bioresource Technology, 101*, 9594–9599.

Rabaey, K., Lissens, G., Siciliano, S. D., & Verstraete, W. (2003). A microbial fuel cell capable of converting glucose to electricity at high rate and efficiency. *Biotechnology Letters, 25*, 1531–1535.

Raghavulu, S. V., Mohan, S. V., Goud, R. K., & Sarma, P. N. (2009). Effect of anodic pH microenvironment on microbial fuel cell (MFC) performance in concurrence with aerated and ferricyanide catholytes. *Electrochemistry Communications, 11*, 371–375.

Ringeisen, B. R., Henderson, E., & Jones-Meehan, J. M. (2006). High power density from a miniature microbial fuel cell using *Shewanella oneidensis* DSP10. *Environmental Science & Technology, 40*, 2629–2634.

Rosenbaum, M., Aulenta, F., Villano, M., & Angenent, L. T. (2011). Cathodes as electron donors for microbial metabolism: Which extracellular electron transfer mechanisms are involved? *Bioresource Technology, 102*, 324–333.

Schroder, U. (2007). Anodic electron transfer mechanisms in microbial fuel cells and their energy efficiency. *Physical Chemistry Chemical Physics, 9*, 2619–2629.

Sotres, A., Díaz-Marcos, J., et al. (2015). Microbial community dynamics in two-chambered microbial fuel cells: Effect of different ion exchange membranes. *Journal of Chemical Technology and Biotechnology, 90*, 1497–1506.

Sun, Y., Wei, J., Liang, P., & Huang, X. (2012). Microbial community analysis in biocathode microbial fuel cells packed with different materials. *AMB Express, 2*, 21.

Tender, L. M., Gray, S. A., et al. (2008). The first demonstration of a microbial fuel cell as a viable power supply: Powering a meteorological buoy. *Journal of Power Sources, 179*, 571–575.

Varanasi, J. L., Roy, S., Pandit, S., & Das, D. (2015). Improvement of energy recovery from cellobiose by thermophillic dark fermentative hydrogen production followed by microbial fuel cell. *International Journal of Hydrogen Energy, 40*, 8311–8321.

Varanasi, J. L., Nayak, A. K., Sohn, Y., Pradhan, D., & Das, D. (2016). Improvement of power generation of microbial fuel cell by integrating tungsten oxide electrocatalyst with pure or mixed culture biocatalysts. *Electrochimica Acta, 199*, 154–163.

Venkata Mohan, S., Veer Raghavulu, S., & Sarma, P. N. (2008). Biochemical evaluation of bioelectricity production process from anaerobic wastewater treatment in a single chambered microbial fuel cell (MFC) employing glass wool membrane. *Biosensors & Bioelectronics, 23*, 1326–1332.

Wrighton, K. C., Agbo, P., & Coates, J. D. (2008). A novel ecological role of the Firmicutes identified in thermophilic microbial fuel cells. *The ISME Journal, 2*, 1146–1156.

Wrighton, K. C., Thrash, J. C., & Coates, J. D. (2011). Evidence for direct electron transfer by a gram-positive bacterium isolated from a microbial fuel cell. *Applied and Environmental Microbiology, 77*, 7633–7639.

Yong, Y.-C., Cai, Z., Yu, Y.-Y., Chen, P., et al. (2013). Increase of riboflavin biosynthesis un erlies enhancement of extracellular electron transfer of Shewanella in alkaline microbial fuel cells. *Bioresource Technology, 130*, 763–768.

You, S. J., Zhao, Q. L., Jiang, J. Q., Zhang, J. N., & Zhao, S. Q. (2006a). Sustainable approach for leachate treatment: Electricity generation in microbial fuel cell. *Journal of Environmental Science and Health, Part A, Toxic and Hazardous Substance Environmental Engineering, 41*, 2721–2734.

You, S., Zhao, Q., Zhang, J., Jiang, J., & Zhao, S. (2006b). A microbial fuel cell using permanganate as the cathodic electron acceptor. *J. Power Sources, Special issue, 162*, 1409–1415.

Zhang, T., Cui, C., Chen, S., Ai, X., Yang, H., Shen, P., & Peng, Z. (2006). A novel mediatorless microbial fuel cell based on direct biocatalysis of *Escherichia coli*. *Chemical Communications (Cambridge), 21*, 2257–2259.

Zhi, W., Ge, Z., He, Z., & Zhang, H. (2014). Methods for understanding microbial community structures and functions in microbial fuel cells: A review. *Bioresource Technology, 171*, 461–468.

Zhou, L., Deng, D., Zhang, D., Chen, Q., Kang, J., Fan, N., & Liu, Y. (2016). Microbial electricity generation and isolation of exoelectrogenic bacteria based on petroleum hydrocarbon-contaminated soil. *Electroanalysis, 28*(7), 1510–1516.

Chapter 4
Microbial Ecology of Anodic Biofilms: From Species Selection to Microbial Interactions

Jean-Clément Flayac, Eric Trably, and Nicolas Bernet

4.1 Introduction to Electroactive Biofilms

Microorganisms have two main lifestyles: planktonic or sessile. In the planktonic mode, bacteria live in a bulk phase with erratic movements according to hydrodynamics. Advantages of this way of life are the ability to reach new ecological niches and easy access to dissolved substrates. When a planktonic cell attaches to a surface, it becomes sessile. If the bacterial cell multiplies and secretes a polysaccharide matrix on the surface, it forms a structure called a biofilm. Biofilms have many advantages including increase of resistance to antimicrobial agents and the ability of microbes to cooperate for nutrients and/or substrates (Simões et al. 2010). According to the type of interactions existing between microbial biofilms and the surface, the support can be classified as inert (silica), nutritious (hydrocarbons) or artificial (electrode). Biofilms that develop on conductive materials and exchange electrons with them are called "electroactive biofilms". In the case of anode respiring biofilms, i.e. biofilms transferring electrons to the conductive material, the terminal electron acceptor is not chemical (O_2, NO_3^-, Fe(III), Mn(III)) but physical and the bacteria are called electroactive bacteria (EABs).

Three main electron transfer modes between microorganisms and electrodes have been identified: (1) Electroactive bacteria can transfer electrons to an electrode by direct contact due to the presence of redox-active proteins such as *c*-type cytochromes on the outer cell surface; (2) Some microorganisms have the ability to produce electronic shuttles to promote electron transfer to an electrode. The known electronic shuttles are flavins, riboflavins or polysaccharides. They can be reduced by redox-active proteins of the outer membrane, such as *c*-type cytochromes. This

J.-C. Flayac • E. Trably (✉) • N. Bernet
INRA-LBE UR050 Laboratoire de Biotechnologie de Environment, Avenue des Etangs, 11100 Narbonne, France
e-mail: eric.trably@supagro.inra.fr

D. Das (ed.), *Microbial Fuel Cell*,
https://doi.org/10.1007/978-3-319-66793-5_4

is an indirect transfer; and (3) Finally, the third mode is achieved via conductive pili also called nanowires. This is a direct transfer to long range. These nanowires allow electron transfer to the electrode. Furthermore, they can also rearrange and create electrical networks to make the whole biofilm connected, facilitating inter-species electron transfers (Lovley 2012).

4.2 Breakdown of Fermentation Mix End Products

Commonly, simple organic molecules are generated by fermentative biodegradation of more complex molecules. These by-products can be further oxidized in BioElectrochemical Systems (BESs) to produce either electricity (MFCs), hydrogen (MECs), or high value molecules formed at the cathode by an electrochemical reduction process or electro-fermentation (Moscoviz et al. 2016a).

Performances of BESs mainly depend on the type of microbial communities involved in the degradation process. A better understanding of the microbial selection and the ecological interactions existing between microorganisms is primordial prior to the development of BESs at larger scales. In this part, the community structure of exoelectrogenic microbial consortia fed with different fermentation end products is described as well as syntrophic processes that improve the biodegradation kinetics.

4.2.1 Acetate

Acetate is a fermentation end product from acetic acid fermentation. This reaction is carried out by acetic bacteria from carbohydrates, primary alcohols, polyhydric alcohols or aldehydes. In BESs, acetate is widely used as a model substrate in most lab-scale studies with pure or mixed cultures. The highest recorded power density obtained with this substrate is 4.3 Wm^{-2}, with a novel cloth-electrode assembly (CEA) MFC design (Fan et al. 2012). There are two main ways to convert acetate in current, either directly by EABs or through hydrogen production (Fig. 4.1).

The direct way is performed by EABs. Indeed, the acetate oxidation is thermodynamically favourable under standard conditions (Table 4.1). In pure culture, the model bacterium in microbial electrochemistry is *Geobacter sulfurreducens* PCA. *G. sulfurreducens* PCA uses acetate as an electron donor and an electrode as electron acceptor (Caccavo et al. 1994; Bond and Lovley 2003).

In mixed culture, it is important to determine the carbon flow from substrate in the ecosystem to understand the trophic chain. This was achieved by using Stable Isotopic Probing and DGGE in MFC. As expected, *Geobacter* sp. was the most abundant bacteria in the communities (28.3 ± 7.3%) and was also dominant in heavy fractions indicating that this is a key genus in acetate uptake (Song et al. 2015). Other EABs able to use acetate have been described in the literature. Most of

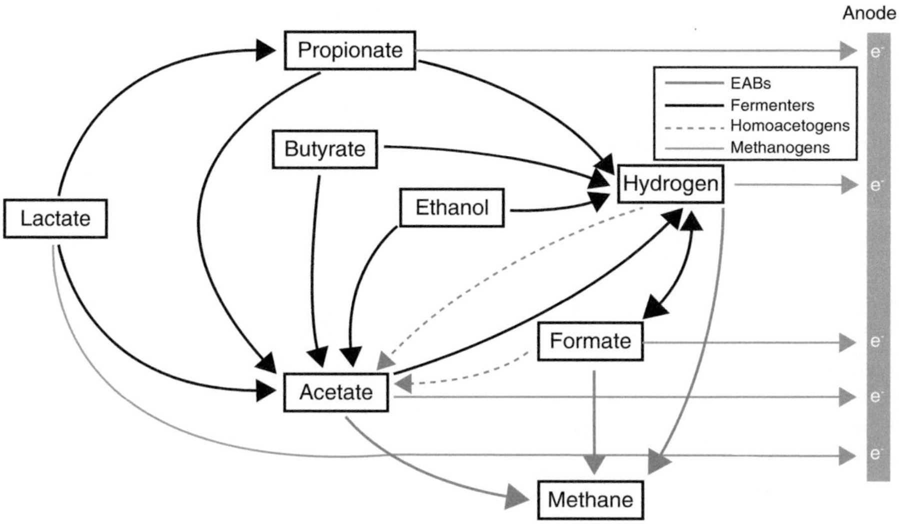

Fig. 4.1 Possible theoretical pathways of electron flow in the anode compartment of BESs from different fermentation end products

Table 4.1 Summary of biochemical reactions during mix end products degradation into BESs

Substrates	Products	$\Delta G^{\circ\prime}$ (kJ mol^{-1})	References
Acetate +4 H_2O	2 HCO_3^- + 9 H^+ + 8 e^-	−35.5	Hari et al. (2016)
Acetate + H^+ + H_2O	2 CO_2 + 4 H_2	+104.6	Thauer et al. (1977)
Formate + H_2O	HCO_3^- + 2 H^+ + 2 e^-	−49.6	Hari et al. (2016)
Formate + H_2O	HCO_3^- + H_2	+1.3	Sun et al. (2012)
Lactate + H_2O	Acetate + CO_2 + 2 H_2	−8.8	Pankhania et al. (1988)
4 Formate + H^+	Acetate +2 HCO_3	−99.1	Hari et al. (2016)
3 Lactate	2 Propionate + Acetate + HCO_3^- + H^+	−164.8	Schink (1997)
Propionate + 5 H_2O	2 CO_2 + HCO_3^- + 14 H^+ + 14 e^-	−72.95	Hari et al. (2016)
Propionate + 3 H_2O	Acetate + H^+ + HCO_3^- + 3 H_2	+76.1	Oh and Logan (2005)
H_2	2 H^+ + 2 e^-	−34.9	Hari et al. (2016)
Propionate + 2 HCO_3^-	Acetate +3 Formate + H^+	+72.2	Hari et al. (2016)
Butyrate +2 H_2O	2 Acetate + H^+ + 2 H_2	+48.1	Thauer et al. (1977)
Ethanol + H_2O	Acetate + H^+ + 2 H_2O	+9.6	Thauer et al. (1977)

them are from the *Geobacteraceae* family, including *Geobacter metallireducens* and *G. anodireducens* (Logan 2009).

The second method for converting acetate to current is through hydrogen production (Table 4.1). Hydrogen scavengers appear to play a key role in the electron flow from acetate towards anode because access to acetate is limited as compared to dissolved chemical electron acceptors which are present in the solution and easily accessible to all microorganisms. Therefore, close to the anode, acetate oxidation is thermodynamically favourable. When the biofilm becomes thicker, electrode becomes unavailable and hydrogen tends to accumulate, increasing its partial pressure. In this case, *G. sulfurreducens* cannot ferment acetate to H_2 because the reaction becomes endergonic (Table 4.1). This reaction can occur if the hydrogen partial pressure is low enough, i.e. in the presence of hydrogen scavengers. Hydrogen scavengers include three different metabolic groups: homo-acetogenic bacteria, hydrogenotrophic methanogens and EABs consuming H_2. Their syntrophic interactions are based on interspecies H_2 transfer (Nath and Das 2004). When methanogenesis was allowed, hydrogenotrophic methanogens (*Methanobacteriales*) were the H_2 scavengers. On the contrary, when this metabolic group was inhibited (with BES), homo-acetogens (*Acetoanaerobium* sp.) became the main channel for electron flow to convert H_2 to current (Parameswaran et al. 2009). Concerning EABs, when hydrogen-consuming partner as *Hydrogenophaga* sp. AR20 is added, acetate can be oxidized by *G. sulfurreducens* since *Hydrogenophaga* sp. AR20 uses hydrogen via NADH and $FADH_2$ and transfers its electrons to the anode by long-range electronic shuttle. This example shows the syntrophic cooperation between EABs and hydrogen-consuming partners which appears to be important in these ecosystems where the electron acceptor is solid (Kimura and Okabe 2013).

In bioelectrochemical systems with mixed microbial cultures, methanogenesis is a strong competitor to current production. Methanogens can indeed metabolize acetate and hydrogen to produce methane, resulting in a trophic competition with EABs towards these two substrates and a strong decrease of coulombic efficiency. In general case, trophic competitions in anaerobic mixed cultures (e.g. dark fermentation) can be controlled by parameters such as pH or hydraulic retention time (HRT). Regarding BESs, the ability to vary resistance is an additional way to control competition pathways. To better understand the resistance effects on anodic microbial communities, Jung and Regan (2011)) varied this parameter and measured methane production. In lower external resistance systems, the substrate consumption rate was higher due to increased rates of electrogenesis (327 μeq/day), low methane production (13 μmol) and high coulombic efficiency (67%). In contrast, high external resistance induces low rates of electrogenesis (43 μeq/day), increase in methanogenic activity (28 μmol of CH_4) and decrease coulombic efficiency (CE 25%). Bacterial 16S rDNA genes were analysed by DGGE. The methanogenic community was dominated by *Methanosaetaceae* (acetoclastic methanogenesis) and *Methanomicrobiales* (hydrogenotrophic methanogenesis), two methanogens families which could interact with EABs to recover electrons (instead of hydrogen) to form methane (Jung and Regan 2011).

4.2.2 Formate

Formic acid is a short chain organic acid resulting from glucose fermentation. In practice, the maximum power density reported with formic acid was 924 mWm^{-2} (Sun et al. 2012). Microbial community profiles in MFC powered with formate suggest three different mechanisms to convert this substrate in current: direct conversion by EABs, indirect conversion through homoacetogenesis or hydrogen (Fig. 4.1).

The direct oxidation of formate by EABs is exergonic under standard condition (Table 4.1). It was observed after addition of an electron-carrying mediator such as humic acids or the humate analogue anthraquinone-2,6-disulfonate (AQDS), by the Gram-positive, spore-forming bacterium *Desulfitobacterium hafniense* DCB2. The current generated was up to 400 mWm^{-2} of cathodic surface in a single-chamber MFC, with a platinum-containing air-fed cathode (Milliken and May 2007). This conversion is also possible without addition of electronic shuttles by *Paracoccus denitrificans* found at 39% abundance in an anodic biofilm of a formate-fed MFC. This bacterium is capable of producing hydrogen from formate and also transfer electrons via flavin shuttle to anode (Kiely et al. 2010).

The second probable way to convert formate to electrical current is a first conversion of formic acid to acetic acid by homoacetogens, an exergonic reaction under standard conditions (Table 4.1). *Acetobacterium* sp. was found at ~25% abundance in anodic communities able to use formate and produce acetate (Balch et al. 1977). Therefore, it is an energetically interesting way for microorganisms. Acetate can be further converted to current by EABs such as *Geobacter sulfurreducens* which represents ~50% of the anodic communities (Sun et al. 2012).

The third method is the conversion of formate to hydrogen, an endergonic reaction which implies an obligatory syntrophy between microorganisms to keep the hydrogen partial pressure low (Table 4.1). In a formic acid-fed MFC, *Paracoccus denitrificans* dominated the biofilm (26–36%) by using a formate dehydrogenase to oxidize formate to carbon dioxide and hydrogen. As hydrogen scavenger, *Geobacter sulfurreducens* converted hydrogen to current and kept the partial pressure of hydrogen very low (Sun et al. 2012).

4.2.3 Lactate

Lactate is a major product of glucose fermentation. Power density from lactate of 474 mW m^{-2} was obtained in MFC, approximately 15% less than using acetate (Kiely et al. 2011b). In BESs, lactate can be converted to electrical current by two main pathways. The first one is the direct conversion by EABs and in the second, lactate is fermented to acetate and propionate that are further used by EABs (Fig. 4.1).

Pure culture experiments showed the direct oxidation of lactate by EABs without electronic shuttle. The first reaction is the intracellular conversion of lactate to acetate, followed by acetate and hydrogen oxidation (Table 4.1). The overall reaction is highly exergonic. More particularly, *Geobacter sulfurreducens* PCA oxidizes lactate to pyruvate, further converted to acetate that can be oxidized at the anode (Call and Logan 2011).

The use of mixed culture helped to highlight the second conversion mode of lactate to current through fermentation process. A 1-year experiment showed the significant presence of lactic acid fermenters. *Pelobacter propionicus* has been found at 39% in an anodic biofilm with lactate as sole fuel. *P. propionicus* oxidizes lactate in propionate and acetate in a 2:1 molar ratio (Table 4.1). *Geobacter sulfurreducens* was present at 7%, suggesting a fermentation pathway to acetate and consequently a possible syntrophy with *P. propionicus* (Kiely et al. 2011a, b).

4.2.4 Propionate

Propionate is a common product in anaerobic ecosystems such as methanogenic systems. Power density produced from propionate is not very high (~115 mW m^{-2}) in MFC as compared to acetate (De Cárcer et al. 2011). Its oxidation involves complex microbial interactions that are currently little known. Based on analysis of microbial communities, three oxidation pathways of propionate were characterized. The electron flow can be direct to anode by EABs or through fermentation depending on the microbial communities present in the electroactive biofilm (Fig. 4.1).

The first degradation pathway is the direct exergonic oxidation of propionate by EABs (Table 4.1). Tests in pure culture revealed that *Geothrix fermentans* is able to oxidize propionate with two electron transfer modes: with or without electronic shuttles. This EAB can oxidize other substrates of interest such as acetate, lactate, malate and succinate (Bond and Lovley 2005).

In mixed culture, DGGE analysis revealed the presence of *Geovibrio ferrireducens* in microbial fuel cell fed with propionate as sole electron donor. This bacterium has *c*-type cytochromes for a direct flow of electrons to the anode (Freguia et al. 2010).

The second pathway is through a first fermentation step to acetate and hydrogen. However, this reaction is endergonic in standard conditions and a syntrophy between fermenters and hydrogen scavengers is necessary. A bacterial strain with 96% similarity with *Propionibacteriaceae* has been identified in a propionate-fed MFC. These bacteria can ferment propionate to acetate (Yu et al. 2012). In MEC reactors fed with different propionate concentrations (4.5 and 36 mM), *Rhodocyclaceae*, *Clostridiales* and *Bacteroidetes* were the probable propionate fermenters (Hari et al. 2016). They were reported to oxidize propionate and were dominant in propionate- and acetate-fed MECs (Chauhan et al. 2004; Kragelund et al. 2008; Hesselsoe et al. 2009; Ruiz et al. 2014).

Hydrogen can be used by three metabolic groups (Fig. 4.1). Concerning EABs, OTUs close to *Geobacter sulfurreducens* PCA and *G. sulfurreducens* subsp. *ethanolicus* were observed in propionate-fed MFC and are both capable of using hydrogen to produce current. In addition, some methane was detected (1.2–2.3%), suggesting that EABs or homo-acetogens were effective as compared to methanogens. In this study, the major electron sink was the anode (60–80% of total electrons) (Hari et al. 2016). The second co-product, acetate, was used more effectively by EABs as compared to methanogens. *G. sulfurreducens* can outcompete acetoclastic methanogens because of kinetic benefits with K_s 0.64 mg COD/L and 177–427 mg COD/L respectively (Gao et al. 2014). *G. sulfurreducens* subsp. *ethanolicus* seems capable of converting hydrogen into electricity (Hari et al. 2016). Contrary to the preceding genus, *Pseudomonas sp.* uses electronic shuttles (pyocyanines) for electron transfer and was found at 99% similarity with band sequence in DGGE analysis of propionate-MFC (Yu et al. 2012).

The third pathway is through a first fermentation step in acetate and formate. Propionate oxidation to acetate and formate is endergonic with $\Delta G^{\circ\prime} = +72.2\ \mathrm{kJmol^{-1}}$ (Table 4.1). Microorganisms able to achieve this step are identical to those of the second pathway acetate/H_2 (*Propionibacteriaceae, Rhodocyclaceae*, *Clostridiales* and *Bacteroidetes*). Likewise, formate oxidation is carried out by the same EABs oxidizing hydrogen (*G. sulfurreducens* PCA, *G. sulfurreducens* subsp. *ethanolicus*) (Hari et al. 2016).

It is not established which is the preferred fermentation pathway (acetate/H_2 or acetate/formate). In anaerobic digestion of propionate, it was shown that interspecies electron transfer through formate was the main mechanism and another study suggested that hydrogen transfer is the preferable way in propionate degrading consortia (Boone et al. 1989; Schmidt and Ahring 1995). Another hypothesis is that hydrogen transfer is efficient at short distance as compared to formate which is dominant way to inter-species electron transfer over a long distance (Bok et al. 2004). More recently, biochemical and genome analysis of pure culture of syntrophic volatile acid showed that degraders use electron transfer via hydrogen and formate simultaneously (Müller et al. 2010). An important parameter which influences the production of hydrogen or formate is pH. The shift from formate to hydrogen during glucose fermentation is due to a pH decrease (Temudo et al. 2007). Formate conversion to hydrogen is endergonic ($\Delta\ G^{\circ\prime} = +1.3\ \mathrm{kJmol^{-1}}$) but very close to the thermodynamic equilibrium (Table 4.1). This reaction is catalyzed by the formate hydrogen lyase complex that is reversible. At 25 °C, the pK_a of carbonate is 6.37, so it is stable above this value and inhibit formate splitting into carbonate and hydrogen (Moscoviz et al. 2016a, b). It is possible that pH has also an influence on the choice of propionate degradation pathway with the way acetate/H_2 when pH medium is below 6.37 and acetate/formate when pH is above. Therefore, pH changes influence the fermentation products from propionate and may also influence certain EABs (hydrogenotrophic or formate oxidizers) and finally on the reactor performances.

Thus, electron transfer mechanisms in propionate fermentation are poorly understood. A better knowledge of propionate fermenters and their interactions with EABs could help to understand the laws governing the electron flow from propionate in BESs.

4.2.5 Butyrate

Butyrate is one of the main compounds generated during the fermentation of organic matter in anoxic environments (Hatamoto et al. 2008). In such environment, further butyrate degradation is performed by syntrophic interactions of butyrate-oxidizing bacteria and hydrogen-scavengers because of thermodynamic constrains in standard conditions (Schink 1997). In BESs, the power produced by butyrate (1000 mg L^{-1}, 305 mWm^{-2}) was 66% lower than that fed with acetate (800 mg L^{-1}, 506 mW m^{-2}) (Liu et al. 2005). The microbial ecology studies in butyrate-fed BESs with mixed cultures reveal only one major pathway for complete oxidation to current with a first fermentation step and a second one to convert products into current (Fig. 4.1).

The first step is butyrate oxidation to acetate and hydrogen. This reaction is endergonic with $\Delta G^{\circ\prime} = +48.1$ $kJmol^{-1}$ and must involve syntrophic interactions to decrease the hydrogen partial pressure and make the reaction possible (Table 4.1). The most abundant fermenter is *Pelomonas saccharophila*, found as a major band in DGGE analysis from MFC. This bacteria has a high metabolic versatility and may use more than 40 carbon sources and, therefore, appears to be involved in butyrate oxidation (Xie and Yokota 2005; Freguia et al. 2010).

The second step is the oxidation of acetate and hydrogen to current. These two reactions are thermodynamically favourable and can be performed by *Geobacter sulfurreducens* PCA present at 13.6% in anodic biofilm of butyrate-fed MFC. Organisms from *Geobacteraceae* family were also found in MEC (9.1%). It seems that these EABs can grow in syntrophy with butyrate oxidizers. Their relative low abundance could be due to the presence of methanogenic competitors (54%), explaining the low coulombic efficiencies of BESs powered with butyrate (Chae et al. 2009; Popov et al. 2016).

As observed in different studies, the conversion of butyrate into current is poorly efficient with only ~10% of electrons collected as electricity. Methanogenesis seems to be the main explanation for this loss of efficiency. Strategies should be found to inhibit this electron flux towards methane and increase the syntrophy between butyrate-oxidizers and exoelectrogenic hydrogen-scavengers, which should result in redirecting the electron flow to current.

4.2.6 Ethanol

Ethanol is a common product of glucose fermentation at neutral pH and is a key breakdown product of cellulose fermentation (Temudo et al. 2007; Lalaurette et al. 2009). Power density produced from ethanol is ~800 mW m^{-2} in MFC, about five time less compared to acetate (Kiely et al. 2011a, b). Mixed culture experiments reveal that ethanol is first fermented to acetate in BESs (Fig. 4.1). This reaction is not thermodynamically favourable with $G^{\circ\prime} = +9.6$ $kJmol^{-1}$ which implies again syntrophy with hydrogen scavengers (Table 4.1). *Pelobacter* species could be involved in this reaction. According to community profiles, *Pelobacter* was present at 85–98% in solution and 35–43% in electroactive biofilm of microbial electrolysis cells fed with ethanol (Parameswaran et al. 2010).

Current was produced in an MFC using a coculture of *P. carbinolicus* and *G. sulfurreducens* with ethanol as the fuel. *P. carbinolicus* can oxidize ethanol but is unable to produce current. *G. sulfurreducens* alone cannot metabolize ethanol but oxidizes acetate and hydrogen from ethanol oxidation by *P. carbinolicus*, utilizing the anode as an electron acceptor. In this syntrophy up to 83% of the electrons available in ethanol were collected as electricity and acetate (Richter et al. 2007).

Acetate is then converted to current. Among the *Geobacteraceae* family, *G. metallireducens* was found in marine sediments fuel cell and can directly oxidize ethanol to current via *c*-type cytochromes but are unable to oxide hydrogen (Bond et al. 2002; Lovley et al. 1993). EABs other than *Geobacter* sp. can interact with *Pelobacter propionicus* to convert acetate to current as *Rhodopseudomonas palustris* found at 27% on anode biofilm (with 38% of *Geobacter* sp.) (Kiely et al. 2011a, b). This purple non-sulphur photosynthetic bacterium could generate high current densities (2.2 $W.m^{-2}$) without light and transfer electron through anode via *c*-type cytochromes (Cheng and Regan 2008).

Hydrogen produced from ethanol fermentation can be used by three metabolic groups as described in Fig. 4.1. Therefore, when ethanol is the only electron donor in BES, they are a three-way syntrophy among ethanol fermenters, acetate-oxidizing anode-respiring bacteria and H_2 scavengers. With this information, it would be interesting to find solution in order to promote homo-acetogenic bacteria against methanogenic archaea to increase the coulombic efficiency (Parameswaran et al. 2010). For example, it is possible to oxidize ethanol with co-culture without hydrogen production in micro-electrolysis cell. *Acetobacter aceti* oxidizes ethanol to acetate and uses electronic shuttles (Ferrocyanide) to regenerate NADH in NAD^+ instead of producing hydrogen. Acetate is then oxidized in carbon dioxide by *Escherichia coli* salting-out its electrons by the same electronic shuttle than *A. aceti*. This method enables high coulombic efficiency 87% (10.4 electrons out of 12 from ethanol molecule converted in current) but requires artificial electronic shuttles. This cascade reaction through two bacteria can be an effective ethanol-converting path in current without H_2 production, thus avoiding competition with H_2-oxidizing methanogens but with addition of artificial electronic shuttle (Matsumoto et al. 2015).

4.3 Breakdown of Glucose

Conversion of glucose to current is an inefficient process regarding coulombic efficiency as compared to the substrates with lower molecular weight. It is probably due to fermentable nature of this substrate (Rabaey et al. 2003). For example, in mixed culture, the coulombic efficiency is ~4.8 times lower with glucose as compared to acetate in MFC (Chae et al. 2009). Indeed, glucose may be oxidized

Table 4.2 Summary of biochemical reactions during glucose degradation into BESs

Substrates	Products	$\Delta G^{\circ\prime}$ ($kJ.mol^{-1}$)	References
Glucose +6 H_2O	6 CO_2 + 24 H^+ + 24 e^-	−1438	Freguia et al. (2008)
Glucose +2 H_2O	2 Acetate +2 CO_2 + 4 H_2 + 3 H^+	−216	Freguia et al. (2008)
Glucose	2 Pyruvate +2 H_2 + 2 H^+	−112.1	Thauer et al. (1977)
Glucose	2 Lactate +2 H^+	−197	Freguia et al. (2008)
Pyruvate + H_2	Lactate	−43.2	Pankhania et al. (1988)
Pyruvate + H_2	Propionate + H_2O	−123	Thauer et al. (1977)
Glucose + HCO_3^-	Succinate + Acetate + Formate	−144	Thiele (1991)
Glucose	2 Ethanol +2 CO_2	−235	Freguia et al. (2008)

to many different molecules, which engages competitive processes between microorganisms for obtaining these products (Table 4.2). EABs producing high current densities have limited metabolic versatility, which could prevent them from converting all the fermentation products from glucose to current and allow the possibility of producing methane, the main concurrent route by diverting the electrons flow (Call and Logan 2008). Microbial ecology studies provide insight into different possible ways of transferring glucose electrons to the anode. All mentioned microorganisms were observed in glucose-fed BESs. This conversion seems to take place in two major steps: fermentation of glucose to pyruvate (glycolysis) then mix of end product and conversion of these compounds in current with EABs through syntrophic interactions (Selembo et al. 2009).

4.3.1 Direct Conversion of Glucose to Current

Most direct path is the conversion of glucose into current (Table 4.2). This reaction releases a large amount of energy with $\Delta G^{\circ\prime} = -1438$ $kJmol^{-1}$ with anodic potential of +200 mV vs SHE (Freguia et al. 2008).

Electron transfer from glucose to anode can be achieved through membrane proteins. *Rhodoferax ferrireducens* carries out this transfer and does not need an electron-shuttling mediator to get energy from oxidation process itself during long-term production (Chaudhuri and Lovley 2003). This direct transfer is also made by *Aeromonas* sp. Strongly dominant in the anodic biofilm (~51.6%), this genus is an electroactive bacterium. In the same condition but with open-circuit MFC, the abundance of this bacterium decreases to 10 times, suggesting that the extracellular electron transfer is the main metabolism employed and gives to *Aeromonas* sp. an ecological advantage to dominate all other species in this ecosystem. The isolated strain ISO2–3 from this study could ferment glucose and likely transfer electron to the anode through membrane-associated compounds (most likely *c*-type cytochromes) which require intimate contact with the anode surface. ISO2–3 can also produce anchor-like filamentous appendages (probably pili) allowing a long-range electron transfer across the thick biofilm (Chung and Okabe 2009).

Unlike previous EABs, other bacteria are using electronic shuttles to utilize anode as terminal electron acceptor. Electronic shuttles of *Klebsiella* genus have been well characterized in glucose-fed MFC. Among this genus two species and one strain are able to oxidize glucose to current: *K. pneumoniae, K. oxytoca* and *Klebsiella* sp. ME17. The first one was characterized in MFC where the anode was coated with microfiltration membrane (0.22 μm) to eliminate biofilm influence mechanisms in order to better understand electron transfer. The maximum voltage outputs achieved were 316.2 and 427.2 mV after 270 and 120 h, respectively. Cyclic voltammetry measurements demonstrated the presence of an electronic shuttle produced by *K. pneumoniae* L17 identified to be 2,6-di-tert-butyl-p-benzoquinon (2,6-DTBBQ), as a recyclable mediator able to transfer electrons between *K. pneumoniae* L17 and the electrode (Lifang et al. 2010). This bacterium has been observed in BESs operated with mixed culture at ~6.45% and 0.02%, suggesting that it may have an interesting role for efficient conversion of glucose (Lu et al. 2012; Chung and Okabe 2009). The second, *K. oxytoca,* was found in glucose-fed MFC at ~7.52 whereas it was only at ~3.15% in open-circuit MFC (without electron flux). This result suggests that *K. oxytoca* is favoured by the extracellular electron transfer mechanism. *Klebsiella* sp. ME17 was used in an H-type MFC with glucose as electron donor. The maximum power density produced by this strain was 1.209 mW m^{-2} with 340 and 1.47 mA of maximum current. The polarization curves illustrated that the strain produced electronic shuttles to promote extracellular electron transfer. Based on cyclic voltammogram, the supernatant was very likely to contain quinone-like substances. However, even after replacing the medium, current was always produced, suggesting that *Klebsiella* sp. ME17 could also transfer electrons through direct contact with electrode (Xia et al. 2010).

Some EABs use exogenous chemical mediators to convert glucose to current such as *Shewanella* genus. Thus, *S. algae* was observed in MFC exposed to light and can use glucose, ribose and fructose as carbon and energy sources (Xing et al. 2009). It is a facultative anaerobe able to produce hydrogen sulfide and reduce Fe (III), making it very corrosive on metal surface (Holt et al. 2005). The iron reduction feature and membership in the *Shewanella* genus known to be exoelectrogen, suggests that *S. algae* could be an exoelectrogen whose function is the conversion of glucose to current (Xing et al. 2009). Similarly, *Lactococcus lactis* uses exogenous mediators but does not convert all glucose electrons into current, but also in acetate. This Gram positive bacterium is normally homolactic fermenter able to produce several kinds of membrane associated quinones and to mediate electron transfer with extracellular electron acceptors such as Fe(III), Cu^{2+} and hexacyanoferrate. *L. lactis* can also perform extracellular electron transfer to anode with two soluble redox mediators suggested to be 2-amino-3-dicarboxy-1,4-naphthoquinone (ACNQ). In classic fermentation, *L. lactis* performs the homolactic metabolism while during electro-fermentation acetate and pyruvate are excreted in stoichiometric amounts with respect to the electrical current. This is the first Gram positive bacteria described capable of extracellular electron transfer (Freguia et al. 2009).

4.3.2 Glucose Fermentation to Mixed End Products

4.3.2.1 Glucose to Acetate and Hydrogen

In BES, glucose can be first fermented into acetate and hydrogen. This reaction is thermodynamically favourable with $\Delta G^{\circ\prime} = -216$ kJ mol^{-1} and is the main route of fermentation (Table 4.2). A clone close to *Acetobacterium paludosum* (90% of similarity) was found as dominant in glucose-fed MFC (Choo et al. 2006). This acetogenic bacteria can ferment different sugars to acetate (Kotsyurbenko et al. 1995). Found in minority as compared to the previous genus, DGGE analysis shows a band corresponding to *Clostridium* (9% in abundance). Among these genus, *C. chromiireducens* found at ~10% in MFC against ~3% in MFC with open circuit suggest that these bacteria are in interaction with EABs for acetate production. This bacterium is able to ferment over fifteen carbohydrates to acetate but also butyrate, formate and lactate. In the same MFC, *Enterobacter asburiae* was present at ~7.5% against 4% in open circuit. *E. asburiae* can ferment glucose and several other compounds (Brenner et al. 1986).

4.3.2.2 Glucose to Lactate

Lactic acid fermentation is a biochemical process by which glucose is converted in two steps to cellular energy and lactate (Luedekingt and Piret 1959). The first step is pyruvate production ($\Delta G^{\circ\prime} = -112.1$ kJmol^{-1}) and is able to ferment glucose to lactate but also two α-galactosides, raffinose and melibiose as well as lactose (β-galactosides) and galactose, making it a fermenter of interest in the case of fermentation of complex substrates (e.g. dairy industry) (Boucher et al. 2003). In the same study, *Dysgonomonas* sp. was found at 15.6%. This genus is able to oxidize approximately fifteen carbohydrates including glucose with production of acid but no gas (Hofstad et al. 2000). However, it was isolated from human clinical specimens with a maximum growth temperature from 25 to 37 °C. Its presence in a biofilm at 4 °C could be attributed to complex syntrophic interactions in this biofilm which could accelerate the metabolic rate of *Dysgonomonas* to oxidize substrates in psychrophilic conditions (Lu et al. 2012). So, there is a difference between conventional lactic acid bacteria (*Lactobacillus, Streptococcus* and *Leuconostoc*) and those present with EABs.

4.3.2.3 Glucose to Propionate

Glucose can be fermented to propionate through the intermediate pyruvate. Pyruvate conversion to propionate is thermodynamically favourable with $\Delta G^{\circ\prime} = -123$ kJ mol^{-1} (Table 4.2). The overall energy balance of propionate oxidation from glucose is $\Delta G^{\circ\prime} = -358.1$ kJ mol^{-1}. The fermenter *Paludibacter propionicigenes* produces propionate and acetate from glucose at a ratio of 2:1 in MFC at ~6% in abundance (Lu et al. 2012). This bacterium produces two different metabolites and should be in interaction with EABs able to convert them in current.

4.3.2.4 Glucose to Succinate, Acetate and Formate

Glucose may be fermented to succinate, acetate and formate concomitantly. Succinate is an intermediate metabolite before propionate. The overall energy balance is exergonic with $\Delta G^{\circ\prime} = -144.0$ kJ mol^{-1} (Table 4.2). Therefore, bacteria are common with the path of acetate and hydrogen fermentation. *Enterobacter asburiae* can produce many molecules from glucose including succinate, acetate and formate which could be in syntrophic relationship with electroactive bacteria because its abundance is affected by the anodic availability (Chung and Okabe 2009).

4.3.2.5 Glucose to Butyrate

Butyrate can be produced from glucose. After producing pyruvate, it is converted to butyrate by classical routes of acetyl-S coenzyme A condensation (Miller and Wolin 1996). The reaction is exergonic with $\Delta G^{\circ\prime} = -264.8$ kJ mol^{-1} (Table 4.2). Two microorganisms found in MFC are able to perform this reaction: *Clostridium chromiireducens* and *Enterobacter asburiae*, present at ~10% and ~7.5% respectively in an anodic biofilm (Chung and Okabe 2009). The abundance of the bacteria is 50% less in the case of an open circuit (no electron transfer through anode) which shows a possible interaction with EABs. These two bacteria can also produce acetate which would thus preferentially choose this metabolite for direct syntrophy with EABs to have greater energy efficiency. No EAB is able to convert directly butyrate to current, suggesting that this pathway is not present in the reactor.

4.3.2.6 Glucose to Ethanol

Glucose can be also converted to ethanol through alcoholic fermentation. In this pathway, one glucose molecule produces two pyruvates. The two molecules produced are then broken down into two acetaldehydes. Then, the two acetaldehydes are converted to two ethanol utilizing protons from NADH regenerating in NAD^+. Glucose into ethanol is thermodynamically favourable with $\Delta G^{\circ\prime} = -235$ kJ mol^{-1} (Table 4.2). *Escherichia coli* was detected in an MFC at ~1% in electroactive biofilm. Its low abundance appears due to the presence of other glucose fermenting organisms such as *Clostridium* (~18% in abundance) (Xing et al. 2009). *E. coli* can generate ethanol with the key enzyme: alcohol dehydrogenase. However, this bacterium can also produce other molecules from glucose such as acetate. Its low abundance and production of many other products appears to minimize ethanol production. No known EAB able to produce electricity from ethanol was found in the biofilm. The possible way to metabolize ethanol is through formate and hydrogen which are easily converted into electricity, even though competitors such as methanogens may emerge.

At this stage, glucose is converted into various fermentation products. These products can then be converted into current by EABs or fermented again to give for example acetate and hydrogen, which may be converted to current. The next sections will describe microorganisms in glucose-fed BESs able to use products from the glucose breakdown to better understand the electroactive ecosystem functioning.

4.3.2.7 Pyruvate to Mixed End Products

Pyruvate is a platform metabolite produced at the end of glycolysis and its fermentation results in the production of mixed end products. Thus, *Sedimentibacter hydroxybenzoicus* was found in microbial fuel cell in minority (~1%) (Xing et al. 2009). This bacterium can live in strict anaerobic condition and growth is supported by the fermentation of pyruvate with acetate, butyrate, isovalerate and isobutyrate produced without hydrogen production (Breitenstein et al. 2002). DGGE analysis showed *Desulfovibrio alcoholovorans* in MFC (Choo et al. 2006). This strict anaerobic and sulphate-reducing organism can use pyruvate as electron donor. However, *D. alcoholovorans* can also use other fermentation end products (H_2, formate, lactate, ethanol) (Qatibi et al. 1991). In view of its metabolic versatility, this bacterium could have different roles in the glucose breakdown. Finally, pyruvate fermentation to propionate and acetate can be carried out by *Pelobacter propionicus*. This bacterium was detected on DGGE band in a glucose-fed MFC (Jung and Regan 2011). In this study, the main exoelectrogen was *Geobacter sulfurreducens* which cannot oxidize propionate. Therefore it seems that for a syntrophic interaction acetate is the major metabolite produce.

4.3.2.8 Lactate to Mixed End Products

Lactate products from lactic acid fermentation may then be converted into simpler molecular product such as acetate or hydrogen. This reaction is exergonic with $\Delta G^{\circ\prime} = -4.2$ kJ mol^{-1} (Table 4.2). As free energy is close to zero, the acidity and hydrogen partial pressure are two major parameters on the reaction balance. Two microorganisms capable of such reaction have been characterized in bioelectrochemical systems. One is *Petrimonas sulfuriphila* able to ferment glucose, but also lactate and more than a dozen other carbohydrate substrates (Grabowski et al. 2005). The second, *Acetobacterium paludosum*, can use lactate as electron donor (Kotsyurbenko et al. 1995). It is interesting to note that these two strains may be present from the beginning to the end of glucose oxidation. Indeed, they can convert glucose to acetate and hydrogen but are in competition with other glucose fermenters. Secondly, when glucose is completely consumed and intermediates are formed as lactate, they may use it as carbon and energy source. Since both products of this strain can be easily metabolized by exoelectrogens such as *Geobacter*, they can be continually in syntrophic mode in order to decrease feedback inhibition and get better metabolic performance than other fermenters. This trophic mode gives them a strong ecological advantage in electroactive biofilms.

Propionate, butyrate and ethanol fermenters in BESs have not been characterized in literature. Further studies including high-throughput sequen-cing should be performed for better detection of these. Then, fermentation end products formed can then be converted into current as described in the next section.

4.3.3 Mixed End Products to Current

As mentioned before, the fermentation end products can be converted to current by EABs. It is interesting to compare microbial communities when single substrate is available against those presents in glucose-fed BESs. Among EABs in direct contact with anode, *Geobacter* is present in glucose-fed reactors (~15% of abundance). It is represented by these different species (*G. sulfurreducens* PCA, *G. metallireducens*, and *G. pickeringii*) (Lu et al. 2012; Jung and Regan 2007). *Geobacter pickeringii*, which was not found in BESs powered with single substrates, has been found in MFC in presence of light. It is probably exoelectrogen and can oxidize many fermentation end products such as ethanol, lactate, or butyrate with Fe(III), S° and may be electrode as electrons acceptors (Shelobolina et al. 2007). Its metabolic capabilities could make it an interesting exoelectrogen when several end products are simultaneously present. Like the previous genus, *Aeromonas* can directly convert glucose to current via *c*-type cytochromes and hydrogen to generate power which can give a strong advantage to establish syntrophies with fermenters (Chung and Okabe 2009). *c*-type cytochromes are also used by *Rhodopseudomonas palustris* found at 27% on anode biofilm of ethanol-fed MFC and present at 32% with glucose, suggesting that it could have an important role as EAB such as converting acetate to electricity (Cheng and Regan 2008; Chae et al. 2009). Electron transfer via *c*-type cytochromes can also be performed by minority EABs as *Desulfovibrio desulfuricans*. This bacterium was observed in two glucose-fed BESs at 1% (Lu et al. 2012; Jung and Regan 2011). Its low abundance suggests that this bacterium has no significance influence on biofilm metabolism. However, this EAB is able to use lactate as electron donor (Logan 2009). This microbe was also observed in lactate-fed MFC and is interesting in the electroactive biofilms in the case of lactic acid fermentation (Zhao et al. 2008). Therefore, many EABs found in single substrate systems are also present in glucose-fed MFC. Ability to use several electron donors may explain that some EABs are present only with glucose.

Although the direct contact with anode (nanowires, *c*-type cytochromes) for electron transfer is present in glucose-fed BESs, EABs with lesser-known transfer mode were observed in majority in the electroactive ecosystem as *Arcobacter* sp. This bacterium was found at 80% in DGGE band in formate-fed MFC. *A. butzleri,* present in 3.2% in the anodic community of glucose-fed MFC, can convert acetate and lactate to current. Concerning electron transfer, this bacterium have lateral appendages which could be used like nanowires (Fedorovich et al. 2009). Even less known, *Desulfuromonas chlorethnica* was not found in BESs powered with single substrate but with glucose. *D. chloroethnica* can use acetate as electron donor and Fe(III) as electron acceptor and could, therefore, be an exoelectrogen (Jung and Regan 2011).

4.4 Microbial Communities for Wastewater Substrates Degradation

BESs can be used in wastewater treatment to remove dissolved organic matter with low energy requirement as compared to conventional aerobic treatment. MFCs yield 50–90% less solids to be disposed of, reducing restatements cost (Du et al. 2007). BESs are more appropriate than anaerobic digestion to treat low-strength wastewater such as municipal wastewater. In addition, BESs allow direct harvesting of electricity in contrast to anaerobic digestion which requires a two-step process: methane generation followed by burning in an internal combustion engine (Tremouli et al. 2016). The multitude of organic compounds present in wastewater can fuel BESs. Hydrogen (MEC) or electricity (MFC) can be produced from fermentation of wastewater substrates. COD removal can reach 95% with double cell MFC which is competitive process as compared to activated sludge (Oh and Logan 2005). The wastewater compositions are very different, depending on their origins (industry or domestic). For example, wastewater from grain industry has a high content of ethanol (221 mg L^{-1}) as compared to that of lagoon (33.6 mg L^{-1}) while the latter has high propionate content (245 mg L^{-1}) as compared to the grain industry (10.6 mg L^{-1}) (Oh and Logan 2005). Microbial ecosystems in wastewater-fed BESs are very complex. Two main steps to convert wastewater substrates into energy are fermentation that will produce low molecular weight compounds (volatile fatty acids, alcohols) and their conversion into current by EABs.

Microorganisms that ferment wastewater substrates are very diverse in view of the diversity of compounds to be oxidized. *Parabacteroides* has been found in 4.7% in wastewater-fed MEC. A strain isolated from human faeces, can oxidize many carbohydrates (L-arabinose, L-rhamnose, glucose, lactose, maltose, D-mannose, D-raffinose, sucrose, D-trehalose and D-xylose) to form acetate or lactate (Sakamoto et al. 2007). This wide range of carbohydrate substrates is an ecological benefit in this ecosystem. In the same biofilm *Pelobacter* sp. was detected at 4.7% of abundance. As described above, this microorganism can ferment alcohols to propionate and acetate. So, the electroactive ecosystem fed with many substrates has two main fermenters, each specialized in a range of molecules. This distribution could decrease the inter-species competition, thus reducing energy lost in the competitive process (synthesis of inhibitor molecules) and stabilizing the ecosystem in long term. The common point between these two microorganisms is the molecules production like acetate or propionate. Concerning EABs, this biofilm is composed of 67% of *Geobacter* with *G. metallireducens* (23%), *G. uraniireducens* (16%), *G. sulfurreducens* (14%) and *G. lovleyi* (14%). The only known exoelectrogen that does not belong to the *Geobacter* genus found in this biofilm at 4.7% in abundance, *Alkaliphilus*, uses the long range transfer mode with flavins and can use acetate and lactate as electron donor (Kumar et al. 2016). The exoelectrogens represent more than two thirds of the biofilm, which prevents fermenters against feedback inhibition by reloading NAD^+, essential for the fermentation process and high COD removal (Cusick et al. 2010; Kiely et al. 2011a, b). It is interesting to note that EABs present in BESs fuelled with single substrates, such as *Geobacter* sp., are also present and dominant with wastewater. This could be due to two reasons: (1) There is a strong selection pressure by anodic material. Microorganisms are able to do extracellular electron transfer to the anode as final electron acceptor. EABs will have a great energetic advantage against other bacteria. (2) Fermentation end products from wastewaters are mainly acetate and propionate, which also restricts the number of EABs able to convert it into current; this favours specific EABs genus and could decrease the bacterial diversity.

4.5 Conclusion

Electroactive biofilms are complex structures capable of converting organic matter into electric current. This is possible through the establishment of syntrophic interactions among the microorganisms that oxidize organic matter to various end-products, and EABs that convert these simple molecules in current. The microbe-electrode exchange is the result of an evolutionary process developed over billions of years, resulting in the development of different electron transfer modes. Among the many EABs, *Geobacter* is the most studied. It is also often the most abundant in the anodic community and has the ability to develop nanowires allowing direct-interspecies electron transfer, a new mode of interaction. Degradations pathways of fermentation end products are specific to each substrate, with a common final point, acetate and hydrogen; both products enable a rapid transfer through EABs. The main concurrent organisms are methanogens, able to use efficiently hydrogen as compared to electroactive bacteria, causing a decrease of the coulombic efficiency. Glucose fermentation produces many compounds, each with its own oxidation pathway. Many microorganisms are commonly present with glucose, including specific EABs developing syntrophy with fermentative organisms. Finally, wastewaters which have a wide range of organic substrates can be effectively used in BESs. The omnipresence of *Geobacter* shows how this genus is important at both small and large scale. However, there is still a large unexplained part with regard to the role of microorganisms in the electroactive biofilm.

To improve stability, performance and coulombic efficiency, it is necessary to better understand the ecology of electroactive biofilms, the different interaction modes (direct, indirect) according to the substrate and use this knowledge to promote the EAB-fermenter interaction to prevent methanogenesis and enable this technology to emerge in the future.

References

Balch, W. E., Schoberth, S., & Tanner, R. S. (1977). *Acetobacterium*, a new genus of hydrogen-oxidizing, carbon dioxide-reducing, anaerobic bacteria. *International Journal of Systematic Bacteriology, 27*(4), 355–361.

Bok, F. D., Plugge, C. M., & Stams, A. J. M. (2004). Interspecies electron transfer in methanogenic propionate degrading consortia. *Water Research, 38*, 1368–1375.

Bond, D. R., & Lovley, D. R. (2003). Electricity production by Geobacter sulfurreducens attached to electrode. *Applied and Environmental Microbiology, 69*, 1548–1555.

Bond, D. R., & Lovley, D. R. (2005). Evidence for involvement of an electron shuttle in electricity generation by *Geothrix fermentans*. *Applied and Environmental Microbiology, 71*(4), 2186–2189.

Bond, D. R., Holmes, D. E., Tender, L. M., & Lovley, D. R. (2002). Electrode-reducing microorganisms that harvest energy from marine sediments. *Science, 295*, 483–486.

Boone, D. R., Johnson, R. L., & Liu, Y. (1989). Diffusion of the interspecies electron carriers H_2 and formate in methanogenic ecosystems and its implications in the measurement of Km for H_2 or formate uptake. *Applied and Environmental Microbiology, 55*(7), 1735–1741.

Boucher, I., Vadeboncoeur, C., & Moineau, S. (2003). Characterization of genes involved in the metabolism of alpha-galactosides by *Lactococcus raffinolactis*. *Applied and Environmental Microbiology, 69*(7), 4049–4056.

Breitenstein, A., Wiegel, J., Haertig, C., Weiss, N., Andreesen, J. R., & Lechner, U. (2002). Reclassification of *Clostridium hydroxybenzoicum* as *Sedimentibacter hydroxybenzoicus* gen. nov., comb. nov., and description of *Sedimentibacter saalensis* sp. nov. *International Journal of Systematic and Evolutionary Microbiology, 52*, 801–807.

Brenner, D. J., McWhorter, A. C., Kai, A., Steigerwalt, A. G., & Farmer, J. J. (1986). *Enterobacter asburiae* sp. nov., a new species found in clinical specimens, and reassignment of *Erwinia dissolvens* and *Erwinia nimipressuralis* to the genus *Enterobacter* as *Enterobacter Femalfetion*. *Journal of Clinical Microbiology, 23*(6), 1114–1120.

Caccavo, F., Lonergan, D. J., Lovley, D. R., Davis, M., Stolz, J. F., & McInerney, M. J. (1994). *Geobacter sulfurreducens* sp. nov., a hydrogen- and acetate-oxidizing dissimilatory metal-reducing microorganism. *Applied and Environmental Microbiology, 60*(10), 3752–3759.

Call, D., & Logan, B. E. (2008). Hydrogen production in a single chamber microbial electrolysis cell lacking a membrane. *Environmental Microbiology Report, 42*(9), 3401–3406.

Call, D. F., & Logan, B. E. (2011). Lactate oxidation coupled to iron or electrode reduction by *Geobacter sulfurreducens* PCA. *Applied and Environmental Microbiology, 77*(24), 8791–8794.

Chae, K. J., Choi, M. J., Lee, J. W., Kim, K. Y., & Kim, I. S. (2009). Effect of different substrates on the performance, bacterial diversity, and bacterial viability in microbial fuel cells. *Bioresource Technology, 100*(14), 3518–3525.

Chaudhuri, S. K., & Lovley, D. R. (2003). Electricity generation by direct oxidation of glucose in mediatorless microbial fuel cells. *Nature Biotechnology, 21*(10), 1229–1232.

Chauhan, A., Ogram, A., & Reddy, K. R. (2004). Syntrophic-methanogenic associations along a nutrient gradient in the Florida Everglades. *Applied and Environmental Microbiology, 70*(6), 3475–3484.

Cheng, S., & Regan, J. M. (2008). Electricity generation by *Rhodopseudomonas palustris*. *Environmental Science and Technology, 42*(11), 4146–4151.

Choo, Y. F., Lee, J., Chang, I. S., & Kim, B. H. (2006). Bacterial communities in microbial fuel cells enriched with high concentrations of glucose and glutamate. *Journal of Microbiology and Biotechnology, 16*(9), 1481–1484.

Chung, K., & Okabe, S. (2009). Characterization of electrochemical activity of a strain ISO2-3 phylogenetically related to *Aeromonas* sp. isolated from a glucose-fed microbial fuel cell. *Biotechnology and Bioengineering, 104*(5), 901–910.

Cusick, R. D., Kiely, P. D., & Logan, B. E. (2010). A monetary comparison of energy recovered from microbial fuel cells and microbial electrolysis cells fed winery or domestic wastewaters. *International Journal of Hydrogen Energy, 35*(17), 8855–8861.

De Cárcer, D. A., Ha, P. T., Jang, J. K., & Chang, I. S. (2011). Microbial community differences between propionate-fed microbial fuel cell systems under open and closed circuit conditions. *Applied Microbiology and Biotechnology, 89*(3), 605–612.

Du, Z., Li, H., & Gu, T. (2007). A state of the art review on microbial fuel cells: A promising technology for wastewater treatment and bioenergy. *Biotechnology Advances, 25*, 464–482.

Fan, Y., Han, S.-K., & Liu, H. (2012). Improved performance of CEA microbial fuel cells with increased reactor size. *Energy & Environmental Science, 5*, 8273–8280.

Fedorovich, V., Knighton, M. C., Pagaling, E., Ward, F. B., Free, A., & Goryanin, I. (2009). Novel electrochemically active bacterium phylogenetically related to *Arcobacter butzleri*, isolated from a microbial fuel cell. *Applied and Environmental Microbiology, 75*(23), 7326–7334.

Freguia, S., Rabaey, K., Yuan, Z., & Keller, J. (2008). Syntrophic processes drive the conversion of glucose in microbial fuel cell anodes. *Environmental Science and Technology, 42*(21), 7937–7943.

Freguia, S., Masuda, M., Tsujimura, S., & Kano, K. (2009). *Lactococcus lactis* catalyses electricity generation at microbial fuel cell anodes via excretion of a soluble quinone. *Bioelectrochemistry, 76*(1–2), 14–18.

Freguia, S., Teh, E. H., Boon, N., Leung, K. M., Keller, J., & Rabaey, K. (2010). Microbial fuel cells operating on mixed fatty acids. *Bioresource Technology, 101*(4), 1233–1238.

Gao, Y., Ryu, H., Santo Domingo, J. W., & Lee, H. S. (2014). Syntrophic interactions between H_2-scavenging and anode-respiring bacteria can improve current density in microbial electrochemical cells. *Bioresource Technology, 153*, 245–253.

Grabowski, A., Tindall, B. J., Bardin, V., Blanchet, D., & Jeanthon, C. (2005). *Petrimonas sulfuriphila* gen. nov., sp. nov., a mesophilic fermentative bacterium isolated from a biodegraded oil reservoir. *International Journal of Systematic and Evolutionary Microbiology, 55*, 1113–1121.

Hari, A. R., Katuri, K. P., Gorron, E., Logan, B. E., & Saikaly, P. E. (2016). Multiple paths of electron flow to current in microbial electrolysis cells fed with low and high concentrations of propionate. *Applied Microbiology and Biotechnology, 100*(13), 5999–6011.

Hatamoto, M., Imachi, H., Yashiro, Y., Ohashi, A., & Harada, H. (2008). Detection of active butyrate-degrading microorganisms in methanogenic sludges by RNA-based stable isotope probing. *Applied and Environmental Microbiology, 74*(11), 3610–3614.

Hesselsoe, M., Füreder, S., Schloter, M., Bodrossy, L., Iversen, N., Roslev, P., Nielsen, P. H., Wagner, M., & Loy, A. (2009). Isotope array analysis of *Rhodocyclales* uncovers functional redundancy and versatility in an activated sludge. *ISME Journal, 3*, 1349–1364.

Hofstad, T., Olsen, I., Eribe, E. R., Falsen, E., Collins, M. D., & Lawson, P. A. (2000). *Dysgonomonas* gen. nov. to accommodate *Dysgonomonas gadei* sp. nov., an organism isolated from a human gall bladder, and *Dysgonomonas capnocytophagoides* (formerly CDC group DF-3). *International Journal of Systematic and Evolutionary Microbiology, 3*, 2189–2195.

Holt, H. M., Gahrn-Hansen, B., & Bruun, B. (2005). *Shewanella algae* and *Shewanella putrefaciens*: Clinical and microbiological characteristics. *Clinical Microbiology and Infection, 11*, 347–352.

Jung, S., & Regan, J. M. (2007). Comparison of anode bacterial communities and performance in microbial fuel cells with different electron donors. *Applied Microbiology and Biotechnology, 77*, 393–402.

Jung, S., & Regan, J. M. (2011). Influence of external resistance on electrogenesis, methanogenesis, and anode prokaryotic communities in microbial fuel cells. *Applied and Environmental Microbiology, 77*(2), 564–571.

Kiely, P. D., Call, D. F., Yates, M. D., Regan, J. M., & Logan, B. E. (2010). Anodic biofilms in microbial fuel cells harbor low numbers of higher-power-producing bacteria than abundant genera. *Applied Microbiology and Biotechnology, 88*(1), 371–380.

Kiely, P. D., Rader, G., Regan, J. M., & Logan, B. E. (2011a). Long-term cathode performance and the microbial communities that develop in microbial fuel cells fed different fermentation end-products. *Bioresource Technology, 102*(1), 361–366.

Kiely, P. D., Regan, J. M., & Logan, B. E. (2011b). The electric picnic: Synergistic requirements for exoelectrogenic microbial communities. *Current Opinion in Biotechnology, 22*(3), 378–385.

Kimura, Z., & Okabe, S. (2013). Acetate oxidation by syntrophic association between *Geobacter sulfurreducens* and a hydrogen-utilizing exoelectrogen. *ISME Journal, 7*(8), 1472–1482.

Kotsyurbenko, O. R., Simankova, M. V., Nozhevnikova, A. N., Zhilina, T. N., Bolotina, N. P., Lysenko, A. M., & Osipov, G. A. (1995). New species of psychrophilic acetogens: *Acetobacterium bakii* sp. nov., *A. paludosum* sp. nov., *A. fimetarium* sp. nov. *Archives of Microbiology, 163*(1), 29–34.

Kragelund, C., Levantesi, C., Borger, A., Thelen, K., Eikelboom, D., Tandoi, V., Kong, Y., Krooneman, J., Larsen, P., Thomsen, T. R., & Nielsen, P. H. (2008). Identity, abundance and ecophysiology of filamentous bacteria belonging to the *Bacteroidetes* present in activated sludge plants. *Microbiology, 154*(3), 886–894.

Kumar, R., Singh, L., & Zularisam, A. W. (2016). Exoelectrogens: Recent advances in molecular drivers involved in extracellular electron transfer and strategies used to improve it for microbial fuel cell applications. *Renewable and Sustainable Energy Reviews, 56*, 1322–1336.

Lalaurette, E., Thammannagowda, S., Mohagheghi, A., Maness, P. C., & Logan, B. E. (2009). Hydrogen production from cellulose in a two-stage process combining fermentation and electrohydrogenesis. *International Journal of Hydrogen Energy, 34*(15), 6201–6210.

Lifang, D., FangBai, L., ShunGui, Z., DeYin, H., & JinRen, N. (2010). A study of electron-shuttle mechanism in *Klebsiella pneumoniae* based microbial fuel cells. *Environmental Science and Technology, 55*(1), 99–104.

Liu, H., Cheng, S., & Logan, B. (2005). Production of electricity from acetate or butyrate using a single-chamber microbial fuel cell. *Environmental Science and Technology, 39*(2), 658–662.

Logan, B. E. (2009). Exoelectrogenic bacteria that power microbial fuel cells. *Nature Reviews Microbiology, 7*(5), 375–381.

Lovley, D. R. (2012). Electromicrobiology. *Annual Reviews in Microbiology, 66*(1), 391–409.

Lovley, D. R., Giovannoni, S. J., White, D. C., Champine, J. E., Phillips, E. J., Gorby, Y. A., & Goodwin, S. (1993). *Geobacter metallireducens* gen. nov. sp. nov., a microorganism capable of coupling the complete oxidation of organic compounds to the reduction of iron and other metals. *Archives of Microbiology, 159*(4), 336–344.

Lu, L., Xing, D. F., Ren, N. Q., & Logan, B. E. (2012). Syntrophic interactions drive the hydrogen production from glucose at low temperature in microbial electrolysis cells. *Bioresource Technology, 124*, 68–76.

Luedekingt, R., & Piret, E. L. (1959). A kinetic study of the lactic acid fermentation: Batch process at controlled pH. *Journal of Biochemical and Microbiological Technology and Engineering, 1*(4), 393–412.

Matsumoto, R., Fujita, S., & Goto, Y. (2015). Cascade reaction through different microorganisms for electrochemical oxidation of ethanol. *ChemElectroChem, 2*(12), 1888–1891.

Miller, T. L., & Wolin, M. J. (1996). Pathways of acetate, propionate, and butyrate formation by the human fecal microbial flora. *Applied and Environmental Microbiology, 62*(5), 1589–1592.

Milliken, C. E., & May, H. D. (2007). Sustained generation of electricity by the spore-forming, Gram-positive, *Desulfitobacterium hafniense* strain DCB2. *Applied Microbiology and Biotechnology, 73*(5), 1180–1189.

Moscoviz, R., Toledo-Alarcón, J., Trably, E., & Bernet, N. (2016a). Electro-fermentation: How to drive fermentation using electrochemical systems. *Trends in Biotechnology*. doi:https://doi.org/10.1016/j.tibtech.2016.04.009.

Moscoviz, R., Trably, E., & Bernet, N. (2016b). Consistent 1,3-propanediol production from glycerol in mixed culture fermentation over a wide range of pH. *Biotechnology for Biofuels, 9*, 32.

Müller, N., Worm, P., Schink, B., Stams, A. J. M., & Plugge, C. M. (2010). Syntrophic butyrate and propionate oxidation processes: From genomes to reaction mechanisms. *Environmental Microbiology Reports, 2*, 489–499.

Nath, M. K., & Das, D. (2004). Improvement of fermentative hydrogen production: Various approaches. *Applied and Environmental Microbiology, 65*, 520–529.

Oh, S., & Logan, B. E. (2005). Hydrogen and electricity production from a food processing wastewater using fermentation and microbial fuel cell technologies. *Water Research, 39*, 4673–4682.

Pankhania, I. P., Spormann, A. M., Hamilton, W. A., & Thauer, R. K. (1988). Lactate conversion to acetate, CO_2 and H_2 in cell suspensions of *Desulfovibrio vulgaris* (Marburg): Indications for the involvement of an energy driven reaction. *Archives of Microbiology, 150*(1), 26–31.

Parameswaran, P., Torres, C. I., Lee, H. S., Krajmalnik-Brown, R., & Rittmann, B. E. (2009). Syntrophic interactions among anode respiring bacteria (ARB) and non-ARB in a biofilm anode: Electron balances. *Biotechnology and Bioengineering, 103*(3), 513–523.

Parameswaran, P., Zhang, H. S., Torres, C. I., Rittmann, B. E., & Krajmalnik-Brown, R. (2010). Microbial community structure in a biofilm anode fed with a fermentable substrate: The significance of hydrogen scavengers. *Biotechnology and Bioengineering, 105*(1), 69–78.

Popov, A. L., Michie, I. S., Kim, J. R., Dinsdale, R. M., Guwy, A. J., Esteves, S. R., & Premier, G. C. (2016). Enrichment strategy for enhanced bioelectrochemical hydrogen production and the prevention of methanogenesis. *International Journal of Hydrogen Energy, 41*(7), 4120–4131.

Qatibi, A. I., Nivière, V., & Garcia, J. L. (1991). *Desulfovibrio alcoholovorans* sp. nov., a sulfate-reducing bacterium able to grow on glycerol, 1,2- and 1,3-propanediol. *Archives of Microbiology, 155*(2), 143–148.

Rabaey, K., Lissens, G., Siciliano, S. D., & Verstraete, W. (2003). A microbial fuel cell capable of converting glucose to electricity at high rate and efficiency. *Biotechnology Letters, 25*(18), 1531–1535.

Richter, H., Lanthier, M., Nevin, K. P., & Lovley, D. R. (2007). Lack of electricity production by *Pelobacter carbinolicus* indicates that the capacity for Fe (III) oxide reduction does not necessarily confer electron transfer ability to fuel cell anodes. *Applied and Environmental Microbiology, 73*(16), 5347–5353.

Ruiz, V., Ilhan, Z. E., Kang, D. W., Krajmalnik-Brown, R., & Buitron, G. (2014). The source of inoculum plays a defining role in the development of MEC microbial consortia fed with acetic and propionic acid mixtures. *Journal of Biotechnology, 182*, 11–18.

Sakamoto, M., Kitahara, M., & Benno, Y. (2007). *Parabacteroides johnsonii* sp. nov., isolated from human faeces. *International Journal of Systematic and Evolutionary Microbiology, 57*(1), 293–296.

Schink, B. (1997). Energetics of syntrophic cooperation in methanogenic degradation. *Microbiology and Molecular Biology Reviews, 61*, 262–280.

Schmidt, J. E., & Ahring, B. K. (1995). Interspecies electron transfer during propionate and butyrate degradation in mesophilic, granular sludge. *Applied and Environmental Microbiology, 61*(7), 2765–2767.

Selembo, P. A., Perez, J. M., Lloyd, W. A., & Logan, B. E. (2009). High hydrogen production from glycerol or glucose by electrohydrogenesis using microbial electrolysis cells. *International Journal of Hydrogen Energy, 34*(13), 5373–5381.

Shelobolina, E. S., Nevin, K. P., Blakeney-Hayward, J. D., Johnsen, C. V., Plaia, T. W., Krader, P., Woodard, T., Holmes, D. E., Vanpraagh, C. G., & Lovley, D. R. (2007). *Geobacter pickeringii* sp. nov., *Geobacter argillaceus* sp. nov. and *Pelosinus fermentans* gen. nov., sp. nov., isolated from subsurface kaolin lenses. *International Journal of Systematic and Evolutionary Microbiology, 57*(1), 126–135.

Simões, M., Simões, L. C., & Viera, M. J. (2010). A review of current and emergent biofilm control strategies. *Food Science and Technology, 43*(4), 573–583.

Song, Y., Xiao, L., Jayamani, I., He, Z., & Cupples, A. M. (2015). A novel method to characterize bacterial communities affected by carbon source and electricity generation in microbial fuel cells using stable isotope probing and Illumina sequencing. *Journal of Microbiological Methods, 108*, 4–11.

Sun, D., Call, D. F., Kiely, P. D., Wang, A., & Logan, B. E. (2012). Syntrophic interactions improve power production in formic acid fed MFCs operated with set anode potentials or fixed resistances. *Biotechnology and Bioengineering, 109*(2), 405–414.

Temudo, M. F., Kleerebezem, R., & van Loosdrecht, M. (2007). Influence of the pH on (open) mixed culture fermentation of glucose: A chemostat study. *Biotechnology and Bioengineering, 98*(1), 69–79.

Thauer, R. K., Jungermann, K., & Decker, K. (1977). Energy conservation in chemotrophic anaerobic bacteria. *Bacteriological Reviews, 41*(1), 100–180.

Thiele, J. H. (1991). Mixed-culture interactions in methanogenesis. In J. G. Zeikus & E. A. Johnson (Eds.), *Mixed cultures in biotechnology* (pp. 261–292). New York: McGraw-Hill.

Tremouli, A., Martinos, M., Bebelis, S., & Lyberatos, G. (2016). Performance assessment of a four-air cathode single-chamber microbial fuel cell under conditions of synthetic and municipal wastewater treatments. *Journal of Applied Electrochemistry, 46*(4), 515–525.

Xia, X., Cao, X. X., Liang, P., Huang, X., Yang, S. P., & Zhao, G. G. (2010). Electricity generation from glucose by a *Klebsiella* sp. in microbial fuel cells. *Applied and Environmental Microbiology, 87*(1), 383–390.

Xie, C., & Yokota, A. (2005). Reclassification of *Alcaligenes latus* strains IAM 12599T and IAM 12664 and *Pseudomonas saccharophila* as *Azohydromonas lata* gen. nov., comb. nov., *Azohydromonas australica* sp. nov. and *Pelomonas saccharophila* gen. nov., comb. nov., respectively. *International Journal of Systematic and Evolutionary Microbiology, 55*(6), 2419–2425.

Xing, D. F., Cheng, S. A., Regan, J. M., & Logan, B. E. (2009). Change in microbial communities in acetate- and glucose-fed microbial fuel cell in the presence of light. *Biosensors & Bioelectronics, 25*(1), 105–111.

Yu, J., Park, Y., Cho, H., Chun, J., Seon, J., Cho, S., & Lee, T. (2012). Variations of electron flux and microbial community in air-cathode microbial fuel cells fed with different substrates. *Water Science & Technology, 66*(4), 748–754.

Zhao, F., Rahunen, N., Varcoe, J. R., Chandra, A., Avignone-Rossa, C., Thumser, A. E., & Slade, R. C. T. (2008). Activated carbon cloth as anode for sulfate removal in a microbial fuel cell. *Environmental Science and Technology, 42*(13), 4971–4976.

Chapter 5
Anodic Electron Transfer Mechanism in Bioelectrochemical Systems

Sanath Kumar Kondaveeti, Jai Sankar Seelam, and G. Mohanakrishna

5.1 Introduction

The applicability of bacteria in transporting electron beyond their cell wall and ability to electrically interact with electrode has been nearly over a century (Potter 1911). This concept was developed into bioelectrochemical systems with active localization of electron transfer and transport through solid surfaces under controlled operating conditions. Microbial fuel cells are growing bioelectrochemical systems that use bacteria as a catalyst and generate bioelectricity using organic matter. The bacteria act as powerhouse at the anode of MFC and oxidize organic matter to CO_2 by generating electrons and protons (Kondaveeti et al. 2014). These electrons move from anode to cathode and get reduced as water by using oxygen as an electron acceptor. Scientific reports suggested that the generated electrons move from the bacteria/biofilm to an anodic surface occurs through two key mechanisms. These mechanisms are broadly classified as follows: (1) direct electron transfer (DET), (2) indirect/mediated electron transfer using mediators (MET) and (3) inter-special electron transfer (Fultz and Durst 1982; Kondaveeti

S.K. Kondaveeti
Department of Environmental Science and Engineering, Kyung Hee University, Gyeonggi-do 446-701, South Korea

J.S. Seelam
Wetsus Academy, P.O. Box 1113, 8900 CC Leeuwarden, The Netherlands

G. Mohanakrishna (✉)
Separation and Conversion Technologies, VITO – Flemish Institute for Technological Research, Boeretang 200, 2400 Mol, Belgium
e-mail: gmohanak@yahoo.com

D. Das (ed.), *Microbial Fuel Cell*,
https://doi.org/10.1007/978-3-319-66793-5_5

and Min 2015). The natural mediators such as flavins which are secreted by bacteria or other active complexes such as *c*-type chromosomes present on outer cell membranes can shuttle electrons. Up to date the metal reducing bacterial species such as *Geobacter* and *Shewanella* have been widely noticed in MFC technology, due to their external electron transfer mechanism and for synthesis of natural mediators (riboflavins), which can be a rival for other exoelectrogens (Logan 2008). The external insoluble shuttles such as neutral red, and methyl viologen etc. were used in microbial fuel cells for electron transfer from the bacterial cell wall to electrodes. The initial studies in addition of exogenous mediators to MFC were pursued (Cohen 1931; Schroder 2007). In this study low current generation in MFC might be due to lack of electromotive oxidation and reductive force. These were resurfaced in 1980 by Bennetto and coworkers and it was further investigated by many other researchers. In the present chapter, the electron transfer mechanisms such as direct electron transfer, mediated electron transfer and interspecies electron transfer mechanisms with electroactive anode bacteria are discussed.

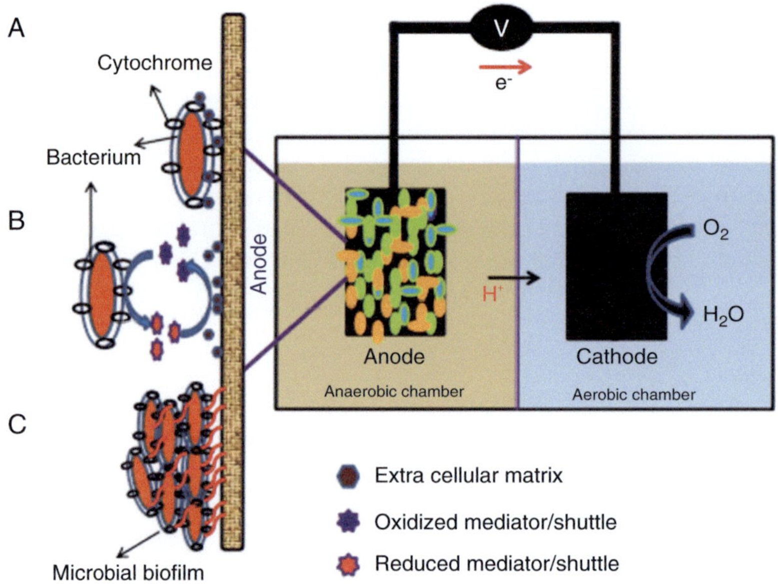

Fig. 5.1 Types of anodic electron transfer mechanism in microbial fuel cells

5.2 Electron Transfer Mechanisms

The process occurs through collaborative microbiological and electrochemical interactions in an electro-redox environment (Fig. 5.1). An overview of each of these electron transfer mechanism has been included.

5.2.1 Direct Electron Transfer

The DET mechanism predominantly occurs through direct physical contact between the bacterial microorganism cell wall and solid electrode without the aid of any redox mediators. These electroactive bacterial species transport the electrons with the help of membrane bound electron transport proteins (cytochromes) and/or electrically conductive cellular appendages (pili or nanowires). These components assist in transferring the electrons from intracellular complex to an external solid acceptor (anode). In general, *c*-type outer-membrane cytochromes and multi heme-proteins which are especially evolved in bacterial species such as *Geobacter, Shewanella*, and *Rhodoferax*. The synthesis of these proteins often depend on the nature of solid electron acceptors (for example iron oxides) in their natural environment (Choi et al. 2003). Anodic electrode also functions as a viable alternative for a solid electron acceptor (Huang et al. 2008). As DET requires physical contact between the anode material and bacterial cell, cytochromes trigger the first monolayer of bacterial cells at the anode to be electrochemically active which consumes energy. Thus limiting the maximum obtainable current densities to 0.6 μA cm^{-2}, 3 μA cm^{-2} and 6.5 μA cm^{-2} by using *Shewanella putrefaciens*, *Rhodeoferax ferrireducens* and *Geobacter sulfurreducens,* respectively (Chaudhuri and Lovley 2003; Kim et al. 2002).

Bacterial species such as *Geobacter* sp. and *Shweanella* sp. can produce electrically conductive molecular pili or nanowires which facilitate the transfer of electrons from cell wall to a solid electron acceptor (anode). These nanowires are connected to the membrane bound cytochromes and microbial proteins (type IV pili). This provides an interface for the transfer of external electron transfer from the cell (Ross et al. 2011). A scientific study presented the pivotal role of outer membrane *c*-cyts and expression of conductive pili in *G. sulfurreducens*. These actively aid in formation of multilayered electroactive (thicker) biofilm thereby leading to enhanced anodic performance (Malvankar and Lovley 2012). Reguera et al. (2006) reported a tenfold increase in current generation by using *Geobacter sulfurreducens* with nanowires. On the other hand, *Geobacter* sp. without omcZ (gene encoding *c*-type cytochromeZ) and PilA (gene encoding type IV pili) had inhibited biofilm formation and consequently lead to drop in output current. Type IV pili primarily constitutes PilA monomers with two types of isoforms such as long and short PilA (Reguera et al. 2006). A study performed by Sandler and Weis using strains exhibiting short PilA noticed disrupted biofilm formation on graphite electrode surface indicating the importance of long PilA during biofilm formation. However, they observed that the presence of short PilA plays a crucial role in biofilm conductivity as it mediates the *c*-type cytochrome, OmcZ towards outer membrane of bacterial cell (Richter et al. 2012). Some pilA mutant studies had pointed that the pili had no important role in electron transport in the cell respiring near the electrode. However, they assist in cell-cell aggregations and optimum biofilm formation (Cologgi et al. 2014).

Electrochemical and energy analysis of DET can provide informative insights on their extended application. Diverse studies observed the relaying of DET mechanism on electrode but they have not evaluated the redox potentials of involved species. Bond and Lovley (2003)) operated MFC and noted open circuit anode potential (−0.17 V) as the redox potential by using the *G. sulfurreducens* colonized electrode growing on acetate. However, in open circuit, the redox potentials of anaerobic electroactive biofilms will shift considerably towards negative potentials due to variation in the concentration terms in the Nernst equation towards reduced species (Logan 2008). Thus the reported redox potentials and open circuit potentials of the *G. sulfurreducens* may not be equal to the potential of cytochromes based electron transfer (Logan et al. 2006). Furthermore, very few microorganisms are capable of performing DET fed upon complex substrates such as glucose. *Rhodeoferax ferrireducens* has the capability to utilize glucose whereas other electroactive species such as *Shewanella* and *Geobacter* strains cannot use complex substrates. They are pertained to simple substrates such as lower molecular weight organic acids and alcohols. This often leads to substantial decline in overall energy conversion efficiency. The assessment of this energy efficiency in DET is complicated and scarce information is available in this respect.

5.2.2 Mediated Electron Transfer

Most researchers primarily preferred the application of DET mechanism for efficient current generation and enhanced electrical performance of an MFC. However, the magnitude of current and power densities obtained using only DET principle were drastically lower in comparison to the use/presence of mediator electron transfer (MET) (Choi et al. 2003). The MET mechanism represents an effective means of electron transfer from the bacterial cell wall to an MFC anode. This mechanism of electron transfer has further been sub-classified based on the use/presence of supplementing redox species/mediators (Fultz and Durst 1982; Park and Zeikus 2000; Rahimnejad et al. 2011). Table 5.1 enlists the performance under application of several mediating agents promoting MET.

Table 5.1 An overview of MFC investigations employing different types of mediators

Type of mediator	Operational mode	Substrate	Power/Current	References
Neutral Red	Double chamber	Glucose	4.5 mA	Park and Zeikus (2000)
Safranine	Double chamber	Glucose	0.09 mA	Choi et al. (2003)
Methylene Blue	Double chamber	Glucose	12.3 μW	Rahimnejad et al. (2011)
Methyl Viologen	Double chamber	Glucose	400 mV	Mohan et al. (2008)
Thionine	Double chamber	Glucose	0.6 mA	Choi et al. (2003)
Resorufin	Double chamber	Glucose	12 μW cm^{-2}	Ganguli and Dunn (2009)
Pyocyanine	Double chamber	Glucose	3977 μW m^{-2}	Rabaey et al. (2005)
Ferric chelate complex	Double chamber	Glucose	0.22 mW	Vega and Fernández (1987)
2-Hydroyl-1,4 napthoquinone	Double chamber	Glucose	0.35 mA	Lee et al. (2002)
Humic acid	Double chamber	Glucose	0.248 mW	Thygesen et al. (2009)
		Acetate	0.510 mW	
Anthraquinone-2,6-disulfonate (AQDS)	Miniature MFC	Lactate	0.8 mW	Ringeisen et al. (2006)
Natural mediator	Single chamber	Waste water	240 μW cm^{-2}	Logan et al. (2006)
Mn (IV) and Mn (II)	Single chamber	Acetate	200 A m^{-3}	Clauwaert et al. (2007)

5.2.2.1 MET via Exogenous Mediators

The usage of exogenous and artificial mediators with anaerobic bacterial culture was first noticed by Cohen. He initially observed that anaerobic growth of bacteria could generate strong negative redox potentials. However, the amount of extractable current for practical purposes was small. His report attributed the smaller current produced might be due to the absence of "electromotively active oxidation-reduction products" (Cohen 1931). In further extension of this research to amplify the current generation, other research teams proposed the utility of potassium ferricyanide or benzoquinone. These supplementary chemicals enhance the electron flow from the bacterial cell wall to solid electron acceptor (anode electrode) (Little et al. 2008; Wei et al. 2012). Later in the early 1980, diverse scientific studies were aimed at investigating alternative compounds were for their suitability and viability in MFC operation as mediators. Chemical agents such as neutral red, methylene blue, safranine, toludine blue, thionine, resorufin and anthraquinone-2, 6-disulfonate were tested as viable options for increasing the electrical output (Kim and Lee 2015; Logan 2008). The major drawback of redox mediator application was that only a fractional increase of current densities was observed. Regular addition of these exogenous compounds was necessary which further escalates the costs. This external influx of chemicals proved to be technologically unfeasible and environmentally unfriendly. This opened up an eco-friendly avenue for wastewater treatment besides conventional bioelectricity generation in an MFC (Jang et al. 2004).

5.2.2.2 MET via Endogenous Mediators

In off time, bacterial species proliferate under conditions where neither a solid electron acceptor nor soluble electron acceptor are readily available for direct reach (DET). For an instance, the electroactive-microbial complex with thick biofilm is limited by the oxygen transport across the film thickness. This oxygen in-depth diffusion is limited primarily due to higher propagation distance (biofilm thickness) and the cell on the outer rim of the film are not in direct contact with (solid/soluble) electron acceptors. In such cases, electroactive microbial communities employs alternative electron transfer pathways by using external (exogenous) electron mediator compounds like humic acids or metal chelates such as Fe (Song et al. 2016). It was noticed in few studies that the self-synthesis of low molecular weight electron mediators/shuttles by secondary electron metabolite pathways was possible. The frequently found self-secreted electron mediators include quinines (2-amino-3-carbox-1, 4-napthoquinone), pycocyanin (pycocyanine/phenazine) and flavin (Schroder 2007). The secreted mediators interact synchronically with the solid electron acceptor (anode) or soluble electron acceptor and bacterial cytochromes for effective electron exchange. Bacterial species also tend to approach different strategies and metabolic pathways for the generation of helpful endogenous mediator compounds (Park and Zeikus 2000). For instance, *G. sulfurreducens* secrete riboflavin-like compounds from its monolayers. Also, in the case of *S. oneidensis,* MR-1 interacts with outer membrane–Cyts and plays a crucial role in electron transfer mechanism (Bond and Lovley 2005; Kim et al. 2002).

The secretion of endogenous mediators has garnered greater interest in MFC applicability. The presence/synthesis of these compounds drives the electron transfer independent of the exogenous (externally supplied) redox mediators. The electron shuttles serve as a reversible electron acceptor in carrying the electrons from bacterial cell wall to either solid electron acceptor or aerobic layer (top layers) of the biofilm and becomes re-oxidized. This to and fro movement enables their availability for further redox process (Kondaveeti and Min 2015). One molecule of endogenous mediator can trigger thousands of redox cycles, so the presence of these compounds in smaller quantities ably assists the bacteria in disposing of electron at sufficiently higher rates (Choi et al. 2012; Rahimnejad et al. 2011). Exclusively, the synthesis of endogenous mediators in batch cultures facilitates higher electron transfer by alternative pathways and subsequently increases the efficiency of current generation. But, the identification of these extracellular secreted mediators is found to be challenging. Pyocyanine and phenazine-1-carboxamide were present in MFC operation with *Pseudomonas aeruginosa* (Rabaey et al. 2005). In other studies, it has been discussed that *S. oneidensis* MR-1 secreted quinonine type of mediators which supports long range electron transfer. However, this mechanism conflicts and overlaps with electron transfer mechanisms of DET using *c*-type cytochromes and/or electrically conductive pili (Newman and Kolter 2000).

The redox potentials of the secreted electron shuttles were more negative than most of outer membrane cytochromes exhibiting higher amount of obtainable current/power densities in comparison to operation with DET (Schroder 2007). The major drawback in use of mediators is that the high efficiency and energy output was confined to batch mode of MFC operation. Because in continuous mode, the system is prone to loss of mediators/shuttles and which can decrease the columbic and energy efficiency (Chaudhuri and Lovley 2003).

5.3 Interspecies Electron Transfer Through Conductive Minerals

Across aerobic and anaerobic biota, the synergistic exchange of electrons between two bacterial species directed at establishing a direct transfer can also be considered for a DET (Kato et al. 2012; Rotaru et al. 2012; Zhao et al. 2015). These reactions are prominently noticed in anaerobic digesters and sediment MFCs (SMFCs) catalyzed by mixed electroactive consortia and fed on complex substrates (Zhao et al. 2015). This process of inter-special extra cellular electron exchange via direct contact can also be referred as IET (direct interspecies electron transfer). IET can also occur with the assistance of natural conductive minerals (Kato et al. 2012). In natural environment, the presence of conductive minerals such as pyrite can facilitate the IET (Cheng and Call 2016). Recent advances on biocathode application revealed that electroactive bacteria can perform the roles of both electron donors and acceptors with subsequent half-cell reactions. For example, *Shwenella* sp. and *Geobacter* sp. can oxidize the substrate in MFC anode and reduce the nitrate at cathode of microbial fuel cells. Based on this principle, a study suggested spatially separated electrodes in SMFC, with anodic sulphide oxidation and cathodic oxygen reduction, generated electric currents in the absence of cross-membrane diffusion species (Nielsen et al. 2010). Though they did not identify the actual electron conduits, discussion on the possibility of electron transfer using nanowires, *c*-type cytochromes and natural conductive mineral such as pyrite was presented (Nielsen et al. 2010). Other types of naturally available conductive minerals such as magnetite and hematite can help SMFC operation and enhance inter-special electron flow (Cheng and Call 2016). Another research group analyzed the participation of conductive minerals using soil bacteria like *Geobacter sulfurreducens* and *Thiobacillus denitrificans* in the presence and absence of iron-oxide minerals for acetate oxidation and nitrate reduction (Kato et al. 2012). They reported maximum current densities in a range of 100 $\mu W\ cm^{-2}$ using iron oxide minerals whereas the values were lower (20 μW cm^{-2}) in the absence of the conductive minerals.

5.4 Factors Influencing Electron Transfer Mechanism

The anodic electron transfer to solid electron accepting surfaces are principally driven by different microbiological (electroactive microbes) and electrochemical (carbon or metal-based electrodes) factors (Schroder 2007). Specific understanding of these governing factors has always been of keen interest for several research groups to predict and optimize the performance of an MFC. These factors predominantly dictate the reaction kinetics of electron propagation, characteristic bio (electro)chemical and metabolic reactions involved during the electron-transfer. A short introduction on biological and design parameters affecting the electron transfer mechanism is presented below.

5.4.1 Biofilm Integrity

The heterogeneous pathway of electron transfer (DET and MET) depends on the position of electroactive cells within the biofilm and their mono/multi layered arrangement (Torres et al. 2010). The arrangement and integrity of these microbes also define the diffusion of substrate molecules which further modulates the biological activity both temporally and spatially (Marcus et al. 2007; Logan et al. 2006). DET mechanism is predominant in microbes resting on electrode surface whereas indirect transfer requires electron transport across the film thickness, subsequently affecting the kinetics. The kinetics of these transfer mechanisms are limited and influenced by several factors. Electroactive species promoting DET are often prone to physical and chemical stress due to diffusion limited transport of waste and nutrients. This substrate feed is readily available for cells present on the outer rim of the biofilm which would prefer indirect exchange. But, indirect electron transfer is limited by the integrity of the biofilm. Biofilm can ably provide a conductive matrix of variable composition for effective transport of soluble and mobile carriers (Torres et al. 2010).

The integrity of biofilm is based on the distribution of electrochemically active (anaerobic respiration microbes) and inactive microbial communities (fermentative organisms and non-electroactive microbes) (Fig. 5.2). Electroactive species in the film facilitate electron exchange and enhance the energy output whereas non-electroactive counterparts degrade and consume complex organics and other electron donors/acceptors. Elimination of these non-electroactive microbes including methanogens, aerobic organisms, hydrogen scavengers and nitrate reducers can effectively improve the facility for electron transfer (Seelam et al. 2015). Bioaugmented MFC systems also support effective electron transfer due to active

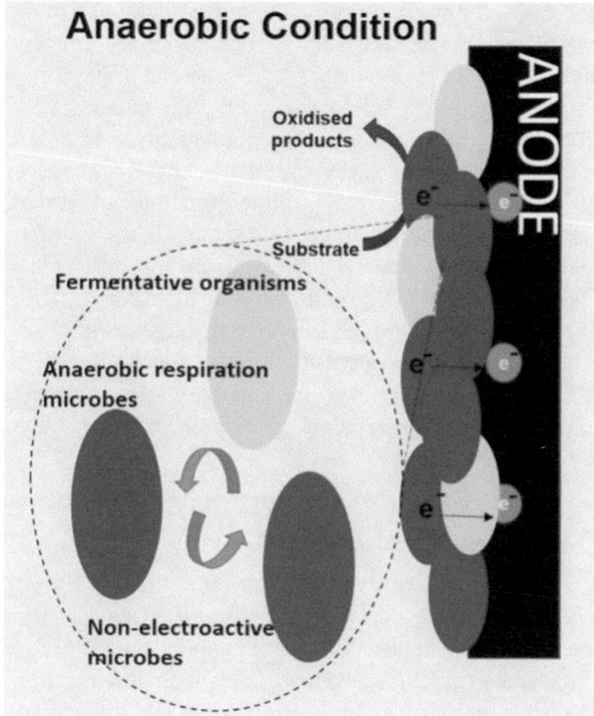

Fig. 5.2 The schematic representation of the biofilm integrity of microbial consortia on anode surface (electroactive and non-electroactive microbes) and electron transfer to the electrode

inter-special and synergistic interactions. Species like *Pseudomonas* sp. and *Shewanella* sp. functioned as augmented agents to deliver redox mediators necessary for electron exchange (Pham et al. 2008; Veer Raghavulu et al. 2012). Furthermore, the electrocatalytic performance of an electroactive biofilm depends on the source, nature (mixed or consortia) and type (Gram positive or negative) of inoculum, conditions of enrichment, mode of operation (pH and temperature) and system architecture (electrodes).

5.4.2 Electrodes Structure

Diverse array of materials in different configurations has been actively applied for application as electrodes in MFCs. Noble metals, metal-based and carbonaceous electrodes functioned as anodic electrode component. The anodic structure and built plays a crucial role in biofilm stability and electron transport across electrode and electroactive microbes. These solid, porous or granular entities provide physical support for biofilm attachment, substrate oxidation, electron transfer and provide active sites for electron exchange. Based on configuration, the electrodes are classed into flat, packed, stuffed and brush models with variable surface properties. Properties like surface roughness or microbially accessible surface area affect the true surface area for electron propagation (Dumas et al. 2008). Specially, the flow-over or flow-through electrodes are preferred for MFC application due to facilitation of pH gradients and high substrate concentration. These electrodes also aid in washout of suspended and non-electroactive microbes. Other electrode complexes like graphite fibre and carbon nanostructures display excellent electron exchange characteristics due to high surface area to volume ratio and viably support bacterial biofilm proliferation (Kumar et al. 2013; Sharma et al. 2008). Graphite fiber built consists of fibrils which are formed by winding fine slices of carbon fibres in the form of brush to improve the accessible surface area. Alternatively, nano-modifications and increase of pore properties of electrodes were applied to improve biofilm formation, substrate diffusion and subsequent maximization of interface for swift electron transfer.

5.4.3 Catalyzed Electrodes

The nature of the electrode materials, their composition and properties also govern the biofilm formation during MFC operation. These properties consequently affect the electron transfer occurring at anode but the transfer kinetics are limited by electrode properties. In the recent years, the electrodes are often catalyzed by modifications which have still better properties like improved biocompatible surface, high conductivity, long-term chemical stability, non-oxidative and non-self-destructive activity (Srikanth et al. 2011; Rosenbaum et al. 2007; Guo et al. 2014). Methods like pulse electro-deposition, impregnation, coating and immobilization were primarily applied to modify anodes for effective electron transfer (Wei et al. 2011; Hubenova and Mitov 2015). These modified electrode materials attain properties to enhance extracellular electron exchange and also function as current collectors (Hubenova and Mitov 2015). Additionally, the impact of electrode on electron discharge varies with the operational conditions. A report suggested that biofilm formation and electron exchange were affected by the electrode characteristic only under open circuit but no changes were presented when operated under closed conditions.

5.4.4 Electrolyte and Electron Carriers

The composition and concentration of substrate, availability and presence of electron carriers in the electrolyte impacts the electrons transport post-substrate oxidative degradation (Schroder 2007). Super-saturated substrates promote shift of metabolic pathway towards non-electroactive reactions (acetogenesis or methanogenesis) which declines the electrocatalytic performance. This shift is also triggered when low-resistance path to an electron sink is absent. On the other hand, electron transfer promoting carriers or mediators in the electrolyte present an enhancing effect on the MFC performance. These mediators/carriers are either metabolically produced (Cheng et al. 2009; Clauwaert and Verstraete 2009) or externally added (Biffinger et al. 2007; Lee and Rittman 2010). They promote electrogenic activity and can accelerate the process of electron exchange. But, excess addition of mediators displayed cytotoxic effect and further influenced the economics of operation due to additional chemical costs (Hubenova and Mitov 2015). Moreover, synthetic carriers often tend to have an impending effect due to toxicity and instability.

5.5 Conclusions

The anodic electron transfer pathway and mechanism offer to be the most critical and decisive players in predicating the overall efficiency and electrical performance of an MFC. This chapter summarized the fundamentals and principle aspects of anodic electron transfer into direct electron transfer, mediated electron transfer and inter-special electron transfer. The factors influencing electron transfer mechanisms presented here highlights the domains of MFC research which needs further improvements. The magnitude of bioelectricity produced by MFC technology is still a major point of concern. To realize large scale and long-term applicability of MFCs, the utility of biofilm-forming electroactive species which can ably facilitate DET is necessary. Further investigations should be focused at developing systems which form and sustain effective conductive biofilm matrix to facilitate high electron rates between electrode and microorganisms. This strategy can actively assist in realizing MFC as a competitive and commercial technology.

References

Biffinger, J. C., Pietron, J., Ray, R., Little, B., & Ringeisen, B. R. (2007). A biofilm enhanced miniature microbial fuel cell using Shewanella oneidensis DSP10 and oxygen reduction cathodes. *Biosensors & Bioelectronics, 22*(8), 1672–1679.

Bond, D. R., & Lovley, D. R. (2003). Electricity production by geobacter sulfurreducens attached to electrodes. *Applied and Environmental Microbiology, 69*(3), 1548–1555.

Bond, D. R., & Lovley, D. R. (2005). Evidence for involvement of an electron shuttle in electricity generation by geothrix fermentans. *Applied and Environmental Microbiology, 71*(4), 2186–2189.

Chaudhuri, S. K., & Lovley, D. R. (2003). Electricity generation by direct oxidation of glucose in mediatorless microbial fuel cells. *Nature Biotechnology, 21*(10), 1229–1232.

Cheng, Q., & Call, D. F. (2016). Hardwiring microbes via direct interspecies electron transfer: Mechanisms and applications. *Environmental Science: Processes and Impacts, 18*, 968–980.

Cheng, S., Xing, D., Call, D. F., & Logan, B. E. (2009). Direct biological conversion of electrical current into methane by electromethanogenesis. *Environmental Science & Technology, 43*(10), 3953–3958.

Choi, Y., Kim, N., Kim, S., & Jung, S. (2003). Dynamic behaviors of redox mediators with in the hydrophobic layers as an important factor for effective microbial fuel cell operation. *Bulletin of the Korean Chemical Society, 24*(4), 437–440.

Choi, O., Um, Y., & Sang, B.-I. (2012). Butyrate production enhancement by Clostridium tyrobutyricum using electron mediators and a cathodic electron donor. *Biotechnology and Bioengineering, 109*(10), 2494–2502.

Clauwaert, P., & Verstraete, W. (2009). Methanogenesis in membraneless microbial electrolysis cells. *Applied Microbiology and Biotechnology, 82*(5), 829–836.

Clauwaert, P., van der Ha, D., Boon, N., Verbeken, K., Verhaege, M., Rabaey, K., & Verstraete, W. (2007). Open air biocathode enables effective electricity generation with microbial fuel cells. *Environmental Science and Technology, 41*(21), 7564–7569.

Cohen, B. (1931). The bacterial culture as an electrical half-cell. *Journal of Bacteriology, 21*, 18–19.

Cologgi, D. L., Speers, A. M., Bullard, B. A., Kelly, S. D., & Reguera, G. (2014). Enhanced uranium immobilization and reduction by Geobacter sulfurreducens biofilms. *Applied and Environmental Microbiology, 80*(21), 6638–6646.

Dumas, C., Basseguy, R., & Bergel, A. (2008). DSA to grow electrochemically active biofilms of Geobacter sulfurreducens. *Electrochimica Acta, 53*(7), 3200–3209.

Fultz, M. L., & Durst, R. A. (1982). Mediator compounds for the electrochemical study of biological redox systems: A compilation. *Analytica Chimica Acta, 140*(1), 1–18.

Ganguli, R., & Dunn, B. S. (2009). Kinetics of anode reactions for a yeast-catalysed microbial fuel cell. *Fuel Cells, 9*(1), 44–52.

Guo, K., Donose, B. C., Soeriyadi, A. H., Prévoteau, A., Patil, S. A., Freguia, S., Gooding, J. J., & Rabaey, K. (2014). Flame oxidation of stainless steel felt enhances anodic biofilm formation and current output in bioelectrochemical systems. *Environmental Science & Technology, 48*(12), 7151–7156.

Hubenova, Y., & Mitov, M. (2015). Extracellular electron transfer in yeast-based biofuel cells: A review. *Bioelectrochemistry, 106*(Part A), 177–185.

Huang, L., Zeng, R. J., & Angelidaki, I. (2008). Electricity production from xylose using a mediator-less microbial fuel cell. *Bioresource Technology, 99*(10), 4178–4184.

Jang, J. K., Pham, T. H., Chang, I. S., Kang, K. H., Moon, H., Cho, K. S., & Kim, B. H. (2004). Construction and operation of a novel mediator- and membrane-less microbial fuel cell. *Process Biochemistry, 39*(8), 1007–1012.

Kato, S., Hashimoto, K., & Watanabe, K. (2012). Microbial interspecies electron transfer via electric currents through conductive minerals. *Proceedings of the National Academy of Sciences of the United States of America, 109*(25), 10042–10046.

Kim, S. K., & Lee, C. G. (2015). *Marine bioenergy: Trends and developments*. Boca Raton: CRC Press.

Kim, H. J., Park, H. S., Hyun, M. S., Chang, I. S., Kim, M., & Kim, B. H. (2002). A mediator-less microbial fuel cell using a metal reducing bacterium, Shewanella putrefaciens. *Enzyme and Microbial Technology, 30*(2), 145–152.

Kondaveeti, S., Lee, J., Kakarla, R., Kim, H. S., & Min, B. (2014). Low-cost separators for enhanced power production and field application of microbial fuel cells (MFCs). *ElectrochimicaActa, 132*(0), 434–440.

Kondaveeti, S., & Min, B. (2015). Bioelectrochemical reduction of volatile fatty acids in anaerobic digestion effluent for the production of biofuels. *Water Research, 87*, 137–144.

Kumar, G. G., Sarathi, V. G. S., & Nahm, K. S. (2013). Recent advances and challenges in the anode architecture and their modifications for the applications of microbial fuel cells. *Biosensors and Bioelectronics Elsevier, 43*, 461–475.

Lee, H. S., & Rittman, B. E. (2010). Significance of biological hydrogen oxidation in a continuous single-chamber microbial electrolysis cell. *International Journal of Hydrogen Energy, 35*, 920–927.

Lee, S. A., Choi, Y., Jung, S., & Kim, S. (2002). Effect of initial carbon sources on the electrochemical detection of glucose by Gluconobacter oxydans. *Bioelectrochemistry, 57*(2), 173–178.

Little, D. M., Agricultural, C. U., & Engineering, B. (2008). *Investigation of a photosynthetically aerated cathode in a MFC*. Clemson University.

Logan, B. E. (2008). *Microbial fuel cells*. Hoboken: Wiley and Sons Inc.

Logan, B. E., Hamelers, B., Rozendal, R., Schroder, U., Keller, J., Freguia, S., Aelterman, P., Verstraete, W., & Rabaey, K. (2006). Microbial fuel cells: Methodology and technology. *Environmental Science & Technology, 40*(17), 5181–5192.

Malvankar, N. S., & Lovley, D. R. (2012). Microbial nanowires: A new paradigm for biological electron transfer and bioelectronics. *ChemSusChem, 5*(6), 1039–1046.

Marcus, A. K., Torres, C. I., & Rittmann, B. E. (2007). Conduction-based modeling of the biofilm anode of a microbial fuel cell. *Biotechnology and Bioengineering, 98*(6), 1171–1182.

Mohan, Y., Manoj Muthu Kumar, S., & Das, D. (2008). Electricity generation using microbial fuel cells. *International Journal of Hydrogen Energy, 33*(1), 423–426.

Newman, D. K., & Kolter, R. (2000). A role for excreted quinones in extracellular electron transfer. *Nature, 405*(6782), 94–97.

Nielsen, L. P., Risgaard-Petersen, N., Fossing, H., Christensen, P. B., & Sayama, M. (2010). Electric currents couple spatially separated biogeochemical processes in marine sediment. *Nature, 463*(7284), 1071–1074.

Park, D. H., & Zeikus, J. G. (2000). Electricity generation in microbial fuel cells using neutral red as an electronophore. *Applied and Environmental Microbiology, 66*(4), 1292–1297.

Pham, T. H., Boon, N., Aelterman, P., Clauwaert, P., De Schamphelaire, L., Vanhaecke, L., De Maeyer, K., Hofte, M., Verstraete, W., & Rabaey, K. (2008). Metabolites produced by Pseudomonas sp. enable a Gram-positive bacterium to achieve extracellular electron transfer. *Applied Microbiology and Biotechnology, 77*, 1119–1129.

Potter, M. C. (1911). Electrical effects accompanying the decomposition of organic compounds. *Proceedings of the Royal Society of London. Series B, Containing Papers of a Biological Character, 84*(571), 260–276.

Rabaey, K., Boon, N., Höfte, M., & Verstraete, W. (2005). Microbial phenazine production enhances electron transfer in biofuel cells. *Environmental Science and Technology, 39*(9), 3401–3408.

Rahimnejad, M., Najafpour, G. D., Ghoreyshi, A. A., Shakeri, M., & Zare, H. (2011). Methylene blue as electron promoters in microbial fuel cell. *International Journal of Hydrogen Energy, 36*(20), 13335–13341.

Reguera, G., Nevin, K. P., Nicoll, J. S., Covalla, S. F., Woodard, T. L., & Lovley, D. R. (2006). Biofilm and nanowire production leads to increased current in geobacter sulfurreducens fuel cells. *Applied and Environmental Microbiology, 72*(11), 7345–7348.

Richter, L. V., Sandler, S. J., & Weis, R. M. (2012). Two isoforms of geobacter sulfurreducens pila have distinct roles in pilus biogenesis, cytochrome localization, extracellular electron transfer, and biofilm formation. *Journal of Bacteriology, 194*(10), 2551–2563.

Ringeisen, B. R., Henderson, E., Wu, P. K., Pietron, J., Ray, R., Little, B., Biffinger, J. C., & Jones-Meehan, J. M. (2006). High power density from a miniature microbial fuel cell using Shewanella oneidensis DSP10. *Environmental Science and Technology, 40*(8), 2629–2634.

Rosenbaum, M., Zhao, F., Quaas, M., Wulff, H., Schroder, U., & Scholz, F. (2007). Evaluation of catalytic properties of tungsten carbide for the anode of microbial fuel cells. *Applied Catalysis B: Environmental, 74*, 261–269.

Ross, D. E., Flynn, J. M., Baron, D. B., Gralnick, J. A., & Bond, D. R. (2011). Towards electrosynthesis in shewanella: Energetics of reversing the Mtr pathway for reductive metabolism. *PloS One, 6*(2), e16649.

Rotaru, A.-E., Shrestha, P. M., Liu, F., Ueki, T., Nevin, K., Summers, Z. M., & Lovley, D. R. (2012). Interspecies Electron transfer via hydrogen and formate rather than direct electrical connections in cocultures of pelobacter carbinolicus and geobacter sulfurreducens. *Applied and Environmental Microbiology, 78*(21), 7645–7651.

Schroder, U. (2007). Anodic electron transfer mechanisms in microbial fuel cells and their energy efficiency. *Physical Chemistry Chemical Physics, 9*(21), 2619–2629.

Seelam, J. S., Pant, D., Patil, S. A., & Kapadnis, B. P. (2015). Biological electricity production from wastes and wastewaters. In *Microbial factories* (pp. 155–183). New Delhi: Springer India.

Sharma, T., Reddy, L. M. A., Chandra, T. S., & Ramaprabhu, S. (2008). Development of carbon nanotubes and nanofluids based microbial fuel cell. *International Journal of Hydrogen Energy, 33*, 6749–6754.

Song, H.-L., Zhu, Y., & Li, J. (2016). Electron transfer mechanisms, characteristics and applications of biological cathode microbial fuel cells – A mini review. *Arabian Journal of Chemistry*. https://doi.org/10.1016/j.arabjc.2015.01.008.

Srikanth, S., Pavani, T., Sarma, P. N., & Venkata Mohan, S. (2011). Synergistic interaction of biocatalyst with bio-anode as a function of electrode materials. *International Journal of Hydrogen Energy, 36*, 2271–2280.

Thygesen, A., Poulsen, F. W., Min, B., Angelidaki, I., & Thomsen, A. B. (2009). The effect of different substrates and humic acid on power generation in microbial fuel cell operation. *Bioresource Technology, 100*(3), 1186–1191.

Torres, A. K., Marcus, H. S., Lee, P., Parameswaran, R. K.-B., & Rittmann, B. E. (2010). A kinetic perspective on extracellular electron transfer by anode-respiring bacteria. *FEMS Microbiology Reviews, 34*, 3–17.

Veer Raghavulu, S., Suresh Babu, P., Kannaiah Goud, R., Venkata Subhash, G., Srikanth, S., & Venkata Mohan, S. (2012). Bioaugmentation of an electrochemically active strain to enhance the electron discharge of mixed culture: Process evaluation through electro-kinetic analysis. *RSC Advances, 2*, 677–688.

Vega, C. A., & Fernández, I. (1987). Mediating effect of ferric chelate compounds in microbial fuel cells with Lactobacillus plantarum, Streptococcus lactis, and Erwinia dissolvens. *Bioelectrochemistry and Bioenergetics, 17*(2), 217–222.

Wei, J., Liang, P., & Huang, X. (2011). Recent progress in electrodes for microbial fuel cells. *Bioresource Technology, 102*(20), 9335–9344.

Wei, L., Han, H., & Shen, J. (2012). Effects of cathodic electron acceptors and potassium ferricyanide concentrations on the performance of microbial fuel cell. *International Journal of Hydrogen Energy, 37*(17), 12980–12986.

Zhao, Z., Zhang, Y., Wang, L., & Quan, X. (2015). Potential for direct interspecies electron transfer in an electric-anaerobic system to increase methane production from sludge digestion. *Scientific Reports, 5*, 11094.

Chapter 6
Development of Suitable Anode Materials for Microbial Fuel Cells

Thi Hiep Han, Sandesh Y. Sawant, and Moo Hwan Cho

6.1 Introduction

Microbial fuel cells (MFCs) and related bioelectrochemical systems (BESs) have shown impressive developments for many purposes over the past decade (Kalathil et al. 2012; Han et al. 2013, 2014, 2016). Even with the noticeable improvements in power density, the large-scale application of MFCs is still limited due to the low power generation and high cost (Wei et al. 2011). To take this technology from laboratory-scale research to commercial applications, the cost and the performance of these systems need to be optimized further. The anode electrode plays an important role in the performance and cost of MFCs. The electrode materials in MFCs have some general and individual characteristics. In general, electrode materials must have good conduction, excellent biocompatibility, good chemical stability, high mechanical strength and low cost. The anode material design has attracted an enormous number of studies over the past decade.

The present chapter summarizes and discusses the recent advances in anode materials and their configurations.

6.2 Essential Requirements of Anode Materials

6.2.1 Surface Area and Porosity

The surface texture and morphology of an anode are important parameters in MFCs. Normally, a rough anode surface facilitates the adhesion of bacteria;

T.H. Han • S.Y. Sawant • M.H. Cho (✉)
School of Chemical Engineering, Yeungnam University, Gyeongsan 712-749, South Korea
e-mail: mhcho@ynu.ac.kr

D. Das (ed.), *Microbial Fuel Cell*,
https://doi.org/10.1007/978-3-319-66793-5_6

hence, the power density of a rough anode is significantly higher than that of a smooth anode (Michaelidou et al. 2011). The microorganisms in an MFC anode are a few micrometres in size. The porosity and surface area ensures the accessibility of microorganisms to the electrode. In general, extended porosity results in a high surface area of the electrode materials. A high surface area provides more space for the microorganisms to immobilize effectively on the anodes. An increase in the surface area of electrodes can minimize the ohmic losses and internal resistance of the MFC system (Kumar et al. 2013), hence increase the performance of MFCs. Therefore, recent studies have focused on the development of a three-dimensional open-porous scaffold anode structure, such as carbon foam (Han et al. 2016) or nickel foam (Wang et al. 2013a), which provides remarkable porosity and surface area for biofilm development.

6.2.2 Fouling and Poisoning

Fouling or clogging is actually the buildup of microorganisms on the electrode surface. After the MFCs operate for a certain time, the microorganism population increases within the confined region of the electrodes which leads to poisoning and fouling. During the development of a biofilm on an electrode surface, microorganisms excrete extra polymer substances. Fouling occurs due to the prolonged use of the electrodes, i.e., the buildup of a thick biofilm layer and extra polymer substances that can foul the electrode surface. To control fouling and poisoning, a high void volume of the electrode and large surface area per volume are needed (Liu et al. 2005).

6.2.3 Electronic Conductivity

The electronic conductivity of anode materials is an important factor that determines the overall MFC performance. Microorganisms oxidize the substrate and release electrons that are required to be transferred effectively to the external circuit via the anode. The excellent conductivity of the anode ensures that the MFCs operate continuously and efficiently. The low conductivity results in large ohmic losses. Only a few ohms of added internal resistance can greatly reduce the level of power generation (Logan 2008). Therefore, the anodes should be excellent conductors to allow the free flow of electrical current. Compared to the carbon-based anode, the metal-based anode is a better conductor with a lower resistance (copper, 0.1 Ω/Cm vs. carbon paper, 0.8 Ω/Cm). On the other hand, their performance is not as good as the carbon based anode. Even good conductivity may not be enough, the material should match many other requirements, such as non-corrosive, high surface area, high porosity, and biocompatibility.

6.2.4 Biocompatibility

Because microorganisms are inoculated directly on the surface of the anodes, the biocompatibility of the anode electrode with a biocatalyst is a critical factor that determines the MFC power generation. The biocompatibility of an electrode allows the microorganisms to adhere and spread over the electrode surface and form an electroactive biofilm. The coarsened surface of the fabricated anodes assists in the inoculation of biomass, which in turn, increases the operation cycle of MFCs. A few electrode materials are cytotoxic to the inoculated microorganisms and might inhibit the growth of microorganisms. The significant potential and power losses that occur in the anodes are due to the non-compatibility of microorganisms with the electrodes and certain fabrication strategies to increase the compatibility are in progress, such as the development of a rougher, porous surface, replacement of the cytotoxic material, and increasing the hydrophobicity of the high surface area.

6.2.5 Stability and Long Durability

For the real applications of MFCs, the durability of electrodes needs to be satisfied so that the replacements can be tailored to the minimum. The electrode materials should be fulfilled by the high chemical and physical stability under an aqueous environment. The anodes employed in the MFCs always make contact with the aqueous environment, which usually lead to the swelling and decomposition of the material. Therefore, hydrophobic electrode materials are preferred as electrode components. The stability of the electrodes is hindered by the molecules present in the pores of the electrode materials, reducing the thriving space for the microorganisms. The electrode surface is preferred to be rough to allow the detachment of the water molecules and provide more space for the sustainability of microorganisms. On the other hand, high surface roughness may result in an increase in fouling, which may decrease the long-term performance of MFCs. Therefore, an optimized roughness is needed to increase the power performance of MFCs.

6.2.6 Electrode Cost and Availability

Reducing the cost of cathode materials is critical for the practical applications of MFCs. Although MFCs are not as expensive as hydrogen-based fuel cells due to the sustainable fuels, the cost of the electrode materials account for a major part of the capital cost of MFCs. The availability of electrode materials for anodes is an essential requirement for improving the commercialization of MFC. Precious metal electrodes are very expensive because of their limited availability, which impedes their utilization in MFCs. Low cost materials with non-precious metals as well as new binders that are less expensive than Nafion, are currently in strong demand. Recent research has focused on carbonaceous or stainless steel mesh electrodes applicable for MFCs. The productions of carbonaceous materials derived from natural resources are on the top of the scales and have opened up the possibility for use as anode materials in MFCs. This effectively replaces the precious metal anodes due to their abundant sources, cost efficiency, prompt conductivity, and chemical inertness (Rozendal et al. 2008).

6.3 Anode Materials Employed in MFCs

6.3.1 Carbonaceous Electrodes

The overall output of MFCs is strongly dependent on its electrode performance, especially the anode. Therefore, considerable effort has been made to develop anode materials with enhanced performance (Pham et al. 2009). The mass transfer, ohmic losses, activation losses, biofilm growth and electron-quenching reactions are the key factors that determine the performance of the anode in MFCs (Pham et al. 2009). Ohmic loss, which depends on the internal resistance of the electrode, and biofilm growth are related directly to the anode material and its structural properties. The surface functionality, reflecting the hydrophilic and hydrophobic nature of the anode, are also very important to the growth of biofilms (Guo et al. 2013). Therefore, several studies have reported the development of the anode material in regard to the higher conductivity and active surface area. Among the different anode materials, carbonaceous anode materials have been explored extensively in MFCs owing to their good electrical conductivity, chemical and thermal stability, high mechanical strength, and most importantly, their comparatively low cost. The high surface area to volume ratio along with the rough surface property of carbonaceous anodes provides more space and more favourable conditions for bacteria growth, which results in better anode performance in MFCs. An increase in the biofilm, i.e., active biomass formation, was reported to increase the anodic current (Reguera et al. 2006). The non-corrosive and biocompatible nature of carbonaceous anode was also found to be advantageous over metal/metal oxide electrodes (Kumar et al. 2013). A large number of carbonaceous anodes have been reported, including graphite rods, graphite plates, carbon cloth, carbon paper, graphite brush, graphite felt, etc. Since the last one and half decades, there has been considerable research in carbon-based anode materials used in MFCs (Fig. 6.1). This part of the chapter focuses mainly on the different types of carbonaceous anodes and their performances based on their structural and chemical properties.

6.3.1.1 Types of Carbonaceous Anode

The geometrical properties of the anode were reported to have a drastic effect on its performance (Chaudhuri and Lovley 2003; Logan et al. 2007). Carbonaceous

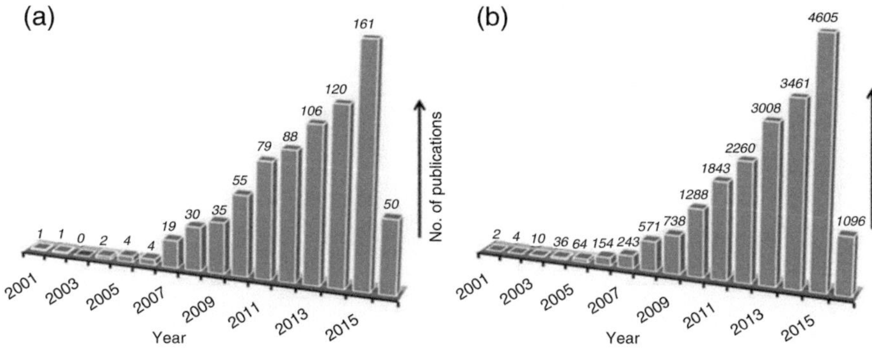

Fig. 6.1 Research trends in the field of carbonaceous anodes in MFC applications (2001 to April 2016)

electrodes can be classified into two categories based on the physical architecture: plane or two (2D) and three dimensional (3D) carbonaceous anodes. A pictorial view of various 2D and 3D carbonaceous anodes used in MFCs are presented in Table 6.1.

6.3.1.2 Plane or 2D Carbonaceous Anodes

2D carbon materials including carbon paper, carbon cloth, graphite foil or sheets have been the most commonly explored anode materials in MFCs. Carbon paper was popular as an anode material in the early stages of the research in MFCs due to the easy connection and precise quantitative analysis of biofilm growth. Owing to the brittle structure of carbon paper, the graphite sheets or plates were found to be favourable owing to their higher strength. The low surface area and comparatively smooth surface of both materials are major disadvantages associated with biofilm formation, which resulted in the low current density and power generation. Forming a rough surface on a graphite plate is an alternative way of enhancing the current density under similar conditions and it can perform better than the metal anode. The favourable bacterial attachment along with low charge transfer resistance are the key points for the better current density generation of a rough graphite plate as compared to a flat graphite plate (ter Heijne et al. 2008). Rabaey et al. (2003) reported a power density of 3.6 W m^{-2} using a glucose-mediated MFC with plain graphite as the anode material. Compared to carbon paper, carbon cloth provides considerable flexibility and a higher surface area but its comparatively high cost makes it a poor candidate as an anode in MFCs (Zhang et al. 2010). Wang et al. (2008) reported a power density of 376 mW m^{-2} using a carbon cloth anode with a surface area of 7 cm^2 from beer brewery wastewater. Activated carbon cloth is an alternative 2D carbonaceous anode material with a higher surface area. Zhao et al. (2008) reported the higher performance of activated carbon cloth as an anode in sulfate-based MFCs as compared to graphite foil and carbon fibre veil owing to its

Table 6.1 Photographs, cost of different 2D and 3D carbonaceous anode materials applied to power production in MFCs

2D electrodes	Cost ($)/size (cm)	3D electrode	Cost ($)/size (cm)
Carbon paper	130/40 × 40[a]	Graphite/Carbon felt	46/40 × 40[a]

Graphite plate	175/30 × 30 × 0.5[a]	Graphite/Carbon granules	6–8/kg[b]

Carbon cloth	79/45 × 40[a]	Reticulated vitreous carbon (RVC)	1.5/1 × 1 × 2.5[c]
		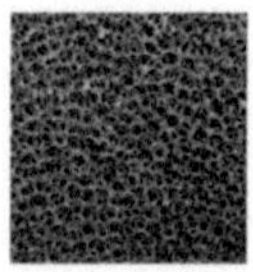	
Carbon mesh	6–40/100 × 100[b]	Carbon monolith	Not available

Carbon or graphite rod	1–10/1 × 20[b]	Carbon brush	Not available

All photographs reprinted with permission from Ref. (Wei et al. 2011) Copyright 2011 by Elsevier
[a]http://fuelcellstore.com
[b]http://www.alibaba.com
[c]http://www.ergaerospace.com/RVC-properties.htm

high surface area and sulfide oxidation property. Carbon cloth has been demonstrated as an anode material in yeast fuel cells and achieved a maximum power density of 1.03 W/m^2 (Haslett et al. 2011). The sheet-like structure obtained by the winding of carbon fibre on two carbon rods was reported to be an anode in MFCs (Wen et al. 2010). Carbon mesh is also available and has been utilized as a porous

2D anode material for power production using MFC and the performance could be enhanced by a heating and ammonia pre-treatment (Wang et al. 2009). The use of a carbon rod as an anode in MFCs was restricted owing to its low porosity but it served as a current collector in many cases (Rhoads et al. 2005; Deng et al. 2010; Pisciotta et al. 2012). The use of a number of carbon rods as an anode was also reported in the case of a single chamber MFC with a maximum power density of 26 mW m^{-2} and 80% of the COD removal (Liu et al. 2004).

6.3.1.3 3D Carbonaceous Anodes

Owing to the low porosity/surface area and sometimes fragile structure of 2D carbonaceous anodes, researchers have focused on the development of 3D architecture of carbon-based anodes for MFC applications. The surface area enhancement of the anode provides a substantial boost in power generation. A porous anode provides a more accessible surface area to microbes, without altering the geometrical area of the electrode. The application of a 3D anode in MFCs is a more practical way of scaling-up and developing real world applications of this technology (Jiang et al. 2011). Chaudhuri and Lovley (2003) reported ~2.4 times higher current production in glucose and *Rhodoferax ferrireducens*-mediated MFCs using porous graphite foam (74 mA m^{-2}) instead of a graphite rod (31 mA m^{-2}) under similar conditions. The idea of the utilization of a 3D electrode system was first reported in 1989 and then by Sell et al. (1989) using a packed bed of granular graphite. Therefore, 3D carbonaceous anode materials can be classified further into two sub-categories: packed anode (sometimes also called as stuffed or filled) and 3D configured. In the packed type of electrode, powdered or granules of electrode materials (activated carbon or graphite) are filled tightly in the anodic chamber and a current collector is inserted externally. The graphite or carbon granule-packed 3D anodes are used either on a laboratory scale (Kalathil et al. 2011) or on a relatively large scale in continuously operated MFCs (You et al. 2007). In contrast, the 3D configured electrode possesses their own 3D geometry, such as carbon felt and reticulated vitreous carbon (RVC) foam. Graphite/carbon felt is used extensively as a traditional 3D–based anode in MFCs. Deng et al. (2010) reported a power production of 784 mW m^{-2} using activated carbon felt as the electrode during a study of activated carbon felt as an anode in variation with different cathode materials in an up-flow type MFC. Han et al. (2016) reported a maximum power generation of 96 W m^{-3} using a 3D nitrogen-doped carbon foam as an anode material with an enhanced surface area, less internal resistivity, and excellent microorganism attachment.

The 3D porous carbon materials derived from natural renewable and abundant sources showed excellent performance as an anode material in MFCs. A simple direct carbonization, low cost and environmentally friendly technique is normally used to obtain such materials. The reticulated carbon foam obtained from Pomelo peel (Fig. 6.2a, b) produced a projected current density (j_p) of 4.02 mA cm^{-2}, which is five times higher than that of commercial RVC and 2.5 times higher than

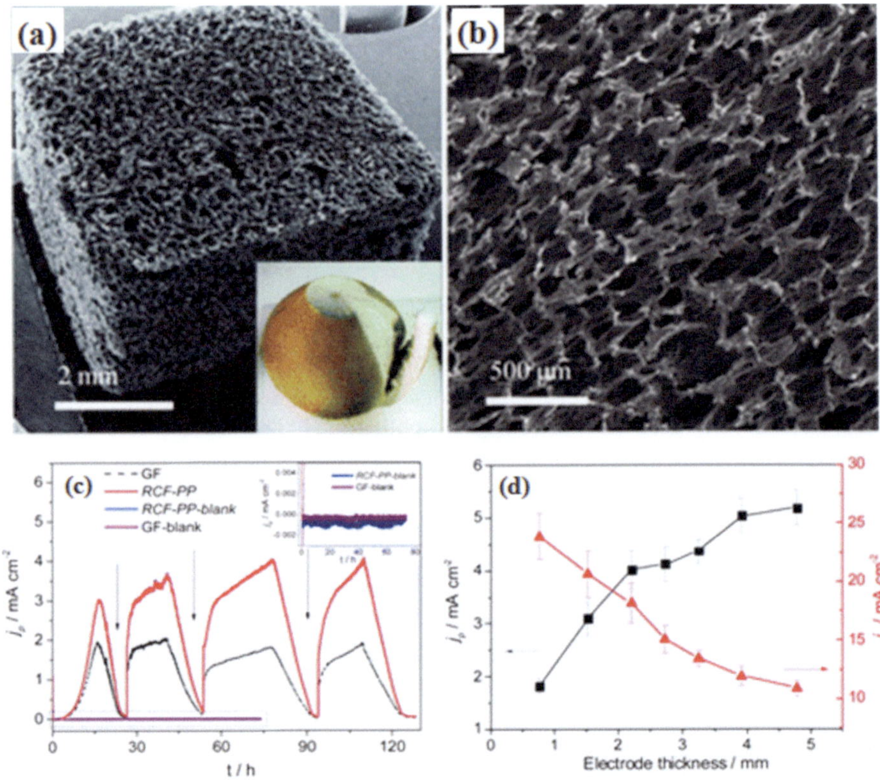

Fig. 6.2 SEM images (**a**) and (**b**) at different magnification and current generation; (**c**) reticulated carbon foam obtained from Pomelo peel (The inset of (*a*) is a digital image of a peeled Pomelo and (*c*) is the current response in sterile PBS solution containing acetate substrate; blank); (**d**) Effect of the electrode thickness on the projected (j_p) and volumetric (j_v) current densities (Reprinted with permission from Chen et al. (2012); Copyright 2012 by the Royal Society of Chemistry)

that of graphite felt with a similar electrode size (Fig. 6.2c) and can be increased further by increasing the electrode size (Fig. 6.2d) (Chen et al. 2012). Yuan et al. (2013) reported the fabrication and modification of a three-dimensional carbon anode using a natural loofah sponge. The enhanced bacterial loading and extracellular electron transfer due to the macro scale porous structure and carbon coating, respectively, resulted in a higher power output of 1090 mW m^{-2} as compared to similarly sized traditional 3D anodes. The 3D carbon material obtained from Kenaf (*Hibiscus cannabinus*) stems has also been reported to be an excellent anode material for MFCs (Chen et al. 2012).

The need for a large surface area with efficient current collection has resulted in the development of a special type of anode electrode in MFCs i.e. carbon brush electrode. The brush electrode was first developed and applied by Garshol and Hasvold (1995) for galvanic seawater cell. Logan et al. (2007) implemented this design to fabricate a carbon brush electrode in MFCs. The carbon brush electrode is very simple and similar to household brushes, which can be fabricated easily by winding graphite fibres into twisted titanium wires as a non-corrosive current collector.

The high surface-volume ratio and low resistance, coupled with a good dispersed distribution of the filaments in the brushes, make them ideal as anodes for small or large scale MFC applications. The confined arrangement of graphite fibres near the connection of the current collector wire limits the microbe-electrode interactions (Xie et al. 2015). The large space required by the carbon brush electrode is one of the major limiting factors in MFC design. To overcome this limitation, the carbon brush electrode was applied as a half-circle brush electrode in membrane based MFCs (Hays et al. 2011; Ahn et al. 2014).

6.3.2 *Non-carbonaceous Electrodes*

Metal materials are much more conductive than carbon materials; for example, copper has a specific conductivity of 58×10^6 S m^{-1}, which is approximately 920 times higher than that of polycrystalline graphite (Baudler et al. 2015). On the other hand, they are not widely applicable as carbon materials in MFCs because the smooth surface of metals does not facilitate the adhesion of bacteria (Wei et al. 2011). The limits of the electrochemical stability can cause corrosion. The antimicrobial property can prevent bacteria colonizing the surface of metals (Baudler et al. 2015). Although many metals have been ruled out because of these reasons, there are many metals reported to be suitable for anode materials in MFCs and related bio-electrochemical systems. Basically, there are two groups of metals that can be used as an anode material in a bioelectrochemical system: electrochemically noble metals or electrochemically passivated metals or their alloys (stainless steel) (Baudler et al. 2015).

6.3.2.1 Noble Metal Materials

Noble metals, such as gold, platinum and silver, belong to the first group of metals. These noble metals are potentially attractive anode materials for MFC applications because they are highly conductive and highly versatile for electrode manufacture (Crittenden et al. 2006). Richter et al. (2008) reported that *Geobacter sulfurreducens* biofilms could grow on gold anodes up to 40 μm thick, producing currents almost as effectively as in graphite anodes. Their high cost, however, prevents them from being used in large technical systems. Interestingly, copper and silver, which are commonly reported to have antimicrobial properties, were shown to achieve comparable performance to graphite (Baudler et al. 2015). Although the tolerance of the electrochemically active bacteria to the antimicrobial effects of copper and silver metals has not been well explored, copper is a highly promising anode material, suitable for applications in high-performance bioelectrochemical systems.

6.3.2.2 Non-noble Metal Materials

In addition to noble metals, nickel, cobalt, titanium and copper have been studied systematically for their suitability as anode materials for MFCs and related bioelectrochemical systems (Baudler et al. 2015). Titanium is a common metal-based anode used in MFCs. As titanium is a non-corrosive, highly stable and biocompatible metal, it is used regularly as a current collector and anode electrode in MFCs. On the other hand, the application of bare titanium in MFCs as an anode electrode results in a clear limiting current as compared to graphite (ter Heijne et al. 2008). It was concluded that bare titanium was unsuitable as an anode material in MFCs. The performance of a titanium electrode can be increased to better than graphite by modifying its surface with Pt (ter Heijne et al. 2008) or platinum-iridium composites or tantalum-iridium composites (Michaelidou et al. 2011).

As an inexpensive base metal and large industrial availability, stainless steel is the most common metal electrode used in MFCs. Stainless steel provides a compact oxide layer (passivation layer) that protects the metal from further oxidation. Hence, it has good mechanical properties and long-term resistance to corrosion (Dumas et al. 2008). As a first attempt to use a non-corrosive material in the anode of MFCs, the stainless steel plate failed to achieve higher power densities as compared to carbon materials (Dumas et al. 2007). In another study, Dumas et al. (2008) reported that the stainless steel plate anode was less efficient than the graphite one at a similar experimental procedure, in which biofilms were grown with a pure culture of *Geobacter sulfurreducens* on anodes polarized at +0.20 V vs. Ag/AgCl. The less obvious successes obtained with the stainless steel plate anodes in benthic fuel cells were certainly linked to the low surface roughness (0.29 μm), which affect biofilm growth on the electrode surface. Therefore, the free evolution of the passive layer may have been a cause of the low current densities obtained previously. In contrast, when a stainless steel plate was replaced with a higher specific area electrode, such as a stainless steel grid, the power densities achieved were much higher than with carbon materials. Erable and Bergel (2009) proved that a stainless steel grid anode produced a current density of 8.2 A m^{-2}, which is 2.5 times higher than that of plain graphite under a constant potential of -100 mV vs. the saturated calomel electrode. This shows that stainless steel is an efficient support for microbial anodes. It would represent a promising solution to the scaling-up of industrial MFCs.

6.3.2.3 3D and Composites Metal-Based Electrodes

Many attempts have been made to increase the energy production efficiency of MFC by the development of a metal-based foam structure of electrodes. Mapelli

Fig. 6.3 (**a**) Sponge anode electrical connection and (**b**) foam porosity filled by bacteria (Reprinted with permission from Mapelli et al. (2013); Copyright 2013 by Wiley Online Library)

et al. (2013) tested the cast iron open cell foam by a high-precision micro-cutter and laser welded technique (Fig. 6.3). The output is comparable to the typical outputs of MFCs systems; however, it is unsuitable for the long time operation of MFCs due to the corrosion that compromises the electrical conductivity of the sponge electrodes. Nickel foam has also attracted considerable interest as a base substrate for the MFC anode because the porous structure can act as a good support for the incorporation of the composite materials with graphene (Wang et al. 2013a, b), or conductive polymers (Karthikeyan et al. 2016).

6.4 Surface Treatment

The modification of electrode materials is an effective way of improving the performance of MFCs because it alters the physical and chemical properties to allow better microbial attachment and electron transfer (Zhou et al. 2011). A huge number of modifications of anode materials have been achieved. This section focuses on some conventional techniques for the modification of anode materials. Generally, the modification methods include surface treatments by physical (heat treatment)/chemical methods (ammonia treatment, acid treatment, electrochemical oxidation), or coating the surface with highly conductive or electroactive composites.

6.4.1 Heat Treatment

As mentioned above, the surface characteristics of anode materials are one of the deciding factors that affect bacterial attachment and electrical connections between the bacteria and electrode surface. The studies that examined the surface treatment of anode materials covering heat, ammonia, acid treatment and electrochemical oxidation are described in Table 6.2.

Table 6.2 Some previous works pertained with the surface modification of anodes

Modification method	Anode materials	Biocatalyst	MFC configuration	Improvement in MFC efficiency	References
Heat treatment	Carbon mesh	Domestic wastewater	Single chamber Air–cathode MFC	Power density increases by 3% to 922 mW m^{-2}	Wang et al. (2009)
	Carbon fibre brush	Domestic wastewater	Single chamber Air–cathode MFC	Power density increases by 25.5% to 1280 mW m^{-2}	Feng et al. (2010)
Ammonia treatment	Carbon cloth	Domestic wastewater	Single chamber Air-cathode MFC	Power density increases by 20.1% to 1970 mW m^{-2} Reduce acclimation time by 50%	Cheng and Logan (2007)
	Carbon mesh	Anaerobic sludge	Single chamber Air-cathode MFC	Power density increases by 33% to 736 mW m^{-2}	Zhou et al. (2012)
Ammonium peroxydisulphate treatment	Graphite felt	Sulphate reducing bacteria	H-type MFC	Power density increases by 25.4% to 355 mW m^{-2}	Du et al. (2016)
Ethylenediamine treatmnet	Graphite felt	Sulphate reducing bacteria	H-type MFC	Power density increases by 92.6% to 545 mW/m^2	Du et al. (2016)
Methylene blue treatment	Graphite felt	Sulphate reducing bacteria	H-type MFC	Power density increases by 80.2% to 510 mW m^{-2}	Du et al. (2016)
Acid treatment	Carbon mesh	Anaerobic sludge	Single chamber Air–cathode MFC	Power density increases by 43% to 792 mW m^{-2}	Zhou et al. (2012)
	Carbon fibre brush	Domestic wastewater	Single chamber Air–cathode MFC	Power density increases by 7.8% to 1100 mW m^{-2}	Feng et al. (2010)
Acid and heat treatment	Carbon fibre brush	Domestic wastewater	Single chamber Air–cathode MFC	Power density increases by 34.3% to 1370 mW m^{-2}	Feng et al. (2010)
	Graphite felt	Brewery wastewater diluted with domestic wastewater	Single chamber Air–cathode MFC	Power density increases twofold to 28.4 mW m^{-2}	Scott et al. (2007)

(continued)

Table 6.2 (continued)

Modification method	Anode materials	Biocatalyst	MFC configuration	Improvement in MFC efficiency	References
Electrochemical oxidation	Carbon mesh	Anaerobic sludge	Single chamber Air–cathode MFC	Power density increased by 43% to 792 mW m^{-2}	Zhou et al. (2012)
	Carbon cloth	Domestic wastewater	Single chamber Air–cathode MFC	Current density increased by 41% (from 4.79×10^{-4} to 6.76×10^{-4} Am^2)	Liu et al. (2014)
	Graphite felt	Mixed bacterial Culture	Two-chamber MFC	Current density increased by 39.5% (from 0.81 mA to 1.13 mA)	Tang et al. (2011)

6.4.1.1 Treatment of Anode Materials

Thermal treatment is one of the effective and cost efficient methods to modify anode materials. During the heat treatment, the cracks were generated and promoted the surface area. It is reported that the heat treatment promoted the actual surface area by 6.94 times to 49.3 m^2 g^{-1} as compared to untreated fibres (Feng et al. 2010), which facilitated the adhesion and inoculation of microorganisms over the electrodes. The thermal treatment of carbon mesh anodes resulted in a 3% increase in the overall power density of MFCs (Wang et al. 2009). High temperature treatment of carbon fibre brush anodes in air (450 °C for 30 min) exhibited an increase in power density up to 15% as compared to its unmodified anodes (Feng et al. 2010).

6.4.1.2 Chemical Treatment

Ammonia/Acid Treatment

Because the microorganisms are negatively charged, the accumulation of microbes depends solely on the surface charges of the electrodes. An increase in the adhesion of microorganisms increases the probability of the facile and direct transfer of electrons to the electrodes. The basis of an ammonia/acid treatment is to increase the adhesion of microorganisms onto the anode interface by enhancing the positive charge of the electrode surface.

The ammonia treatment improvises the positively charged functional groups over the electrode surfaces, facilitates bacterial adhesion and increases electron transfer to the anode surface. Ammonium treatment of carbon electrodes was widely applied with different anode materials such as carbon cloth (Cheng and Logan 2007), graphite brush (Logan et al. 2007), and carbon mesh (Wang et al. 2009). Ammonia gas treatment on carbon cloth was conducted using continuous flow of ammonia in a thermogravimetric analyzer at 700 °C. The acclimation time of MFCs was reduced greatly by 50% upon treatment. The ammonia-treated carbon cloth exhibited a power density of 1970 mW m^{-2}, which was 1.5 times higher than the untreated anode yielded (Cheng and Logan 2007). Moreover, nitrogenous compounds such as ethylenediamine and methylene blue were also applied to modify graphite felt anodes in MFCs and exhibited the great improvement in MFC performances. The maximum power density of the MFC with modified anode was 545 and 510 mW/m^2, respectively, which was larger than the un-modified anode (283 mW/m^2). The increase of power density was correlated with increase in nitrogen content, which could make bacterial adhesion more favourable to the anode, and facilitated the electrons' transfer from bacteria to anode (Du et al. 2016).

Oxidation of the anode surface using an acid is another effective way to modify the surface of electrodes. It has been achieved by impregnating the electrodes in concentrated acid solutions. This increases the native surface area of the anodes and facilitates the protonation of functional groups over the anodes, thereby increasing the positive charge on the electrode surface. Graphite modified with anthraquinone-1,6-disulphonic acid produce five-times greater power (Lowy et al. 2006). A method involving HNO_3 treatments of Ketjen Black carbon supports and followed by heat treatment up to 900 °C has been investigated to modify graphite felt anode. The performance of MFC with that modified graphite felt anode increases about twofold (Scott et al. 2007). The increase in performance of above modified anodes was attributed to the increase in surface area and biocompatibility by introducing the quinone group (Scott et al. 2007). In addition, the combination of acid and thermal treatments (Feng et al. 2010) or acid treatment and ammonia treatment (Wang et al. 1999) gives higher electrocatalytic activity and increases MFC power generation than either treatment alone. Although high power generation was guaranteed by using heat/acid/ammonia treatment, the special demands for a sophisticated environment, complicated apparatus, and high temperature, long treatment time of these strategies can increase the capital cost and limit their potential scale applications.

Electrochemical Oxidation

Electrochemical oxidation is known as a convenient, effective and practical method for anode modification to improve the performance of MFCs (Tang et al. 2011; Zhou et al. 2012; Liu et al. 2014). Electrochemical oxidation changes the electrochemical properties of anode, increases the electrochemical active surface area by 2.9 times (from 11.2 to 44.1 cm^2), and improves the exchange current density by 41% (Liu et al. 2014). This process results in the generation of new native functional groups, such as carboxyl (Tang et al. 2011) or amide groups (Liu et al. 2014) on to the electrode surface, which enhances electron transport from the microorganisms to the anode surface. The presence of functional groups facilitated the formation of peptide bonds between the electrode surface and microorganisms, acting as highways for effective electron transfer. The graphite felt modified by electrochemical oxidation exhibited a 39.5% increase in the current density compared to the untreated anodes (Tang et al. 2011). In the same manner, power density of MFCs with carbon mesh modified by electrochemical oxidation method increased by 43% to 792 mW m^{-2} (Zhou et al. 2012). Compared with other electrode modification techniques, this electrochemical approach is much simpler and quicker, operating at ambient temperature with the cheap apparatus of power supply.

6.4.2 Advanced Nanostructure Modification of Anodes

6.4.2.1 Modification of Anodes by Carbon Nanotubes (CNT) and Its Composites

In a recent period, advanced carbon nanostructures, including CNT, graphene, carbon nanofibres, etc., possessing a high surface area and inherent conductivity with easily tailored surface functionalities, were introduced as an efficient anode in MFCs. The carbon nanostructures could improve the extracellular electron transfer through the surface functional groups (Huang et al. 2011). The most exploited and easiest way to use the carbon nanostructures as 2D or 3D anode materials is by coating them on conducting 2D or 3D supports, such as stainless steel mesh, carbon paper, nickel foam, etc. (Guo et al. 2014; Hou et al. 2014; Qiao et al. 2014). Sun et al. (2010) have reported that an up to 20% increase in the power density of MFCs could be achieved with a simple coating of CNTs on carbon paper. Interestingly, Liang et al. (2011) showed that the total output voltage could be increased with a shortened MFC startup time by the addition of CNTs in an anodic medium. Mink and Hussain (2013) developed micron-sized MFC (75 μL) (Fig. 6.4) on silicon using CMOS-compatible processes, in which they used multiwall carbon nanotubes (MWCNTs) as the on-chip anode. Surprisingly, the MWCNTs device produced approximately a 20 times higher initial current (880 mA m^{-2}) than with gold (29 mA m^{-2}) and nickel (37 mA m^{-2}) anodes. This proves that the bacteria could grow more rapidly and transfer electrons to the MWCNT anode than the others due to the increased surface area and exceptional conductivity of MWCNTs. As per authors understanding, the oxygen intrusion into the device, bacterial clogging of the anode and non-optimal contact with the MWCNT anode were the main reasons for the decrease in the performance of micron-sized MFC comprised of MWCNTs.

The 3D CNT textile anode was also fabricated by a simple dipping-drying process and compared that with carbon cloth (Xie et al. 2011). The high

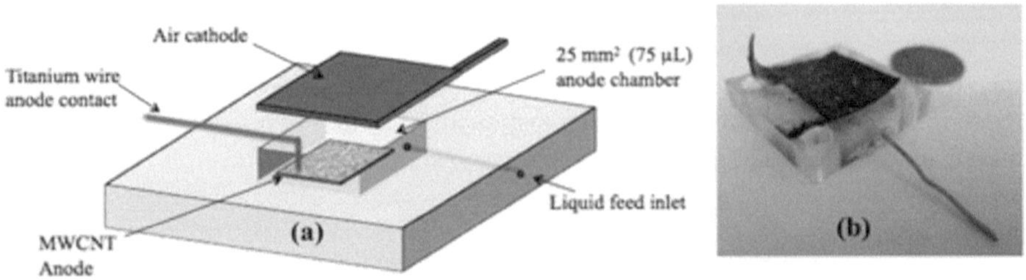

Fig. 6.4 (**a**) Schematic diagram of the microsized (75 µL) MFC with MWCNTs on a silicon chip anode and air cathode; (**b**) Photograph of the MWCNTs on a silicon chip MFC in a plastic encasing with a titanium wire contact visible as well as the black air cathode compared to a U.S. penny (Reprinted with permission from Mink and Hussain (2013); Copyright 2013 by the American Chemical Society)

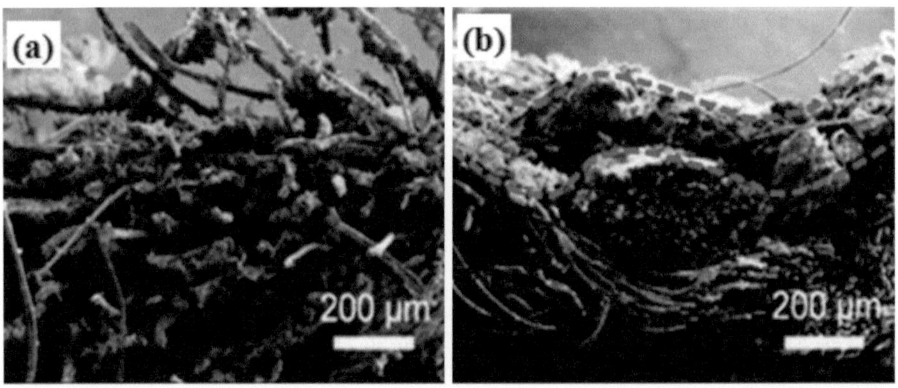

Fig. 6.5 SEM images of microbial growth on CNT-textile and carbon cloth. (**a**) Cross section of the CNT-textile anode illustrating internal colonization. A microbial biofilm wraps around each CNT-textile fibre, including the exterior and interior fibres; (**b**) Cross section of the carbon cloth anode. The biofilm is largely restricted to the outer surface of the carbon cloth anode (area between two broken lines) with a few microorganisms present on the interior fibres (Reprinted with permission from Xie et al. (2011); Copyright 2011 by the American Chemical Society)

conductivity (50 S/cm) was achieved by a ~ 200 nm thick CNT coating. SEM showed that the open macroscale porous structure of the CNT-textile provided sufficient substrate transport inside the CNT-textile anode to maintain internal colonization (Fig. 6.5a). On the other hand, microbial colonization was restricted largely to the outer surface of the carbon cloth anode (Fig. 6.5b). The MFCs operated with the CNT-textile anode produced a 68% higher maximum power density than those operated with the carbon cloth anode (1098 vs. 655 mW m^{-2}). Moreover, 141% more energy was produced using the CNT-textile anode than the carbon cloth anode utilizing the same amount of glucose.

6.4.2.2 Modification of Anodes by Graphene and Its Composites

When combined with graphene, the nickel foam electrode enhances the surface area, chemical stability and electrical conductivity significantly. Therefore, the performance of a graphene/nickel foam electrode in MFCs is quite impressive. Wang et al. (2013a, b)) developed a 3D electrode by coating reduced graphene oxide (rGO) sheets on Ni foam for MFC devices (Fig. 6.6). This MFC device with a

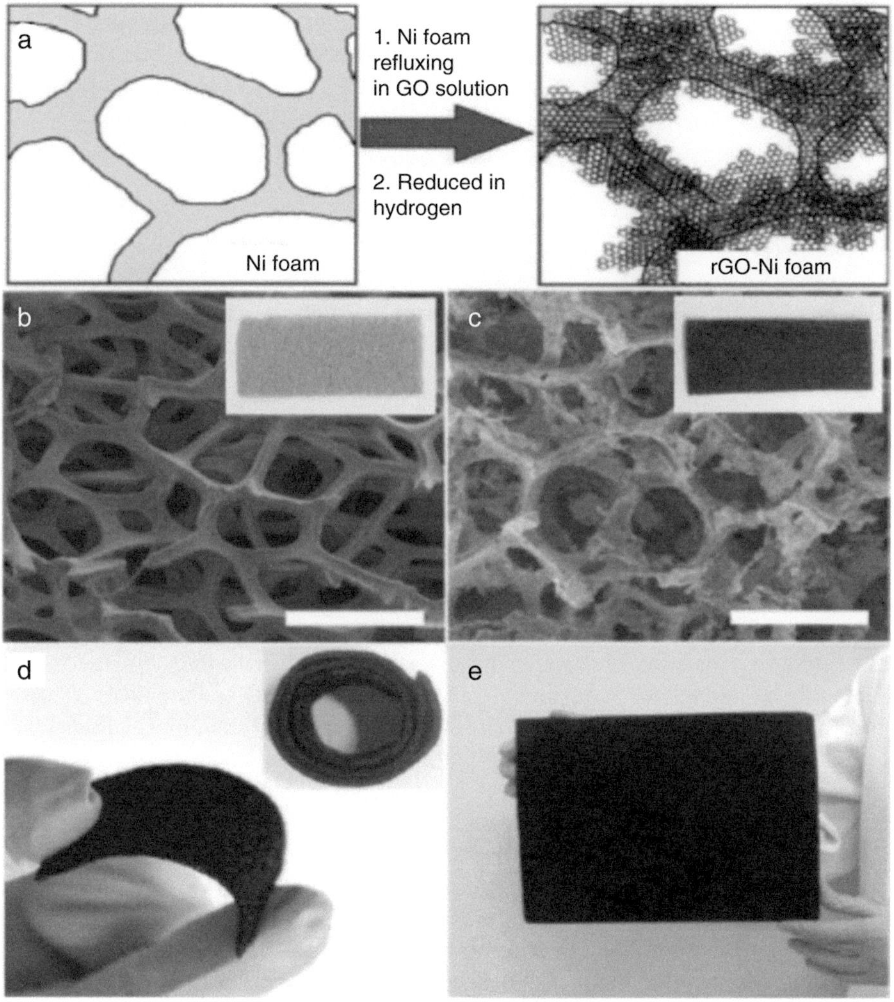

Fig. 6.6 (**a**) Schematic diagram of the preparation of rGO-Ni anode; (**b** and **c**) SEM images and digital pictures (insets) of plain nickel foam and rGO-Ni foam. Scale bars are 200 @ m; (**d**) Digital image of a curved rGO-Ni foam. Inset: rGO-Ni foam rolled up into a cylindrical shape; (**e**) Digital image of a 25 cm × 20 cm rGO-Ni foam (Reprinted with permission from Wang et al. (2013a, b); Copyright 2013 by the Royal Society of Chemistry)

flexible rGO-Ni electrode achieved a noticeable volumetric power density of 27 W m^{-3} (based on the volume of the anode chamber), which is 26 times and 16.7 times larger than the values gained by carbon felt and carbon cloth anodes, respectively.

3D scaffolds of CNTs and graphene possessing a hierarchical and open porous structure with good electrical conductivity are favourable for microbial colonization and have been studied as promising candidates for MFC anodes. 3D carbon nanostructures can be synthesized either by coating on a 3D support or an aerogel technique (Dumitru et al. 2008). Carbon nanostructured aerogels are biocompatible (Gutierrez et al. 2007) and easy to fabricate using a range of methods, such as a template (Antonietti et al. 2014), 3D printing (Zhu et al. 2015), and sol-gel using polysaccharides (e.g. chitosan, chondroitin sulphate) (Katuri et al. 2011), protein-like polymers (e.g. gelatin) (Nardecchia et al. 2013), organic agents (ethylene diamine) (Hu et al. 2013) as a chelating agent. Xie et al. (2012a, b)) developed 3D flexible sponges with a repeated coating of CNTs and graphene on polyurethane sponges as a low cost anode for MFCs and achieved a maximum power density of 182 W m^{-3} and 394 W m^{-3}, respectively.

6.4.2.3 Modification of Anodes by Conductive Polymer and Its Composites

Using a conductive polymer composite to improve the conductivity is an efficient approach to improving the conductivity, specific surface area of the anode, and the MFC output (Yong Zhao et al. 2010). Long polypyrrole chains could penetrate into the bacterial cell membrane and bring out the electrons via a metabolic pathway (Yuan and Kim 2008). The weak compatibility of precious metal anodes with the microorganisms was attributed to the absence of surface roughness and is improved by the modification of a conductive polymer (Kumar et al. 2013). The maximum MFC power density derived for the polyaniline modified titanium anode was 2317 mW m^{-3} and the observed power was correlated with their enhanced biocompatibility and nature of the inoculated bacterial culture (Benetton et al. 2010). A maximum power density of 18.8 W m^{-3} (2.3 times higher than nickel foam) was obtained in the case of the anode modified with a composite of chitosan, polyaniline and titanium carbide. The backbone of polyaniline with activated carbon forms a high surface area; TC enhances the coulombic efficiency; and the positive charge of CT leads to the higher adhesion of bacteria (Karthikeyan et al. 2016).

According to Qiao et al. (2007), a composite of polyaniline with CNTs provides protection and reduces the cellular toxicity of CNTs as well as enhances the surface area and contributes to the high power output of 42 mW m^{-2} with a cell voltage of 450 mV. In the case of MnO_2/CNT, however, electron conduction was facilitated by the enhanced electron transfer between the microorganisms and the anode material due to Mn^{4+} (Kalathil et al. 2013). The capacitive property of the composite was also found to be favourable for power production using MFCs. Roh and Woo (2015) explored the polypyrrole/CNT coated on carbon felt as an anode for the *Shewanella oneidensis* catalyzed MFC and reported that polypyrrole/CNT coated on carbon felt showed 38% improvement in power production as compared to plain carbon felt anode. The modification of the graphene-based anode with positively charged ionic liquids was favourable for the electrostatic interactions between the microbes and anode (Zhao et al. 2013). The decoration of conducting polymer through electro-polymerization also provided a larger active surface area and lower charge-transfer resistance by acting as an electric bridge between graphene and the support (Wang et al. 2013a, b).

6.5 Challenge and Outlook

At the current state of MFC applications, low efficiency and high cost are major obstacles. In this chapter, besides the considerable research on MFC design and application, particularly with the anode electrode, none of the current anode electrodes fully satisfy the needs of either the performance or cost. Carbonaceous materials can provide a great platform to develop an anode owing to their excellent conductivity, bio-compatibility and large surface area. As noted in this chapter, carbonaceous materials possess a robust surface, which can be modified easily by surface functionalization and composites with metal oxide or polymers. 3D carbon anodes are more promising towards large scale applications. Nano sized carbon materials, such as graphene and CNT, are good alternatives, which can be used either directly or as additives in anode materials.

On the other hand, despite the considerable progress in the field of carbonaceous anode in MFCs, the following areas are essential to be developed in prospective MFC commercialization: (1) As most of the work describes the initial results of the anode performance, the stability of the anode material is still a large issue in commercialization and long term applications of MFCs; (2) most of the synthesis methods of excellent anode materials are tedious and lengthy; (3) cost-effective production of anode materials; (4) the development of 3D carbon anodes with a suitable pore size to provide an enhanced surface area without clogging; and (5) the *in-situ* modification of anode materials either by metal/metal oxide doping or carbon nanomaterials to enhance the conductivity without compromising the bio-compatibility. Carbonaceous materials are expected to be low cost and durable anode materials with satisfactory performance in MFCs.

6.6 Conclusion

Electrode designs are the greatest challenge in manufacturing MFCs as a cost-effective technology. For their superior performance in MFCs, the most significant properties of anode electrodes include the surface area, morphology, biocompatibility, conductivity and stability. A range of carbon and metal materials have been explored to develop anodes and cathodes, and several electrode modification methods have been developed to improve power generation. The electrode configuration has evolved from a planar to a 3D structure together with many modification strategies. To date, a number of modifications are underway, such as surface treatment, coating with nanomaterials and composites. On the other hand, all these studies were conducted on a laboratory scale. The power generation and electrode cost have not reached the level for commercial use. Further studies on more effective anode materials and optimization of the configuration are expected to address these challenges.

References

Ahn, Y., Hatzell, M. C., Zhang, F., & Logan, B. E. (2014). Different electrode configurations to optimize performance of multi-electrode microbial fuel cells for generating power or treating domestic wastewater. *Journal of Power Sources, 249*, 440–445.

Antonietti, M., Fechler, N., & Fellinger, T.-P. (2014). Carbon aerogels and monoliths: Control of porosity and nanoarchitecture via sol–gel routes. *Chemistry of Materials, 26*, 196–210.

Baudler, A., Schmidt, I., Langner, M., Greiner, A., & Schroder, U. (2015). Does it have to be carbon? Metal anodes in microbial fuel cells and related bioelectrochemical systems. *Energy & Environmental Science, 8*, 2048–2055.

Benetton, X. D., Navarro-Ávila, S. G., & Carrera-Figueiras, C. (2010). Electrochemical evaluation of Ti/TiO2-polyaniline anodes for microbial fuel cells using hypersaline microbial consortia for syntheticwastewater treatment. *Journal of New Materials for Electrochemical Systems, 13*, 1–6.

Chaudhuri, S. K., & Lovley, D. R. (2003). Electricity generation by direct oxidation of glucose in mediatorless microbial fuel cells. *Nature Biotechnology, 21*, 1229–1232.

Chen, S., Liu, Q., He, G., Zhou, Y., Hanif, M., Peng, X., Wang, S., & Hou, H. (2012). Reticulated carbon foam derived from a sponge-like natural product as a high-performance anode in microbial fuel cells. *Journal of Materials Chemistry, 22*, 18609–18613.

Cheng, S., & Logan, B. E. (2007). Ammonia treatment of carbon cloth anodes to enhance power generation of microbial fuel cells. *Electrochemistry Communications, 9*, 492–496.

Crittenden, S. R., Sund, C. J., & Sumner, J. J. (2006). Mediating electron transfer from bacteria to a gold electrode via a self-assembled monolayer. *Langmuir, 22*, 9473–9476.

Deng, Q., Li, X., Zuo, J., Ling, A., & Logan, B. E. (2010). Power generation using an activated carbon fiber felt cathode in an upflow microbial fuel cell. *Journal of Power Sources, 195*, 1130–1135.

Du, H., Bu, Y., Shi, Y., Zhong, Q., & Wang, J. (2016). Effect of an anode modified with nitrogenous compounds on the performance of a microbial fuel cell. *Energy Sources, Part A: Recovery, Utilization, and Environmental Effects, 38*, 527–533.

Dumas, C., Mollica, A., Féron, D., Basséguy, R., Etcheverry, L., & Bergel, A. (2007). Marine microbial fuel cell: Use of stainless steel electrodes as anode and cathode materials. *Electrochimica Acta, 53*, 468–473.

Dumas, C., Basseguy, R., & Bergel, A. (2008). Electrochemical activity of Geobacter sulfurreducens biofilms on stainless steel anodes. *Electrochimica Acta, 53*, 5235–5241.

Dumitru, A., Morozan, A., Ghiurea, M., Scott, K., & Vulpe, S. (2008). Biofilm growth from wastewater on MWNTs and carbon aerogels. *Physica Status Solidi (a), 205*, 1484–1487.

Erable, B., & Bergel, A. (2009). First air-tolerant effective stainless steel microbial anode obtained from a natural marine biofilm. *Bioresource Technology, 100*, 3302–3307.

Feng, Y., Yang, Q., Wang, X., & Logan, B. E. (2010). Treatment of carbon fiber brush anodes for improving power generation in air–cathode microbial fuel cells. *Journal of Power Sources, 195*, 1841–1844.

Garshol, T., & Hasvold, O. (1995). *Galvanic seawater cell*. Google Patents.

Guo, K., Freguia, S., Dennis, P. G., Chen, X., Donose, B. C., Keller, J., Gooding, J. J., & Rabaey, K. (2013). Effects of surface charge and hydrophobicity on anodic biofilm formation, community composition, and current generation in bioelectrochemical systems. *Environmental Science & Technology, 47*, 7563–7570.

Guo, W., Cui, Y., Song, H., & Sun, J. (2014). Layer-by-layer construction of graphene-based microbial fuel cell for improved power generation and methyl orange removal. *Bioprocess and Biosystems Engineering, 37*, 1749–1758.

Gutierrez, M. C., Garcia-Carvajal, Z. Y., Hortiguela, M. J., Yuste, L., Rojo, F., Ferrer, M. L., & del Monte, F. (2007). Biocompatible MWCNT scaffolds for immobilization and proliferation of E. coli. *Journal of Materials Chemistry, 17*, 2992–2995.

Han, T. H., Khan, M. M., Kalathil, S., Lee, J., & Cho, M. H. (2013). Simultaneous enhancement of methylene blue degradation and power generation in a microbial fuel cell by gold nanoparticles. *Industrial & Engineering Chemistry Research, 52*, 8174–8181.

Han, T. H., Cho, M. H., & Lee, J. (2014). Indole oxidation enhances electricity production in an *E. coli*-catalyzed microbial fuel cell. *Biotechnology and Bioprocess Engineering, 19*, 126–131.

Han, T. H., Sawant, S. Y., Hwang, S.-J., & Cho, M. H. (2016). Three-dimensional, highly porous N-doped carbon foam as microorganism propitious, efficient anode for high performance microbial fuel cell. *RSC Advances, 6*, 25799–25807.

Haslett, N. D., Rawson, F. J., Barriëre, F., Kunze, G., Pasco, N., Gooneratne, R., & Baronian, K. H. R. (2011). Characterisation of yeast microbial fuel cell with the yeast Arxula adeninivorans as the biocatalyst. *Biosensors and Bioelectronics, 26*, 3742–3747.

Hays, S., Zhang, F., & Logan, B. E. (2011). Performance of two different types of anodes in membrane electrode assembly microbial fuel cells for power generation from domestic wastewater. *Journal of Power Sources, 196*, 8293–8300.

Hou, J., Liu, Z., Yang, S., & Zhou, Y. (2014). Three-dimensional macroporous anodes based on stainless steel fiber felt for high-performance microbial fuel cells. *Journal of Power Sources, 258*, 204–209.

Hu, H., Zhao, Z., Wan, W., Gogotsi, Y., & Qiu, J. (2013). Ultralight and highly compressible graphene aerogels. *Advanced Materials, 25*, 2219–2223.

Huang, Y.-X., Liu, X.-W., Xie, J.-F., Sheng, G.-P., Wang, G.-Y., Zhang, Y.-Y., Xu, A.-W., & Yu, H.-Q. (2011). Graphene oxide nanoribbons greatly enhance extracellular electron transfer in bio-electrochemical systems. *Chemical Communications, 47*, 5795–5797.

Jiang, D., Curtis, M., Troop, E., Scheible, K., McGrath, J., Hu, B., Suib, S., Raymond, D., & Li, B. (2011). A pilot-scale study on utilizing multi-anode/cathode microbial fuel cells (MAC MFCs) to enhance the power production in wastewater treatment. *International Journal of Hydrogen Energy, 36*, 876–884.

Kalathil, S., Lee, J., & Cho, M. H. (2011). Granular activated carbon based microbial fuel cell for simultaneous decolorization of real dye wastewater and electricity generation. *New Biotechnology, 29*, 32–37.

Kalathil, S., Lee, J., & Cho, M. H. (2012). Efficient decolorization of real dye wastewater and bioelectricity generation using a novel single chamber biocathode-microbial fuel cell. *Bioresource Technology, 119*, 22–27.

Kalathil, S., Nguyen, V. H., Shim, J.-J., Khan, M. M., Lee, J., & Cho, M. H. (2013). Enhanced Performance of a microbial fuel cell using CNT/MnO_2 nanocomposite as a bioanode material. *Journal of Nanoscience and Nanotechnology, 13*, 7712–7716.

Karthikeyan, R., Krishnaraj, N., Selvam, A., Wong, J. W.-C., Lee, P. K. H., Leung, M. K. H., & Berchmans, S. (2016). Effect of composites based nickel foam anode in microbial fuel cell using *Acetobacter aceti* and *Gluconobacter roseus* as a biocatalysts. *Bioresource Technology, 217*, 113–120.

Katuri, K., Ferrer, M. L., Gutierrez, M. C., Jimenez, R., del Monte, F., & Leech, D. (2011). Three-dimensional microchanelled electrodes in flow-through configuration for bioanode formation and current generation. *Energy & Environmental Science, 4*, 4201–4210.

Kumar, G. G., Sarathi, V. G. S., & Nahm, K. S. (2013). Recent advances and challenges in the anode architecture and their modifications for the applications of microbial fuel cells. *Biosensors and Bioelectronics, 43*, 461–475.

Liang, P., Wang, H., Xia, X., Huang, X., Mo, Y., Cao, X., & Fan, M. (2011). Carbon nanotube powders as electrode modifier to enhance the activity of anodic biofilm in microbial fuel cells. *Biosensors and Bioelectronics, 26*, 3000–3004.

Liu, H., Ramnarayanan, R., & Logan, B. E. (2004). Production of electricity during wastewater treatment using a single chamber microbial fuel cell. *Environmental Science & Technology, 38*, 2281–2285.

Liu, H., Cheng, S., & Logan, B. E. (2005). Power generation in fed-batch microbial fuel cells as a function of ionic strength, temperature, and reactor configuration. *Environmental Science & Technology, 39*, 5488–5493.

Liu, J., Liu, J., He, W., Qu, Y., Ren, N., & Feng, Y. (2014). Enhanced electricity generation for microbial fuel cell by using electrochemical oxidation to modify carbon cloth anode. *Journal of Power Sources, 265*, 391–396.

Logan, B., Cheng, S., Watson, V., & Estadt, G. (2007). Graphite fiber brush anodes for increased power production in air-cathode microbial fuel cells. *Environmental Science & Technology, 41*, 3341–3346.

Logan, B. E. (2008). *Microbial fuel cells* (p. 53). Hoboken: Wiley.

Lowy, D. A., Tender, L. M., Zeikus, J. G., Park, D. H., & Lovley, D. R. (2006). Harvesting energy from the marine sediment–water interface. II: Kinetic activity of anode materials. *Biosensors and Bioelectronics, 21*, 2058–2063.

Mapelli, C., Mapelli, V., Olsson, L., Mombelli, D., Gruttadauria, A., & Barella, S. (2013). Viability study of the use of cast iron open cell foam as microbial fuel cell electrodes. *Advanced Engineering Materials, 15*, 112–117.

Michaelidou, U., ter Heijne, A., Euverink, G. J. W., Hamelers, H. V. M., Stams, A. J. M., & Geelhoed, J. S. (2011). Microbial communities and electrochemical performance of titanium-based anodic electrodes in a microbial fuel cell. *Applied and Environmental Microbiology, 77*, 1069–1075.

Mink, J. E., & Hussain, M. M. (2013). Sustainable design of high-performance microsized microbial fuel cell with carbon nanotube anode and air cathode. *ACS Nano, 7*, 6921–6927.

Nardecchia, S., Carriazo, D., Ferrer, M. L., Gutierrez, M. C., & del Monte, F. (2013). Three dimensional macroporous architectures and aerogels built of carbon nanotubes and/or graphene: Synthesis and applications. *Chemical Society Reviews, 42*, 794–830.

Pham, T. H., Aelterman, P., & Verstraete, W. (2009). Bioanode performance in bioelectrochemical systems: Recent improvements and prospects. *Trends in Biotechnology, 27*, 168–178.

Pisciotta, J. M., Zaybak, Z., Call, D. F., Nam, J.-Y., & Logan, B. E. (2012). Enrichment of microbial electrolysis cell biocathodes from sediment microbial fuel cell bioanodes. *Applied and Environmental Microbiology, 78*, 5212–5219.

Qiao, Y., Li, C. M., Bao, S.-J., & Bao, Q.-L. (2007). Carbon nanotube/polyaniline composite as anode material for microbial fuel cells. *Journal of Power Sources, 170*, 79–84.

Qiao, Y., Wu, X.-S., Ma, C.-X., He, H., & Li, C. M. (2014). A hierarchical porous graphene/nickel anode that simultaneously boosts the bio- and electro-catalysis for high-performance microbial fuel cells. *RSC Advances, 4*, 21788–21793.

Rabaey, K., Lissens, G., Siciliano, S. D., & Verstraete, W. (2003). A microbial fuel cell capable of converting glucose to electricity at high rate and efficiency. *Biotechnology Letters, 25*, 1531–1535.

Reguera, G., Nevin, K. P., Nicoll, J. S., Covalla, S. F., Woodard, T. L., & Lovley, D. R. (2006). Biofilm and nanowire production leads to increased current in Geobacter sulfurreducens fuel cells. *Applied and Environmental Microbiology, 72*, 7345–7348.

Rhoads, A., Beyenal, H., & Lewandowski, Z. (2005). Microbial fuel cell using anaerobic respiration as an anodic reaction and biomineralized manganese as a cathodic reactant. *Environmental Science & Technology, 39*, 4666–4671.

Richter, H., McCarthy, K., Nevin, K. P., Johnson, J. P., Rotello, V. M., & Lovley, D. R. (2008). Electricity generation by geobacter sulfurreducens attached to gold electrodes. *Langmuir, 24*, 4376–4379.

Roh, S.-H., & Woo, H.-G. (2015). Carbon nanotube composite electrode coated with polypyrrole for microbial fuel cell application. *Journal of Nanoscience and Nanotechnology, 15*, 484–487.

Rozendal, R. A., Hamelers, H. V. M., Rabaey, K., Keller, J., & Buisman, C. J. N. (2008). Towards practical implementation of bioelectrochemical wastewater treatment. *Trends in Biotechnology, 26*, 450–459.

Scott, K., Rimbu, G. A., Katuri, K. P., Prasad, K. K., & Head, I. M. (2007). Application of modified carbon anodes in microbial fuel cells. *Process Safety and Environmental Protection, 85*, 481–488.

Sell, D., Krämer, P., & Kreysa, G. (1989). Use of an oxygen gas diffusion cathode and a three-dimensional packed bed anode in a bioelectrochemical fuel cell. *Applied Microbiology and Biotechnology, 31*, 211–213.

Sun, J.-J., Zhao, H.-Z., Yang, Q.-Z., Song, J., & Xue, A. (2010). A novel layer-by-layer self-assembled carbon nanotube-based anode: Preparation, characterization, and application in microbial fuel cell. *Electrochimica Acta, 55*, 3041–3047.

Tang, X., Guo, K., Li, H., Du, Z., & Tian, J. (2011). Electrochemical treatment of graphite to enhance electron transfer from bacteria to electrodes. *Bioresource Technology, 102*, 3558–3560.

ter Heijne, A., Hamelers, H. V. M., Saakes, M., & Buisman, C. J. N. (2008). Performance of non-porous graphite and titanium-based anodes in microbial fuel cells. *Electrochimica Acta, 53*, 5697–5703.

Wang, H., Côté, R., Faubert, G., Guay, D., & Dodelet, J. P. (1999). Effect of the pre-treatment of carbon black supports on the activity of fe-based electrocatalysts for the reduction of oxygen. *The Journal of Physical Chemistry B, 103*, 2042–2049.

Wang, X., Feng, Y. J., & Lee, H. (2008). Electricity production from beer brewery wastewater using single chamber microbial fuel cell. *Water Science and Technology, 57*, 1117–1121.

Wang, X., Cheng, S., Feng, Y., Merrill, M. D., Saito, T., & Logan, B. E. (2009). Use of carbon mesh anodes and the effect of different pretreatment methods on power production in microbial fuel cells. *Environmental Science & Technology, 43*, 6870–6874.

Wang, H., Wang, G., Ling, Y., Qian, F., Song, Y., Lu, X., Chen, S., Tong, Y., & Li, Y. (2013a). High power density microbial fuel cell with flexible 3D graphene-nickel foam as anode. *Nanoscale, 5*, 10283–10290.

Wang, Y., Zhao, C.-E., Sun, D., Zhang, J.-R., & Zhu, J.-J. (2013b). A graphene/poly (3,4-ethylenedioxythiophene) hybrid as an anode for high-performance microbial fuel cells. *ChemPlusChem, 78*, 823–829.

Wei, J., Liang, P., & Huang, X. (2011). Recent progress in electrodes for microbial fuel cells. *Bioresource Technology, 102*, 9335–9344.

Wen, Q., Wu, Y., Zhao, L.-X., Sun, Q., & Kong, F.-Y. (2010). Electricity generation and brewery wastewater treatment from sequential anode-cathode microbial fuel cell. *Journal of Zhejiang University SCIENCE B, 11*, 87–93.

Xie, X., Hu, L., Pasta, M., Wells, G. F., Kong, D., Criddle, C. S., & Cui, Y. (2011). Three-dimensional carbon nanotube−textile anode for high-performance microbial fuel cells. *Nano Letters, 11*, 291–296.

Xie, X., Ye, M., Hu, L., Liu, N., McDonough, J. R., Chen, W., Alshareef, H. N., Criddle, C. S., & Cui, Y. (2012a). Carbon nanotube-coated macroporous sponge for microbial fuel cell electrodes. *Energy & Environmental Science, 5*, 5265–5270.

Xie, X., Yu, G., Liu, N., Bao, Z., Criddle, C. S., & Cui, Y. (2012b). Graphene-sponges as high-performance low-cost anodes for microbial fuel cells. *Energy & Environmental Science, 5*, 6862–6866.

Xie, X., Criddle, C., & Cui, Y. (2015). Design and fabrication of bioelectrodes for microbial bioelectrochemical systems. *Energy & Environmental Science, 8*, 3418–3441.

You, S., Zhao, Q., Zhang, J., Jiang, J., Wan, C., Du, M., & Zhao, S. (2007). A graphite-granule membrane-less tubular air-cathode microbial fuel cell for power generation under continuously operational conditions. *Journal of Power Sources, 173*, 172–177.

Yuan, Y., & Kim, S.-H. (2008). Improved performance of a microbial fuel cell with polypyrrole/carbon black composite coated carbon paper anodes. *Bulletin of the Korean Chemical Society, 29*, 1344–1348.

Yuan, Y., Zhou, S., Liu, Y., & Tang, J. (2013). Nanostructured macroporous bioanode based on polyaniline-modified natural *Loofah* sponge for high-performance microbial fuel cells. *Environmental Science & Technology, 47*, 14525–14532.

Zhang, F., Saito, T., Cheng, S., Hickner, M. A., & Logan, B. E. (2010). Microbial fuel cell cathodes with poly(dimethylsiloxane) diffusion layers constructed around stainless steel mesh current collectors. *Environmental Science & Technology, 44*, 1490–1495.

Zhao, F., Rahunen, N., Varcoe, J. R., Chandra, A., Avignone-Rossa, C., Thumser, A. E., & Slade, R. C. T. (2008). Activated carbon cloth as anode for sulfate removal in a microbial fuel cell. *Environmental Science & Technology, 42*, 4971–4976.

Zhao, Y., Watanabe, K., Nakamura, R., Mori, S., Liu, H., Ishii, K., & Hashimoto, K. (2010). Three-dimensional conductive nanowire networks for maximizing anode performance in microbial fuel cells. *Chemistry - A European Journal, 16*, 4982–4985.

Zhao, C., Wang, Y., Shi, F., Zhang, J., & Zhu, J.-J. (2013). High biocurrent generation in *Shewanella*-inoculated microbial fuel cells using ionic liquid functionalized graphene nanosheets as an anode. *Chemical Communications, 49*, 6668–6670.

Zhou, M., Chi, M., Luo, J., He, H., & Jin, T. (2011). An overview of electrode materials in microbial fuel cells. *Journal of Power Sources, 196*, 4427–4435.

Zhou, M., Chi, M., Wang, H., & Jin, T. (2012). Anode modification by electrochemical oxidation: A new practical method to improve the performance of microbial fuel cells. *Biochemical Engineering Journal, 60*, 151–155.

Zhu, C., Han, T. Y.-J., Duoss, E. B., Golobic, A. M., Kuntz, J. D., Spadaccini, C. M., & Worsley, M. A. (2015). Highly compressible 3D periodic graphene aerogel microlattices. *Nature Communications, 6*, 6962.

Chapter 7
Performances of Separator and Membraneless Microbial Fuel Cell

Ruchira Rudra, Vikash Kumar, Arpita Nandy, and Patit Paban Kundu

7.1 Introduction

Microbial fuel cell (MFC) is a typical bio-electrochemical system (BES) that produces green electricity from organic wastes by mimicking bacterial interactions under anoxic conditions. This bacteria catalysed system cannot only provide a renewable alternative clean source of energy without the need for fossil fuels, but also become as a remote power source in the field of domestic wastewater treatment, breweries, hydrogen production, remote sensing, desalination plants, pollution remediation and so on (Pant et al. 2010). Though MFC technology belongs to its infancy stage in our modern day, but it has several advantages over other conventional systems. Firstly, it can eliminate pollution caused by burning fossil fuels in a mild reaction condition. Maintenance is quite simple and economically viable. It can also be operated without any noise and has no memory effect while getting refuelled. It can provide high quality DC power output and the power densities are very higher than expected. To set up the basis of an MFC, the bacteria are allowed to grow on the electrode (placed in anode), commonly known as biofilm in order to bio-catalyse the reaction. The anode is separated from the cathode compartment through the ion exchange membrane, salt bridge, sediment, etc. Another electrode without bio-film is placed in the cathode compartment. These

R. Rudra • V. Kumar • A. Nandy
Advanced Polymer Laboratory, Department of Polymer Science & Technology, University of Calcutta, 92, A.P.C. Road, Kolkata 700 009, India

P.P. Kundu (✉)
Advanced Polymer Laboratory, Department of Polymer Science & Technology, University of Calcutta, 92, A.P.C. Road, Kolkata 700 009, India

Department of Chemical Engineering, Indian Institute of Technology Roorkee, Roorkee 247667, India
e-mail: ppk923@yahoo.com

D. Das (ed.), *Microbial Fuel Cell*,
https://doi.org/10.1007/978-3-319-66793-5_7

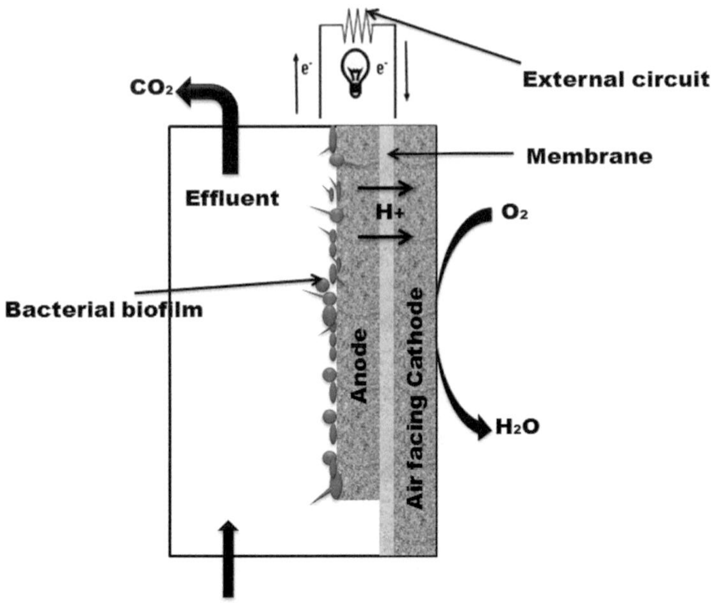

Fig. 7.1 Simplified diagram of a single chambered microbial fuel cell (Pant et al. 2010)

two electrodes are connected by an external conductive wire. During the conversion of organic materials by the metabolic activity of bacteria, the generated electrons are transferred from anode to cathode compartment through the wire. The protons produced in this process also flow from anode to cathode through separator in order to maintain the charge difference. At the cathode side an oxidant (generally oxygen) is reduced. Thus, electrons and protons are consumed with oxygen to form water on the cathode side (Perez et al. 2012). This has been pictorially represented in Fig. 7.1.

The biofilm growth on the electrode and the electrons produced by the exoelectrogens from the organic wastes are governed by the mediators or electron shuttlers that are externally added to the system (e.g. thionene, methyl viologen, methylene blue, humic acid, neutral red etc.). These redox metal complexes are quite expensive as well as toxic in nature which get degraded with the progression of time (Rabaey et al. 2005; Logan 2007). Pure culture as well as mixed cultures are used in mediatorless MFC in which electrochemically active bacteria (EAB) is used to transfer electrons to the electrode. They can directly transfer the electrons to the electrode through an array of enzymes in the bacterial respiratory enzymes, like *Shewanella putrefaciens*, *Aeromonas hydrophila* etc. The higher power output is achieved by using mixed bacterial cultures over pure cultures. It was found that a group of EAB cultures comprising *Geobacteraceae*, *Desulfuromonas*, *Alcaligenesfaecalis*, *Enterococcus faecium*, *Pseudomonas aeruginosa*, *Proteobacteria*, *Clostridia*, *Bacteroides* and *Aeromonas* are used in

Table 7.1 Performances of mediatorless MFC

Micro-organism	Substrate	Anode	Current density (mA)	Power density (mW m^{-2})	References
Shewanellaputrefaciens	Lactate	Woven graphite	0.031	0.19	Kim et al. (2002)
Geobacter sulfurreducens	Acetate	Graphite	0.40	13	Bond and Lovley (2003)
Rhodoferax ferrireducens	Glucose	Graphite	0.2	8	Chaudhuri and Lovley (2003)
	Glucose	Woven graphite	0.57	17.4	
	Glucose	Porous graphite	74	33	
Mixed seawater culture	Acetate	Graphite	0.23	10	Bond et al. (2002)
	Sulphide/ acetate		60	32	Tender et al. (2002)
Mixed active sludge culture	Acetate	Graphite	5	–	Lee et al. (2003)
	Glucose	Graphite	30	3600	Rabaey et al. (2003)
	Sewage	Woven graphite	0.2	8	Kim et al. (2004)
Mixed culture containing *P. aeruginosa*	Glucose	Graphite	30.9	4310	Rabaey et al. (2004)

Rabaey et al. (2005)

separator-less MFC and wastewater treatment plants having the advantages of higher resistances against the process disturbances, higher substrate consumption rate, smaller substrate specificity, and overall higher power generation. Table 7.1 summarizes the maximum power output delivered by the mediatorless MFC.

$$\text{Anode}: C_6H_{12}O_6 + 6H_2O \rightarrow 6CO_2 + 24H^+ + 24e^- \tag{7.1}$$

$$\text{Cathode}: 6O_2 + 24H^+ + 24e^- \rightarrow 12H_2O \tag{7.2}$$

7.2 Parameters Used in MFC Performance

MFCs are utilized for current generation through the biodegradation of organic matter. Now there are some parameters which can determine the performances of MFC. These are summerized in Table 7.2.

Table 7.2 Parameters used in determining MFC performance

Parameters	Unit	Description
Loading rate	kg $m^{-3}d^{-1}$	An amount of organic wastes loaded to MFC normalized with respect to net anode volume and time.
Effluent quality	kg m^{-3}	Concentration of organic waste in an effluent discharged from the anode compartment.
Chemical oxygen demand (COD)	mg L^{-1}	Final concentration of organic compound
Treatment efficiency	%	COD removal efficiency obtained by dividing the COD concentration in the effluent with respect to that of effluent.
Power density (per volume or per area)	W m^{-3} or W m^{-2}	A power output is normalized to an anode volume (or area) or a sum of anode and cathode volume (area) as obtained from the maximum peak of a polarization curve.
Current density	A m^{-2}	A current generated is normalized to an anode area. This may be an index related to the total metabolic activity of bacteria in the anode chamber.
Open circuit voltage (OCV)	V	A voltage between the anode and cathode measured in the absence of current. A difference between the total electromotive force (emf; the potential difference between the cathode and anode) and OCV is regarded as the total potential loss.
Internal resistance (R_{int})	Ω	This is obtained from the slope of the polarization curve and is helpful to evaluate the total internal loss in an MFC process.
Coulombic efficiency (CE)	%	This is defined as the ratio of Coulombs measured as the current to the total Coulombs contained in substrates (estimated from the total COD value). If there are alternative electron acceptors present in an anode chamber, this value diminishes.
Energy efficiency (EE)	%	This is calculated as the ratio of power produced by MFC to the heat energy obtained by combustion of substrates added, and is considered to be the most important to evaluate an MFC process as an energy-recovery process.
Voltammetry studies (e.g. cyclic voltammetry, linear sweep voltammetry)	mV S^{-1}	This is useful in calculating the standard electrode potential, electrochemical behaviour of microbial strains, extent of mediator production during biofilm growth, and test the performances of novel cathode material.
Oxygen diffusivity or oxygen crossover or oxygen permeability	cm S^{-1}	This is a mass transfer coefficient as characterized by the oxygen permeability from cathode to anode chamber with respect to time.

Watanabe (2008)

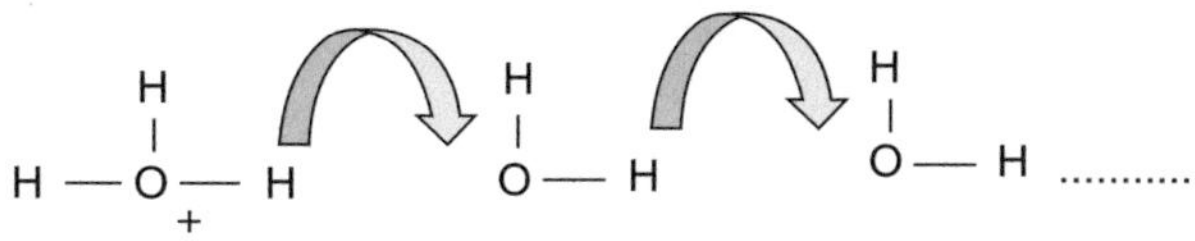

Fig. 7.2 The Grotthuss mechanism with the consecutive proton transfer indicated by the arrows

7.2.1 Proton Transport Mechanism in a PEM

Protons (H^+) are transported via the Grotthuss mechanism through the ion exchange membrane during fully hydrated condition. In this mechanism proton can hop from one water molecule to another, forcing the excess hydrogen atom at the next water molecule to hop to another molecule thereby restarting the chain of events (Fig. 7.2).

7.3 Advantages and Disadvantages of Separator and Separatorless MFC

The overall features of MFC depend not only on the microbial activity, but also the hardware and configuration of the fuel cell. Thus the electricity generation is guided by: (a) catalytic activity of microbes; (b) anode performance developed with the electron transfer from microbes to the electrodes; (c) cathode performance that is a reduction of electron acceptors at the cathode; and (d) proton transfer from anode to cathode (Li et al. 2011). In separatorless MFC, higher proton transfer rate as well as higher power density (264–494 W m^{-2}) was reported earlier (Liu and Logan 2004; Yang et al. 2009). Moreover, a simpler configuration was achieved in a cost effective way with the drawbacks of increased oxygen and substrate permeability which would be rather responsible for deteriorated MFC performances caused by biofouling and, therefore, decreasing the columbic efficiency (CE) as well as decreasing bioelectrocatalytic activity of microbes (Liu et al. 2005). To minimize the difficulties of the separatorless MFC, certain electrode spacing would be recommended, which will result in higher internal resistance. Thus, to optimize the critical situation, a separator is required to make sustainable power generation. Separator plays a very important role in the MFC design as it greatly enhances the power production and CE. An ideal separator material should have high proton transfer coefficient to ensure that the substrate does not hinder proton transfer from anode to cathode and lower oxygen diffusivity improves CE and it must be non-biodegradable.

7.4 Type of Separators and Their Performance in MFC

7.4.1 Ion-Exchange Membranes

Cation exchange membranes (CEM) such as Nafion, Ultrex CMI 7000 (Membranes Inc., USA), Hyflon (Solvay-Solexis, Italy), Zirfon etc. are used to facilitate the proton conduction. Nafion has good proton conductivity due to the presence of negatively charged hydrophilic sulphonate group in the hydrophobic polymeric chain. It was reported that the MFC using thinner Nafion 112 membrane has a higher power density of 31.32 mW m^{-2} (maximum voltage 670 mV and current density 150.6 mA m^{-2}) than thicker Nafion 117 membrane which would produce 9.95 mW m^{-2} (maximum voltage 668 mV and current density 60.28 mA m^{-2}) due to the lower ohmic resistance of the former (Leong et al. 2013). But they can transport other cations like Na^+, K^+, Ca^{2+}, Mg^{2+} and NH_4^+ at the concentration rate of 10^5 higher than protons leading to a severe problem called pH splitting, higher ohmic resistances etc. This could hamper the biocatalytic activity of microorganism as well as biofouling. Greater oxygen permeability limits the practical application of the high cost CEMs in MFCs. Zirfon, an ultrafiltration composite membrane where an inorganic filler ZrO_2 is incorporated into the asymmetric polysulphone structure, have lower ionic resistance (2727 Ω cm) than Nafion 117 (17,000 Ω cm) but it has higher oxygen diffusivity (1.9×10^{-3} cm s^{-1}) than the later (2.8×10^{-4} cm s^{-1}).

The deficiency of CEM can be overcome to some extent by using anion exchange membrane (AEM). AEM can transfer protons by the carriers like phosphate, carbonate and pH buffer. MFC with an AEM, AMI-7001 have been reported to generate 25% higher power density (610 mW m^{-2}) than that of a CEM (480 mW m^{-2}). Such improvement was also found in case of membrane electrode assembly (MEA) based on AEM (13.1 W m^{-3}) than that of CEM based MEA (8.3 W m^{-3}). Another type of AEM is AFN, which shows lower resistance and higher current density (0.38 mA cm^{-2}) followed by AM-1 (0.28 mA cm^{-2}) and ACS (0.21 mA cm^{-2}) membrane (Ji et al. 2011). However, AEM has higher substrate diffusivity and susceptible to deform readily than that of CEM and thus leading to higher internal resistance when used in a single chambered MFC. All of the disputes of both CEM and AEM can be rectified when both of the two ion exchange membranes are mounted together, called a bipolar membrane (Harnisch and Schroder 2009). The ion selective nature of the membrane efficiently migrates and transports protons, generated by water splitting at the interface between CEM and AEM (Fig. 7.3). However, this type of membrane was unable to solve the problem of pH splitting and the ions would deposit at the interface of the membranes generating an ion flux for charge balance. Such type of membrane can be applicable in microbial desalination cell and high salt containing water treatment.

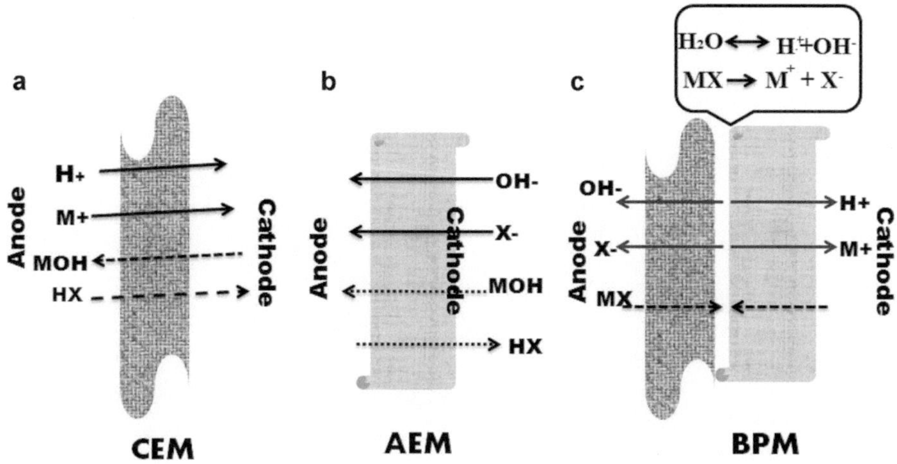

Fig. 7.3 Schematic ions and mass transport across: (**a**) cation exchange membrane (CEM); (**b**) anion exchange membrane (AEM); and (**c**) Bipolar membranes

Drawbacks of ion exchange membranes:

- Resistivity of membranes lowers the power density.
- Leakage of other cations lowers the energy efficiency.
- Membranes need to be cleaned and replaced due to biofouling.

7.4.2 *Salt Bridge*

A salt bridge is basically made of glass tubes filled with electrolytes such as saturated KCl solution and phosphate buffer solution for ion conduction (Fig. 7.4). Agar is added in order to separate the two liquids. Thus the oxygen diffusion is almost unpredictable in an agar salt bridge while used in MFC (Min et al. 2005). But it can develop lower power density and higher internal resistance (19,920 Ω) at a much lower peak voltage (0.1 V) as compared with an ion exchange membrane (0.2 V) (Liu and Li 2007). It was also reported that the salt bridge using MFC can develop a maximum power density of 1.5 Wm^{-2} as compared to a CEM-MFC (0.94 Wm^{-2}) in a chocolate industry wastewater. The internal resistance can be lowered by decreasing the electrode spacing between the electrodes and the power density can be improved by increasing the surface area of the salt bridge in contact with the cathode and anode chamber.

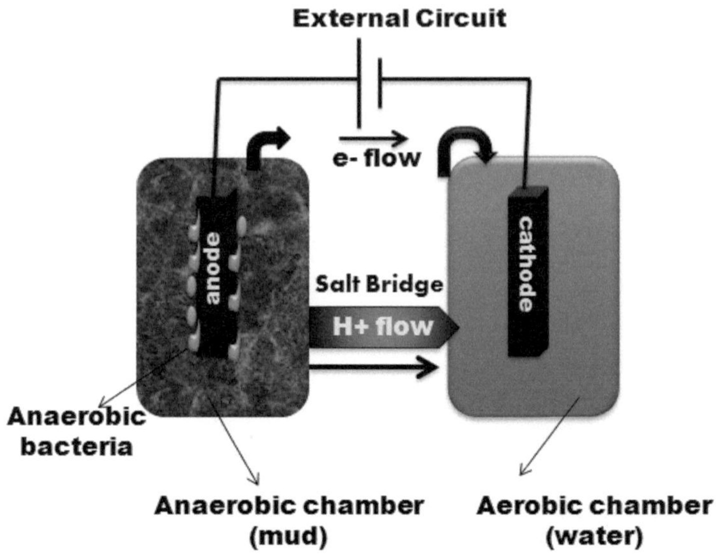

Fig. 7.4 A dual chambered MFC with a salt bridge

7.4.3 Porous Membrane

The concept of porous membranes or size selective separators was introduced in order to minimize the pH splitting and the ion selectivity nature for transferring protons. Based on their filtration performance, the porous membranes can be divided into two categories, namely microporous membrane such as microfiltration membrane (MFM) and ultra-filtration membrane (UFM) and coarse pore filters. In the first type of membrane, proton can be transferred very efficiently by passing through the ions into the pores of the membranes. It was reported that the conducting and catalytic layer coated UFM can generate 17.7 Wm^{-3} of power density than that of a CEM (6.6 Wm^{-3}) (Zuo et al. 2007). They are much more durable and cost effective than any other membrane and thus can be used as sludge separators for wastewater treatment. However, their oxygen and substrate diffusivity is very high. Moreover, increasing mass transport, ohmic resistance and internal resistance can pose a bigger issue while using in MFC.

Coarse pore filters such as fabrics, glass fibre, non-woven cloths, nylon mesh, and cellulose fibres, etc. are reported as separators in an MFC application. Glass fibre mats, J-cloth produce a current density of 790 mW m^{-2} as compared with a membrane of an identical thickness of CEM (Zhang et al. 2009). The columbic efficiencies of non-biodegradable glass fibre mats is 81% higher than that of J-cloths. It was also reported that the increasing mesh size of nylon fibres increases the power density significantly, but similar phenomenon is not found in case of glass fibres due to the narrower pore size of glass fibres (Zhang et al. 2010). Porous ceramic and biodegradable bag (biobag), and biodegradable natural rubber were also used as porous separator and comparable with CEM (Winfield et al. 2013b). Virgin natural rubber is impermeable to proton transfer, but it can develop voltage during degradation with respect to time. Though its MFC performance is generally lower than any CEM material, it can be applied to improve the circuit potential when both CEM and AEM performance deteriorate after 11 months and aid to generate steady power with time. Another cheap and porous separator is a non-woven fabric filter (NWF) which can produce 97 mW m^{-2} power density at lower internal resistance, higher voltage operation. Thus NWF can give a stable MFC performance over 300 days operation, but the water permeability of this material is quite high, whereas its oxygen crossover is almost comparable to Nafion117 or any other PEM material. The higher proton conductivity and lower ohmic losses of NWF makes it as a suitable candidate in MFC performance. It is reported that by applying a proton transfer conductor such as poly [2,5-benzimidazole] (ABPBI) on the NWF can effectively enhance its ion conductivity and power density (twice than that of PEM) during MFC application (Choi et al. 2013).

Another kind of cheapest separator is earthen pot. Terracotta separator can be used extensively as porous separator. An MFC utilizing this along with a stainless steel mesh cathode and $KMnO_4$ as the oxidant generates fourfold higher maximum power of 70.48 W m^{-3} at twofolds higher CE of 64.5% than another MFC having an earthen pot separator but with a graphite plate cathode (Behera et al. 2010). An MFC using an earthen pot separator treating rice mill wastewater generates fourfold higher power density of 2.3 W m^{-3} at 4–6% higher removal efficiencies of 96.5% for chemical oxygen demand (COD) as compared to an MFC using a Nafion117 separator (Behera et al. 2010). Tubular-type up-flow MFC using earthen pot separator generates 46% higher power density of 14.59 W m^{-3} than that produced in an identical tubular MFC with a Nafion117 membrane (Jana et al. 2010). Similarly, MFCs using cylindrical earthen pot produce 75% higher current and 33% more power than MFCs with terracotta separator (Winfield et al. 2013a). The higher power density is probably due to the lower internal resistance of the MFC with cylindrical earthen pot separator compared with the MFC having Nafion and terracotta separators (Jana et al. 2010; Winfield et al. 2013a). The thickness of the earthen separator also affects MFC performance (Behera and Ghangrekar 2011). The reactor with the thinnest separator performed excellent to obtain 60% higher power density of 24.32 mW m^{-2}, volumetric power of 1.04 W m^{-3} and CE of 7.7%. Although the volumetric power density is greater, but the CE is much lower due to substrate consumption by anaerobic bacteria such as methanogenic microorganisms or sulphate-reducing bacteria as reported earlier (He et al. 2005).

7.4.4 *Polymer Electrolyte Membrane and Composite Membranes*

In recent years MFC separators based on polymeric materials have been studied and utilized as novel proton exchange membrane in MFC performances. The major disputes of Nafion117 like high cost, oxygen diffusivity, difficulty in processing and bio-fouling can limit its MFC application. To replace it, sulphonated polymers like poly ether ether ketone (SPEEK) and polystyrene-ethylene-butylene-polystyrene (SPSEBS) membranes were developed as cost effective material (Lim et al. 2012; Ayyaru et al. 2012). Moreover, the presence of aromatic backbone in the macromolecules makes it durable for longer period of time and the presence of sulphonating group can ease proton conduction. Sometimes addition of some inorganic nano-silica into the polymer can effectively enhance the power density. Ayyaru and Dharmalingam (2015a) reported that with the addition of sulphonated nano silica into SPEEK, the maximum power density was found to be 1008 mW m^{-2} as compared to the SiO_2-SPEEK (802 mW m^{-2}) and SPEEK (680 mW m^{-2}). MFC using SPSEBS membrane also generates 600 mW m^{-2} which is much higher than Nafion membrane. But with the addition of 7.5% sulphonated silica, sulphonated TiO_2 into SPSEBS produces a maximum of 1209 $\pm$ 17 mW m^{-2}, 1345 $\pm$ 17 mW m^{-2} power density having a very much low oxygen diffusivity (Ayyaru and Dharmalingam 2015b; Ayyaru et al. 2016). Organic conductive nano-filler like graphene oxide shows a little bit of reduced power density (902 mW m^{-2}) when intercalated with SPEEK membrane (Leong et al. 2015). Low conducting polymeric membrane such as PES when blended with high conducting 5 wt% SPEEK, 170 mW m^{-2} power density was obtained.

On incorporation of iron oxide nano-particle into PES polymer, a maximum power density of 20 mW m^{-2} was achieved. This value is much lower than the above studies (Rahimnejad et al. 2012). BPSH membranes having higher degree of sulphonation become susceptible to proton transfer due to its hydrophilic nature and can reduce biofouling problem (power density 126 mWm^{-2}). It is also reported that utilizing conducting polymer such as polyaniline, polypyrrole, or polythiophene can drastically enhance the performance of MFC (Dutta and Kundu 2014). It has been observed that the Nafion 112/polyaniline composite membrane can drastically enhance power density (124.03 mW m^{-2}) than pristine Nafion112 (13.98 mW m^{-2}). In addition, activated carbon nano fibre (CNF)/Nafion composite membrane was also reviewed, showing a maximum of 57.64 mW m^{-2} power density due to higher porosity of the membrane. Chitosan-carbon nano tube (CNT) membranes were also used as a polymer electrolyte separator in MFC (Venkatesan and Dhamalingam 2013). The oxygen crossover was reduced and the mechanical as well as chemical stability were improved by using CNT. Recent advances on the various polymeric separators based single chambered MFC are discussed in terms of power generation in Table 7.3.

Nowadays, a new and novel approach of ionic liquid membrane supported PEM is utilized in the field of wastewater treatment (Hernández-Fernández et al. 2015, 2016). Ionic liquid primarily is composed of an organic salt like 1-octyl-3-methylimidazolium tetrafluoroborate, $[omim^+][BF_4^-]$, 1-butyl-3-methylimidazolium hexafluorophosphate, $[bmim^+]$ $[PF_6^-]$, 1-octyl-3-methylimidazolium hexafluorophosphate, $[omim^+][PF_6^-]$, 1-butyl-3-methylimidazolium bis{(trifluoromethyl)sulfonyl}imide, $[bmim^+][NTf_2^-]$, 1-octyl-3-methylimidazolium bis-{(trifluoromethyl)

Table 7.3 A comparative study of MFCs in terms of power generation using different polymer electrolyte membrane

Separator used	Maximum power density ($mW\ m^{-2}$)	Maximum current density ($mA\ m^{-2}$)	Open circuit potential (mV)	References
Partially sulphonated PVDF-co-HFP	290.176 ± 15	1390.866 ± 70	621 ± 30	Kumar et al. (2015a)
Crosslinked IPN of sulphonated styrene and sulphonated PVDF-co-HFP	447.42 ± 22	1729.63 ± 87	722 ± 20	Kumar et al. (2015b)
Nano-alumina filled sulphonated PVDF-co-HFP/Nafion blend	541.52 ± 27	1900 ± 95	741 ± 20	Kumar et al. (2016a)
PBI/PVP blend membrane	231.38 ± 12	1242 ± 62	607 ± 30	Kumar et al. (2016b)
PVA separator crosslinked by acid catalysed gluteraldehyde	119.13 ± 6	447.81 ± 18	501 ± 18	Rudra et al. (2015)
Graphene oxide impregnated PVA-STA composite membrane	86.7	425	773	Khilari et al. (2013)
Sulphonated polybenzimidazole	86.8	318	460	Singha et al. (2016)

sulfonyl}imide, $[omim^+][NTf_2^-]$, in which one part is organic cation and another is polyatomic inorganic anion.

7.5 Separatorless MFC

The major drawbacks associated with membranes are bio-fouling, lower proton transfer rate, higher internal resistance, membrane and MFC operational cost etc. In case of separatorless MFC, the electrode spacing is minimized, thus internal resistance is lowered. But, the high proton transfer rate combines with higher oxygen diffusivity towards the anode and also associated with the columbic efficiency which would deteriorate the overall performance of MFC as compared with the separator containing MFC; even high power density is achieved in the former case. The tendency of forming biofilm on the cathode is much easier as no hindrance is found to occur in case of separatorless MFC. The biofilm retards oxygen diffusion to the cathode and hence performance of fuel cell is hampered. Thus it is not suitable for long term operation. One example of separatorless MFC is sediment MFC (Fig. 7.5) where anode is placed in anaerobic sediment and cathode is immersed into overlying water containing dissolved oxygen (Zabihallahpoor et al. 2015). The high salinity of sea water develops ionic potential between the electrodes and the sediment is full

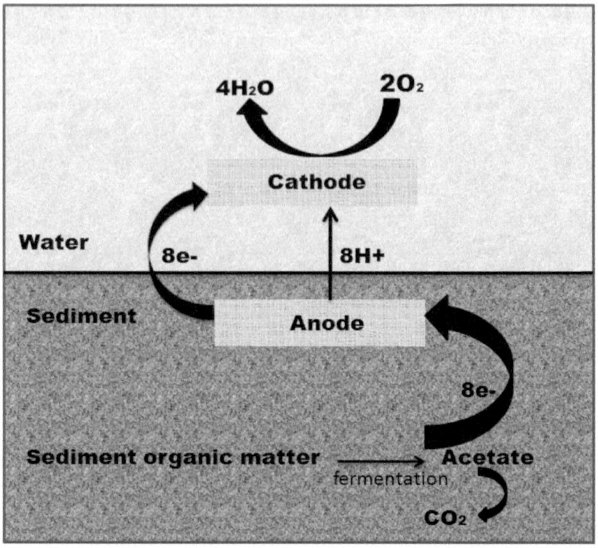

Fig. 7.5 Sediment microbial fuel cell

of organic matter from which bacteria produces electricity (maximum power output of 50 mW m^{-2}). Chen et al. (2013) reported that activated carbon cathode MFC lacking a separator produces 860 ± 10 mW m^{-2} power density (CE 20%–55%) than that of inexpensive PVA separator containing MFC (840 ± 42 mW m^{-2}, CE 43%–89%) having similar activated carbon cathode. A new approach for separatorless MFC design avoids oxygen permeability by forcing electrolytes to flow simultaneously from the anode chamber to the cathode chamber, which accelerates proton transfer but limits oxygen diffusion in the opposite direction. The COD removal efficiency in this MFC could reach as high as 90.5%, but in most cases cathode chamber is occupied by aerobic bacteria (Du et al. 2011).

7.6 Current Status

MFC deals with several technical limitations, where bacterial role with employed membranes have been given enormous attention in terms of energy recovery as a whole. One of the biggest associated problem is applied separating material between the two terminals. Earlier, commercially available PEM/CEMs were commonly used as they were relatively better proton conductors. But, their high cost and relatively higher mass transport served as a major limiting factor in overall systemic performance. Alternatively, several other materials e.g., polystyrene, polyether ether ketone (PEEK), poly (arylene ether sulphone), phenylated polysulphone, polyphosphazenes, polyimides, polybenzimidazole (PBI) and polypropylene (PP) have been tested and optimized as low cost separators in MFC application. Relatively, other porous materials (e.g., glass wool, earthen pot and ceramics) are widely studied, including different inorganic and organic materials. The main challenge resides in maintaining uniform porosity for higher efficiency. For storage, MFCs are usually stacked to enhance the respective voltages and currents, where voltage reversal becomes the major problem due to uneven performance of the individual stacks. Nevertheless, its effect can be minimized with efficient circuit design using capacitors in parallel. Different authors (Oh and Logan 2007; Winfield et al. 2013b) reported that the more porous and less iron-rich earthenware is the better material for power production. It would be interesting to study such low cost separators with elementary modifications that can be controlled to enhance the overall transversal of ionic conductivity in the system (Choi and Chae 2013).

7.7 Conclusion

MFC being a green technology, needs profound investigation in such diverse areas of possible low cost seperators, microbes and reactor fabrication that would make it practically more feasible. Despite of various materials employed, its commercialization still has many limitations and that needs to be reduced. Factors like ionic conductivity, uniform porosity and lower mass transfer are three major areas, signifying the role of employed separators in MFC. Further, unit design and used microbes are the other possible areas to explore in total systemic behaviour. In advance, with long term stability, feasible constraints associated with the separator development also calls for such challenges to be addressed equally, prior to commercialization.

References

Ayyaru, S., & Dharmalingam, S. (2015a). Influence of sulfonated SiO_2 in sulfonated polyether ether ketone nanocomposite membrane in MFC. *Fuel, 159*, 689–696.

Ayyaru, S., & Dharmalingam, S. (2015b). A study of influence on nanocomposite membrane of sulfonated TiO_2 and sulfonated polystyrene-ethylene-butylene-polystyrene for MFC application. *Energy, 88*, 202–208.

Ayyaru, S., Letchoumanane, P., Dharmalingam, S., & Stanislaus, A. R. (2012). Performance of sulfonated polystyrene-ethylene-butylene-polystyrene membrane in MFC for bioelectricity production. *Journal of Power Sources, 217*, 204–208.

Ayyaru, S., Dharmalingam, S., & Young-Ho, A. (2016). Nanocomposite membranes based on sulfonated polystyrene ethylene butylenes polystyrene (SSEBS) and sulfonated SiO_2 for MFC application. *Chemical Engineering Journal, 289*, 442–451.

Behera, M., & Ghangrekar, M. M. (2011). Electricity generation in low cost MFC made up of earthenware of different thickness. *Water Science and Technology, 64*, 2468–2473.

Behera, M., Jana, P. S., More, T. T., & Ghangrekar, M. M. (2010). Rice mill waste water treatment in MFCs fabricated using proton exchange membrane and earthen pot at different pH. *Bioelectrochemistry, 79*, 228–233.

Bond, D. R., Holmes, D. E., Tender, L. M., & Lovley, D. R. (2002). Electrodereducing microorganisms that harvest energy from marine sediments. *Science, 295*, 483–485.
Bond, D. R., & Lovley, D. R. (2003). Electricity production by *Geobacter sulfurreducens* attached to electrodes. *Applied and Environmental Microbiology, 69*, 1548–1555.
Chaudhuri, S. K., & Lovley, D. R. (2003). Electricity generation by direct oxidation of glucose in mediatorless microbial fuel cells. *Nature Biotechnology, 21*, 1229–1232.
Chen, G., Zhang, F., Logan, B. E., & Hickner, M. A. (2013). Poly(vinyl alcohol) separators improve the coulombic efficiency of activated carbon cathodes in MFCs. *Electrochemistry Communications, 34*, 150–152.
Choi, S., & Chae, J. (2013). Optimal biofilm formation and power generation in a micro-sized MFC (MFC). *Sensors and Actuators A: Physical, 195*, 205–212.
Choi, S., Kim, J. R., Cha, J., Kim, Y., Premier, G. C., & Kim, C. (2013). Enhanced power production of a membrane electrode assembly microbial fuel cell (MFC) using a cost effective poly [2,5-benzimidazole] (ABPBI) impregnated non-woven fabric filter. *Bioresource Technology, 128*, 14–21.
Du, F., Xie, B., Dong, W., Jia, B., Dong, K., & Liu, H. (2011). Continuous flowing membraneless MFCs with separated electrode chambers. *Bioresource Technology, 102*, 8914–8920.
Dutta, K., & Kundu, P. P. (2014). A review on aromatic conducting polymer-based catalyst supporting matrices for application in MFC. *Polymer Reviews, 54*, 401–435.
Harnisch, F., & Schroder, U. (2009). Selectivity versus mobility: Separation of anode and cathode in microbial bioelectrochemical systems. *Chemistry and Sustainability, Energy and Materials, 2*, 921–926.
He, Z., Minteer, S. D., & Angenent, L. T. (2005). Electricity generation from artificial wastewater using an upflow MFC. *Environmental Science and Technology, 39*, 5262–5267.
Hernández-Fernández, F. J., de los Ríos, A. P., Mateo-Ramírez, F., Godínez, C., Lozano-Blanco, L. J., et al. (2015). New application of supported ionic liquids membranes as proton exchange membranes in MFC for waste water treatment. *Chemical Engineering Journal, 279*, 115–119.
Hernández-Fernández, F. J., de los Ríos, A. P., Mateo-Ramírez, F., Juarez, M. D., et al. (2016). New application of polymer inclusion membrane based on ionic liquids as proton exchange membrane in MFC. *Separation and Purification Technology, 160*, 51–58.
Jana, P. S., Behera, M., & Ghangrekar, M. (2010). Performance comparison of up-flow MFCs fabricated using proton exchange membrane and earthen cylinder. *International Journal of Hydrogen Energy, 35*, 5681–5686.
Ji, E., Moon, H., Piao, J., Ha, P. T., An, J., Kim, D., et al. (2011). Interface resistances of anion exchange membranes in MFCs with low ionic strength. *Biosensors and Bioelectronics, 26*, 3266–3271.
Khilari, S., Pandit, S., Ghangrekar, M. M., Pradhan, D., & Das, D. (2013). Graphene oxide-impregnated PVA−STA composite polymer electrolyte membrane separator for power generation in a single-chambered MFC. *Industrial and Engineering Chemistry Research, 52*, 11597–11606.
Kim, H. J., Park, H. S., Hyun, M. S., Chang, I. S., Kim, M., & Kim, B. H. (2002). A mediator-less microbial fuel cell using a metal reducing bacterium, *Shewanella putrefaciens. Enzyme and Microbial Technology, 30*, 145–152.
Kim, B. H., Park, H. S., Kim, H. J., Kim, G. T., Chang, I. S., Lee, J., & Phung, N. T. (2004). Enrichment of microbial community generating electricity using a fuel-cell-type electrochemical cell. *Applied Microbiology and Biotechnology, 63*, 672–681.
Kumar, V., Nandy, A., Das, S., Salahuddin, M., & Kundu, P. P. (2015a). Performance assessment of partially sulfonated PVdF-co-HFP as polymer electrolyte membranes in single chambered MFCs. *Applied Energy, 137*, 310–321.
Kumar, V., Kumar, P., Nandy, A., & Kundu, P. P. (2015b). Crosslinked inter penetrating network of styrene and sulfonated PVdF-co-HFP as sulfonated electrolytic membrane in a single chamber MFC. *RSC Advances, 5*, 30758–30767.

Kumar, V., Kumar, P., Nandy, A., & Kundu, P. P. (2016a). A nanocomposite membrane composed of incorporated nano-alumina within sulfonated PVDF-co-HFP/Nafion blend as separating barrier in a single chambered MFC. *RSC Advances, 6*, 23571–23580.

Kumar, V., Mondal, S., Nandy, A., & Kundu, P. P. (2016b). Analysis of polybenzimidazole and polyvinylpyrrolidone blend membranes as separating barrier in single chambered MFCs. *Biochemical Engineering Journal, 111*, 34–42.

Lee, J., Phung, N. T., Chang, I. S., Kim, B. H., & Sung, H. C. (2003). Use of acetate for enrichment of electrochemically active microorganisms and their 16S rDNA analyses. *FEMS Microbiology Letters, 223*, 185–191.

Leong, J. X., Daud, W. R. W., Ghasemi, M., Liew, K. B., & Ismail, M. (2013). Ion exchange membranes as separators in MFCs for bioenergy conversion: A comprehensive review. *Renewable and Sustainable Energy Reviews, 28*, 575–587.

Leong, J. X., Daud, W. R. W., Ghasemi, M., Ahmad, A., Ismail, M., & Liew, K. B. (2015). Composite membrane containing graphene oxide in sulfonated polyether ether ketone in MFC applications. *International Journal of Hydrogen Energy, 40*, 11604–11614.

Li, W. W., Sheng, G. P., Liu, X. W., & Yu, H. Q. (2011). Recent advances in the separators for MFCs. *Bioresource Technology, 102*, 244–252.

Lim, S. S., Daud, W. R. W., Jahim, J. M., Ghasemi, M., Chong, P. S., & Ismail, M. (2012). Sulfonated poly (ether ether ketone)/poly (ether sulfone) composite membranes as an alternative proton exchange membrane in MFCs. *International Journal of Hydrogen Energy, 37*, 11409–11424.

Liu, Z. D., & Li, H. R. (2007). Effects of bio- and abio-factors on electricity production in a mediatorless MFC. *Biochemical Engineering Journal, 36*, 209–214.

Liu, H., & Logan, B. E. (2004). Electricity generation using an air-cathode single chamber MFC in the presence and absence of a proton exchange membrane. *Environmental Science and Technology, 38*, 4040–4046.

Liu, H., Cheng, S., & Logan, B. E. (2005). Power generation in a feed-batch microbial fuel cells as a function of ionic strength, temperature and reactor configuration. *Environmental Science and Technology, 39*, 5488–5493.

Logan, B. E. (2007). *Microbial fuel cells*. New York: Wiley.

Min, B., Cheng, S., & Logan, B. E. (2005). Electricity generation using membrane and salt bridge MFCs. *Water Research, 39*, 1675–1686.

Oh, S. E., & Logan, B. E. (2007). Voltage reversal during MFC stack operation. *Journal of Power Sources, 167*, 11–17.

Pant, D., Bogaert, G. V., Diels, L., & Vanbroekhoven, K. (2010). A review of the substrates used in microbial fuel cells (MFCs) for sustainable energy production. *Bioresource Technology, 101*, 1533–1543.

Perez, N. A., Coenen A. M. J., Mabesoone, M. F. J., Ripken, A. C.,& Roosen, J. (2012). *Bacteria, membranes and energy MFCs and their applications in Society*. Radboud Honours Academy. www.ru.nl/publish/pages/687330/eind.pdf.

Rabaey, K., Lissens, G., Siciliano, S. D., & Verstraete, W. (2003). A microbial fuel cell capable of converting glucose to electricity at high rate and efficiency. *Biotechnology Letters, 25*, 1531–1535.

Rabaey, K., Boon, N., Siciliano, S. D., Verhaege, M., & Verstraete, W. (2004). Biofuel cells select for microbial consortia that self-mediate electron transfer. *Applied and Environmental Microbiology, 70*, 5373–5382.

Rabaey, K., Lissens, G., & Verstraete, W. (2005). Biofuels for fuel cells: Renewable energy from biomass fermentation. In *MFCs: Performances and perspectives, Chapter 20* (pp. 377–399). London: IWA Publishing.

Rahimnejad, M., Ghasemi, M., Najafpour, G. D., Ismail, M., Mohammad, A. W., Ghoreyshi, A. A., & Hassan, S. H. A. (2012). Synthesis, characterization and application studies of self-made Fe_3O_4/PES nanocomposite membranes in MFC. *Electrochimica Acta, 85*, 700–706.

Rudra, R., Kumar, V., & Kundu, P. P. (2015). Acid catalysed cross-linking of poly vinyl alcohol (PVA) by glutaraldehyde: Effect of crosslink density on the characteristics of PVA membranes used in single chambered MFCs. *RSC Advances, 5*, 83436–83447.

Singha, S., Jana, T., Modestra, J. A., Kumar, A. N., & Venkata Mohan, S. (2016). Highly efficient sulfonated polybenzimidazole as a proton exchange membrane for MFCs. *Journal of Power Sources, 317*, 143–152.

Tender, L. M., Reimers, C. E., Stecher, H. A., Holmes, D. E., Bond, D. R., Lowy, D. A., Pilobello, K., Fertig, S. J., & Lovley, D. R. (2002). Harnessing microbially generated power on the seafloor. *Nature Biotechnology, 20*, 821–825.

Venkatesan, P. N., & Dhamalingam, S. (2013). Characterization and performance study on chitosan-functionalized multi walled carbon nano tube as separator in MFC. *Journal of Membrane Science, 435*, 92–98.

Watanabe, K. (2008). Recent developments in MFC technologies for sustainable bioenergy. *Journal of Bioscience and Bioengineering, 106*, 528–536.

Winfield, J., Greenman, J., Huson, D., & Ieropoulos, I. (2013a). Comparing terracotta and earthenware for multiple functionalities in MFCs. *Bioprocess and Biosystems Engineering, 36*, 1913–1921.

Winfield, J., Chambers, L. D., Rossiter, J., & Ieropoulos, I. (2013b). Comparing the short and long term stability of biodegradable, ceramic and cation exchange membranes in MFCs. *Bioresource Technology, 148*, 480–486.

Yang, S. Q., Jia, B. Y., & Liu, H. (2009). Effects of the Pt loading side and cathode-biofilm on the performance of a membrane-less and single-chamber MFC. *Bioresource Technology, 100*, 1197–1202.

Zabihallahpoor, A., Rahimnejad, M., & Talebnia, F. (2015). Sediment MFC as a new power source of renewable and sustainable energy: Present status and future prospects. *RSC Advances, 5*, 94171–94183.

Zhang, X. Y., Cheng, S., Wang, X., Huang, X., & Logan, B. E. (2009). Separator characteristics for increasing performance of MFCs. *Environmental Science and Technology, 43*, 8456–8461.

Zhang, X., Cheng, S., Huang, X., & Logan, B. E. (2010). Improved performance of single-chamber MFCs through control of membrane deformation. *Biosensors and Bioelectronics, 25*, 1825–1828.

Zuo, Y., Cheng, S., Call, D., & Logan, B. E. (2007). Tubular membrane cathodes for scalable power generation in MFCs. *Environmental Science and Technology, 41*, 3347–3353.

Chapter 8
Role of Cathode Catalyst in Microbial Fuel Cell

Santimoy Khilari and Debabrata Pradhan

8.1 Introduction

Electrodes (cathode and anode) are the primary components of fuel cells. Oxidation of fuel at anode surface produces electrons and oxidized by-products. Subsequently, the electrons travel to cathode through an external circuit (Lu and Li 2012). At the cathode surface, appropriate terminal electron acceptors (TEA) are reduced by the incoming electrons and complete the cathodic half-cell reaction. The performance of a microbial fuel cell (MFC) strongly depends on the efficiency of the cathode. Generally, the efficacy of a cathode is determined from the reduction kinetics of the TEA (Rismani-Yazdi et al. 2008). The cathodic reduction is a surface electrochemical phenomenon. Thus, the surface of cathode plays a crucial role on the reduction kinetics of TEA. However, the widely used carbon-based cathodes show very poor reduction kinetics and limit the fuel cell performance (Rismani-Yazdi et al. 2008). This obstacle prompts the researchers to develop electrocatalysts for the enhanced cathodic reduction. The main purpose of the present chapter is to discuss different types of cathode catalysts and their influence on the bioelectricity generation in MFC. An effective cathode catalyst has to qualify several crucial requirements for practical applicability. First, it must be capable to deliver high intrinsic catalytic activity. Second, catalyst should show long term durability. Moreover, the electrocatalyst should have good electrical conductivity to minimize the resistive losses. In addition, the catalyst must be inexpensive, abundant and can easily be synthesized in large scale. In past, efforts have been made to find appropriate cathode catalysts. The findings are reviewed by several authors (Lu and Li 2012; Rismani-Yazdi et al. 2008; Kannan 2016; Liew et al. 2014).

S. Khilari • D. Pradhan (✉)
Materials Science Centre, Indian Institute of Technology Kharagpur, Kharagpur, West Bengal 721302, India
e-mail: deb@matsc.iitkgp.ernet.in

D. Das (ed.), *Microbial Fuel Cell*,
https://doi.org/10.1007/978-3-319-66793-5_8

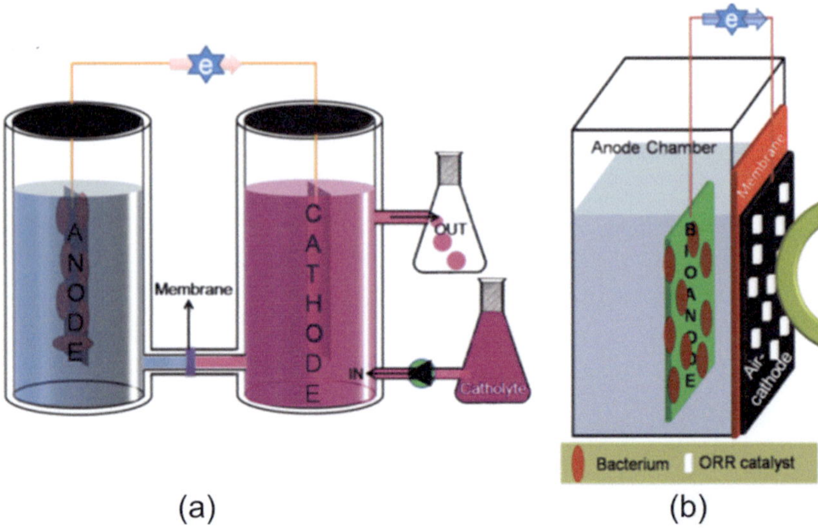

Fig. 8.1 Schematic of (**a**) double chamber MFC; and (**b**) single chamber MFC

Among the different configurations of MFC reported in past decades, dual-chambered MFC (dMFC) and single chambered MFC (sMFC) are the most common. A dMFC consists of two chambers i.e. anode and cathode separated by an ion exchange membrane (Fig. 8.1a). On the other hand, sMFC has one chamber in which anode is placed and the cathode is fixed to one wall of the chamber (Fig. 8.1b) attached through a membrane called membrane cathode assembly. The dMFC with separate cathode chamber allows different types of TEAs to be used (Pandit et al. 2011). On the other hand, sMFC always utilizes atmospheric oxygen (O_2) as TEA (Khilari et al. 2014). The open circuit potential (OCP) of a fuel cell is determined from the redox potential of anodic and cathodic half-cell reactions. The TEA redox potential mainly governs the cathode half-cell potential. Beside O_2, several other organic or inorganic TEAs such as ferricyanide, permanganate, dichromate, metal ion and nitrate have been studied in dMFC (Wei et al. 2012; Pandit et al. 2011; Lu and Li 2012). A high theoretical reduction potential of TEA offers better cathodic half-cell potential and thus improves the overall OCP. The cathodic reduction of various TEAs and their corresponding theoretical potentials are listed in Table 8.1. These TEAs also have several shortcomings that make them ineffective for practical application in MFC. Regeneration of non-O_2 TEAs is usually difficult in MFC. So, replacement of permanganate and other non-O_2 TEA is required at frequent intervals that add extra cost to MFC operation. In addition, some of the TEAs generate insoluble end product ($KMnO_4$ reduce to MnO_2) that can hinder cathodic reduction process and hamper the sustainability of MFC. In this context, being abundant and cost-free TEA, O_2 is desirable. Moreover, water being the end product of O_2, reduction is sustainable to the environment. However, the reduction of O_2 is an uphill electrochemical process with high activation energy.

Table 8.1 Different terminal electron acceptors and their half-cell reactions

Electron acceptor	Cathodic reduction reaction	E^0	E	Reaction conditions
$Fe(CN)_6^{3-}$	$Fe(CN)_6^{3-} + e^- \rightarrow Fe(CN)_6^{4-}$	0.36	0.36	$[Fe(CN)_6^{3-}] = [Fe(CN)_6^{4-}]$
MnO^{4-}	$MnO^{4-} + 3e^- + 2H^+ \rightarrow MnO_2 + 2H_2O$	1.7	1.1	$[MnO^{4-}] = 5$ mM, pH = 7
O_2	$O_2 + 4e^- + 4H^+ \rightarrow 2H_2O$	1.23	0.8	$pO_2 = 0.2$, pH = 7
O_2	$O_2 + 2e^- + 2H^+ \rightarrow 2H_2O_2$	0.69	0.33	$pO_2 = 0.2$, $[H_2O_2] = 5$ mM, pH = 7
NO^{3-}	$2NO^{3-} + 10e^- + 12H^+ \rightarrow N_2 + 6H_2O$	1.25	0.73	$[NO^{3-}] = 5$ mM, $pN_2 = 0.2$, pH = 0.7
NO^{3-}	$2NO^{3-} + 2e^- + 2H^+ \rightarrow NO^{2-} + H_2O$	0.83	0.42	$[NO^{3-}] = [NO^{2-}]$, pH = 0.7
$S_2O_8^{2-}$	$S_2O_8^{2-} + 2e^- \rightarrow 2SO_4^{2-}$	1.96	1.96	$[S_2O_8^{2-}] = [SO_4^{2-}] = 5$ mM
ClO^{4-}	$ClO^{4-} + 8e^- + 8H^+ \rightarrow Cl^- + 4H_2O$	1.29	0.87	$[ClO^{4-}] = [Cl^-]$, pH = 7
$Cr_2O_7^{2-}$	$Cr_2O_7^{2-} + 6e^- + 14H^+ \rightarrow 2Cr^{3+} + 7H_2O$	1.36	0.42	$[Cr_2O_7^{2-}] = [Cr^{3+}] = 5$ mM, pH = 7
Cu^{2+}	$Cu^{2+} + 2e^- \rightarrow Cu(s)$	0.34	0.27	$[Cu^{2+}] = 5$ mM
VO_2^{2+}	$VO_2^{2+} + 2e^- + 2H^+ \rightarrow VO^{2+} + H_2O$	1.00	0.17	$[VO_2^{2+}] = [VO^{2+}]$, pH = 7
CO_2	$CO_{-2} + 12e^- + 12H^+ \rightarrow C_6H_{12}O_6$ (biomass) + $3O_2$	N.A.	N.A.	N.A.

E^0 standard reduction potential of the TEA, *E* half-cell potential under operational condition

To reduce the activation energy, suitable catalysts are generally employed (Nørskov et al. 2004). The use of an effective catalyst can induce a low energy intermediate and improve the cathodic O_2 reduction reaction (ORR) kinetics. Efforts have been made to elaborate the fundamentals of ORR mechanism and catalyst design in the present chapter. Further, the application of various developed cathode catalysts in MFC is discussed.

8.2 Non-oxygen Terminal Electron Acceptors

Several studies have been devoted to find suitable electron acceptor for dMFC. A number of studies used potassium ferricyanide-based catholyte as electron acceptor. Wei et al. (2012) found higher power generation with potassium ferricyanide (426 mV, 181.48 mW m^{-3}) than dissolved oxygen (DO) catholyte (150 mV, 22.5 mW m^{-3}). Although ferricyanide is cheaper and good electron acceptor, it undergoes major side reaction at low pH generating toxic and hazardous HCN (Forrestal et al. 2014). The inefficiency of DO-based system is realized due to the limited solubility of molecular O_2 as compared to other TEAs. In a different study, very high OCP (1.04 V) and improved power density (7.29 mW m^{-2}) was reported with $KMnO_4$ based catholyte as compared to both potassium ferricyanide (0.71 V, 0.92 mW m^{-2}) and potassium dichromate (0.56 V, 0.79 mW m^{-2}) containing catholyte (Guerrero-Rangel et al. 2010). Pandit et al. reported slight lower performance with potassium persulphate (1.1 V, 101.7 mW m^{-2}) than that of $KMnO_4$ (1.11 V, 116.2 mW m^{-2}) based catholyte. However, the OCP for persulfate-based cathode was stable for longer duration as compared to $KMnO_4$. Additionally, an external ariation on the catholyte improves the performance which arises from the synergetic electron uptake by both O_2 and non-O_2 TEA (Pandit et al. 2011). A separate study conducted on persulphate and fericyanide reports that persulphate is capable to produce high power after a few cycles (Li et al. 2009). Very recently hypochloride and sodium bromate were tested as TEA in dMFC (Jadhav et al. 2014; Dai et al. 2016a). The studies on TEA in dMFC cathode signify that the OCP is primarily governed by the standard reduction potential of the TEA, concentration of TEA, and pH of the catholyte.

8.3 Oxygen Reduction Reaction (ORR) at Cathode: Fundamentals

The catalytic reduction of one O_2 molecule requires four electrons ($4e^-$) and four protons to produce water as represented in the following equation

$$O_2 + 4e^- + 4H^+ \rightarrow 2H_2O, E^0 = 1.23 \text{ V (vs. RHE)} \qquad (8.1)$$

In real experiment, ORR needs extra bias to the equilibrium potential (E^0 = 1.23 V). The extra bias voltage is called overpotential (Gewirth and Thorum 2010). Thus, the overpotential of ORR determines the activity of the catalyst. Lowering of overpotential by introducing more active material is the prime requirement for ORR catalyst development. The overall ORR process is an integration of several discrete electrons coupled or decoupled proton transfer steps (Katsounaros et al. 2014). The well-established ORR mechanism is schematically represented in Fig. 8.2. Initiation of reduction process involves the diffusion of DO molecule to the electrode surface and form an adsorbed O_2 molecule (O_2*, where * signifies the active site on the catalyst surface). The adsorbed O_2 undergoes sequential bond cleavage and reduction. Depending on the mode of bond cleavage to form O*, three different pathways can be considered. The first pathway is termed as dissociative pathway in which an adsorbed O_2 molecule directly dissociates to produce O*. The O* is then reduced by the electron and proton to OH* and H_2O successively. The second pathway is known as associative pathway where OOH* is first generated from O_2*. Subsequently, the O–O bond of the intermediate OOH* is cleaved to produce OH* and O* intermediates.

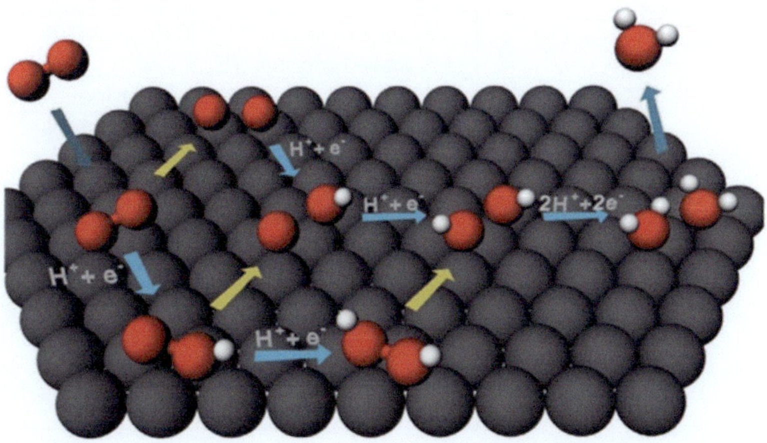

Fig. 8.2 Schematic representation of proposed ORR catalysis mechanism on a catalyst surface. *Black*, *red* and *white/grey* balls represent catalytic surface, oxygen and hydrogen, respectively (Adapted from Xia et al. 2016 with permission from Wiley publisher)

The third pathway is known as peroxo pathway (also known as second associative pathway) where O_2* is reduced to OOH* and form HOOH* before O–O bond breaking. The selection of reduction pathway on a catalyst depends on the physicochemical characteristics of catalyst surface as well as surrounding environment. It is always preferable to avoid the third pathway for fruitful catalysis as it liberates corrosive peroxide (H_2O_2) intermediate. H_2O_2 formation not only hampers the catalytic process but also degrades the polymeric membrane by generating reactive radicals (Khilari et al. 2013). For the design of an efficient catalyst, the free energy change on the catalyst surface during catalysis is another important factor. Generally, density functional theory (DFT) is used to calculate the free energy change on a catalyst surface suggesting the ORR pathway. The DFT study reveals that dissociative mechanism is predominate at low O_2 coverage whereas high O_2 coverage prefers associative pathway (Nørskov et al. 2004). However, the selectivity towards feasible pathway significantly depends on the characteristics of the electrode material.

The ORR kinetics at an electrode surface is strongly correlated with the adsorption of intermediates (O*, OH* and OOH*) on the catalyst surface. Thus, it is essential to experimentally probe these intermediates to identify the exact reaction mechanism. However, monitoring of the intermediates is difficult (Shao et al. 2006). The progress of computational studies helps to estimate the surface interaction energies with sufficient accuracy. A DFT study suggests that the adsorption of O_2 on the electrode is responsible for the origin of overpotential (Nørskov et al. 2004). Generally, O_2 is strongly adsorbed on the electrode surface at high potential. Thus, proton and electron can't be transferred to the highly stable adsorbed O_2 molecule. Further, upon switching the potential to a lower value weakens the stability of adsorbed O_2 and reaction becomes feasible. Additionally, the binding

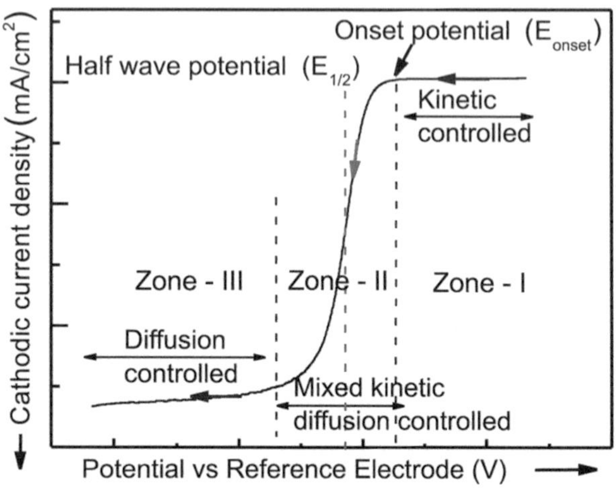

Fig. 8.3 Characteristic regions of a representative ORR polarization plot

affinity toward oxygenated species determines the catalytic activity of the catalyst. Thus, an efficient catalyst should have an optimum binding strength to the intermediates. A very weak binding of adsorbed O_2 or oxygenated species on catalyst surface limits the electron and proton transfer, thereby ORR kinetics. On the other hand, too strong binding of O* and OH* restricts desorption of H_2O and blocks the active sites for further O_2 adsorption. The binding energy of intermediates varies from catalyst to catalyst. Generally, the intrinsic binding characteristic of a catalyst depends on its electronic structure (Stamenkovic et al. 2006). A high energy metal *d*-state relative to Fermi energy offers stronger interaction of the electrode surface to the intermediates (Nørskov et al. 2009). Thus, the catalyst with optimal electronic structure can offer appropriate intermediates binding energy and boost ORR catalysis significantly.

8.3.1 Evaluation of ORR Catalysts: Figure of Merits

The ORR catalytic activity is generally evaluated by employing a catalyst on the cathode and comparing with benchmark Pt under identical condition. The ORR mechanism is studied from the half-cell reaction in a standard electrochemical cell. The standard half-cell characterization of an ORR catalyst is carried out with a thin-film catalyst on a rotating disk electrode (RDE) or rotating ring disk electrode (RRDE). The cathodic polarization plot is recorded by employing linear sweep voltammetric technique at steady state or with a known rotation of RDE or RRDE. A typical ORR polarization plot of a catalyst consists of three characteristic regions as shown in Fig. 8.3. The first zone (Zone I) is termed as kinetic-control zone where the ORR rate is quite slow and cathodic current density increases very slowly. The second region (Zone II) corresponds to mixed kinetics and diffusion-controlled zone where a significant increase of current density with potential drop is noticed. The third zone (Zone III) refers to diffusion controlled region (Khilari et al. 2014). In this region the electrode's current density is governed by the diffusion rate of reactants reaching to the rotating electrode surface. In addition to these three characteristic zones, other two most important parameters often used to evaluate a catalyst performance are (i) onset potential (E_{onset}) and (ii) half-wave potential ($E_{1/2}$). The more positive value of these two potentials infers better catalytic activity of the catalyst. However, the definition of E_{onset} differs from article to article. Some literature considers E_{onset} as the potential required to achieve 5% of the diffusion-limited current density (J_{L}) (Zhou et al. 2014). Another definition states that E_{onset} is the potential where the current density exceeds the threshold value of 0.1 mA cm^{-2} (Daems et al. 2014). Thus, it is important to mention the definition of E_{onset} used in the ORR study of a catalyst and should be compared to Pt with E_{onset} evaluated at similar conditions.

The mechanistic pathway of ORR can be extracted from the kinetic limiting current density (J_{K}). In this connection, mass-transport corrected polarization plot is employed to find kinetics parameters using Koutechky-Levich equation (Khilari et al. 2014).

$$1/J = 1/J_L + 1/J_K \tag{8.2}$$

where J is the measured current density, J_{K} and J_{L} represent the kinetic limiting and Levich current, respectively. The measured current density must be the catalytic current density which generally interferes with capacitive current density. This type of interference is remarkable in porous earth-abundant oxides and carbon materials due to the electrical double layer formation by adsorption of ions on the catalyst surface. Generally, the capacitive interference is minimized by employing very slow scan rate during polarization study of the electrode and background correction. The background current of the electrode can be eliminated by subtracting the current density recorded in a N_2 saturated electrolyte experiment from the O_2 saturated measurement.

The average number of electron transfer (n) to each O_2 molecule and amount of H_2O_2 intermediate generation during ORR are the key points to figure out an efficient ORR catalyst. These two critical factors can be evaluated from RDE and RRDE polarization study. Following equations are used to estimate the 'n' value of a catalyst:

$$\text{B} = 0.2nFC_0(D_0)^{2/3}\upsilon^{-1/6} \tag{8.3}$$

$$4I_{\text{d}} = n(I_{\text{d}} + I_{\text{r}}/N) \tag{8.4}$$

where n, F, C_0, D_0 and υ are assigned as the number of electron involved in the reaction, Faraday constant (96,485 C mol^{-1}), bulk O_2 concentration, diffusion coefficient of O_2 in the electrolyte, and kinematic viscosity of the electrolyte, respectively. Further, I_{d}, I_{r}, and N represent the disk current, ring current and current collection efficiency of the Pt ring, respectively (Su et al. 2013). The efficiency of an ORR catalyst can be estimated from the H_2O_2 generation. The H_2O_2 generated at the catalyst surface coated on a disk electrode diffused to the Pt ring and successively reduced. Thus, the ring current measurement can provide the quantitative estimation of H_2O_2 generation during ORR. Following equation is used to estimate H_2O_2 intermediate generation on a disk electrode (Liu et al. 2013):

$$\%H_2O_2 = 200 \times (I_r/N)/(I_d + I_r/N) \tag{8.5}$$

The efficient ORR catalysis needs to follow $4e^-$ pathway whereas $2e^-$ pathway leads to formation of H_2O_2 which reduces the performance as well as the stability of the cathode. Apart from the 'n' value and % H_2O_2 production, the electron transfer coefficient is another important parameter that affects the catalyst efficiency. The electron transfer coefficient of ORR catalyst can be estimated from the Tafel slope in a linear portion of Tafel plot. Tafel plot of a catalyst is plotted by using Tafel equation as given below (Liew et al. 2014).

$$\eta = (RT/2.303n\alpha F) \log (i/i_0) \tag{8.6}$$

where η, R, T and α represent overpotential, universal gas constant, temperature in absolute scale and electron transfer coefficient, respectively. Further, i and i_0 refer to measured ORR current density and exchange current density, respectively. The electron transfer coefficient can be extracted from Tafel slope ($RT/2.303n\alpha F$). Thus, the smaller value of Tafel slope reflects higher transfer coefficient indicating the increment of current density resulted with a slower increase of overpotential. To obtain Tafel plot from RDE study the mass transport correction is essential on the recorded polarization current. Equation (8.7) has been used to find out the mass transport corrected current of a polarization study (Singh et al. 2014a).

$$J_k = J \times J_L/(J_L - J) \tag{8.7}$$

8.4 Cathode Catalysts

Electrochemical reduction of molecular O_2 at the cathode surface is an important reaction for the power generation in different types of fuel cells including MFC (Khilari et al. 2014). However, ORR is sluggish with graphite or other extensively used carbon electrodes (Kannan 2016; Xia et al. 2016). Thus, electrocatalysts are introduced to improve the cathode performance. The noble metals and especially Pt are extensively used as cathode catalyst in MFC (Wang et al. 2015). However, the high cost and catalyst poisoning are the major issues associated with Pt for its practical application (Xia et al. 2016). Numerous efforts have been devoted to develop non-Pt catalysts as alternate. Different approaches have thus been considered to overcome this problem by developing more efficient catalysts. The typical approaches made on this purpose are:

(i) Tuning of size, shape, exposed facets and structure of Pt nanostructures.
(ii) Designing Pt-based multimetallic (PtM or PtM_1M_2) nanostructures.
(iii) Building superior ORR catalysts with earth abundant elements.

In last few decades, Pt and Pt-based complex systems are well studied for ORR catalysis—both experimentally and theoretically (Stamenkovic et al. 2006; Bing et al. 2010; Stephens et al. 2012). The critical factors such as surface structure, electronic property, geometry, and lattice strain affect the ORR activity of Pt-based materials. In recent time, paramount research interests involve rational designing of durable non-Pt nanomaterials with high ORR activity. Combination of different earth-abundant active elements in a single molecule or nanostructure with high surface area, more active sites, and special structure has been considered to enhance ORR activity (Nie et al. 2015; Chen et al. 2011). However, the intrinsic catalytic activity of non-Pt catalysts remains lower than that of Pt (Xia et al. 2016; Nie et al. 2015). The ORR activity of Pt-free cathode can be partially improved by higher catalyst loading. However, higher loading reduces the mass transport onto the electrode and makes the device heavier. Thus it is essential to increase the intrinsic catalytic sites by other suitable methods such as doping and/or forming micropore/mesopore to facilitate the mass transport. Significant progress in material science and nanotechnology has strengthened the research in designing and fabricating highly efficient non-precious metals and related compounds (oxides, sulphides, metal complexes etc.) with functional nanostructures and precise composition for ORR catalysts. In next section, the role of several Pt- and non-Pt based ORR catalysts in the cathode of MFC is discussed.

8.4.1 Pt and Pt-based ORR Catalysts

The ORR is an irreversible electrochemical phenomenon which results in significant amount of energy loss. Moreover, the high standard reduction potential (1.23 V vs. SHE) is the major limiting factor of ORR for most of the catalysts. Except a few noble metals, most of the materials are unstable at high potential that restricts the applicability of non-noble metal-based ORR catalysts (Singh et al. 2014b). The electrocatalytic study executed with different noble metals (Hg, Au, Ag, Cu, Pd, Pt etc.) revealed that Pt follows a quasi $4e^-$ ORR pathway with less H_2O_2 generation as compared to other metals (Sánchez-Sánchez and Bard 2009). This signifies the importance of Pt as a state-of-the-art ORR catalyst. In acidic environment, metal catalysts catalyze ORR via two most favourable pathways: (i) peroxide pathway and (ii) metal oxide (MO) pathway. The former pathway is initiated by generation of a superoxide ion (O_2^-) which successively combines with proton and electron to produce H_2O_2. Further, the H_2O_2 can either be disproportionate or reduced to form

H_2O as end product. The proposed mechanism can be presented by following set of reactions (Eqs. 8.8, 8.9 and 8.10)

$$O_2 + 2e^- + 2H^+ \rightarrow H_2O_2, E^0 = 0.67 \text{ V (vs. RHE)} \tag{8.8}$$

$$H_2O_2 + 2e^- + 2H^+ \rightarrow 2H_2O, E^0 = 1.77 \text{ V (vs. RHE)} \tag{8.9}$$

$$H_2O_2 \rightarrow 2H_2O + \tfrac{1}{2}\, O_2 \tag{8.10}$$

The MO pathway originates from the formation of corresponding MO by reaction with molecular O_2 and subsequently MO is reduced to initial metal (M). Here, MO serves as an active intermediate for ORR. Equations 8.11 and 8.12 can be used to represent the MO-based ORR pathway.

$$2\,M + O_2 \rightarrow 2MO \tag{8.11}$$

$$MO + 2e^- + 2H^+ \rightarrow M + 2H_2O \tag{8.12}$$

The proposed pathways simultaneously proceed on the catalyst surface and thus termed as mixed ORR pathway. The ways of combination (series or parallel) of different elementary reactions lead to different amount of H_2O_2 production as well as different '*n*' values for overall ORR. Further, combination of above two elementary reactions (Eqs. 8.8, 8.9, 8.10, 8.11, and 8.12) can give rise to an overall $4e^-$ reduction pathway. Thus, MO-based ORR pathway proceeds via efficient $4e^-$ reduction of O_2 and exclude formation of corrosive H_2O_2. Moreover, the probability of MO pathway depends on (i) MO formation trend and (ii) effective reduction of MO to M. The tendency to form MO depends on the O_2 adsorption ability on the metal surface and subsequent reaction. Moreover, the reduction of MO to M solely depends on the thermodynamics of reduction process. The optimal O_2 adsorption energy and high reduction potential of Pt^{2+} contributes to its superior ORR catalytic activity. Nørskov et al. showed that the activity of metallic ORR catalysts mainly depends on the adsorption energies of O_2 and different intermediates. On the basis of DFT calculation they reported excellent ORR activity of Pt than other metals (Nørskov et al. 2004). Figure 8.4a shows the ORR activity of different metals as a function of their oxygen binding energy, which clearly states the superiority of Pt among the metals. Liu and Logan (2004) developed a Pt catalyst (0.5 mg cm^{-2}) coated carbon cloth for air cathodic ORR. Recently, Peng et al. (2013) utilized Pt in a membraneless MFC that delivered power density of 693 ± 20 mW m^{-2} with a FeOOH modified bioanode.

Commercially available Pt supported on activated carbon (AC) is the most commonly used standard cathode catalyst in MFC as well as several other type fuel cells. However, high cost and scarcity of Pt limits its large scale application. In addition, Pt suffers from its poor stability (dissolution, agglomeration and sintering) during long term uses (Xia et al. 2016). Alloying of Pt with earth-abundant metals not only lowers the cost by reducing Pt content but also significantly improves the catalytic activity as compared to virgin Pt. Alloying of metals also can lead to change in the atomic and electronic structure of material along with creation of

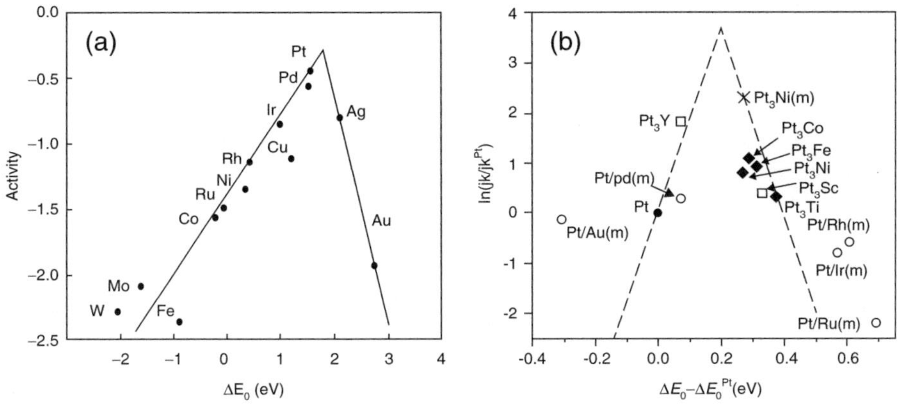

Fig. 8.4 (**a**) Trends of ORR catalytic activity of different metals as a function of oxygen binding energy; and (**b**) Volcano plots for the ORR on Pt-based bimetallic alloys (Adapted from Ref. Nørskov et al. 2004 (**a**) with permission from the ACS publisher and Xia et al. 2016 and (**b**) with permission from Wiley publisher)

more active sites. Moreover, binding affinity to the adsorbed species can be easily tuned with proper composition design of alloy catalysts. Figure 8.4b represents a volcano plot of different Pt-based alloys as a function of O_2 adsorption energies (Xia et al. 2016). Several bimetallic catalysts were tested in MFC with higher or comparable catalytic activity to Pt/C. A high mass activity is achieved with Pt_2-Ni/C catalyst (1724 mW m^{-2}) among different compositions of Pt and Ni metals while outperforming the Pt/C (1413 mW m^{-2}) (Wang et al. 2015). The Pt-Co/graphene composite with much low Pt content (15%) delivers comparable power density of 1378 mW m^{-2} to Pt(20%)/C (1406 mW m^{-2}) (Yan et al. 2013). The Fe-Pt catalysts exhibit high ORR catalytic activities in neutral as well as in acid medium. The best performance was recorded with Pt_3-Fe/C catalyst (1680±15 mW m^{-2}) which delivered 18% high power than Pt/C (1422±18 mW m^{-2}) (Yan et al. 2014). These studies demonstrate that the efficiency of bimetallic catalysts is strongly correlated to the chemical composition of the alloys.

8.4.2 *Pt-free ORR Catalysts in MFC*

8.4.2.1 Metals and Multimetallics

The high cost, poor earth-abundance and surface poisoning are the demerits associated with Pt, which drive scientific community to explore Pt-free alternate catalysts. Many scientific and engineering strategies have been made to lower the cost-to-performance ratio of MFC which include design of an appropriate electrode structure with more accessible O_2 adsorption sites and easy mass transfer (Rismani-Yazdi et al. 2008). In this concern, electrode with three dimensional

(3D) architecture and well-defined porosity has been introduced to promote the mass transport. The 3D structured foam composed nano Pd-Cu and activated carbon-Ni foam (1240 mW m^{-2}) have shown good ORR catalytic activity (Xiong et al. 2013; Cheng and Wu 2013). Very recently, Hou et al. (2016) reported a CoNi alloy nanoparticle encapsulated bamboo-like CNT with several graphene (GR) layers on its inner cavity that catalyzes ORR with '*n*' value of 3.63, comparable to Pt/C (3.9) and generates maximum power density ($P_{d,max}$) of 2.0±0.1 W/m^2 close to the benchmark Pt/C (2.6±0.2 W/m^2). The improvement of electrocatalytic activity is believed to be originated from the synergetic contribution of nitrogeneous functional groups of support material and bimetallic catalyst.

8.4.2.2 Metal Oxide-Based ORR Catalysts

Towards the development of noble metal-free ORR catalysts, the low cost metal oxides (MO) have generated considerable attention as alternate to Pt/C in MFC. Different types of transition MOs were studied including simple MOs, spinel type MOs, and hybrids. The simple synthesis protocol, low cost, excellent ORR catalytic activity, and environmental benignity are the important factors that attract scientific community to use MO-based catalysts. Owing to the low cost, high abundance and multiple oxidation states, manganese oxides (MnO_x) have been studied extensively as cathode catalysts in MFC. The theoretical and experimental studies revealed that Mn^{III} metal centre serves as active species for ORR on MnO_x surface. Thus the concentration of Mn^{III} species on the catalyst surface determines the efficiency of the MnO_x. A proposed ORR mechanism on MnO_2 surface can be represented by following set of reactions (Khilari et al. 2013):

$$Mn^{IV}O_2 + H_2O + e^- \rightarrow Mn^{III}OOH + OH^- \tag{8.13}$$

$$2Mn^{III}OOH + O_2 \rightarrow 2Mn^{IV}OOHO^- \tag{8.14}$$

$$Mn^{IV}OOHO^- + e^- \rightarrow Mn^{IV}O_2 + OH^- \tag{8.15}$$

Morphology, crystal structure and specific surface area play important role in controlling the activity of MO-based electrocatalysts. The catalytic activity of different polymorphs of MnO_2 is reported to be in the order β-<λ-<γ-<α- MnO_2 (Khilari et al. 2013). Additionally, the chemical composition of catalyst affects the catalytic activity. Among different MnO_x, the ORR activity follow the order of Mn_5O_8<Mn_3O_4<Mn_2O_3<MnOOH~MnO_2 (Khilari et al. 2013). The α-MnO_2 nanotubes modified MFC delivered a $P_{d,max}$ of 2100 mW m^{-2} which is higher than α-MnO_2 nanorods (1850 mW m^{-2}) confirming the role of morphology (Gnanakumar et al. 2014). Very recently Liu et al. (2015) reported the morphology-dependent catalytic performance of electrodeposited Cu_xO (with tuneable porosity) cathode catalyst in sMFC. The improved ORR catalytic activity was owing to high surface area and well defined microporosity that provides large number of active sites. Several other transition MOs such as Fe_2O_3 and NiO have

also been tested in MFC (Martin et al. 2011; Huang et al. 2015). The power generation follow the order of carbon black (8 W m^{-3})<Fe_2O_3 (15 W m^{-3})< Mn_2O_3 (32 W m^{-3})<Pt/C (90 W m^{-3}) (Martin et al. 2011). Impurity doping is another way to boost the ORR activity of MOs. The cryptomelane-type octahedral molecular sieve MnO_2 has been doped with Co, Cu and Ce, which shows significant improvement in performance to the parent MnO_2 (Li et al. 2010). The doping of lower valent cation on to Mn^{4+} sites in MnO_2 generates O_2 vacancies to maintain the charge neutrality. This disordered structure is mainly responsible for the improved catalytic activity.

Spinel type MOs are a special class of materials with chemical formula AB_2O_4 where A and B are either same or different transition metals (Khilari et al. 2014). Several studies have been carried out with spinel type catalysts such as $MnCo_2O_4$, Fe_3O_4, $MnFe_2O_4$, Co_3O_4 and $NiCo_2O_4$ (Khilari et al. 2014, 2015; Ma et al. 2014; Ge et al. 2015; 2016). The multiple valent metal centres in spinel play major role in improving the ORR activity by offering more active sites. Further, the tailoring of chemical composition 'x' in $A_{1+x}B_{2-x}O_4$ resulted in many intriguing electrocatalysts with varied performance (Cheng et al. 2011) due to change in their properties. The tailoring of chemical composition is adopted to create disorder for enhancing the catalytic activity by minimizing the activation barrier at the defect sites.

Low electronic conductivity is a demerit of most of the MO-based catalysts that increases the internal resistance of fuel cell. To improve the electronic conductivity of MO-based cathode, electrically conductive supports such as AC, graphitized carbon, CNTs, graphene (GR) and conducting polymer are generally employed. MnO_2/CNT and NiO/CNT were evaluated as ORR catalyst in MFC (Liew et al. 2014; Huang et al. 2015). Similarly, graphene and heteroatom doped GR supported MOs (MnO_2 NTs/GR, MnO_2/GO, Co_3O_4/N-doped GR) were tested (Kannan 2016). Among different compositions, 77% NiO/CNT was reported to be the best ORR catalyst with $n = 3.5$ which is almost two times higher to virgin NiO ($n = 1.9$) in neutral PBS medium. In addition, 77% NiO/CNT modified sMFC offers $P_{d,max}$ of 670 mW m^{-2} and an OCV of 0.772 V (Huang et al. 2015, Fig. 8.5). Moreover, Gnanakumar et al. reported a significant increment of power generation by integrating α-MnO_2 nanotubes with graphene oxide (GO). A composite catalyst with 2:1 ratio of MnO_2:GO generated $P_{d,max}$ of 3359 mW m^{-2} which is quite close to Pt/C (Gnanakumar et al. 2014). A number of recent review articles discussed the role of graphene/MO hybrid catalysts with detail insight (Kannan 2016; Liew et al. 2014).

8.4.2.3 Metal Macrocycles-Based ORR Catalysts

Metal complexes are known to be the efficient electrocatalysts for many electrochemical processes. The macrocyclic ligand bounded metal complex such as metal-phthalocyanines and tetramethoxyphenylporphyrins were paid much attention as low cost ORR catalyst in recent years (HaoYu et al. 2007). Figure 8.6 shows a metal

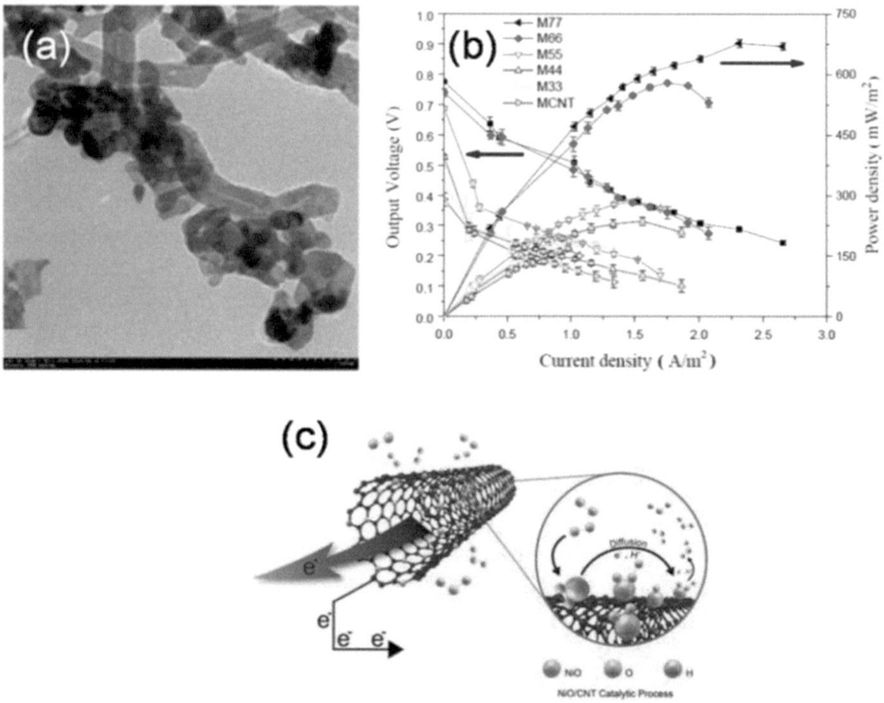

Fig. 8.5 (**a**) TEM image of 44% NiO/CNT composite; (**b**) power density and cell voltage as a function of current density obtained with different NiO catalysts; and (**c**) schematic of ORR pathway in NiO/CNT composite (Adapted from Huang et al. 2015 with permission from Elsevier publisher)

Fig. 8.6 Molecular structure of a metal pthalocyanine (MPc) complex

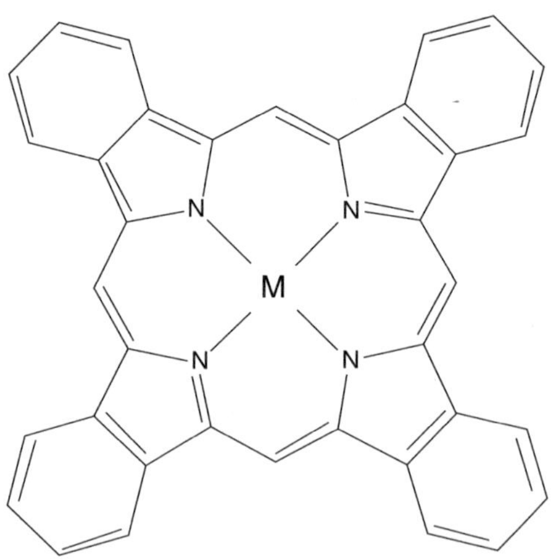

pthalocyanine (MPc) complex with a central metal ion (M) coordinated to four nitrogen (N) atoms of cyclic ligand such as phthalocyanine. Although metal phthalocyanines can be used in neutral, alkaline and acidic environment, they show limited stability below pH 3 due to demetalation of the macrocyclic ring. Thus, metal macrocycles are supported on AC by pyrolysis and this has been found to be effective to improve the stability and catalytic activity in acidic medium (HaoYu et al. 2007; Zhao et al. 2005). Moreover, the pyrolysis process introduces metal and chelating N onto graphitic planes and form different covalently bonded MN_x sites (e.g. $FeN_2C_2^-$, $FeN_3C_2^-$) (Lalande et al. 1996). Furthermore, the heat treatment and pyrolysis of metal macrocycles with catalyst support was reported to improve the catalytic activity by controlling the particle size, surface morphology and dispersibility. In such metal macrocycles, ligand acts as a bridge for the electron transfer from metal centre to the catalyst support and simultaneously protect the central metal ion (Kim et al. 2011). However, a few reports demonstrate that the unpyrolized macrocycles are more active in neutral medium as compared to pyrolized analogue (Yuan et al. 2011a, b). Thus further work is needed in this direction. Apart from metal macrocycles, studies have been conducted on different metal chelated complexes as ORR catalyst in MFC including pyrolized iron-ethylenediamine tetraacetic acid complex/activated carbon (FeEDTA/AC) (Liew et al. 2014). Another study with FeEDTA/AC suggests that the ORR catalytic performance depends on the amount of AC (Xia et al. 2013).

8.4.2.4 Carbon-Based ORR Catalysts

The carbon-based materials have been found to be crucial to many applications. The introduction of nanostructured carbon materials such as AC, CNT, GR, nanohorn, nanofibre, and nanoonion add extra dimension to carbon material's applicability. The excellent electrical conductivity, chemical stability, mechanical strength and high surface area made the nanocarbons a superior choice for variety of electrochemical applications. Moreover, the carbon materials have been found as active catalyst for ORR. AC with a PTFE gas diffusion layer cold pressed to a nickel current collector (1220 mW/m^2) delivers higher power than a Pt modified carbon cloth cathode (1060 $mW\ m^{-2}$) (Zhang et al. 2009). The performance of AC cathode was reported to be improved by addition of carbon black (CB) and the best ORR performance achieved with 10% of AC:CB composition. This enhancement was attributed to the reduction of charge transfer resistance upon CB blending (Zhang et al. 2014). Pristine GR and CNTs lack positive charge density and defects for facile adsorption and cleavage of molecular O_2. Thus, the ORR occurs at very high overpotential on GR and CNTs. Surface functionalization of pristine catalysts with hetero atoms (e.g. N, B and S) creates asymmetric spin densities, atomic charge densities and defects that improve the adsorption and effective cleavage of O=O bond (Kannan 2016). Recently, attention has been given on the development of N doped CNT (N-CNT), carbon fibres, and GR (N-GR) as metal-free ORR catalysts. The bamboo-shaped N-CNT grown (Fig. 8.7) via a chemical vapour

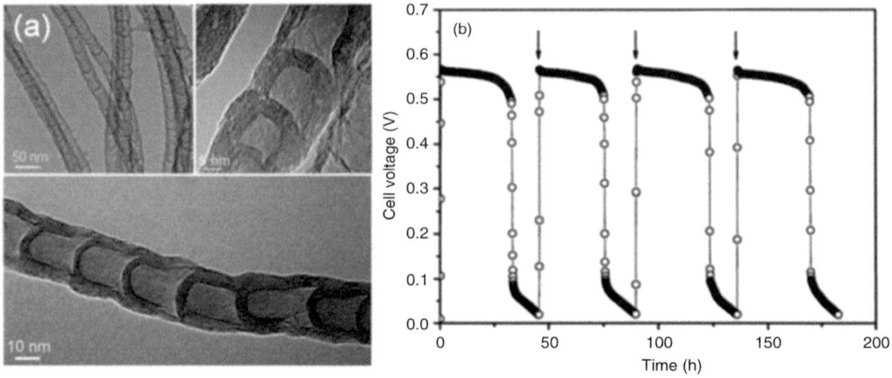

Fig. 8.7 (**a**) TEM images of bamboo-like multi-walled N-CNT structures; and (**b**) Cell voltage vs. time profile of N-CNT based MFC during first four cycles (Adapted from Feng et al. 2011b with permission from RSC)

deposition was found to be a superior ORR catalyst ($4e^-$ pathway in neutral pH, lower internal resistance and higher cathode potential) in MFC than that of Pt/C (Feng et al. 2011b). The DFT calculation reveals that the high electron affinity of N and substantial positive charge density on the adjacent carbon atoms favour easy O_2 adsorption. Moreover, the parallel diatomic adsorption of O_2 on N-CNT weakens the O−O bond which results in a direct reduction to OH^- via a $4e^-$ reduction pathway (Gong et al. 2009).

A number of studies have been conducted on N-GR as ORR catalyst for MFC cathode. The metal-free N-GR modified air cathode sMFC produced $P_{d,max}$ of 1350±15 mW m^{-2} which is comparable to Pt/C (1420±25 mW m^{-2}) (Feng et al. 2011a). Considering the catalytic activity and electronic conductivity of heteroatom doped carbon, several hybrid catalysts were developed for ORR catalysis in MFC. The reported hybrid includes Co/Co_9S_8/NPGC (1156±18 mW m^{-2}) (Pt/C-1023±13 mW m^{-2}) (Li et al. 2016) and Co_3O_4/N-GR (1340±10 mW m^{-2}) (Pt/C-1470±10 mW m^{-2}) (Su et al. 2013). In recent past, Ag/Fe/N/C composite (1791 mW m^{-2}) derived from carbonization of Ag and Fe chelated melamine complex is found to be a superior cathode catalyst than benchmark Pt/C (1192 mW m^{-2}) in sMFC (Dai et al. 2016a, b). The authors claimed that Fe-N sites and different nitrogeneous functionalities were responsible for high ORR activity. More detail study on N-GR and hybrid ORR catalysts is reported in recently published review articles (Kannan 2016; Liew et al. 2014).

8.4.2.5 Metal Carbides as ORR Catalysts

Metal carbides (MCs) are emerging as a promising class of material for fuel cell application due to their high chemical and mechanical stability, and resistance to corrosion during the electrochemical reactions. The MCs have several advantages over the parent metal in terms of catalytic activity, selectivity and resistance to poisoning (Ham and Lee 2009). However, the MCs can form different chemical composition of M_xC_y with different reactivity towards electrocatalysis. Thus, selection of appropriate composition is important to enhance the ORR activity (Ham and Lee 2009). Since, the discovery of metal-like behaviour of tungsten carbide (WC) by Levy and Boundart in 1973, it has been widely used for fuel cell applications (Levy and Boudart 1973). Gong et al. (2013) reported Ag-WC/C hybrid ORR catalyst for cathode modification in a dMFC. The polarization study and K-L plot analysis suggested a quasi-4e$^-$ ($n = 3.86$) ORR pathway on Ag-WC/C whereas parent WC and WC/C offers the n value of 2.23 and 2.8, respectively. The synergy between Ag and WC is responsible for improved ORR activity of Ag-WC/C ($P_{d,max}$ of 20.62 W m^{-3}) in comparison to Pt/C (21.4 W m^{-3}). The thermodynamically favourable O_2 adsorption followed by O–O cleavage on WC surface reduces the activation barrier and high positive reduction potential of Ag–O facilitates the reduction process. Wen et al. (2012) reported N-Fe/Fe_3C@C catalyst with 'n' value of 3.98 in a neutral electrolyte which is comparable to Pt/C ($n = 4.0$). The performance with N-Fe/Fe_3C@C (4.2 W/m^3) as cathode catalyst in dMFC outperforms the state-of-the-art Pt/C (3.98 W/m^3).

8.4.2.6 Electronically Conductive Polymer Catalysts

In recent time, electronically conducting polymers (ECPs) are investigated for diverse applications. The high electronic conductivity and environmental stability make this class of materials unique over many others. Flexibility and large scale electrode fabrication with conducting polymer offer huge opportunity to build practically applicable devices. Among different ECPs, PANI, polypyrrole (PPy), poly (3-methyl)thiophene (PMeT) and poly(3,4-ethylenedioxythiophene) (PEDOT) were found to be more active ORR catalyst. However, the transformation of PANI from emeralidine to leucoemeraldine form during ORR reduces its catalytic activity. Additionally, the long-term electrochemical stability of ECPs is poor. So, these polymers were generally used as ORR catalysts along with another material such as AC, CNTs, GR, metal complexes and MOs. The ECPs weaken the molecular O–O bond of chemiadsorbed O_2 at polymer surface which lowers the activation barrier for ORR. Figure 8.8 displays the bridge model of O_2 adsorption on PANI and PPy (Khomenko et al. 2005). Considering this unique adsorption profile of PPy and PANI, CB was integrated to them forming hybrids for enhanced performance than AC. Similarly, GR and CNT-based ECPs hybrids were also tested as ORR catalyst in MFC. The pthalocyanine, anthraquinone and prussian blue based EPC hybrid ORR catalysts were reviewed (Yuan et al. 2011a; Li et al. 2014). Recently, many MO/PANI or PPy composite catalysts were also used as cathode catalyst in MFC. The reported MOs-based composites are $MnCo_2O_4$/PPy (6.11 W m^{-3}) (Khilari et al. 2014), V_2O_5/PANI (79.26 mW m^{-2}) (Ghoreishi et al. 2014), MnO_2/PANI (37.6 mW m^{-2}) (Ansari et al. 2016), and $MnFe_2O_4$/PANI (6.49 W m^{-3}) (Khilari et al. 2015). In these cases, enhancement in ORR activity is recorded by incorporating ECP.

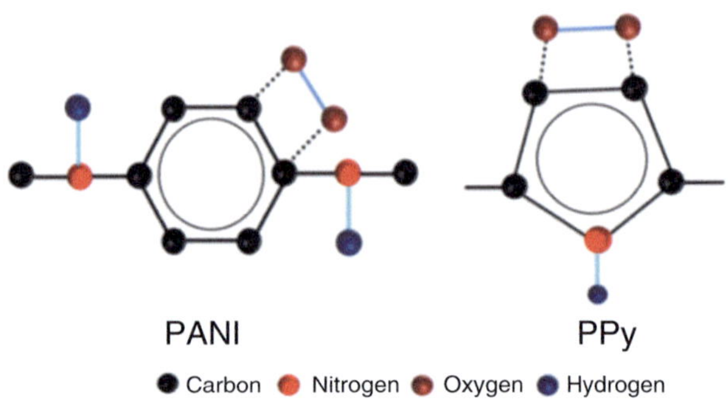

Fig. 8.8 Bridge model of O_2 adsorption on PANI and PPy

8.4.2.7 Biocatalysts for Cathodic Reduction

The abiotic cathode catalysts have several limitations for practical applicability of MFC which includes high cost, complex fabrication protocol, and most of the catalysts are environmentally unfriendly. Thus significant efforts have been made to develop biotic cathode to overcome the limitations of abiotic counterpart. Biocathode consists of a biofilm that can be used to catalyze the cathodic reduction making the MFC a totally bioelectrochemical system. The advantages of biocathode include environmental benignity, self-sustainable catalyst regeneration, and low operational cost. Generally, two types of biocathodes are reported in literature. One type catalyzes the ORR in aerobic medium. Moreover, the ORR process at biocathode is mediated by both Mn and Fe (He and Angenent 2006). Another type of biocathodes catalyzes cathodic reduction of sulphate and nitrate in an anaerobic environment. The performance of biocathode primarily depends on the electrochemical activity of bacteria anchored to the cathode (Zhang et al. 2012). Pure culture such as *Sphingobacterium* sp., *Acinetobacter* sp. and *Letothrix discophora* sp., were utilized in the biocathode as cathode catalyst (Liew et al. 2014; He and Angenent 2006).

Previous reports suggest that the $P_{d,max}$ achieved with an isolated *Sphingobacterium* sp. (49 mW m^{-2}) is two times higher than that of isolated *Acinetobacter* sp. (24 mW m^{-2}) and thrice of bare cathode (15 mW m^{-2}) (Rabaey et al. 2008). In addition to pure culture, reports are available on mixed culture-based biocathodes (Liew et al. 2014; Rabaey et al. 2008). The efficacy of a biocathode depends on the number of active sites which is directly related to the bacterial population on the cathode surface. Thus, an efficient cathode surface must possess a populated biofilm for dense active sites. In this regard, different strategies have been adopted to create suitable cathode structure and modification of cathode surface with functional materials such as GR, CNT etc. (Zhang et al. 2013). The progress of biocathode application in MFC is reviewed in literature (Liew et al. 2014; He and Angenent 2006).

8.5 Conclusions

Last few years witnessed significant progress and several breakthrough advancements in the MFC application. In particular, efforts have been made to overcome the challenges for cost-effective bioelectricity generation using non-noble metal catalysts in the MFC. Although, till date efficiency of a few cathode catalysts approaches close to Pt/C in MFC, their commercial uses remain to be seen. We have discussed here on the improvement of MFC performance while focusing the development on different types of cathode catalysts. Further improvement on catalytic activity is required and theoretical investigations such as DFT study and molecular dynamics can assist to design superior functional cathode catalyst. The hybrid catalysts show promising ORR catalytic activity and have potential to compete the state-of-the-art Pt/C. In order to increase the efficiency of waste to bioelectricity generation using MFC, low-cost functional supports for suitable ORR catalyst such as metal oxides and metal complex or metal is also important. In addition, most of the existing catalysts were studied in a lab scale. Thus, the necessity of cost-effective cathode catalysts for MFC scale up is to be resolved. Scale up of MFC from lab scale to industrial scale can be attained with proper synchronization of engineering, scientific skill and development of effective catalyst materials.

References

Ansari, S. A., Parveen, N., Han, T. H., Ansari, M. O., & Cho, M. H. (2016). Fibrous polyaniline@ manganese oxide nanocomposites as supercapacitor electrode materials and cathode catalysts for improved power production in microbial fuel cells. *Physical Chemistry Chemical Physics, 18*, 9053–9060.

Bing, Y., Liu, H., Zhang, L., Ghosh, D., & Zhang, J. (2010). Nanostructured Pt-alloy electrocatalysts for PEM fuel cell oxygen reduction reaction. *Chemical Society Reviews, 39*, 2184–2202.

Chen, Z., Higgins, D., Yu, A., Zhang, L., & Zhang, J. (2011). A review on non-precious metal electrocatalysts for PEM fuel cells. *Energy & Environmental Science, 4*, 3167–3192.

Cheng, S., & Wu, J. (2013). Air-cathode preparation with activated carbon as catalyst, PTFE as binder and nickel foam as current collector for microbial fuel cells. *Bioelectrochemistry, 92*, 22–26.

Cheng, F., Shen, J., Peng, B., Pan, Y., Tao, Z., & Chen, J. (2011). Rapid room-temperature synthesis of nanocrystalline spinels as oxygen reduction and evolution electrocatalysts. *Nature Chemistry, 3*, 79–84.

Daems, N., Sheng, X., Vankelecom, I. F. J., & Pescarmona, P. P. (2014). Metal-free doped carbon materials as electrocatalysts for the oxygen reduction reaction. *Journal of Materials Chemistry A, 2*, 4085–4110.

Dai, Y., Chan, Y., Jiang, B., Wang, L., Zou, J., Pan, K., & Fu, H. (2016a). Bifunctional Ag/Fe/N/C Catalysts for Enhancing Oxygen Reduction via Cathodic Biofilm Inhibition in Microbial Fuel Cells. *ACS Applied Materials & Interfaces, 8*, 6992–7002.

Dai, H., Yang, H., Liu, X., Zhao, Y., & Liang, Z. (2016b). Performance of sodium bromate as cathodic electron acceptor in microbial fuel cell. *Bioresource Technology, 202*, 220–225.

Feng, L., Chen, Y., & Chen, L. (2011a). Easy-to-operate and low-temperature synthesis of gram-scale nitrogen-doped graphene and its application as cathode catalyst in microbial fuel cells. *ACS Nano, 5*, 9611–9618.
Feng, L., Yan, Y., Chen, Y., & Wang, L. (2011b). Nitrogen-doped carbon nanotubes as efficient and durable metal-free cathodic catalysts for oxygen reduction in microbial fuel cells. *Energy & Environmental Science, 4*, 1892–1899.
Forrestal, C., Huang, Z., & Ren, Z. J. (2014). Percarbonate as a naturally buffering catholyte for microbial fuel cells. *Bioresource Technology, 172*, 429–432.
Ge, B., Li, K., Fu, Z., Pu, L., & Zhang, X. (2015). The addition of ortho-hexagon nano spinel Co_3O_4 to improve the performance of activated carbon air cathode microbial fuel cell. *Bioresource Technology, 195*, 180–187.
Ge, B., Li, K., Fu, Z., Pu, L., Zhang, X., Liu, Z., & Huang, K. (2016). The performance of nano urchin-like $NiCo_2O_4$ modified activated carbon as air cathode for microbial fuel cell. *Journal of Power Sources, 303*, 325–332.
Gewirth, A. A., & Thorum, M. S. (2010). Electroreduction of dioxygen for fuel-cell applications: Materials and challenges. *Inorganic Chemistry, 49*, 3557–3566.
Ghoreishi, K. B., Ghasemi, M., Rahimnejad, M., Yarmo, M. A., Daud, W. R. W., Asim, N., & Ismail, M. (2014). Development and application of vanadium oxide/polyaniline composite as a novel cathode catalyst in microbial fuel cell. *International Journal of Energy Research, 38*, 70–77.
Gnanakumar, G., Awan, Z., Nahm, K. S., & Xavier, J. S. (2014). Nanotubular MnO_2/graphene oxide composites for the application of open air-breathing cathode microbial fuel cells. *Biosensors and Bioelectronics, 53*, 528–534.
Gong, K., Du, F., Xia, Z., Durstock, M., & Dai, L. (2009). Nitrogen-doped carbon nanotube arrays with high electrocatalytic activity for oxygen reduction. *Science, 323*, 760–764.
Gong, X.-B., You, S.-J., Wang, X.-H., Gan, Y., Zhang, R.-N., & Ren, N.-Q. (2013). Silver–tungsten carbide nanohybrid for efficient electrocatalysis of oxygen reduction reaction in microbial fuel cell. *Journal of Power Sources, 225*, 330–337.
Guerrero-Rangel, N., Rodríguez-de la Garza, J. A., Garza-García, Y., Rios-Gonzalez, L. J., SosaSantillán, G. J., De la Garza-Rodríguez, I. M., Martínez-Amador, S. Y., Rodriguez-Garza, M. M., & Rodríguez-Martínez, J. (2010). Comparative study of three cathodic electron acceptors on the performance of mediatorless microbial fuel cell. *International Journal of Electrical and Power Engineering, 4*, 27–31.
Ham, D. J., & Lee, J. S. (2009). Transition metal carbides and nitrides as electrode materials for low temperature fuel cells. *Energies, 2*, 873–899.
HaoYu, E., Cheng, S., Scott, K., & Logan, B. (2007). Microbial fuel cell performance with non-Pt cathode catalysts. *Journal of Power Sources, 171*, 275–281.
He, Z., & Angenent, L. T. (2006). Application of bacterial biocathodes in microbial fuel cells. *Electroanalysis, 18*, 2009–2015.
Hou, Y., Yuan, H., Wen, Z., Cui, S., Guo, X., He, Z., & Chen, J. (2016). Nitrogen-doped graphene/CoNi alloy encased within bamboo-like carbon nanotube hybrids as cathode catalysts in microbial fuel cells. *Journal of Power Sources, 307*, 561–568.
Huang, J., Zhu, N., Yang, T., Zhang, T., Wu, P., & Dang, Z. (2015). Nickel oxide and carbon nanotube composite (NiO/CNT) as a novel cathode non-precious metal catalyst in microbial fuel cells. *Biosensors and Bioelectronics, 72*, 332–339.
Jadhav, D. A., Ghadge, A. N., Mondal, D., & Ghangrekar, M. M. (2014). Comparison of oxygen and hypochlorite as cathodic electron acceptor in microbial fuel cells. *Bioresource Technology, 154*, 330–335.
Kannan, M. V. (2016). Current status, key challenges and its solutions in the design and development of graphene based ORR catalysts for the microbial fuel cell applications. *Biosensors and Bioelectronics, 77*, 1208–1220.

Katsounaros, I., Cherevko, S., Zeradjanin, A. R., & Mayrhofer, K. J. (2014). Oxygen electrochemistry as a cornerstone for sustainable energy conversion. *Angewandte Chemie International Edition, 53*, 102–121.

Khilari, S., Pandit, S., Ghangrekar, M. M., Das, D., & Pradhan, D. (2013). Graphene supported α-MnO_2 nanotubes as a cathode catalyst for improved power generation and wastewater treatment in single-chambered microbial fuel cells. *RSC Advances, 3*, 7902–7911.

Khilari, S., Pandit, S., Das, D., & Pradhan, D. (2014). Manganese cobaltite/polypyrrole nanocomposite-based air-cathode for sustainable power generation in the single-chambered microbial fuel cells. *Biosensors and Bioelectronics, 54*, 534–540.

Khilari, S., Pandit, S., Varanasi, J. L., Das, D., & Pradhan, D. (2015). Bifunctional Manganese Ferrite/Polyaniline Hybrid as Electrode Material for Enhanced Energy Recovery in Microbial Fuel Cell. *ACS Applied Materials & Interfaces, 7*, 20657–20666.

Khomenko, V. G., Barsukov, V. Z., & Katashinskii, A. S. (2005). The catalytic activity of conducting polymers toward oxygen reduction. *Electrochimica Acta, 50*, 1675–1683.

Kim, J. R., Kim, J. Y., Han, S. B., Park, K. W., Saratale, G. D., & Oh, S. E. (2011). Application of Co-naphthalocyanine (CoNPc) as alternative cathode catalyst and support structure for microbial fuel cells. *Bioresource Technology, 102*, 342–347.

Lalande, G., Faubert, G., Cote, R., Guay, D., Dodelet, J. P., Weng, L. T., & Bertrand, P. (1996). Catalytic activity and stability of heat-treated iron phthalocyanines for the electroreduction of oxygen in polymer electrolyte fuel cells. *Journal of Power Sources, 61*, 227–237.

Levy, R. B., & Boudart, M. (1973). Platinum-like behavior of tungsten carbide in surface catalysis. *Science, 181*, 547–549.

Li, J., Fu, Q., Liao, Q., Zhu, X., Ye, D. D., & Tian, X. (2009). Persulfate: A self-activated cathodic electron acceptor for microbial fuel cells. *Journal of Power Sources, 194*, 269–274.

Li, X., Hu, B., Suib, S., Lei, Y., & Li, B. (2010). Manganese dioxide as a new cathode catalyst in microbial fuel cells. *Journal of Power Sources, 195*, 2586–2591.

Li, Y., Liu, L., Liu, J., Yang, F., & Ren, N. (2014). PPy/AQS (9, 10-anthraquinone-2-sulfonic acid) and PPy/ARS (Alizarin Red's) modified stainless steel mesh as cathode membrane in an integrated MBR/MFC system. *Desalination, 349*, 94–101.

Li, R., Dai, Y., Chen, B., Zou, J., Jiang, B., & Fu, H. (2016). Nitrogen-doped Co/Co_9S_8/partly-graphitized carbon as durable catalysts for oxygen reduction in microbial fuel cells. *Journal of Power Sources, 307*, 1–10.

Liew, K. B., Daud, W. R. W., Ghasemi, M., Leong, J. X., Lim, S. S., & Ismail, M. (2014). Non-Pt catalyst as oxygen reduction reaction in microbial fuel cells: A review. *International Journal of Hydrogen Energy, 39*, 4870–4883.

Liu, H., & Logan, B. E. (2004). Electricity generation using an air-cathode single chamber microbial fuel cell in the presence and absence of a proton exchange membrane. *Environmrntal Science and Technology, 38*, 4040–4046.

Liu, Y., Liu, H., Wang, C., Hou, S.-X., & Yang, N. (2013). Sustainable energy recovery in wastewater treatment by microbial fuel cells: Stable power generation with nitrogen-doped graphene cathode. *Environmental Science & Technology, 47*, 13889–13895.

Liu, Z., Li, K., Zhang, X., Ge, B., & Pu, L. (2015). Influence of different morphology of three-dimensional Cu_xO with mixed facets modified air–cathodes on microbial fuel cell. *Bioresource Technology, 195*, 154–161.

Lu, M., & Li, S. F. Y. (2012). Cathode reactions and applications in microbial fuel cells: A review. *Critical Reviews in Environmental Science and Technology, 42*, 2504–2525.

Ma, M., Dai, Y., Zou, J. L., Wang, L., Pan, K., & Fu, H. G. (2014). Synthesis of iron oxide/partly graphitized carbon composites as a high-efficiency and low-cost cathode catalyst for microbial fuel cells. *ACS Applied Materials & Interfaces, 6*, 13438–13447.

Martin, E., Tartakovsky, B., & Savadogo, O. (2011). Cathode materials evaluation in microbial fuel cells: A comparison of carbon, Mn_2O_3, Fe_2O_3 and platinum materials. *Electrochimica Acta, 58*, 58–66.

Nie, Y., Li, L., & Wei, Z. (2015). Recent advancements in Pt and Pt-free catalysts for oxygen reduction reaction. *Chemical Society Reviews, 44*, 2168–2201.

Nørskov, J. K., Rossmeisl, J., Logadottir, A., Lindqvist, L. R. K. J., Kitchin, J. R., Bligaard, T., & Jonsson, H. (2004). Origin of the overpotential for oxygen reduction at a fuel-cell cathode. *Journal of Physical Chemistry B, 108*, 17886–17892.

Nørskov, J. K., Bligaard, T., Rossmeisl, J., & Christensen, C. H. (2009). Towards the computational design of solid catalysts. *Nature Chemistry, 1*, 37–46.

Pandit, S., Sengupta, A., Kale, S., & Das, D. (2011). Performance of electron acceptors in catholyte of a two-chambered microbial fuel cell using anion exchange membrane. *Bioresource Technology, 102*, 2736–2744.

Peng, X., Yu, H., Wang, X., Gao, N., Geng, L., & Ai, L. (2013). Enhanced anode performance of microbial fuel cells by adding nanosemiconductor goethite. *Journal of Power Sources, 223*, 94–99.

Rabaey, K., Read, S. T., Clauwaert, P., Freguia, S., Bond, P. L., Blackall, L. L., & Keller, J. (2008). Cathodic oxygen reduction catalyzed by bacteria in microbial fuel cells. *The ISME Journal, 2*, 519–527.

Rismani-Yazdi, H., Carver, S. M., Christy, A. D., & Tuovinen, O. H. (2008). Cathodic limitations in microbial fuel cells: An overview. *Journal of Power Sources, 180*, 683–694.

Sánchez-Sánchez, C. M., & Bard, A. J. (2009). Hydrogen peroxide production in the oxygen reduction reaction at different electrocatalysts as quantified by scanning electrochemical microscopy. *Analytical Chemistry, 81*, 8094–8100.

Shao, M. H., Liu, P., & Adzic, R. R. (2006). Superoxide anion is the intermediate in the oxygen reduction reaction on platinum electrodes. *Journal of the American Chemical Society, 128*, 7408–7409.

Singh, R. N., Awasthi, R., & Sharma, C. S. (2014a). Review: An overview of recent development of platinum-based cathode materials for direct methanol fuel cells. *Int. J. Electrochem. Sci, 9*, 5607–5639.

Singh, K. P., Song, M. Y., & Yu, J.-S. (2014b). Iodine-treated heteroatom-doped carbon: Conductivity driven electrocatalytic activity. *Journal of Materials Chemistry A, 2*, 18115–18124.

Stamenkovic, V., Mun, B. S., Mayrhofer, K. J., Ross, P. N., Markovic, N. M., Rossmeisl, J., Greeley, J., & Nørskov, J. K. (2006). Changing the activity of electrocatalysts for oxygen reduction by tuning the surface electronic structure. *Angewandte Chemie International Edition, 118*, 2963–2967.

Stephens, I. E. L., Bondarenko, A. S., Grønbjerg, U., Rossmeisl, J., & Chorkendorff, I. (2012). Understanding the electrocatalysis of oxygen reduction on platinum and its alloys. *Energy & Environmental Science, 5*, 6744–6762.

Su, Y., Zhu, Y., Yang, X., Shen, J., Lu, J., Zhang, X., Chen, J., & Li, C. (2013). A highly efficient catalyst toward oxygen reduction reaction in neutral media for microbial fuel cells. *Industrial & Engineering Chemistry Research, 52*, 6076–6082.

Wang, Z., Yan, Z., Wang, M., & Zhao, J. (2015). Glycerol stabilized $NaBH_4$ reduction for preparation of carbon supported Pt-Ni alloy nanoparticles used as oxygen-reduction electrocatalysts for microbial fuel cells. *International Journal of Electrochemical Science, 10*, 1953–1965.

Wei, L., Han, H., & Shen, J. (2012). Effects of cathodic electron acceptors and potassium ferricyanide concentrations on the performance of microbial fuel cell. *International Journal of Hydrogen Energy, 37*, 12980–12986.

Wen, Z., Ci, S., Zhang, F., Feng, X., Cui, S., Mao, S., Luo, S., He, Z., & Chen, J. (2012). Nitrogen-enriched core-shell structured Fe/Fe_3C-C nanorods as advanced electrocatalysts for oxygen reduction reaction. *Advanced Materials, 24*, 1399–1404.

Xia, X., Zhang, F., Zhang, X., Liang, P., Huang, X., & Logan, B. E. (2013). Use of Pyrolyzed Iron Ethylenediaminetetraacetic acid modified activated carbon as Air–Cathode catalyst in microbial fuel cells. *ACS Applied Materials & Interfaces, 5*, 7862–7866.

Xia, W., Mahmood, A., Liang, Z., Zou, R., & Guo, S. (2016). Earth-abundant nanomaterials for oxygen reduction. *Angewandte Chemie International Edition, 55*, 2650–2676.

Xiong, L., Huang, Y.-X., Liu, X.-W., Sheng, G.-P., Li, W.-W., & Yu, H.-Q. (2013). Three-dimensional bimetallic Pd–Cu nanodendrites with superior electrochemical performance for oxygen reduction reaction. *Electrochimica Acta, 89*, 24–28.

Yan, Z., Wang, M., Huang, B., Liu, R., & Zhao, J. (2013). Graphene supported Pt-Co alloy nanoparticles as cathode catalyst for microbial fuel cells. *International Journal of Electrochemical Science, 8*, 149–158.

Yan, Z., Wang, M., Liu, J., Liu, R., & Zhao, J. (2014). Glycerol-stabilized $NaBH_4$ reduction at room-temperature for the synthesis of a carbon-supported Pt_xFe alloy with superior oxygen reduction activity for a microbial fuel cell. *Electrochimica Acta, 141*, 331–339.

Yuan, Y., Ahmed, J., & Kim, S. (2011a). Polyaniline/carbon black composite-supported iron phthalocyanine as an oxygen reduction catalyst for microbial fuel cells. *Journal of Power Sources, 196*, 1103–1106.

Yuan, Y., Zhao, B., Jeon, Y., Zhong, S., Zhou, S., & Kim, S. (2011b). Iron phthalocyanine supported on amino-functionalized multi-walled carbon nanotube as an alternative cathodic oxygen catalyst in microbial fuel cells. *Bioresource Technology, 102*, 5849–5854.

Zhang, F., Cheng, S., Pant, D., Bogaert, G. V., & Logan, B. E. (2009). Power generation using an activated carbon and metal mesh cathode in a microbial fuel cell. *Electrochemistry Communications, 11*, 2177–2179.

Zhang, G., Zhao, Q., Jiao, Y., Wang, K., Lee, D. J., & Ren, N. (2012). Biocathode microbial fuel cell for efficient electricity recovery from dairy manure. *Biosensors and Bioelectronics, 31*, 537–543.

Zhang, Y., Sun, J., Hu, Y., Li, S., & Xu, Q. (2013). Carbon nanotube-coated stainless steel mesh for enhanced oxygen reduction in biocathode microbial fuel cells. *Journal of Power Sources, 239*, 169–174.

Zhang, X., Xia, X., Ivanov, I., Huang, X., & Logan, B. E. (2014). Enhanced activated carbon cathode performance for microbial fuel cell by blending carbon black. *Environmental Science & Technology, 48*, 2075–2081.

Zhao, F., Harnisch, F., Schröder, U., Scholz, F., Bogdanoff, P., & Herrmann, I. (2005). Application of pyrolysed iron (II) phthalocyanine and CoTMPP based oxygen reduction catalysts as cathode materials in microbial fuel cells. *Electrochemistry Communications, 7*, 1405–1410.

Zhou, X., Qiao, J., Yang, L., & Zhang, J. (2014). A review of graphene-based nanostructural materials for both catalyst supports and metal-free catalysts in PEM fuel cell oxygen reduction reactions. *Advanced Energy Materials, 4*, 1301523–1301547.

Chapter 9
Role of Biocathodes in Bioelectrochemical Systems

V. Prakasam, S.G.F. Bagh, S. Ray, B. Fifield, L.A. Porter, and J.A. Lalman

9.1 Introduction

Environmental damage, depleting fossil fuels and energy security are major factors driving intensive research efforts to develop carbon neutral or carbon negative technologies which can be used to produce electricity and chemicals. Technologies under development to achieve this goal include those based on bioelectrochemical, biological, thermal and chemical processes. Evolving technologies employing biological as well as electrochemical principles are grouped in the bioelectrochemical systems (BESs) category. The main focus of this chapter is on biocathodes used in BESs.

A BES such as a microbial fuel cell (MFC) is a two-chamber system consisting of anode and cathode chambers separated by a proton exchange membrane (PEM). This system is configured with an anode chamber containing electrochemically-active microorganisms, while the cathode is abiotic. Microorganisms attached to the anode oxidize electron donors such as glucose to produce electrons which travel to the cathode. Protons are transported through the solution or across a membrane separator. Depending on the application, BESs are configured with abiotic and biotic cathodes (biocathode).

V. Prakasam • S.G.F. Bagh • J.A. Lalman (✉)
Department of Civil and Environmental Engineering, University of Windsor, Windsor, ON, Canada, N9B 3P4
e-mail: lalman@uwindsor.ca

S. Ray
Department of Chemical Engineering, National Institute of Technology, Agartala, Jirania, Tripura, India, 79904

B. Fifield • L.A. Porter
Department of Biological Sciences, University of Windsor, Windsor, ON, Canada, N9B 3P4

D. Das (ed.), *Microbial Fuel Cell*,
https://doi.org/10.1007/978-3-319-66793-5_9

Biocathodes can operate under aerobic, anoxic and anaerobic conditions. The standard reduction potential for O_2 to H_2O is 0.818 V (Thauer et al. 1977). The O_2/H_2O couple is more positive than the standard reduction potentials for sulphate/bisulphide (−0.217 V), nitrate/ammonia (0.360 V) and nitrate/nitrite (0.430 V). In comparison, only nitrate/nitrogen reduction couple (0.760 V) is close to the O_2/H_2O couple. Electrons at the cathode reduce O_2 to H_2O or H_2O_2 in MFCs, or water to H_2 in microbial electrochemical cells (MECs).

Although the anodic oxidation reactions are similar, the different cathode reactions result in MFCs producing electrical energy while in MECs, additional energy is required to drive the overall reaction (Sasaki et al. 2011; Tartakovsky et al. 2009). MFC technology can be classified into a wide array of technologies designated as MXCs, where x is desalination, electrolysis and solar. Microbial desalination cells (MDCs) are used in desalinating brackish water (Kim and Logan 2013) while MECs are used for production of hydrogen (Tartakovsky et al. 2009) and microbial solar cells (MSCs) are used for carbon dioxide sequestration (Pisciotta et al. 2012).

9.2 BES Technology Utilizing Biocathodes

Abiotic cathodes constructed from platinum (Jeremiasse et al. 2010; Rozendal et al. 2008) are commonly used in MECs. However, limiting factors due to cost and reduced catalytic efficiency caused by sulphide poisoning sulphide have forced researchers to seek alternative options. BESs configured with biocathodes catalyze electron transfer from cathodes to electro-positive terminal electron acceptors such as oxygen or nitrate (Clauwaert et al. 2007a, b). Clauwaert et al. (2007a, b) successfully employed BESs designed with bioanodes and biocathodes utilizing inexpensive materials such as carbon or graphite rather than expensive metals such as palladium. Additionally, biocathodes have also been configured into photosynthetic MSCs to reduce carbon dioxide (Cao et al. 2009).

MDCs configured with biocathodes have been reported in several studies (Meng et al. 2014; Wen et al. 2012). According to Meng et al. (2014), a biocathode constructed from a graphite brush embedded with graphite granules was developed for synergistic desalination, electricity generation and biosolids stabilization in an MDC. These researchers reported a maximum power output reaching 3.2 W m^{-3} and an open circuit voltage (OCV) of 1.12 V. Wen et al. (2012) employed a biocathode configured MDC and reported a maximum voltage of approximately 610 mV.

9.3 Electron Acceptors and Microorganisms

Microorganisms accepting electrons directly or indirectly from a cathode are referred to as electrotrophs (Lovley 2011; Logan 2009). BESs are operable using a variety of electron acceptors such as oxygen, nitrate, sulphate, iron, manganese, arsenate, fumarate, or carbon dioxide (Virdis et al. 2010; Freguia et al. 2010; Rabaey et al. 2008; Clauwaert et al. 2007a).

Aerobic biocathodes variable performance and wide range of bacteria involved with electron transfer are major issues under investigation by many researchers. Mixed-community aerobic biocathode biofilms are a suitable substitute for chemical catalysts at the cathode because of cost, robustness and sustainability. A wide range of bacterial species belonging to the *Alphaproteobacteria* (Du et al. 2014; Wang et al. 2013a; Zhang et al. 2011a; Clauwaert et al. 2007b), *Betaproteobacteria* (Du et al. 2014; Wang et al. 2013b; Sun et al. 2012; Zhang et al. 2011b, 2012a; Chen et al. 2008; Rabaey et al. 2008), *Gammproteobacteria* (Wang et al. 2013b, 2015; Du et al. 2014; Strycharz-Glaven et al. 2013; Chung et al. 2011; Zhang et al. 2011a; Chen et al. 2010; Schamphelaire et al. 2010; Rabaey et al. 2008; Clauwaert et al. 2007b; Reimers et al. 2006), *Bacteroidetes* (Wang et al. 2013a; Sun et al. 2012; Xia et al. 2012; Chen et al. 2008, 2010) and other less well-known microorganisms (Rimboud et al. 2015; Blanchet et al. 2014; Du et al. 2014; Wang et al. 2013a) have been identified as dominant in mixed community aerobic biocathodes. Carbajosa et al. (2010) reported acidophilic *Acidithiobacillus ferrooxidans* is involved with oxygen reduction in biocathodes. Other microorganisms such as *Chlorella vulgaris* have been utilized under aerobic conditions (Campo et al. 2013; Wu et al. 2013).

Under anoxic conditions, nitrate and sulphate have been employed as electron acceptors (Nguyen et al. 2015; Clauwaert et al. 2009). Denitrification and sulphate reduction in BESs have been reported in many studies. Cai et al. (2014) reported sulphate-reducing bacteria and homoacetogens were detected in MEC biocathodes. Autotrophic hydrogen and sulphate consumer *Desulfovibrio* and planktonic microbial consortia was detected on a biocathode configured BES (Pozo et al. 2015). Nguyen et al. (2015) and Clauwaert et al. (2009) reported denitrification using microorganisms attached to a cathode. Nitrifying bacteria detected on anoxic biocathodes include *Nitrosomonas* sp., *Nitrospira* sp. and *Nitrobacter* sp. (Du et al. 2014). Chen et al. (2010) reported using nitrate as electron acceptor and the major microbial population detected included *Alphaproteobacteria, Betaproteobacteria, Gammaproteo-bacteria* and *Flavobacteria*.

Jeremiasse et al. (2010) reported operating an MEC with the anode and cathode catalyzing reactions mediated by microorganisms. These authors described the formation of calcium phosphate and production of hydrogen plus methane. In an MEC configured with a biocathode, microorganisms utilize the electrode as an electron source to catalyze combining electrons and protons to produce hydrogen. In other systems, microorganisms were able to produce methane from CO_2 plus an electron donor (Marshall et al. 2012, 2013; Pisciotta et al. 2012). Hydrogen and formate have been detected in mixed microbial communities autotrophic biocathode configured MES reactors (Marshall et al. 2012, 2013; Zaybak et al. 2013; Pisciotta et al. 2012). Zaybak et al. (2013) used a BES configured with a biocathode containing mixed microbial autotrophic community to produce butanol, ethanol, hydrogen, acetate, propionate and butyrate.

Sulphate-reducing bacteria (SRB) as well as homoacetogens have been identified in MEC biocathodes (Kim et al. 2015). Sulphate reduction has been reported using biocathodes with hydrogen as the electron donor (Coma et al. 2013; Yu et al. 2008).

Photo-biocathodes are a promising option for combining photosynthesis into BES systems (Xiao et al. 2015; Xiao and He 2014). BES configured with photo-biocathodes has the potential of attaining the concept of a carbon neutral system. Photo-biocathodes are useful for chemical production, carbon fixation (El-Mekawy et al. 2014) and biomass production. Adding photosynthetic organisms is a means for electron acceptors cathode as well as dissolved oxygen for the oxygen reducing reaction (Campo et al. 2013; Gajda et al. 2013; Berk and Canfield 1964). Using phototrophic biocatalysts in the cathode half-cell reaction is expected to permit meeting the oxygen level requirements for the oxygen reducing reaction (Berk and Canfield 1964) as well as producing biomass which utilized as a fuel for the MFC anode. McCormick et al. (2011) and Xiao et al. (2015) reported cyanobacteria *Leptolyngbya* and the green alga *Acutodesmus* as dominant photoautotrophs in cathode suspension and biofilms.

9.4 Biocathode Materials

Electrode material properties are critical elements for the efficient and economical feasibility for operating MFCs. Suitable materials properties should consider high conductivity, low corrodibility, high specific surface area, suitable for microbial growth and low cost (Wei et al. 2011a). Various materials utilized include graphite felt (Zhang et al. 2012b), graphite granules (Zhang et al. 2011a), polyaniline (Ren et al. 2013; Li et al. 2012), graphite fibre brush (Li et al. 2012; Zhang et al. 2011a, b), poly(aniline-co-o-aminophenol) (Li et al. 2012), graphite plate (Behera et al. 2010; You et al. 2009), poly(aniline-co-2, 4-diaminophenol) (Li et al. 2012), poly (aniline-1,8-diaminonaphthalene (Li et al. 2012), carbon felt (Li et al. 2012; Schamphelaire et al. 2010), carbon paper (Zhang et al. 2012b), carbon fibre brush (Tursun et al. 2016), granular carbon (semi coke) (Wei et al. 2011b) granular activated carbon (Tursun et al. 2016; Sun et al. 2012; Wei et al. 2011b), carbon felt cubes (Wei et al. 2011b), carbon nano tubes (Zhang et al. 2013), graphene nano sheets (Ren et al. 2013) and stainless steel mesh (Zhang et al. 2012b; Zhang et al. 2013).

Tursun et al. (2016) compared graphite granules, activated carbon granules and activated carbon powder and concluded that activated carbon granules show improved power generation, higher chemical oxygen demand removal and coulombic efficiency. Zhang et al. (2012b) evaluated current density, power density and polarization and reported graphite felt was the most effective biocathode when compared to carbon paper and stainless steel mesh. In other studies, Zhang et al. (2011b) compared the performance of graphite brushes, graphite granules and graphite brushes plus graphite granules and concluded the startup time was less for the graphite brushes plus graphite granules cathode configuration when compared to graphite brushes. They also reported a maximum power density achieved with a higher coulombic efficiency for the graphite brushes plus graphite granules cathode.

9.4.1 General Material Characteristics

Electrode characteristics such as surface roughness, surface area, porosity, conductivity, and hydrophobicity are major factors affecting biofilm formation and, hence, the performance of biocathodes.

9.4.1.1 Biocompatibility and Surface Roughness

Surface roughness can significantly impact the heterogeneity nature of biofilms on surfaces. The ability to form and sustain biofilms on biocathode surfaces directly impacts the electron transfer process. The structural heterogeneity and microbial characteristics can influence the biofilm activities as well as the substrate and product mass transfer (Yang et al. 2000). According to Dumas et al. (2008), surface roughness enhances biofilm formation on graphite rather than on stainless steel and the current density was much larger for the stainless steel electrode. Pons et al. (2011) reported the biofilm structure significantly affecting current production with isolated cells and small local colonies providing higher current density than dense colonies.

9.4.1.2 Surface Area and Porosity

High surface area electrodes are affiliated with increasing power production (Logan 2009). Increasing porosity increases the surface area available for biofilm growth (Santoro et al. 2014). Higher porosity reduces the diffusional resistance to the mass transfer of oxygen, in case of aerobic biocathodes (Tursun et al. 2016).

9.4.1.3 Conductivity

Highly conductive materials reduce the resistance to the flow of electrons and increases electron transfer. Higher resistance leads to a loss of energy in the form of heat. Highly conductive materials increase the electron transfer to micro-organisms (Jourdin et al. 2014).

9.4.1.4 Hydrophobicity

The hydrophobic nature of the material can affect the type of microorganisms attached to the surface (Mieke et al. 2013). Many microorganisms are capable of changing their cell surface characteristics (charge and hydrophobicity) depending on the environmental conditions (Marshall et al. 1971; Busscher and Weerkamp 1987). Bacterial adhesion can be enhanced by changing cathode properties such as hydrophobicity as well as changing the pH and conductivity of the catholyte (Rijnaarts et al. 1995; Van Loosdrecht et al. 1989).

9.5 Biofilm Formation

Biofilms are structured microbial communities encased and glued in a matrix of complex extracellular biopolymers. A micro-colony, a basic unit of the biofilm, is enclosed in a dense biopolymer attached to a surface (Costerton et al. 1995). Biopolymer composition and the microbial populations in biofilms are major factors affecting the electron transfer efficiency. A gradual decrease in the power generation was observed with increasing cathode biofilm thickness on graphite plate and stainless steel mesh (Behera et al. 2010). Biofilm formation is affected by numerous factors such as reactor operational conditions, inoculum source, environmental conditions, substrate and surface characteristics.

9.5.1 Biofilm Architecture

Biofilms architecture is also not always a uniform distribution of cells constructed layer upon layer. Homogenous distribution of microorganisms on an electrode is reported in many studies (Huang et al. 2011; Nevin et al. 2010). The availability of nutrients plays an important role in biofilm formation as microorganism position themselves according to the presence of nutrient gradients (Jain et al. 2007; Costerton et al. 1995). Biopolymers can also enhance biofilm formation. Other factors, including physicochemical properties (e.g. pH, temperature, viscosity and flow rate) and the presence of other microorganisms can also affect biofilm development (Patil et al. 2011; Costerton et al. 1995). Convective flow and turbulence, caused by the roughness of the surface, allows for circulation within the biofilm. Establishing biofilm growth on bioanodes and biocathodes by acclimatization of electroactive bacteria (EAB) is a mechanism to increase establishment of micro-colonies on surfaces (Xu et al. 2016; Wang et al. 2009; Kim et al. 2005).

Acclimatization is useful in increasing the electrode performance by enhancing biofilm attachment and growth and/or allowing the biofilm to attain a steady-state open circuit potential (OCP) (Renslow et al. 2011; Larrosa-Guerrero et al. 2010).

9.6 Electron Transfer

Many microorganisms exchange electrons with solid surfaces or mediators. Electrons are transferred between microorganisms, between microorganisms and surfaces and surface to microorganism. Electron acceptors reduction is achieved either by bacteria or direct reduction on a biocathode.

9.6.1 Aerobic and Anaerobic Bacterial Electron Transport Chains

Electron transport is mediated by a series of electron carriers which cascade electrons from an electron donor with a relatively lower redox potential to an electron acceptor with higher redox potential. The carrier system, localized in bacterial plasma membranes, consists of linked electron carriers such as proteins, flavins, cytochromes, non-heme iron components, other smaller non-protein carriers as well as multiple cytochrome oxidases (Kracke et al. 2015).

A generalized electron transport chain is initiated by the oxidation of NADH ($+H^+$) proceeds from flavoprotein dehydrogenase. Electrons are transported into a non-heme iron centre and subsequently to ubiquinone to produce ubiquinol. The non-heme iron centre receive electrons from ubiquinone, in case of aerobes, then to cytochrome b1 and subsequently channelled to multiple cytochrome oxidases such as cytochromes a, c4, c5 or d (Jurtshuk 1996). Configuration of these subsequent carriers depends on the oxygen tension (Kurosu and Begari 2010).

In anaerobes, electrons from ubiquinone are transferred to cytochrome b1, to c1 and to c type cytochromes. Electrons from b1 may transfer to non-heme iron centre or from c type cytochrome to cytochrome d and to the electron acceptor (Jurtshuk 1996).

Geobacter sulfurreducens and *Shewanella oneidensis* have been used as model organisms to examine direct and indirect electron transfer mechanisms between electrode surfaces (Ross et al. 2011; Rosenbaum et al. 2011; Thrash and Coates 2008; Bond and Lovley 2003). Outer-membrane cytochromes have been identified as enabling compounds promoting electron transfer (Breuer et al. 2015; Shi et al. 2007, 2009). However, studies have shown significant differences in the different electron transport chains in microorganisms. For example, soluble electron carriers detected in *Shewanella* sp. are not present in *Geobacter* sp. (Marsili et al. 2008; Holmes et al. 2006).

9.6.2 Electrode-Microbe Electron Transfer Mechanisms

Microbe-anode and cathode-microbe electron transfer are similar but electron flows in the opposite direction (Semenec and Franks 2015; Strycharz et al. 2011). Electrons transfer from electrode to bacteria can be categorized into the following (Choi and Sang 2016):

1. Direct electron transfer (DET)
2. Indirect electron transfer (IDET)

Microorganisms utilize several electron transfer modes such as: direct— involve physical contact or attachment with nanowires, and indirect—use small organic secreted by cells or added exogenously molecules as redox mediators or primary metabolites or other intermediates (Patil et al. 2012).

9.6.2.1 Direct Electron Transfer (DET)

Direct Electron Transfer Through Surface Proteins

Outer bound, inner bound and trans-membrane proteins associated with electron transport chains are involved with a series of redox reactions. Electron transfer systems are not the same in microorganisms; however, these systems essentially perform the same function. Although electron transfer systems in *Geobacter sulfurreducens* can function on both anodes and cathodes (Geelhoed and Stams 2011; Strycharz et al. 2011; Nevin et al. 2009), it is unclear if the same proteins are involved.

Carriers mediating electron transfer reactions include an extracellular site on the outer membrane designated as MtrC, MrtA, a periplasmic c-type cytochrome, CymA, positioned on a site between the inner membrane and the periplasm and OmcA, a protein located on the inner membrane (Choi and Sang 2016). OmcS and OmcF, two of the most abundant proteins from the outer surfaces of intact c-type cytochrome cells, are affiliated with the electron transfer process.

Direct electron transfer from cathode to hydrogenases via cytochromes has been proposed by Rosenbaum et al. (2011). Genomic sequence of *Desulfovibrio vulgaris* has shown to possess genes coded for various c-type cytochromes (Heidelberg et al. 2004). These cytochromes may create a network of heme centres in electron transport chains. Dinh et al. (2004) observed a *Desulfobacterium* isolate and a *Methanobacterium* which can accept electrons from metallic iron. Various types of cytochromes have been detected in the outer membrane of *Desulfovibrio* sp. and their role as redox partners of hydrogenase was confirmed by many researchers (Barton et al. 2007; Guiral-Brugna et al. 2001; Pereira et al. 1996). These findings form the basis for a mechanism of cytochromes accepting electrons from cathode and coupling to hydrogenases producing hydrogen in electron transport chains.

Direct Electron Transfer Through Conducting Pili

This direct electron transfer mechanism relies on conductive pili (nanowires) for electron transfer. Nanowires may function by enhancing the formation of thicker electroactive biofilms which assist with charge transfer. Nanowires have been reported in *Geobacter* and *Shewanella* strains (El-Naggar et al. 2010). In brief, DET occurs when active sites on the membrane/extracellular proteins are bound or close to the electrode surface.

9.6.2.2 Indirect Electron Transfer (IDET)

In microbial systems, electron transfer is mediated through chemical addition (artificially), by-products as well as extracellular polymeric substances (EPS) (Park and Zeikus 1999; Park et al. 1999; Pequin et al. 1994; Kim and Zeikus 1992; White et al. 1987).

Mediator Shuttled Electron Transfer

Indirect electron transfer (IDET) is associated with the electron transport via shuttle chemicals. These mediating chemicals adsorb and facilitate the transfer of electrons from carbon and metal surfaces to microorganisms. Many redox dyes, such as methyl viologen (MV) (Pequin et al. 1994; White et al. 1987), benzyl viologen (White et al. 1987), and neutral red (Kim and Zeikus 1992) with redox potential lower values than NAD can affect biological redox reactions in bacterial systems. For example, methyl viologen, an artificial mediator, was shown to increase ethanol production from acetate at a cathode (Steinbusch et al. 2010).

EPS Mediated Electron Transfer

Extracellular polymeric substances (EPS) have been shown to function as a means for electron transport from cathode surfaces to micro-organisms. This property has become basis for many studies in improving cathode material for maximum power production. Cathode surfaces modified with oligosaccharides or polysaccharides and other polymeric materials have been reported to increase the specific surface area, alter the charge of cathode material and improve efficient electron transfer (Ren et al. 2013; Li et al. 2012).

9.6.2.3 Proteins Affiliated with Extracellular Electron Transfer

In addition to membrane bound cytochromes, other proteins such as ferrodoxin oxidoreductases, hydrogenases coupled with ferrodoxins, methyl transferases, rubredoxin, fumarate reductase, and formate dehydrogenase participate in electron transfer (Choi and Sang 2016).

9.7 Microbial Characterization Methods

Genomic/metagenomic, transcriptomic/metatranscriptomic, proteomic/meta-proteomic, metabolomics/meta-metabomoics, microscopic coupled with flow cytometry are employed to understand microbial interactions as well as metabolism in microorganisms. Combined metagenomic and metaproteomic analyses can be particularly useful for generating information related to the functions of biofilm constituents. Metaproteomics is useful because the data can provide functional information from electrode-grown cells in order to understand biofilm electron transfer pathways.

9.7.1 Biofilm Characterization

A variety of techniques, including biological, microscopic and chemical, are available for characterizing and for community analysis of biofilm on electrode surfaces. Each method has different level of taxonomical resolution and, therefore, a specific application purpose. Each technique is classified into the following three levels of detection: microbial detection, assessing and characterizing the community composition and characterization based on functional genes and microbial activity.

9.7.2 Microorganism Detection

Microbial communities can be analyzed using both culture-dependent and independent approaches. Isolation of specific populations is accomplished using a variety of methods. Cultures are serially diluted and plated on solid medium where single colonies are subsequently cultured further as a pure culture. Further, culture independent techniques are needed for complete detection of populations of microbes.

Another powerful tool used to detect and quantify microbes in a biofilm makes use of fluorescent markers is fluorescence *in situ* hybridization (FISH) (Moter and Göbel 2000). In this technique, fluorescent oligonucleotide probes are designed against specific known sequences such as genus specific sequences. Many variations of FISH have been developed to enhance the signal generated from the sample, which is especially useful if a low microbial population is predicted (Pernthaler et al. 2002). Quantitative real time polymerase chain reaction (q-PCR) is used to rapidly determine and detect specific microbes within a biofilm (White et al. 2009).

9.7.3 Composition and Characterization of Microbial Communities

Detection and determination of the relative abundance of microbes within a community is not sufficient for complete characterization of the population. The gold standard in identifying cultures is 16S rRNA sequencing (Rosselló-Mora and Amann 2001; Amann et al. 1995; Woese et al. 1985). The highly conserved regions of 16S rRNA allows for reliable universal primers, while variable regions can be used for comparative taxonomy (Prakash et al. 2007). Identification of the isolates is examined by comparing the 16S rRNA gene sequence to international public sequence databases (e.g. GenBank). Genetic fingerprinting techniques are commonly employed to characterize the composition of a microbial community. Gel electrophoresis techniques are used to separate the amplicons based on their sequence. Denaturing gradient gel electrophoresis (DGGE) relies on an increasing gradient of denaturants (formamide and urea) to separate amplicons; alternatively non-denaturing electrophoresis is used to separate these fragments while maintaining their secondary structure. Temperature gradient gel electrophoresis (TGGE) uses a temperature gradient for amplicon separation.

An alternative genetic approach includes terminal restriction fragment length polymorphism (T-RFLP). T-RFLP is based on the amplification of target genes, typically the hyper-variable region of 16S rRNA genes from the bulk population (Liu et al. 1997). Another version of RFLP is amplified ribosomal DNA restriction analysis (ARDRA). This technique is another extension of RFLP where conserved regions at the end of the 16S gene are amplified and digested with restriction enzymes (Vaneechoutte et al. 1992; Vaneechoutte et al. 1993).

Single-strand conformation polymorphism (SSCP) analysis is a simple and sensitive technique for examining genetic variations within microbial communities (Widjojoatmodjo et al. 1995; Orita et al. 1989). The SSCP method is based on analyzing an altered conformation in comparison to a defined conformation. A single base substitution leads to altered mobility of the single-stranded DNA under non-denaturing electrophoresis. The differential intra-molecular folding results in the banding pattern obtained from single-strand conformation analysis.

Genomic methods dedicated towards extraction of nucleic acid and its subsequent analysis have been used to characterize microbial community. Metagenomics is used to analyze the collective genome from microorganisms in an ecosystem.

An indirect method to delineate differences between bacterial species is using the percent similarity via DNA-DNA hybridization. With increasing number of sequenced microbial genomes the direct determination of the similarity between entire genome sequences may replace DNA-DNA hybridization (Goris et al. 2007; Vandamme et al. 1996).

Genetic approaches can yield a great deal of information regarding the characterization of biofilms. However, further classification based on phenotypic or chemotaxonomic characterization is required. An example of chemotaxonomy is cellular fatty acid methyl ester (FAME) analysis. FAME analysis is performed using gas chromatography (GC) or liquid chromatography (LC) coupled with mass spectrometry (MS) (Vandamme et al. 1996).

9.7.4 Analysis of Functional Genes and Activity of Microbes

When assessing and characterizing a biofilm, not only is the structure and composition important, the functions these microbes perform and their activity are also crucial to fully understand these populations. To achieve this, techniques based on metatranscriptome and metaproteome are used. Metatranscriptomics can be used to quantify levels of gene expression within complex microbial communities in a high throughput manner. Combining metaproteomic and metagenomic analyses is considered useful for gathering information of constituents and functions of constituents in a biofilm (Wang et al. 2015).

For metatranscriptomic techniques, RNA is isolated from the population and analyzed via real time PCR (RT-PCR) or microarray analysis (Sharkey et al. 2004). Using RT-PCR, changes in gene expression of functional genes in response to varying environmental conditions can be assessed (Sharkey et al. 2004). The microarray based metatranscriptomic analysis allows for a large number of target sequences to be analyzed at one time, provide a rapid, high throughput method of analyzing expression of functional genes.

Metaproteomic data is able to produce similar information as metatranscriptomics. However, metaproteomics can also provide information at the level of protein expression (Wilmes and Bond 2006). Specific functions and activities performed by microbes are known to be associated with expression of particular sets of proteins; thus functional activity of the microbes can be analyzed.

Stable isotope probing (SIP) can be used to identify the active populations within a microbial community (Dumont and Murrell 2005; Radajewski et al. 2000). Stable isotopes (^{13}C or ^{15}N) which have successfully metabolized and incorporated the substrate can be separated by density gradient centrifugation. Isolated populations can then be analyzed using genetic approaches to determine the strain of microbe response for metabolizing the substrate.

9.7.5 Polyphasic Taxonomical Approach

Using the polyphasic approach to classify microbes undoubtedly enhances the accuracy of establishing phylogenetic relationships between microbes, as well as in characterizing new species. This approach uses many of the techniques described in the above sections including complete gene sequencing and comparative analysis by phylogenetic trees, DNA-DNA hybridization, analyses of molecular markers, biochemical assays and microscopic characterization to gather collective information on microorganisms (Prakash et al. 2007; Vandamme et al. 1996). Polyphasic taxonomy is strengthened by further integrating genotypic, phenotypic and phylogenetic microbial data in a stepwise manner.

9.7.6 Microscopic Methods

Light microscopy techniques have dramatically improved visibility and contrast in unstained and living material. Adaptions to microscope hardware have made contrast imaging possible; one example is the Hoffman Modulation Contrast (HMC) which creates a 3D image by placing an amplitude filter in the objective of the microscope. The result is a high resolution three-dimensional image capable of optical sectioning, directionality and control over contrast and coherence (Hoffman 1988).

Scanning electron microscopy (SEM) is a powerful technique capable of examining the surface structure of biological samples and often used for visualization of microorganisms in a biofilm. Environmental SEM (ESEM) is preferred for examining biofilms because of issues related to the water of hydration with the vacuum system. Energy x-ray spectroscopy (EDS) is another useful analytical tool integrated with SEM.

Transmission electron microscopy (TEM) allows quantitative analysis at high resolution across the sections of the biofilms and, thereby, deriving valuable information on the spatial arrangement and cellular ultra-structure.

AFM is a type of near-field scanning probe microscope which uses a sharp probe or tip to map the contours of a sample and is therefore not limited in resolution due to diffraction effects. The AFM technique is very useful since it allows the precise measurement of dimensions of individual bacteria (Surman et al. 1996; Nivens et al. 1995).

Confocal scanning laser microscopy (CSLM) has improved the analytical precision of light microscopy. CSLM is well suited for studying microbial biofilms in that the *in-situ*, nondestructive analysis of living, fully hydrated biofilms is performed without chemical fixation or embedding. Combining confocal microscope capabilities with fluorescence *in situ* hybridization (FISH), which uses fluorescently labelled oligonucleotide probes to identify specific microbial community members, allows for accurate and quantitative analysis of multi-species biofilms in both natural and artificial environments (Almeida et al. 2011; Yang et al. 2011).

9.7.7 Spectroscopic Methods

Fourier transform infrared spectroscopy (FTIR) is used to study the interaction between IR radiation and the vibrational modes of molecules in condensed phases (solids and liquids). Commercially available IR instruments are capable of studying aqueous biofilms. Coupling of attenuated total reflection (ATR) sampling technique with FT-IR spectrometry allows examination of biofilms in aqueous environments; however, the water background signal requires subtracting. Surface-enhanced Raman scattering (SERS) is another technique which is used to gather information on microorganisms (Ivleva et al. 2008).

9.7.8 Nuclear Magnetic Resonance Imaging

Nuclear magnetic resonance imaging (NMRI) works on the principle of tracing the water movement within the biofilm. Intra-biofilm flow is known to profoundly affect mass transport within biofilms. NMRI is useful in understanding molecular diffusion and biofilm process modelling (Costerton et al. 1995).

9.7.9 Flow Cytometry

Flow cytometry as a method for microbial community characterization is beginning to gain more attention in recent years. In brief, samples are stained with fluorescent dyes and then examined using a series of lasers. Detectors are able to determine intensity of the fluorescent signal, which is subsequently used to quantify either the relative abundance of microbes in a particular sample (Hammes et al. 2008), or can also be used to determine the composition of samples (Kerstens et al. 2015; Ksontini et al. 2013).

To enable identification of specific microbial species within a community, combining flow cytometry with FISH, known as flow-FISH, can be utilized (Nettmann et al. 2013; Amann et al. 1990). In this technique, fluorescently labelled probes with sequences specific to 16S rRNA sequences are used to label a microbial community. Samples are then analyzed using a flow cytometer where the expression and quantification of the fluorescent probe can be readily and rapidly evaluated, allowing for quick identification of microbial species in a given community.

9.8 Conclusions

BESs are an evolving technology which remains in its developmental stages. A crucial factor affecting developing and implementing the technology is the cathode. Abiotic cathodes and biocathodes have been evaluated in BESs. Numerous advantages have been attributed to using biocathodes when compared to abiotic cathodes. Biocathodes are a crucial design element affecting efficient electron transfer and, hence, electricity production. Advancing the technology requires a thorough understanding of how material characteristics as well as microbial factors affect electron transfer. Investigating electron transfer mechanisms has been evaluated using numerous biological and chemical tools.

Acknowledgements Dr. Srimanta Ray was a visiting research associate in the Department of Civil and Environmental Engineering, University of Windsor, Ontario, Canada during preparation of this manuscript. Financial contribution for Dr. Ray's associateship was provided by the Department of Biotechnology, Government of India. The Canadian Queen Elizabeth II Diamond Jubilee Scholarship program provided funding to support Mr. Vignesh Prakasam. Funding to support Ms. Fatemeh Bagh was provided by the Ontario Trillium foundation.

References

Almeida, C., Azevedo, N. F., Santos, S., Keevil, C. W., & Vieira, M. J. (2011). Discriminating multi-species populations in biofilms with peptide nucleic acid fluorescence in-situ hybridization (PNA FISH). *PloS One, 6*(3), e14786.

Amann, R. I., Binder, B. J., Olson, R. J., Chisholm, S. W., Devereux, R., & Stahl, D. A. (1990). Combination of 16S rRNA-targeted oligonucleotide probes with flow cytometry for analyzing mixed microbial populations. *Applied and Environmental Microbiology, 56*(6), 1919–1925.

Amann, R. I., Ludwig, W., & Schleifer, K. H. (1995). Phylogenetic identification and in situ detection of individual microbial cells without cultivation. *Microbiological Reviews, 59*(1), 143–169.

Barton, L. L., Goulhen, F., Bruschi, M., Woodards, N. A., Plunkett, R. M., & Rietmeijer, F. J. M. (2007). The bacterial metallome: Composition and stability with specific reference to the anaerobic bacterium *Desulfovibrio desulfuricans*. *Biometals, 20*, 291–302.

Behera, M., Jana, P. S., & Ghangrekar, M. M. (2010). Performance evaluation of low cost microbial fuel cell fabricated using earthen pot with biotic and abiotic cathode. *Bioresource Technology, 101*(4), 1183–1189.

Berk, R. S., & Canfield, J. H. (1964). Bioelectrochemical energy conversion. *Applied Microbiology, 12*, 10–12.

Blanchet, E., Pécastaings, B., Erable, B., Roques, C., & Bergel, A. (2014). Protons accumulation during anodic phase turned to advantage for oxygen reduction during cathodic phase in reversible bioelectrodes. *Bioresource Technology, 173*, 224–230.

Bond, D. R., & Lovley, D. R. (2003). Electricity production by *Geobacter sulfurreducens* attached to electrodes. *Applied Environmental Microbiology, 69*, 1548–1555.

Breuer, M., Rosso, K. M., Blumberger, J., & Butt, J. N. (2015). Multi-heam cytochromes in *Shewanella oneidensis* MR-1: Structures, functions and opportunities. *Journal of Royal Society Interface, 12*(102), 20141117.

Busscher, H. J., & Weerkamp, A. H. (1987). Specific and non-specific interactions in bacterial adhesion to solid substrata. *FEMS Microbiology Letters, 46*(2), 165–173.

Cai, J., Zheng, P., Qaisar, M., Xing, Y. (2014). Effect of operating modes on simultaneous anaerobic sulfide and nitrate removal in microbial fuel cell. *Journal of Industrial Microbiology and Biotechnology, 41*(5), 795–802.

Campo, G. D. A., Cañizares, P., Rodrigo, M. A., Fernández, F. J., & Lobato, J. (2013). Microbial fuel cell with an algae-assisted cathode: A preliminary assessment. *Journal of Power Sources, 242*, 638–645.

Cao, X., Huang, X., Liang, P., Boon, N., Fan, M., Zhang, L., & Zhang, X. (2009). A completely anoxic microbial fuel cell using a photo-biocathode for cathodic carbon dioxide reduction. *Energy and Environmental Science, 2*(5), 498–501.

Carbajosa, S., Malki, M., Caillard, R., Lopez, M. F., Palomares, F. J., Martín-Gago, J. A., et al. (2010). Electrochemical growth of *Acidithiobacillus ferrooxidans* on a graphite electrode. *Biosensors and Bioelectronics, 26*(2), 877–880.

Chen, G. W., Choi, S. J., Lee, T. H., Lee, G. Y., Cha, J. H., & Kim, C. W. (2008). Application of biocathode in microbial fuel cells: Cell performance and microbial community. *Applied Microbiology and Biotechnology, 79*(3), 379–388.

Chen, G. W., Choi, S. J., Lee, T. H., Lee, G. Y., Cha, J. H., & Kim, C. W. (2010). Microbial community dynamics and electron transfer of a biocathode in microbial fuel cells. *Korean Journal of Chemical Engineering, 27*(5), 1513–1520.

Choi, O., & Sang, B.-I. (2016). Extracellular electron transfer from cathode to microbes: Application for biofuel production. *Biotechnology for Biofuels, 9*, 11.

Chung, K., Fujiki, I., & Okabe, S. (2011). Effect of formation of biofilms and chemical scale on the cathode electrode on the performance of a continuous two-chamber microbial fuel cell. *Bioresource Technology, 102*(1), 355–360.

Clauwaert, P., Rabaey, K., Aelterman, P., Schamphelaire, D. L., Pham, H. T., Boeckx, P., et al. (2007a). Biological denitrification in microbial fuel cells. *Environment Science and Technology, 41*, 3354–3360.

Clauwaert, P., van der Ha, D., Boon, N., Verbeken, K., Verhaege, M., & Verstrate, W. (2007b). Open air biocathodes enables effective electricity generation with microbial fuel cells. *Environmental Science and Technology, 41*, 7564–7569.

Clauwaert, P., Desloover, J., Shea, C., Nerenberg, R., Boon, N., & Verstraete, W. (2009). Enhanced nitrogen removal in bio-electrochemical systems by pH control. *Biotechnology Letters, 31*(10), 1537–1543.

Coma, M., Puig, S., Pous, N., Balaguer, M. D., & Colprim, J. (2013). Biocatalysed sulphate removal in a BES cathode. *Bioresource Technology, 198*, 218–223.

Costerton, J. W., Lewandowski, Z., Caldwell, D. E., Korber, D. R., & Lappin-Scott, H. M. (1995). Microbial biofilms. *Annual Review of Microbiology, 49*, 711–745.

Dinh, H. T., Kuever, J., Mussmann, M., Hassel, A. W., Stratmann, M., & Widdel, F. (2004). Iron corrosion by novel anaerobic microorganisms. *Nature, 427*, 829–832.

Du, Y., Feng, Y., Dong, Y., Qu, Y., Liu, J., Zhou, X., & Ren, N. (2014). Coupling interaction of cathodic reduction and microbial metabolism in aerobic biocathode of microbial fuel cell. *RSC Advances, 4*, 34350–34355.

Dumas, C., Basseguy, R., & Bergel, A. (2008). Microbial electrocatalysis with *Geobacter sulfurreducens* biofilm on stainless steel cathodes. *Electrochimica Acta, 53*, 2494–2500.

Dumont, M. G., & Murrell, J. C. (2005). Stable isotope probing-linking microbial identity to function. *Nature Reviews Microbiology, 3*(6), 499–504.

El-Mekawy, A., Hegab, H. M., Vanbroekhoven, K., & Pant, D. (2014). Techno-productive potential of photosynthetic microbial fuel cells through different configurations. *Renewable and Sustainable Energy Reviews, 39*, 617–627.

El-Naggar, M. Y., Wanger, G., Leung, K. M., Yuzvinsky, T. D., Southam, G., Yang, J., et al. (2010). Electrical transport along bacterial nanowires from *Shewanella oneidensis* MR-1. *Proceedings of the National Academy of Sciences U S A, 107*(42), 18127–18131.

Freguia, S., Tsujimura, S., & Kano, K. (2010). Electron transfer pathways in microbial oxygen biocathodes. *Electrochimica Acta, 55*, 813–818.

Gajda, I., Greenman, J., Melhuish, C., & Ieropoulos, I. (2013). Photosynthetic cathodes for microbial fuel cells. *International Journal of Hydrogen Energy, 38*, 11559–11564.

Geelhoed, J. S., & Stams, A. J. M. (2011). Electricity-assisted biological hydrogen production from acetate by *Geobacter sulfurreducens. Environmental Science and Technology, 45*(2), 815–820.

Goris, J., Konstantinidis, K. T., Klappenbach, J. A., Coenye, T., Vandamme, P., & Tiedje, J. M. (2007). DNA-DNA hybridisation values and their relationships to whole-genome sequence similarities. *International Journal of Systematic and Evolutionary Microbiology, 57*, 81–91.

Guiral-Brugna, M., Giudici-Orticoni, M. T., Bruschi, M., & Bianco, P. (2001). Electrocatalysis of the hydrogen production by [Fe] hydrogenase from *Desulfovibrio vulgaris* Hildenborough. *Journal of Electroanalytical Chemistry, 510*(1–2), 136–143.

Hammes, F., Berney, M., Wang, Y., Vital, M., Köster, O., & Egli, T. (2008). Flow-cytometric total bacterial cell counts as a descriptive microbiological parameter for drinking water treatment processes. *Water Research, 42*(1), 269–277.

Heidelberg, J. F., Seshadri, R., Haveman, S. A., Hemme, C. L., Paulsen, I. T., Kolonay, J. F., et al. (2004). The genome sequence of the anaerobic, sulfate-reducing bacterium *Desulfovibrio vulgaris* Hildenborough. *Nature Biotechnology, 22*, 554–559.

Hoffman, R. (1988). Applications of the modulation contrast microscope. *International Laboratory, 18*, 32–39.

Holmes, D. E., Chaudhuri, S. K., Nevin, K. P., Mehta, T., Methé, B. A., Liu, A., Ward, J. E., Woodard, T. L., Webster, J., & Lovley, D. R. (2006). Microarray and genetic analysis of electron transfer to electrodes in *Geobacter sulfurreducens. Environmental Microbiology, 8*, 1805–1815.

Huang, L., Regan, J. M., & Quan, X. (2011). Electron transfer mechanisms, new applications, and performance of biocathode microbial fuel cells. *Bioresource Technology, 102*(1), 316–323.

Ivleva, N. P., Wagner, M., Horn, H., Niessner, R., & Haisch, C. (2008). In-situ surface-enhanced Raman scattering analysis of biofilm. *Analytical Chemistry, 80*(22), 8538–8544.

Jain, A., Gupta, Y., Agrawal, R., Khare, P., & Jain, S. K. (2007). Biofilms – A microbial life perspective: A critical review. *Critical Reviews on Therapeutic Drug Carrier System, 24*, 393–443.

Jeremiasse, A. W., Hamelers, H. V. M., & Buisman, C. J. N. (2010). Microbial electrolysis cell with a microbial biocathode. *Bioelectrochemistry, 78*, 39–43.

Jourdin, L., Freguia, S., Donose, B. C., Chen, J., Wallace, G. G., Keller, J., & Flexer, V. (2014). A novel carbon nanotube modified scaffold as an efficient biocathode material for improved microbial electrosynthesis. *Journal of Materials Chemistry A, 2*(32), 13093.

Jurtshuk, P. Jr. (1996). Chapter 4: Bacterial metabolism. In S. Baron (Ed.), *Medical microbiology* (4th ed.). Galveston: University of Texas Medical Branch at Galveston. Available from: http://www.ncbi.nlm.nih.gov/books/NBK7919/

Kerstens, M., Boulet, G., Clais, S., Lanckacker, E., Delputte, P., Maes, L., & Cos, P. (2015). A flow cytometric approach to quantify biofilms. *Folia Microbiologica, 60*(4), 335–342.

Kim, Y., & Logan, B. E. (2013). Microbial desalination cells for energy production and desalination. *Desalination, 308*, 122–130.

Kim, B. H., & Zeikus, J. G. (1992). Hydrogen metabolism in *Clostridium acetobutylicum* fermentation. *Journal of Microbiology and Biotechnology, 2*, 2771–2776.

Kim, J. R., Min, B., & Logan, B. E. (2005). Evaluation of procedures to acclimate a microbial fuel cell for electricity production. *Applied Microbiology and Biotechnology, 68*, 23–30.

Kim, B. H., Lim, S. S., Daud, W. R., Gadd, G. M., & Chang, I. S. (2015). The biocathode of microbial electrochemical systems and microbially-influenced corrosion. *Bioresource Technology, 190*, 395–401.

Kokabian, B., & Gude, V. G. (2013). Photosynthetic microbial desalination cells (PMDCs) for clean energy, water and biomass production. *Environmental Science: Processes & Impacts, 15* (12), 2178–2185.

Kracke, F., Vassilev, I., & Krömer, L. O. (2015). Microbial electron transport and energy conservation – The foundation for optimizing bioelectrochemical systems. *Frontiers in Microbiology, 6*, 75.

Ksontini, H., Kachouri, F., & Hamdi, M. (2013). Impact of *Lactococcus lactis* spp. *lactis* bio adhesion on pathogenic bacillus cereus biofilm on silicone flowing system. *Indian Journal of Microbiology, 53*(3), 269–275.

Kurosu, M., & Begari, E. (2010). Vitamin K2 in electron transport system: Are enzymes involved in vitamin K2 biosynthesis promising drug targets? *Molecules, 15*, 1531–1553.

Larrosa-Guerrero, A., Scott, K., Katuri, K. P., Godinez, C., Head, I. M., & Curtis, T. (2010). Open circuit versus closed circuit enrichment of anodic biofilms in MFC: Effect on performance and anodic communities. *Applied Microbiology and Biotechnology, 87*, 1699–1713.

Li, C., Ding, L., Cui, H., Zhang, L., Xu, K., & Ren, H. (2012). Application of conductive polymers in biocathode of microbial fuel cells and microbial community. *Bioresource Technology, 116*, 459–465.

Liu, W. T., Marsh, T. L., Cheng, H., & Forney, L. J. (1997). Characterization of microbial diversity by determining terminal restriction fragment length polymorphisms of genes encoding 16S rRNA. *Applied and Environmental Microbiology, 63*(11), 4516–4522.

Logan, B. E. (2008). *Microbial fuel cells*. New York: John Wiley & Sons.

Logan, B. E. (2009). Exoelectrogenic bacteria that power microbial fuel cells. *Nature Reviews Microbiology, 7*(5), 375–381.

Lovley, D. R. (2011). Powering microbes with electricity: Direct electron transfer from electrodes to microbes. *Environmental Microbiology Reports, 3*, 27–35.

Marshall, K. C., Stout, R., & Mitchell, R. (1971). Selective sorption of bacteria from seawater. *Canadian Journal of Microbiology, 17*(11), 1413–1416.

Marshall, C. W., Ross, D. E., Fichot, E. B., Norman, R. S., & May, H. D. (2012). Electrosynthesis of commodity chemicals by an autotrophic microbial community. *Applied and Environmental Microbiology, 78*, 8412–8420.

Marshall, C. W., Ross, D. E., Fichot, E. B., Norman, R. S., & May, H. D. (2013). Long-term operation of microbial electrosynthesis systems improves acetate production by autotrophic microbiomes. *Environment Science and Technology, 47*, 6023–6029.

Marsili, E., Baron, D. B., Shikhare, I. D., Coursolle, D., Gralnick, J. A., & Bond, D. R. (2008). *Shewanella* secretes flavins that mediate extracellular electron transfer. *Proceedings of the National Academy of Sciences U S A, 105*, 3968–3973.

McCormick, A. J., Bombelli, P., Scott, A. M., Philips, A. J., Smith, A. G., Fisher, A. C., & Howe, C. J. (2011). Photosynthetic biofilms in pure culture harness solar energy in a mediatorless bio-photovoltaic cell (BPV) system. *Energy and Environment Science, 4*, 4699–4709.

Meng, F., Jiang, J., Zhao, Q., Wang, K., Zhang, G., Fan, Q., Wei, L., Ding, J., & Zheng, Z. (2014). Bioelectrochemical desalination and electricity generation in microbial desalination cell with dewatered sludge as fuel. *Bioresource Technology, 157*, 120–126.

Mieke, C. A. A., Eerten-Jansen, V., Veldhoen, A. B., Plugge, C. M., Stams, A. J. M., Buisman, C. J. N., & Heijne, A. T. (2013). Microbial community analysis of a methane-producing biocathode in a bioelectrochemical system. *Archaea*. Article ID 481784.

Moter, A., & Göbel, U. B. (2000). Fluorescence in situ hybridization (FISH) for direct visualization of microorganisms. *Journal of Microbiological Methods, 41*(2), 85–112.

Nettmann, E., Fröhling, A., Heeg, K., Klocke, M., Schlüter, O., & Mumme, J. (2013). Development of a flow-fluorescence in situ hybridization protocol for the analysis of microbial communities in anaerobic fermentation liquor. *BMC Microbiology, 13*(278), 1–15.

Nevin, K. P., Kim, B. C., Glaven, R. H., Johnson, J. P., Woodward, T. L., Methé, B. A., et al. (2009). Anode biofilm transcriptomics reveals outer surface components essential for high density current production in *Geobacter sulfurreducens* fuel cells. *PloS One, 4*(5).

Nevin, K. P., Woodard, T. L., Franks, A. E., Summers, Z. M., & Lovley, D. R. (2010). Microbial electrosynthesis: Feeding microbes electricity to convert carbon dioxide and water to multicarbon extracellular organic compounds. *MBio, 1*, e00103–e00110.

Nguyen, V. K., Hong, S., Park, Y., Jo, K., & Lee, T. (2015). Autotrophic denitrification performance and bacterial community at biocathodes of bioelectrochemical systems with either abiotic or biotic anodes. *Journal of Bioscience and Bioengineering, 119*(2), 180–187.

Nivens, D. E., Palmer, R. J., Jr., & Whitez, D. C. (1995). Continuous nondestructive monitoring of microbial biofilms: A review of analytical techniques. *Journal of Industrial Microbiology, 15*, 263–276.

Orita, M., Iwahana, H., Kanazawa, H., Hayashi, K., & Sekiya, T. (1989). Detection of polymorphisms of human DNA by gel electrophoresis as single-strand conformation polymorphisms. *Proceedings of the National Academy of Sciences, 86*(8), 2766–2770.

Park, D. H., & Zeikus, J. G. (1999). Utilization of electrically reduced neutral red by *Actinobacillus succinogenes*: Physiological function of neutral red in membrane-driven fumarate reduction and energy conservation. *Journal of Bacteriology, 181*(8), 2403–2410.

Park, D. H., Laivenieks, M., Guettler, M. V., Jain, M. K., & Zeikus, J. G. (1999). Microbial utilization of electrically reduced neutral red as the sole electron donor for growth and metabolite production. *Applied and Environmental Microbiology, 65*(7), 2912–2417.

Patil, S. A., Harnisch, F., Koch, C., Hübschmann, T., Fetzer, I., Carmona-Martínez, A. A., Müller, S., & Schröder, U. (2011). Electroactive mixed culture derived biofilms in microbial bioelectrochemical systems: The role of pH on biofilm formation, performance and composition. *Bioresource Technology, 102*(20), 9683–9690.

Patil, S. A., Hägerhäll, C., & Gorton, L. (2012). Electron transfer mechanisms between microorganisms and electrodes in bioelectrochemical systems. *Bioanalytical Reviews, 4*, 159–192.

Pequin, S., Delorme, P., Goma, G., & Soucaille, P. (1994). Enhanced alcohol yields in batch cultures of Clostridium using a three-electrode potentiometric system with methyl viologen as electron carrier. *Biotechnology Letters, 16*, 269–274.

Pereira, A. S., Franco, R., Feio, M. J., Pinto, C., Lampreia, J., Reis, M. A., et al. (1996). Characterization of representative enzymes from a sulfate reducing bacterium implicated in the corrosion of steel. *Biochemical and Biophysical Research Communications, 221*, 414–421.

Pernthaler, A., Pernthaler, J., & Amann, R. (2002). Fluorescence in situ hybridization and catalyzed reporter deposition for the identification of marine bacteria. *Applied and Environmental Microbiology, 68*(6), 3094–3101.

Pisciotta, J. M., Zaybak, Z., Call, D. F., Nam, J. Y., & Logan, B. E. (2012). Enrichment of microbial electrolysis cell biocathodes from sediment microbial fuel cell bioanodes. *Applied and Environmental Microbiology, 78*, 5212–5219.

Pons, L., Délia, M. L., & Bergel, A. (2011). Effect of surface roughness, biofilm coverage and biofilm structure on the electrochemical efficiency of microbial cathodes. *Bioresource Technology, 102*(3), 2678–2683.

Pozo, G., Jourdin, L., Lu, Y., Ledezma, P., Kellera, J., & Freguia, S. (2015). *Methanobacterium* enables high rate electricity-driven autotrophic sulfate reduction. *RSC Advance, 5*, 89368–89374.

Prakash, O., Verma, M., Sharma, P., Kumar, M., Kumari, K., Singh, A., et al. (2007). Polyphasic approach of bacterial classification—An overview of recent advances. *Indian Journal of Microbiology, 4*(2), 98–108.

Rabaey, K., Read, S. T., Clauwaert, P., Freguia, S., Bond, P. L., Blackall, L. L., & Keller, J. (2008). Cathodic oxygen reduction catalyzed by bacteria in microbial fuel cells. *ISME Journal, 2*(5), 519–527.

Radajewski, S., Ineson, P., Parekh, N. R., & Murrell, J. C. (2000). Stable-isotope probing as a tool in microbial ecology. *Nature, 403*(6770), 646–649.

Reimers, C. E., Girguis, P., Stecher, H. A., III, Tender, L. M., Ryckelynck, N., & Whaling, P. (2006). Microbial fuel cell energy from an ocean cold seep. *Geobiology, 4*, 123–136.

Ren, Z., Lan, Y., & Wang, Y. (2013). Aligned carbon nanotubes physics, concepts, fabrication and devices. In P. Avouris, B. Bhushan, D. Bimberg, K. V. Klitzing, H. Sakaki, & R. Wiesendanger (Eds.), *Nano science and technology* (pp. 7–44). Berlin/Heidelberg: Springer.

Renslow, R., Donovan, C., Shim, M., Babauta, J., Nannapaneni, S., Schenk, J., & Beyenal, H. (2011). Oxygen reduction kinetics on graphite cathodes in sediment microbial fuel cells. *Physical Chemistry Chemical Physics, 13*, 21573–21584.

Rijnaarts, H. H. M., Norde, W., Bouwer, E. J., Lyklema, J., & Zehnder, A. J. B. (1995). Reversibility and mechanism of bacterial adhesion. *Colloids and Surfaces B, 4*(1), 5–22.

Rimboud, M., Desmond-Le Quemener, E., Erable, B., Bouchez, T., & Bergel, A. (2015). The current provided by oxygen-reducing microbial cathodes is related to the composition of their bacterial community. *Bioelectrochemistry, 102*, 42–49.

Rosenbaum, M., Aulenta, F., Villano, M., & Angenent, L. T. (2011). Cathodes as electron donors for microbial metabolism: Which extracellular electron transfer mechanisms are involved? *Bioresource Technology, 102*(1), 324–333.

Ross, D. E., Flynn, J. M., Baron, D. B., Gralnick, J. A., & Bond, D. R. (2011). Towards electrosynthesis in Shewanella: Energetics of reversing the mtr pathway for reductive metabolism. *PloS One, 6*, e16649.

Rosselló-Mora, R., & Amann, R. (2001). The species concept for prokaryotes. *FEMS Microbiology Reviews, 25*(1), 39–67.

Rozendal, R. A., Jeremiasse, A. W., Hamelers, H. V. M., & Buisman, C. J. N. (2008). Hydrogen production with a microbial biocathode. *Environment Science and Technology, 42*, 629–634.

Santoro, C., Guilizzoni, M., Correa Baena, J. P., Pasaogullari, U., Casalegno, A., Li, B., et al. (2014). The effects of carbon electrode surface properties on bacteria attachment and start up time of microbial fuel cells. *Carbon, 67*, 128–139.

Sasaki, K., Morita, M., Sasaki, D., Hirano, S., Matsumoto, N., Ohmura, N., & Igarashi, Y. (2011). Methanogenic communities on the electrodes of bioelectrochemical reactors without membranes. *Journal of Bioscience and Bioengineering, 111*, 47–49.

Schamphelaire, D. L., Boeckx, P., & Verstraete, W. (2010). Evaluation of biocathodes in freshwater and brackish sediment microbial fuel cells. *Applied Microbiology and Biotechnology, 87*(5), 1675–1687.

Semenec, L., & Franks, A. E. (2015). Delving through electrogenic biofilms: From anodes to cathodes to microbes. *AIMS Bioengineering, 2*(3), 222–248.

Sharkey, F. H., Banat, I. M., & Marchant, R. (2004). Detection and quantification of gene expression in environmental bacteriology. *Applied and Environmental Microbiology, 70*(7), 3795–3806.

Shi, L., Squier, T. C., Zachara, J. M., & Fredrickson, J. K. (2007). Respiration of metal (hydr) oxides by *Shewanella* and *Geobacter*: A key role for multihaem c-type cytochromes. *Molecular Microbiology, 65*, 12–20.

Shi, L., Richardson, D. J., Wang, Z., Kerisit, S. N., Rosso, K. M., Zachara, J. M., & Fredrickson, J. K. (2009). The roles of outer membrane cytochromes of *Shewanella* and *Geobacter* in extra cellular electron transfer. *Environmental Microbiology Reports, 1*, 220–227.

Steinbusch, K. J. J., Hamelers, H. V. M., Schaap, J. D., Kampman, C., & Buisman, C. J. N. (2010). Bioelectrochemical ethanol production through mediated acetate reduction by mixed cultures. *Environmental Science and Technology, 44*(1), 513–517.

Strycharz, S. M., Glaven, R. H., Coppi, M. V., Gannon, S. M., Perpetua, L. A., Liu, A., et al. (2011). Gene expression and deletion analysis of mechanisms for electron transfer from electrodes to *Geobacter sulfurreducens*. *Bioelectrochemistry, 80*(2), 142–150.

Strycharz-Glaven, S. M., Glaven, R. H., Wang, Z., Zhou, J., Vora, G. J., & Tender, L. M. (2013). Electrochemical investigation of a microbial solar cell reveals a nonphotosynthetic biocathode catalyst. *Applied and Environmental Microbiology, 79*, 3933–3942.

Sun, Y., Wei, J., Liang, P., & Huang, X. (2012). Microbial community analysis in biocathode microbial fuel cells packed with different materials. *AMB Express, 2*(1), 21.

Surman, S. B., Walker, J. T., Goddard, D. T., Morton, L. H. G., Keevil, C. W., Weaver, W., et al. (1996). Comparison of microscope techniques for the examination of biofilms. *Journal of Microbiological Methods, 25*, 57–70.

Tartakovsky, B., Manuel, M.-F., Wang, H., & Guiot, S. R. (2009). High rate membraneless microbial electrolysis cell for continuous hydrogen production. *International Journal of Hydrogen Energy, 34*, 672–677.

Thauer, R. K., Jungermann, K., & Decker, K. (1977). Energy conservation in chemotrophic anaerobic bacteria. *Bacteriological Reviews, 41*(1), 100–180.

Thrash, J. C., & Coates, J. D. (2008). Review: Direct and indirect electrical stimulation of microbial metabolism. *Environment Science and Technology, 42*, 3921–3931.

Tursun, H., Liu, R., Li, J., Abro, R., Wang, X., Gao, Y., & Li, Y. (2016). Carbon material optimized biocathode for improving microbial fuel cell performance. *Frontiers in Microbiology, 7*, 1–9.

Van Loosdrecht, M. C. M., Lyklema, J., Norde, W., & Zehnder, A. J. B. (1989). Bacterial adhesion: A physiochemical approach. *Microbial Ecology, 17*(1), 1–15.

Vandamme, P., Pot, B., Gillis, M., De Vos, P., Kersters, K., & Swings, J. (1996). Polyphasic taxonomy: A consensus approach to bacterial systematics. *Microbiological Reviews, 60*, 407–438.

Vaneechoutte, M., Rossau, R., De Vos, P., Gillis, M., Janssens, D., Paepe, N., et al. (1992). Rapid identification of bacteria of the Comamonadaceae with amplified ribosomal DNA-restriction analysis (ARDRA). *FEMS Microbiology Letters, 93*(3), 227–233.

Vaneechoutte, M., De Beenhouwer, H., Claeys, G., Verschraegen, G. E. R. D. A., De Rouck, A., Paepe, N., et al. (1993). Identification of Mycobacterium species by using amplified ribosomal DNA restriction analysis. *Journal of Clinical Microbiology, 31*(8), 2061–2065.

Virdis, B., Rabaey, K., Rozendal, R. A., Yuan, Z., & Keller, J. (2010). Simultaneous nitrification, denitrification and carbon removal in microbial fuel cells. *Water Research, 44*, 2970–2980.

Wang, X., Feng, Y. J., Ren, N. Q., Wang, H. M., Lee, H., Li, N., & Zhao, Q. L. (2009). Accelerated start-up of two-chambered microbial fuel cells: Effect of anodic positive poised potential. *Electrochimica Acta, 54*, 1109–1114.

Wang, X., Feng, Y., Liu, J., Lee, H., Li, C., Li, N., & Ren, N. (2010). Sequestration of CO_2 discharged from anode by algal cathode in microbial carbon capture cells (MCCs). *Biosensors and Bioelectronics, 25*(12), 2639–2643.

Wang, Z., Zheng, Y., Xiao, Y., Wu, S., Wu, Y., Yang, Z., & Zhao, F. (2013a). Analysis of oxygen reduction and microbial community of air-diffusion biocathode in microbial fuel cells. *Bioresource Technology, 144*, 74–79.

Wang, H., Jiang, S. C., Wang, Y., & Xiao, B. (2013b). Substrate removal and electricity generation in a membrane-less microbial fuel cell for biological treatment of wastewater. *Bioresource Technology, 138*, 109–116.

Wang, Z., Leary, D. H., Malanoski, A. P., Li, R. W., Hervey, W. J., Eddie, B. J., et al. (2015). A previously uncharacterized, nonphotosynthetic member of the *Chromatiaceae* is the primary CO_2-fixing constituent in a self-regenerating biocathode. *Applied and Environmental Microbiology, 81*(2), 699–712.

Wei, J., Liang, P., & Huang, X. (2011a). Recent progress in electrodes for microbial fuel cells. *Bioresource Technology, 102*(20), 9335–9344.

Wei, J., Liang, P., Cao, X., & Huang, X. (2011b). Use of inexpensive semicoke and activated carbon as biocathode in microbial fuel cells. *Bioresource Technology, 102*(22), 10431–10435.

Wen, Q., Zhang, H., Chen, Z., Li, Y., Nan, J., & Feng, Y. (2012). Using bacterial catalyst in the cathode of microbial desalination cell to improve wastewater treatment and desalination. *Bioresource Technology, 125*, 108–113.

White, H., Lebertz, H., Thanos, I., & Simon, H. (1987). *Clostridium thermoaceticum* production of methanol from carbon monoxide in the presence of viologen dyes. *FEMS Microbiology Letters, 43*, 173–176.

White, H. K., Reimers, C. E., Cordes, E. E., Dilly, G. F., & Girguis, P. R. (2009). Quantitative population dynamics of microbial communities in plankton-fed microbial fuel cells. *The ISME Journal, 3*(6), 635–646.

Widjojoatmodjo, M. N., Fluit, A. C., & Verhoef, J. (1995). Molecular identification of bacteria by fluorescence-based PCR-single-strand conformation polymorphism analysis of the 16S rRNA gene. *Journal of Clinical Microbiology, 33*(10), 2601–2606.

Wilmes, P., & Bond, P. L. (2006). Metaproteomics: Studying functional gene expression in microbial ecosystems. *Trends in Microbiology, 14*(2), 92–97.

Woese, C. R., Stackebrandt, E., Macke, T. J., & Fox, G. E. (1985). A phylogenetic definition of the major eubacterial taxa. *Systematic and Applied Microbiology, 6*(2), 143–151.

Wu, X. Y., Song, T. S., Zhu, X. J., Wei, P., & Zhou, C. C. (2013). Construction and operation of microbial fuel cell with Chlorella vulgaris biocathode for electricity generation. *Applied Biochemistry and Biotechnology, 171*(8), 2082–2092.

Xia, X., Sun, Y., Liang, P., & Huang, X. (2012). Long-term effect of set potential on biocathodes in microbial fuel cells: Electrochemical and phylogenetic characterization. *Bioresource Technology, 120*, 26–33.

Xiao, L., & He, Z. (2014). Applications and perspectives of phototrophic microorganisms for electricity generation from organic compounds in microbial fuel cells. *Renewable and Sustainable Energy Reviews, 37*, 550–559.

Xiao, L., Young, E. B., Grothjan, J. J., Lyon, S., Zhang, H., & He, Z. (2015). Wastewater treatment and microbial communities in an integrated photo-bioelectrochemical system affected by different wastewater algal inocula. *Algal Research, 12*, 446–454.

Xu, L., Zhao, Y., Doherty, L., Hu, Y., & Hao, X. (2016). Promoting the bio-cathode formation of a constructed wetland-microbial fuel cell by using powder activated carbon modified alum sludge in anode chamber. *Scientific Reports, 6*, 26514.

Yang, X., Beyenal, H., Harkin, G., & Lewandowski, Z. (2000). Quantifying biofilm structure using image analysis. *Journal of Microbiological Methods, 39*(2), 109–119.

Yang, L., Liu, Y., Wu, H., Høiby, N., Molin, S., & Song, Z.-J. (2011). Review: Current understanding of multi-species biofilms. *International Journal of Oral Science, 3*, 74–81.

You, S. J., Ren, N. Q., Zhao, Q. L., Wang, J. Y., & Yang, F. L. (2009). Power generation and electrochemical analysis of biocathode microbial fuel cell using graphite fibre brush as cathode material. *Fuel Cells, 9*(5), 588–596.

Yu, L., Duan, J., Zhao, W., Huang, Y., & Hou, B. (2008). Characteristics of hydrogen evolution and oxidation catalyzed by *Desulfovibrio caledoniensis* biofilm on pyrolytic graphite electrode. *Electrochimica Acta, 54*(1), 29–34.

Zaybak, Z., Pisciotta, J. M., Tokash, J. C., & Logan, B. E. (2013). Enhanced start-up of anaerobic facultatively autotrophic biocathodes in bioelectrochemical systems. *Journal of Biotechnology, 168*, 478–485.

Zhang, G. D., Zhao, Q. L., Jiao, Y., Zhang, J. N., Jiang, J. Q., Ren, N., & Kim, B. H. (2011a). Improved performance of microbial fuel cell using combination biocathode of graphite fiber brush and graphite granules. *Journal of Power Sources, 196*(15), 6036–6041.

Zhang, L., Zhu, X., Li, J., Liao, Q., & Ye, D. D. (2011b). Biofilm formation and electricity generation of a microbial fuel cell started up under different external resistances. *Journal of Power Sources, 196*, 6029–6035.

Zhang, G., Zhao, Q., Jiao, Y., Wang, K., Lee, D. J., & Ren, N. (2012a). Biocathode microbial fuel cell for efficient electricity recovery from dairy manure. *Biosensors and Bioelectronics, 31*(1), 537–543.

Zhang, Y., Sun, J., Hu, Y., Li, S., & Xu, Q. (2012b). Bio-cathode materials evaluation in microbial fuel cells: A comparison of graphite felt, carbon paper and stainless steel mesh materials. *International Journal of Hydrogen Energy, 37*(22), 16935–16942.

Zhang, Y., Sun, J., Hu, Y., Li, S., & Xu, Q. (2013). Carbon nanotube-coated stainless steel mesh for enhanced oxygen reduction in biocathode microbial fuel cells. *Journal of Power Sources, 239*, 169–174.

Chapter 10
Physicochemical Parameters Governing Microbial Fuel Cell Performance

Sanath Kondaveeti, Ramesh Kakarla, and Booki Min

10.1 Introduction

Microbial fuel cell (MFC) performance has been dramatically improved by optimizing physicochemical parameters, especially in the early period of MFC research, for MFC practical applications. The enhancement in power output of MFC can be dependent on several physical and chemical parameters such as electrode material and morphology, catalyst on electrode, reactor design, membrane/separator, temperature, pH, electrolyte conductivity, and types of oxidants and substrates (fuels). The optimized conditions of physical and chemical parameters can enhance performance of MFC by decreasing internal resistance which is the sum of over potentials at the anode and cathode chambers and the separator part, and by increasing Columbic efficiencies. The effects of these parameters on MFC performance are discussed below in details.

10.2 Anode Electrode for MFC

10.2.1 Plain Anode Materials

In most of MFC operations, carbon materials such as graphite rod, graphite plate, graphite fibre brush, carbon cloth, carbon paper and reticulated vitreous carbon (RVC) have been used as an anode electrode. The carbon materials are implemented

---

S. Kondaveeti • R. Kakarla • B. Min (✉)
Department of Environmental Science and Engineering, Kyung Hee University,
1 Seocheon-dong, Giheung-gu, Yongin Si, Gyeonggi-do 446-701, Republic of Korea
e-mail: bkmin@khu.ac.kr

D. Das (ed.), *Microbial Fuel Cell*,
https://doi.org/10.1007/978-3-319-66793-5_10

in MFC due to its excellent electrical conductivity, lower resistance and chemical stability. High-conductivity electrode can lead to decrease in few ohms of internal resistance, which can increase MFC power. Logan (2008) reported the electrical conductivity of graphite fibre, carbon paper and carbon cloth as 1.6, 0.8 and 2.2 Ω cm $^{-1}$, respectively. Carbon paper is often used as the anode electrode in MFC operation, but it is relatively thin and brittle in comparison with other electrodes. It can be easily connected to electrical collector (wire) and sealed with non-conductive resin. The carbon cloth appears to be highly flexible and porous allowing higher surface area for bacterial adhesion and growth. Due to its flexibility and firmness, the carbon cloth materials are often used in hydrogen fuel cell and flat plate MFC operations with a minimum spacing between electrodes to reduce ohmic resistance for enhanced MFC performance. Graphite plates, rods, felt and sheets exhibit high strength and low porosity, except for graphite felt. These electrodes are also often used as the anode material or in combination of activated carbon. MFC operation with graphite felt resulted in 2.4 times amplification in current generation than with graphite rod (Chaudhuri and Lovley 2003). Ahn and Logan (2012) suggested that increase in anode thickness from 0.32 to 1.27 cm of graphite felt exhibited a 23% increase in maximum power density due to its sponge structure. Graphite fibre brushes are made of graphite fibres that are wound using one or more electrically conductive and non-corrosive metal wires such as titanium or stainless steel. It is attractive due to high available surface area in comparison with other 2D conventional electrodes for bacterial growth and low electrical resistance. The graphite fibre brush anode with high specific surface area (18,200 m^2 m^{-3}) could produce the maximum power density of 1430 mW m^{-2}, which was about 2.4 times higher power density than with carbon paper (surface area = 22.5 cm^2; 600 mW m^{-2}) (Logan et al. 2007; Logan 2008). Comparison studies of anode electrode materials in MFC by using carbon paper and carbon brush clearly point the significant enhancement, in terms of power generation (Fig 10.1) (Logan et al. 2007).

Reticulated vitreous carbon (RVC) was used in few studies due to its high conductive nature (200 S cm^{-1}; 5×10^{-3} Ω cm^{-1}) and porosity. Similar to graphite fibre brush, RVC exhibits high surface area which helps in filling the anode chamber, but it is high-cost and brittle material. Few studies have tested

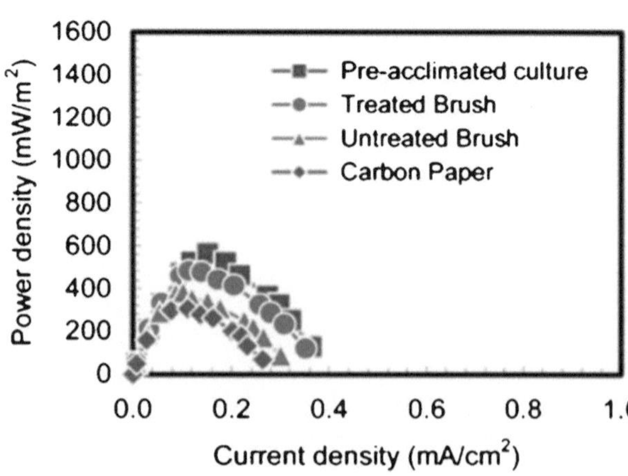

Fig. 10.1 Power density curves in single chamber bottle type MFCs with carbon paper and carbon brush anodes (Logan et al. 2007)

non-carbon materials like stainless steel (SS), titanium, ceramic with carbon, and gold as anode electrode materials (Guerrini et al. 2014; ter Heijne et al. 2008; Thorne et al. 2011). Dumas et al. (2007) operated sediment MFCs using SS as an anode and could generate 4 mW m^{-2} power density. MFCs using gold (7.8 cm^2) could generate current density of 688 mA m^{-2} and with carbon cloth (6.45 cm^2), the current density was 3147 mA m^{-2} (Richter et al. 2008). The higher current value with carbon cloth was possibly due to more biomass on the electrode surface than with smooth and polished electrodes.

10.2.2 Surface Modifications of Anode Electrode

One of the successful modifications of anode electrode was the treatment of carbon cloth anode with 5% NH_3 gas at 700 °C and helium carrier gas. The maximum power density with the NH_3 surface treatment was increased from 1640 to 1970 mW m^{-2}, and the lag period (startup) was reduced up to 50%. The increase in maximum power densities was attributed to increase in positive surface charge from 0.38 to 3.9 meq m^{-2} (Cheng and Logan 2007). The other types of surface treatments include the heating of carbon mesh at 450 °C for 30 min (He et al. 2005) and surface oxidation of carbon electrodes using sulphuric acid or the combination of heat and acid treatment (Feng et al. 2010). The increase in performance with acid or heat treatment can be due to several factors such as: (1) increase in specific surface area which helps in bacterial adhesion on to the anode electrode, (2) high ratio of protonation of N to total N giving more positive charge on electrode, and (3) lower composition of C–O resulting in lesser formation of contaminants on electrode (Feng et al. 2010). Electrochemical oxidation of graphite plate by applying +1.85 V (vs. Ag/AgCl) had gained attention due to its simple operation process and exhibited a similar increase (39–58%) in power density to the operation with heat and acid pretreated anodes in comparison to control (no pretreatment) (Lowy and Tender 2008). The simple pretreatment process of scrubbing the electrode by sandpaper could also increase the surface roughness of graphite rod/sheets/plates and enhance the bacterial adhesion for high power generation (Guo et al. 2015).

Surface coatings of an anode electrode have been tested as the other approach to increase anode performance. The surface coating materials include carbon nanotubes (CNT), conductive polymers (ex: poly aniline), immobilized mediators (neutral red), metals and composites of these materials (polymer/CNTs). CNT or modified forms of CNT with metals are most frequently used materials for increasing power generations. Electrochemical analysis such as cyclic voltammetry and electrochemical impedance spectroscopy revealed that the presence of CNT can increase electron transfer from bacteria to electrode and decrease the internal resistance of MFC as well (Qiao et al. 2007). The presence of poly aniline or polypyrole coating on MFC anodes enhanced maximum power densities, which were presumably due to increase in surface area and bacterial adhesion or due to enhanced current collection (Tsai et al. 2009). Neutral red (NR), methylene blue (MB) and 1,4-napthoquinone (NQ) were used as the mediator to efficiently transfer the electrons from bacterial cell to electrode (Wang et al. 2011a). However, the continuous supplementations of these mediators are needed to maintain their performance and also limited to batch mode of operation due to washout. These problems can be partially overcome by immobilizing the mediators on electrode surface, but in the long-term operation, the immobilized mediators can be deactivated or degraded. The stability of electrodes with immobilized mediators needs to be confirmed in operational practice.

10.3 Cathode Electrode

10.3.1 Cathode Electrode with Catalysts

The cathode catalyst plays a significant role in achieving enhanced cathode performance and decreasing construction costs. The cathode catalyst accounts up to 47% of capital cost in air cathode MFCs (Rozendal et al. 2008). Platinum (Pt) is the most commonly used cathode catalyst for oxygen reduction in both air and aqueous cathode MFCs. However, the Pt catalyst is limited to bench-scale application because of its expensive nature. Few researchers have tried to decrease the loading ratio of platinum without losing its catalytic performance. The deposition of Pt on carbon paper with electron beam evaporation reduced the thickness of Pt layer (1000 A) and reduced the loading with increase in current density (0.42 A m^{-2}) as compared to commercially available Pt-black (0.22 A m^{-2}; Vulcan XC-72, E-Tek) (Park et al. 2007). The combination of the transition metals (Fe, Co, Ni) with Pt can decrease the oxygen binding energies. The Pt-Co alloy at a ratio of 2:1 on cathode electrode could obtain higher power densities (1681 mW m^{-2}) than with the platinum coated carbon paper (1315 mW m^{-2}) (Yan et al. 2014). The MFC with MnO_2 catalysts could produce the maximum power densities ranging 88 to 172 mW m^{-2}, which were smaller than with Pt (268 mW m^{-2}) (Zhang et al. 2009a). Li et al. (2010) observed an increase in current generation (180 mW m^{-2}) by doping MnO_2 with cobalt (only with MnO_2 as control: 86 mW m^{-2}), which was similar to the value with Pt (198 mW m^{-2}). For coating of catalyst (ex. Pt or CoTMPP) on electrode substrate, the binders such as perfluorosulphonic acid (Nafion) and poly tetra fluroethylene (PTFE) are commonly used.

Cheg et al. pursued comparison studies in using of Nafion and PTFE as binders (catalyst: platinum) in single chamber air cathode MFCs. Air cathode operation with Nafion and PTFE exhibited maximum power densities of 480 and 331 mW m^{-2}, respectively. In this study they also suggested that biofilm formation on PTFE cathodes were thin and low in comparison to Nafion as a binder due to variation in hydrophobicity on cathode electrode (Cheng et al. 2006b). In the same study, they have tested CoTMPP coating on cathode with Nafion as a binder and noticed power generations of 369 mW m^{-2}. However, it is virtuous to know that Nafion costs up to 500 times more than PTFE. Zhang et al. (2010) used the PDMS (poly dimethyl siloxane), which were much cheaper than PTFE and could result in high CE of upto 80% and prevent water leakage. As an alternative to expensive catalyst on cathode and pretreatment process, researchers have developed biocathodes for decreasing over potentials. Moreover, the biocathodes have several advantages such as low cost, environmental friendly, regenerative nature, and they are also efficient for wastewater treatment, biogas generation and bioelectrochemical synthesis of fuels. Yo et al. operated MFCs with biotic carbon brush and produced 68.4 W m^{-3} power density, which was higher than with abiotic carbon brush (31.5 W m^{-3}) (Chae et al. 2009). In their study, the biocathodes with graphite plate and graphite granules could generate high power density of 51.4 W m^{-3} and 61.3 W m^{-3}, respectively.

10.3.2 Cathode Electrode Without Catalysts

MFCs using plain carbon paper without catalysts could produce a small power density of 0.04 W m^{-3} (cathode working volume) (Oh et al. 2004). Some researchers have used plain cathode electrodes with high surface area. Freguia et al. (2007) operated air cathode MFC with granular graphite and achieved a maximum power density of 50 W m^{-3} (based on cathode working volume). In comparison to the usage of graphite rods/plates and other plain limited surface area carbon materials (carbon cloth, carbon paper), activated carbon (AC) with high porosity was beneficial in obtaining higher power generation (Freguia et al. 2007). The pretreatment of AC cathodes with various chemicals such as HNO_3, H_3PO_4, KOH and H_2O_2 have been tested for oxygen reduction in comparison with Pt-coated cathodes (Duteanu et al. 2010). Among all the chemicals tested, MFC operation with HNO_3 pretreated AC exhibited a higher power density (170 mW cm^{-2}) than without pretreated AC (51 mW cm^{-2}), but it was relatively smaller than with Pt coated AC (217 mW cm^{-2}) (Duteanu et al. 2010).

10.4 Membranes/Separators Tested in MFC

10.4.1 Ion Exchange Membrane

Cation exchange membrane (CEM) or proton exchange membrane (PEM) is commonly used for protons or cations transfer from anode to cathode chamber in MFC operations. Most commonly used CEM is Nafion-117, which draws interest in hydrophobic fluorocarbon backbone (CF_2-CF_2) and sulphonate (SO_3) groups for diffusion of cations. Other frequently used CEM in MFC is CMI-7000, which is strong acid polymer membrane with gel polystyrene and divinyl benzene structure.

CMI-700 exhibits comparable cation conductivity, sustainability and mechanical durability to Nafion, but it tends to exhibit higher ohmic resistance due to difference in chemical structural and physical morphologies. Several alternative low cost CEMs such as SPEEK, Hyflon (sulphonated polyether ether ketone) were tested and exhibited better power densities, low ohmic resistance and higher conductivity than conventional CEMs (Nafion, CMI-700) (Ayyaru and Dharmalingam 2011; Ayyaru and Dharmalingam 2013; Ieropoulos et al. 2010). All CEM endured several limitations such as biofouling, competition of proton and cation and high cost for practical MFC operations. Competitive proton transport with other cations can lead to decrease of pH at anode, which would obviously increase the internal resistance and varies power densities (Li et al. 2011). AEM allows protons to efficiently diffuse from the anode to cathode by phosphate or carbonate anions as proton carrier and pH buffer.

The diffusion of phosphate anions (buffering effect) also contributes in enhancing performance of MFC. Kim et al. (2007) carried out comparison studies of MFC with between CEM and AEM and noticed maximum power densities of 0.48 and 0.61 W m^{-2}, respectively. However operation of MFC with AEM can be limited due to high substrate cross over and deformation in comparison to CEM. The bipolar membranes (BPM) are composed of two mono polar membranes (AEM and CEM), which helps in simultaneous transport of positive and negative ions. However, the competitive transport of electrolyte ions at the membrane can pose a major flux in charge balance, which can severely affect the performance of MFC by altering pH balance (Krol and Engineering Krol 1997). Low proton transferability of BPM limits the MFC application to wastewater treatment, but it still has interesting nature (two mono layers), which is helpful for desalination (Harnisch et al. 2008).

10.4.2 Size Selective Separators

The size selective separators can be differentiated into two categories based on their pore size: microporous separators (ex: UF and MF) and course pore separators (ex: GF, nylon mesh, fabric materials and cellulose filters). Numerous types of microporous filters were extensively studied in water and wastewater treatments due to their higher durability and performance. Recently these separators have been used in MFC for enhancing proton diffusion. Microporous separators can physically separate the anode and cathode solutions, and are able to diffuse the various charged and neutral species with their efficient pore size. Zuo et al. (2007) pursued a tubular MFC studies by coating UF with conductive and catalytic layer and observed the maximum power density of 17.7 W m^{-3}, which was higher than with CEM alone (6.6 W m^{-3}). Course pore filters such as fabrics, nylon mesh and glass fibres are much cheaper and exhibit a higher potential for practical application. Course pore materials such as NWF and J cloth have been reported to exhibit enhanced power densities due to their high transport capability of proton ions (Choi et al. 2013).

Kondaveeti et al. (2014) carried out comparison studies between size-selective separators and ion exchange membrane. In these studies they have noted increase in power generations with size selective separators (PP100: 117 mW m^{-2}; PPS: 102 mW m^{-2}, S-PPS:190 mW m^{-2}), in comparison to ion exchange membrane (Nafion: 24 mW m^{-2}; CMI-7000: 78 mW m^{-2}), which were attributed to abundant ion transfer, high cathode potentials and low pH at cathode (Moon et al. 2015b). An MFC with non-woven fabric (NWF) separator produced higher power density of 1027 mW m^{-2} than with Nafion (609 mW m^{-2}) (Choi et al. 2013). However, size selective separators are limited with higher diffusions of oxygen and substrate cross over. The larger pore size of course pore filters (NWF) adversely affects the MFC performance due to high permeation of oxygen (7.0×10^{-4} cm/s) in comparison with Nafion (6.7×10^{-4} cm s^{-1}) (Choi et al. 2013). Few researchers have overcome this problem by increasing the distance of electrode. However, with the increase of electrode spacing internal resistance of MFC would be increased (Moon et al. 2015a). Further studies would be needed in optimizing the performance of MFC with size-selective separators.

10.5 Reactor Configurations

Reactor configurations can govern the MFC performance with the variations in reactor volume, electrode spacing, membrane area, oxygen supply and hydraulic flow pattern. The diverse configurations of MFCs are available, which are ranging from microlitres to thousands of litres in size. Among various configurations, double chamber "H" type MFCs have been typically used. The key aspect of this configuration is ion exchange compartment which helps in diffusion of protons and limits the substrate and oxygen crossover (Kondaveeti et al. 2014). However, the relatively longer distance of electrodes can limit high MFC performance due to increase in internal resistance. In order to decrease internal resistance, researchers have operated MFCs by clamping electrodes along with membranes or separators (Choi et al. 2013). By using closer spacing of electrodes in double chamber MFC, Choi et al. noted a smaller internal resistance of 93 Ω (57 mW m^{-2}), in comparison to other studies of MFC using larger spacing 672 Ω (103 mW m^{-2}) (Sun et al. 2009). However, placing the electrodes closer to membrane can result in higher oxygen diffusions from cathode to anode. The difference in internal resistance due to different configurations of MFCs can affect directly the power density, and the unknown power density can be calculated based on the internal and external resistance of the two systems using the following equation (Min et al. 2005; Kondaveeti et al. 2014).

$$P_{\mathrm{a}}=\frac{P_{\mathrm{h}}\times R_{\mathrm{i,h}}^{2}}{R_{\mathrm{e,h}}}\times\frac{R_{\mathrm{e,a}}}{R_{\mathrm{i,a}}^{2}} \tag{10.1}$$

P_a is power density (mW m^{-2}) with advanced configuration of MFC, P_h is power density with an H type MFC, $R_{i,h}$ is internal resistance of an H type MFC, and $R_{i,a}$ is internal resistance of an advanced cell. Assume that external resistance ($R_{e,h}$, $R_{e,a}$) of each cell for maximum power generation is equal to internal resistance. Min et al. reported that MFC configuration with a relatively advanced design for ion transports (membrane; system internal resistance: 1286 Ω) could produce about 17 times higher power generation than with a salt bridge compartment (19,920 Ω internal resistance). They could also estimate the power density value (2.8 mW m^{-2}) using the Eq. 10.1 from the MFC with a salt bridge, which was similar to the measured one (2.2 mW m^{-2}), based on the information about the system resistances (Min et al. 2005). Cube type MFCs produced about 13.5 times higher power generation (514 mW m^{-2}; 84 Ω internal resistance) than using bottle type MFCs (38 mW m^{-2}; 1272 Ω internal resistance) (Kim et al. 2007). It is not necessary to aerate the cathode when using oxygen as final electron acceptor, which leads to develop single chamber MFCs with air cathode assembly. These configurations have gained attention due to several advantages such as simple nature of operation, decrease in internal resistance, enhanced oxygen reduction rate on cathode, increased proton diffusion and reduced electrode spacing. Logan et al. (2007) carried out the comparison study between single chamber cube and bottle types of MFCs and noted a smaller internal resistance (8 Ω) and higher power density with cube MFC (2400 mW m^{-2}) than with bottle type MFCs (50 Ω; 1200 mW m^{-2}). Fan et al. (2007) could produce higher power density (303 W m^{-3}) with sandwich type MFCs with MEA than with MFC (71 W/m^3) having 1.7 cm spacing between anode and cathode electrodes (Fan et al. 2007).

One of the major limitations in using a single chamber MFC is water leakage, evaporation and high oxygen diffusion. Cheng et al. (2006a) found that PTFE diffusion layers on the cathode could decrease oxygen diffusion and water loss, which results in increase in Columbic efficiency and maximum power densities. They could increase CE about 1.7 times higher with four PTFE layers (32%) than without coating (19.1%), and also high power density (766 mW m^{-2}) with PTFE coating was obtained with a noticeable decrease in water loss (without coating, 538 mW m^{-2}). The other configurations with low internal resistance and high power density include packed bed upflow MFC, flat plate MFC, and tubular MFC. He et al. (2005, 2006) operated two different configurations of UMFCs, and one reactor produced the maximum power density of 170 mW m^{-2} (Internal resistance: 84 Ω) and the other one generated 29.2 W m^{-3} (internal resistance: 17 Ω). Rabaey et al. operated a tubular MFC and obtained the maximum power density of 48 mW m^{-2} (internal resistance: 4 Ω). In this study, they have noticed a COD removal of less than 20% and a higher conversion of sulphate to sulphide suggesting the alternate electron acceptor. Min and Logan (2004) developed a flat plate MFC and noticed the maximum power density of 309 mW m^{-2}. Li et al. (2009) operated a wetted wall MFC (WWMFC) and noticed a maximum power density of 18.2 W m^{-3} (420 Ω internal resistance). A submersible MFC with MEA assembly was developed and the maximum power density of 631 mW m^{-2}, and the internal resistance of 82 Ω were observed (Min et al. 2012).

10.6 Effect of Temperature on MFC Performance

Temperature is one of the crucial governing factors in MFC power generation. Temperature variations in MFC system could affect the kinetics (activation energy, mass transfer coefficient, solution conductivity), Gibbs free energy, electrode potential, chemical solubility and distribution and activity of microorganisms on the electrodes (Larrosa-Guerrero et al. 2010). Min et al. (2008) reported that operational temperature of 30 °C would be beneficial in decrease of lag period and increase in power density. Guerrini et al. (2014) found that increase in operational temperature from 4 to 35 °C resulted in an increase in current density (20 to 100 mA m^{-2}) and COD removal rate (16 to 39 mg L^{-1} $hr.^{-1}$) (Larrosa-Guerrero et al. 2010). In this study, they have also noted an increase of initial pH and solution conductivity with an increase in temperature. Liu et al. (2011) operated MFC at different operational temperatures (30, 35, 40 and 45 °C) and observed an increase in the maximum power densities from 764 to 1065 mW m^{-2} with temperature increase from 30 to 45 °C (internal resistance decrease from 177 to 121 Ω). However, with a further increase to 53 °C, no presence of catalytic activity were noticed. The anode and cathode potentials were decreased by 36.2% (−392 to −250 mV) and by 16.6% (246 to 187 mV), respectively, with temperature decrease from 37 to 10 °C (Li et al. 2013).

10.7 Electrolyte pH in Governing MFC Performances

Electrolyte (anolyte and catholyte) pH in MFC is important parameter in governing the growth and metabolism of bacteria on anode, transport of proton from anode to cathode, and oxygen reduction on cathode. The electricity generation in MFC starts with generation of electron and protons (H^+) in the anode which eventually is transferred to cathode for the reduction process (Kakarla et al. 2015; Kakarla and Min 2014a). Most of the studies till now make use of buffer to control the pH changes and prevent potential imbalances in-between anode and cathode (Kakarla and Min 2014b). Without a buffer system, the pH at anode is always acidic condition due to H^+ production from the oxidation of organic substrates by the bacteria preset on anode electrode and, in the cathodic chamber, the pH on the cathode electrode becomes at alkaline condition due to the formation of OH^- ions. The optimum pH conditions, invariably, exhibit a positive effect on MFC current generations as the generation of electron and metabolism of electrogenic bacteria present on the anode electrode. The optimum pH conditions vary among the different bacterial species in an anode chamber (He et al. 2008). Vologni et al.

(2013) reported that the maximum power densities with primary sludge from a wastewater treatment plant were observed at neutral pH conditions rather than acidic conditions. In some other study, a fed batch MFC with anaerobic sewage sludge generated higher maximum power densities at pH 6 (237 mW m^{-2}) and pH 7 (182 mW m^{-2}) than at pH 4 (95 mW m^{-2}) and pH 5 (116 mW m^{-2}), suggesting that at lower pH conditions microbes are prone to acidic shock (Zhang et al. 2009b). He et al. (2008) reported that the anodic polarization resistance was decreased with increase in pH from 5 to 7, but further increase in pH from 7 to 10 the anodic polarization resistance was increased. However, some studies had reported that the maximum MFC performance was observed at more alkaline condition (at pH 10.5) with use of pure culture *Bacillus* strain (Akiba et al. 1987). Similarly, the maximum power densities of 87.2 mW m^{-2} and 107 mW m^{-2} were observed from an air-cathode MFC between pH 8 and pH 10, respectively, whereas this maximum power density was dropped to 21 mW m^{-2} at pH 5 (He et al. 2008).

The MFC Coulombic efficiencies were also greatly influenced by operational pH conditions. It was reported that in a SMFC without pH control the pH of the medium dropped from 6.2 to 4.9 with Coulombic efficiency of 1.06%. However, the CE was increased to 16% at a neutral pH with addition of 50 mM PBS buffer (Vologni et al. 2013). Similar findings were observed with a single chamber air cathode MFC fed with brewery wastewater, in which CE was increased from 10% (pH drop from 5.7 to 4.5 without buffer) to 16% with addition of 50 mM PBS (Yang et al. 2013). Few studies also reported internal resistances of both anode and cathode electrodes at varied pH conditions. He et al. reported that the anodic polarization resistance (Rp^a) decreased from pH 5 to 7 and then increased after reaching a minimum pH of 7 indicating that Rp^a is related to bacterial activity on anode electrode. The lowest Rp^a occurred at pH 7 showing that a neutral pH is beneficial to bacterial activity. The pH values higher or lower than 7 had restricted the bacterial activity on anode electrode. While the cathodic polarization resistance (Rp^c) continuously decreased from pH 5 to 10, indicating that the oxygen reaction increased at a higher pH (He et al. 2008). The redox potential of oxygen reduction on cathode at pH 7 was calculated by the Nernst equation shown in Eqs. 10.2, 10.3. Here, E is half cell reduction potential, $E°$ is standard potential, R is universal gas constant, T is temperature in Kelvin, n is number of electrons transferred and F is Faraday's constant (the number of Coulombs per mole of electrons).

$$E = E^{o} - \frac{RT}{nF}\ln\frac{1}{[O_2]^{1/2}[H^+]^2} \tag{10.2}$$

$$E = E^{o} - \frac{(8.31\,\text{J/mol K})(298.15\ \text{K})}{(2)(96{,}500\ \text{C/mol})}\ln\frac{1}{[0.2]^{1/2}[10^{-7}]^2} = 0.805\text{V} \tag{10.3}$$

Logan et al. (2006) previously reported different oxygen redox potentials of cathode at different pHs, which were 0.805 V (pH 7) and 0.627 V (pH 10), respectively. You et al. (2006) reported that with decreasing the pH from 7 to 3.5 the open circuit potential (OCP) was increased from 0.90 V to 1.38 V, which could be due to high redox potential (1.70 V, Eq. 10.4) of permanganate at acidic conditions. However, an increase in alkaline conditions had shown a negative lead to drop in OCP and cathode potentials as shown in Eq. 10.5.

$$MnO_4^- + 4H + 4e^- \rightarrow MnO_2 + 2H_2O, E^o = 1.70\ V \tag{10.4}$$

$$MnO_4^- + 2H_2O + 3e^- \rightarrow MnO_2 + 4OH^-, E^o = 0.59\ V \tag{10.5}$$

10.8 Electrolyte Conductivity

Electrolyte of MFC can be classified into anolyte (anodic solution) and catholyte (cathodic solution), and both electrolytes play a crucial role in preventing potential imbalances for better MFC performance. The electrolyte internal resistance was increased from 47.3% to 82.2% when the phosphate buffer concentration (conductivity) was decreased from 200 mM to 50 mM (Fan et al. 2008). In some other studies (membrane free single chambered MFC) with addition of supplementary NaCl (100 mM) to 100 mM phosphate buffer had decreased the MFC system internal resistance from 161 Ω to 91 Ω (200 mM) (Liu et al. 2005a). However, the further increase of electrolyte conductivity from 200 mM to 400 mM didn't drop the system internal resistance significantly. The MFC internal resistance including anode, cathode and solution were greatly affected by electrolyte conductivity. Fan et al. (2008) reported that for all MFCs, the electrolyte resistance contributed around 36–78% of total internal resistance, while the resistance of anode contribution was measured to be around 2 to 35%. In the MFCs with equally sized anodes and cathodes (7 cm^2) and 200 mM phosphate buffer, the anode only accounted for 5.4% of the internal resistance, while the cathode and the electrolyte each contributed 47.3%. Moreover, with increasing the phosphate buffer concentration from 50 to 200 mM, the total internal resistance of MFC is decreased from 208 Ω to 85.9 Ω, whereas the electrolyte internal resistance was decreased from 78.2 to 47.3%. Cheng and Logan (2011) observed the increase of power densities of MFC from 16 W m^{-3} to 33 W m^{-3} (107%) with increasing the electrolyte conductivity from 1.7 mS cm^{-1} (10 mM PBS) to 7.8 mS cm^{-1} (50 mM PBS). Further increases in conductivity to 20 mS cm^{-1} (200 mM PBS) increased the power up to 60 m^{-3}. Similarly, in some other studies the MFC maximum power density of 2680 mW m^{-2} were increased to 4000 mW m^{-2} (49% increases) with increase in phosphate buffer concentration from 50 mM to 100 mM. Further increase in buffer concentration to 200 mM resulted in an additional 39% increase in power density (Fan et al. 2008). In some other study the addition of buffer to a primary sludge (50 mM PBS) had increased the average power generation of SMFC from 0.140 mW m^{-2} to 0.204 mW m^{-2}, which was about 1.5 times higher than the value without the phosphate buffer addition (Vologni et al. 2013). Liu et al. (2005a) observed 85% increase of power density when the solution conductivity was

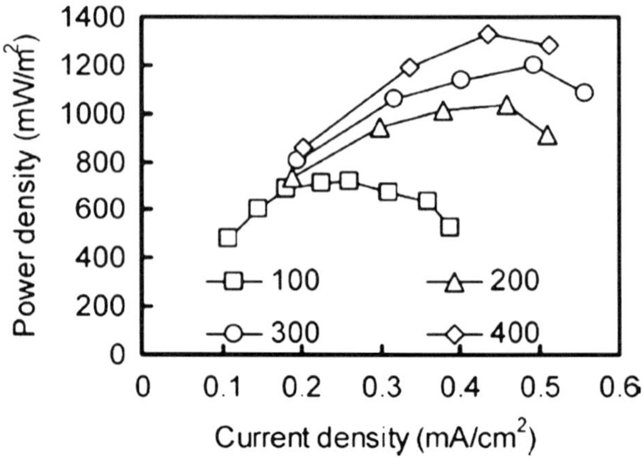

Fig. 10.2 Effect of solution conductivity (100–400 mM) on MFC power generation (Liu et al. 2005a)

increased from 100 mM (720 mW m^{-2}) to 400 mM (1330 mW m^{-2}) in a single chambered air cathode MFC (Fig. 10.2).

Wang et al. (2011b) has tested the effect of NaCl addition (0–250 mM) in air cathode MFC with Pt (platinium) cathode electrode. He found that with increasing the NaCl concentration from 0 to 100 mM the power densities of MFC with Pt cathode were increased by 15% (from 798 mW m^{-2} to 921 mW m^{-2}). However, a slight decrease in the power density was observed with further increase of NaCl from 100 to 250 mM. Vologni et al. (2013) found that addition of 50 mM to a SMFC with primary sludge could increase the coulombic efficiency from 1.07% to 16% due to incresed solution conductivity from 0.8 mS cm^{-1} to 8.4 mS cm^{-1}. Yang et al. also observed that CE was increased from 10 to 16% with buffer increase from 50 to 200 mM PBS in a single chamber air cathode MFC with brewery wastewater (Yang et al. 2013).

10.9 Oxidants in an MFC Cathode

The oxidants used in the MFC cathode play a major role in determining half-cell potential and overall MFC performances. Several oxidants are widely being used in MFC cathode as an electron acceptor including O_2 (dissolved and gaseous), $Fe(CN)_6^{3-}$, NO_3^-, SO_4^{2-}, Fe^{3+}, fumarate, MnO_4^{2-}, ubiquinone, and various types of Cytochromes including Cytochrome a, b and c. Some of the common oxidants in cathode along with reduced forms in MFC cathode are shown in Fig. 10.3.

The most commonly employed oxidant for MFC cathode is oxygen due to its high reduction potential of 0.82 V and availability (Logan 2008). However, in practice cathode reduction potential with oxygen is lower than the theoretical value. General open circuit potential (OCP) of an air cathode is around 0.4 V, with an approximate short circuit potential of 0.25 V having a Pt coated cathode electrode

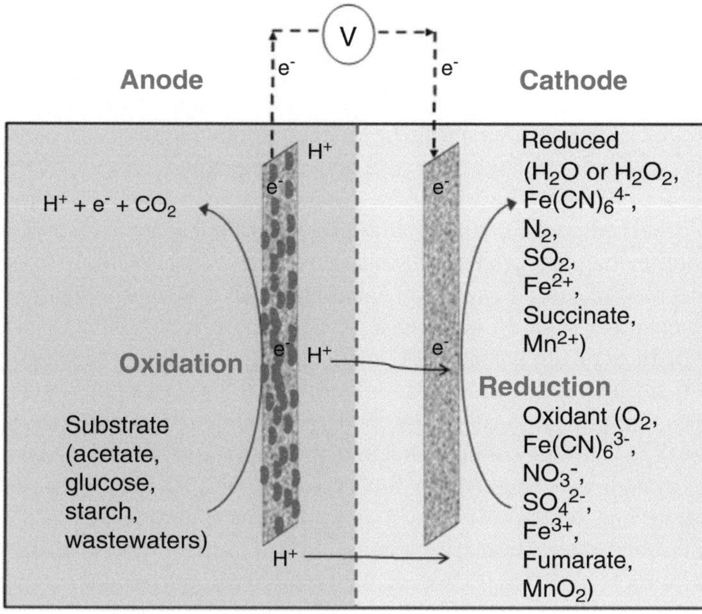

Fig. 10.3 Possible electron acceptors (oxidizers) in MFC cathode chamber

as previously reported (Liu and Logan 2004). The reduction of oxygen to form water needs four electrons (Eq. 10.6). However, this process could be hindered with formation of hydrogen peroxide with use of two electrons. The reduction potential of this hydrogen peroxide formation is a bit lower than the water formation reaction, which is about 0.695 V (Eq. 10.7). The formation of hydrogen peroxide at cathode could result in degradation of cathode materials and membranes of MFC. However, the formation of hydrogen peroxide might act as a disinfectant in keeping the cathode and membrane free from microbial contamination (Logan et al. 2006).

$$O_2 + 4H^+ + 4e^- \rightarrow 2H_2O \tag{10.6}$$

$$O_2 + 2H^+ + 2e^- \rightarrow H_2O_2 \tag{10.7}$$

In some other studies, researchers had tried to use regenerative chemical of ferricyanide [$Fe(CN)_6{}^{3-}$] which has higher reduction potentials than oxygen but are unsustainable. The use of ferricyanide eliminates the necessity to use precious catalysts like platinum with a high cathode reduction potential compared with use of oxygen. This could be due to slight polarization of cathode electrode as reported previously (You et al. 2006).

$$4Fe(CN)_6^{3-} + 4e^- \rightarrow 4Fe(CN)_6^{4-} \tag{10.8}$$

Mohan et al. (2009) observed 11% higher voltage generation (0.651 V) with ferricyanide catholyte (50 mM PBS) than the MFC (0.578 V) with aeration. Similarly observations were found by Wei et al. (2012) in which the MFC catholyte aerated ferricyanide had generated 155% and 111% higher cell voltage than MFCs catholytes with only aeration and ferricyanide without aeration, respectively. You et al. (2006) study reported that with use of permanganate (1.7 V, pH 3.5) the OCP of cathode was found to be higher as compared to oxygen (0.8 V, pH 7). The maximum observed power densities with use of permanganate (115 mW m^{-2}) was 11.3-fold higher than that produced by using oxygen (10.2 mW m^{-2}). In some other studies, enzyme (laccase) was implemented alternative to oxygen and ferricyanide in cathode chamber with an oxygenase, assisted by a redox mediator (Schaetzle et al. 2009). In their study, the MFC cathode with 2,2′-azino-bis (3-ethylbenzothiazoline-6-sulfonic acid) diammonium salt (ABTS)/laccase/O_2 system yielded higher maximum power density of 46 mW m^{-2} compared to with use of ferricyanide (31 mW m^{-2}). This could be due to higher redox potentials (1.1 V) of ABTS/laccase/O_2 system as compared to ferricyanide (0.42 V, pH 5). In some other studies, nitrate and sulphate was used as a final electron acceptor in the cathode through microbial interaction (biocathode) with minimal amount of oxygen concentration (Song et al. 2015). However, the reported cathode reductions potential of cathode with use of biocathode and oxygen was smaller as compared to use of oxygen (Cournet et al. 2010; Freguia et al. 2010). Some of redox couples and respective reduction potentials for MFC cathode were explained (Logan 2008; Logan et al. 2006).

The MFC cell and its cathode potentials can be varied according to concentration of reactant (partial pressures of oxidizers; example use of pure O_2 instead of normal air) in the cathode according to the Nernst's equation (Eq. 10.3). With increase in partial pressures of oxygen in cathode there are three possible effects as reported previously (James Larmine 2003): (1) The increased open circuit potential due to increase in oxygen partial pressure; (2) The decreased activation losses at cathode because of better use of maximum catalyst sites on cathode electrode; and (3). The increase in limiting current with reduced mass transport or concentration overvoltage losses. This type of observations was reported in Kakarla et al. (2015) study in which an air cathode MFC performance and its cathode potentials were evaluated at increased oxygen concentrations, supplied from an externally connected algae bioreactor. In this study, they have reported that open circuit cell voltage (OCV) and cathode (OCP) with use of normal air (20.8% O_2) were 0.715 V (vs. Ag/AgCl) and 0.348 V (vs. Ag/AgCl), respectively. With 39% oxygen supplied by algae bioreactor, the OCV and OCP were increased by 13% (0.806 V) and 54% (0.536 V) compared with normal air. This measured OCP of 0.733 V (vs. NHE) at 39% O_2 level was very close to the standard cathode potential of 0.805 V (Logan 2008).

10.10 Substrates (Fuels) in the MFC Anode Chamber

The substrate is considered as one of the most vital factor in determining the MFC's maximum electricity generation (Li et al. 2009). Up to now a great variety of substrates was successfully used in MFCs for electricity generation, ranging from simple substrates (pure) to complex substrates (mixed). The sources include domestic wastewaters, agriculture wastes, animal and industrial wastewaters, which are different in their organic strength, complexity and microbial community. The complexity, chemical oxygen demand (COD) concentration, pH and bacterial community of wastewaters determine the overall MFC performance in getting its maximum electricity generation. It has been reported that complex substrates are generally difficult to be utilized as a carbon source for electricity generation compared with simple substrate like acetate (Pant et al. 2010). The complex substrates need more time and energy to be first metabolized to simpler molecules before being used as a fuel. The MFC with simple substrates like acetate and glucose could generate higher electricity generation and Coulombic efficiency (CE) than with complex substrates (starch and cellulose). In Chae et al. (2009) study, the double chambered MFC fed with acetate had showed the highest CE of 72.3% as compared to butyrate (43.0%), propionate (36.0%) and glucose (15.0%). This lower CE with glucose was due to non-electrogenic process such as fermentation or methanogenesis. However, the MFC had showed the highest power density of 156 mW m^{-2} with use of glucose as a substrate; whereas with use of acetate, propionate and butyrate substrates the maximum power densities were 64.3 mW m^{-2}, 58.0 mW m^{-2} and 51.4 mW m^{-2}, respectively.

A similar result was reported from a flat plate air cathode MFC (Min and Logan 2004). However, in their study, the highest power density was observed with acetate (309 mW m^{-2}) rather than with glucose (220 mW m^{-2}). Various wastes and wastewaters have been also fed to the anode chamber of several MFCs at the substrate source. Vologni et al. (2013) operated a sedimental MFC (SMFC) fed with primary sludge which exhibited a voltage generation of 0.25 V with a fixed external resistance of 470 Ω. Jiang et al. (2009) operated a double chambered MFC and noticed 0.687 V as voltage generation from primary sludge (sewage sludge) with a 1000 Ω resistor and 50 mM $K_3[Fe(CN)_6]$ catholyte. The voltage generations from a double chambered MFC operated using swine wastewaters produced a cell voltage of 0.357 V with fixed external resistance of 1000 Ω (Min et al. 2005). The substrate concentration has a great effect on power densities and CE of MFCs. Vologni et al. (2013) studied different initial SCOD concentrations ranging from 0.018 to 0.4 g L^{-1}, where the power density was linearly increased up to 0.1 g L^{-1} of SCOD, but the slope of power vs. SCOD was decreased with increased SCOD. A similar observation was reported with increased initial SCOD concentrations (Min and Angelidaki 2008; Zhang et al. 2011). Min et al. (2005) double chambered MFC with use of swine wastewaters had produced a maximum power density of 182 mW m^{-2} with a CE of 8% and SMFC with use of primary sludge without buffer had shown a CE of 1.6% (Vologni et al. 2013). Moreover, with increasing the initial wastewater TCOD concentration from 500 to 4500 mg L^{-1}, CE was dropped from 26% to 10% (Min et al. 2005). A similar type of observation was found in Liu et al. (2005b) study, in which increasing acetate concentration from 80 to 800 mg/l decreased the CE from 28% to 13%.

10.11 Conclusions

The physical and chemical parameters play a major role in governing the MFC performance by affecting the anode and cathode catalytic abilities, the system internal resistance and sustainability. The optimization of physical and chemical factors including MFC architecture, electrode and catalyst materials, membrane/separator, temperature, pH, conductivity and substrate concentration are very crucial for efficient, economic and sustainable MFC power generation. Especially when MFCs are applied to the real field, the main physicochemical parameters affecting the MFC operations should be thoroughly determined and optimized.

References

Ahn, Y., & Logan, B. E. (2012). Altering anode thickness to improve power production in MFCs with different electrode distances. *Energy & Fuels, 27*(1), 271–276.

Akiba, T., Bennetto, H. P., Stirling, J. L., & Tanaka, K. (1987). Electricity production from alkalophilic organisms. *Biotechnology Letters, 9*(9), 611–616.

Ayyaru, S., & Dharmalingam, S. (2011). Development of MFC using sulphonated polyether ether ketone (SPEEK) membrane for electricity generation from waste water. *Bioresource Technology, 102*(24), 11167–11171.

Ayyaru, S., & Dharmalingam, S. (2013). Improved performance of MFCs using sulfonated polyether ether ketone (SPEEK) TiO_2-SO_3H nanocomposite membrane. *RSC Advances, 3*(47), 25243–25251.

Chae, K.-J., Choi, M.-J., Lee, J.-W., Kim, K.-Y., & Kim, I. S. (2009). Effect of different substrates on the performance, bacterial diversity, and bacterial viability in MFCs. *Bioresource Technology, 100*(14), 3518–3525.

Chaudhuri, S. K., & Lovley, D. R. (2003). Electricity generation by direct oxidation of glucose in mediatorless MFCs. *Nat Biotech, 21*(10), 1229–1232.

Cheng, S., & Logan, B. E. (2007). Ammonia treatment of carbon cloth anodes to enhance power generation of MFCs. *Electrochemistry Communications, 9*(3), 492–496.

Cheng, S., & Logan, B. E. (2011). Increasing power generation for scaling up single-chamber air cathode MFCs. *Bioresource Technology, 102*(6), 4468–4473.

Cheng, S., Liu, H., & Logan, B. E. (2006a). Increased performance of single-chamber MFCs using an improved cathode structure. *Electrochemistry Communications, 8*(3), 489–494.

Cheng, S., Liu, H., & Logan, B. E. (2006b). Power densities using different cathode catalysts (Pt and CoTMPP) and polymer binders (Nafion and PTFE) in single chamber MFCs. *Environmental Science & Technology, 40*(1), 364–369.

Choi, S., Kim, J. R., Cha, J., Kim, Y., Premier, G. C., & Kim, C. (2013). Enhanced power production of a membrane electrode assembly MFC (MFC) using a cost effective poly

[2,5-benzimidazole] (ABPBI) impregnated non-woven fabric filter. *Bioresource Technology, 128*, 14–21.

Cournet, A., Bergé, M., Roques, C., Bergel, A., & Délia, M.-L. (2010). Electrochemical reduction of oxygen catalyzed by Pseudomonas aeruginosa. *Electrochimica Acta, 55*(17), 4902–4908.

Dumas, C., Mollica, A., Féron, D., Basséguy, R., Etcheverry, L., & Bergel, A. (2007). Marine MFC: Use of stainless steel electrodes as anode and cathode materials. *Electrochimica Acta, 53*(2), 468–473.

Duteanu, N., Erable, B., Senthil Kumar, S. M., Ghangrekar, M. M., & Scott, K. (2010). Effect of chemically modified Vulcan XC-72R on the performance of air-breathing cathode in a single-chamber MFC. *Bioresource Technology, 101*(14), 5250–5255.

Fan, Y., Hu, H., & Liu, H. (2007). Enhanced Coulombic efficiency and power density of air-cathode MFCs with an improved cell configuration. *Journal of Power Sources, 171*(2), 348–354.

Fan, Y., Sharbrough, E., & Liu, H. (2008). Quantification of the internal resistance distribution of MFCs. *Environmental Science & Technology, 42*(21), 8101–8107.

Feng, Y., Yang, Q., Wang, X., & Logan, B. E. (2010). Treatment of carbon fiber brush anodes for improving power generation in air–cathode MFCs. *Journal of Power Sources, 195*(7), 1841–1844.

Freguia, S., Rabaey, K., Yuan, Z., & Keller, J. (2007). Non-catalyzed cathodic oxygen reduction at graphite granules in MFCs. *Electrochimica Acta, 53*(2), 598–603.

Freguia, S., Tsujimura, S., & Kano, K. (2010). Electron transfer pathways in microbial oxygen biocathodes. *Electrochimica Acta, 55*(3), 813–818.

Guerrini, E., Cristiani, P., Grattieri, M., Santoro, C., Li, B., & Trasatti, S. (2014). Electrochemical behavior of stainless steel anodes in membraneless MFCs. *Journal of the Electrochemical Society, 161*(3), H62–H67.

Guo, K., Prévoteau, A., Patil, S. A., & Rabaey, K. (2015). Engineering electrodes for microbial electrocatalysis. *Current Opinion in Biotechnology, 33*, 149–156.

Harnisch, F., Schröder, U., & Scholz, F. (2008). The suitability of monopolar and bipolar ion exchange membranes as separators for biological fuel cells. *Environmental Science & Technology, 42*(5), 1740–1746.

He, Z., Minteer, S. D., & Angenent, L. T. (2005). Electricity generation from artificial wastewater using an upflow MFC. *Environmental Science & Technology, 39*(14), 5262–5267.

He, Z., Wagner, N., Minteer, S. D., & Angenent, L. T. (2006). An upflow MFC with an interior cathode: Assessment of the internal resistance by impedance spectroscopy. *Environmental Science & Technology, 40*(17), 5212–5217.

He, Z., Huang, Y., Manohar, A. K., & Mansfeld, F. (2008). Effect of electrolyte pH on the rate of the anodic and cathodic reactions in an air-cathode MFC. *Bioelectrochemistry, 74*(1), 78–82.

Ieropoulos, I., Greenman, J., & Melhuish, C. (2010). Improved energy output levels from small-scale MFCs. *Bioelectrochemistry, 78*(1), 44–50.

James Larmine, A. D. (2003). *Fuel cell systems explained*. Chichester: John Wiley and Sons Ltd.

Jiang, J., Zhao, Q., Zhang, J., Zhang, G., & Lee, D.-J. (2009). Electricity generation from bio-treatment of sewage sludge with MFC. *Bioresource Technology, 100*(23), 5808–5812.

Kakarla, R., & Min, B. (2014a). Evaluation of MFC operation using algae as an oxygen supplier: Carbon paper cathode vs. carbon brush cathode. *Bioprocess and Biosystems Engineering, 37*(12), 2453–2461.

Kakarla, R., & Min, B. (2014b). Photoautotrophic microalgae Scenedesmus obliquus attached on a cathode as oxygen producers for MFC (MFC) operation. *International Journal of Hydrogen Energy, 39*(19), 10275–10283.

Kakarla, R., Kim, J. R., Jeon, B.-H., & Min, B. (2015). Enhanced performance of an air-cathode MFC with oxygen supply from an externally connected algal bioreactor. *Bioresource Technology, 195*, 210–216.

Kim, J. R., Cheng, S., Oh, S.-E., & Logan, B. E. (2007). Power generation using different cation, anion, and ultrafiltration membranes in MFCs. *Environmental Science & Technology, 41*(3), 1004–1009.

Kondaveeti, S., Lee, J., Kakarla, R., Kim, H. S., & Min, B. (2014). Low-cost separators for enhanced power production and field application of MFCs (MFCs). *Electrochimica Acta, 132*, 434–440.

Krol, J. J. (1997). *Monopolar and bipolar ion exchange membranes: Mass transport limitations*. Enschede: Department of Chemical Engineering/University of Twente. doc.utwente.nl/9372/1/t0000002.pdf.

Larrosa-Guerrero, A., Scott, K., Head, I. M., Mateo, F., Ginesta, A., & Godinez, C. (2010). Effect of temperature on the performance of MFCs. *Fuel, 89*(12), 3985–3994.

Li, Z., Zhang, X., Zeng, Y., & Lei, L. (2009). Electricity production by an overflow-type wetted-wall MFC. *Bioresource Technology, 100*(9), 2551–2555.

Li, X., Hu, B., Suib, S., Lei, Y., & Li, B. (2010). Manganese dioxide as a new cathode catalyst in MFCs. *Journal of Power Sources, 195*(9), 2586–2591.

Li, W.-W., Sheng, G.-P., Liu, X.-W., & Yu, H.-Q. (2011). Recent advances in the separators for MFCs. *Bioresource Technology, 102*(1), 244–252.

Li, L. H., Sun, Y. M., Yuan, Z. H., Kong, X. Y., & Li, Y. (2013). Effect of temperature change on power generation of MFC. *Environmental Technology, 34*(13–14), 1929–1934.

Liu, H., & Logan, B. E. (2004). Electricity generation using an air-cathode single chamber MFC in the presence and absence of a proton exchange membrane. *Environmental Science & Technology, 38*(14), 4040–4046.

Liu, H., Cheng, S., & Logan, B. E. (2005a). Power generation in fed-batch MFCs as a Function of ionic strength, temperature, and reactor configuration. *Environmental Science & Technology, 39*(14), 5488–5493.

Liu, H., Cheng, S., & Logan, B. E. (2005b). Production of electricity from acetate or butyrate using a single-chamber MFC. *Environmental Science & Technology, 39*(2), 658–662.

Liu, Y., Climent, V., Berná, A., & Feliu, J. M. (2011). Effect of temperature on the catalytic ability of electrochemically active biofilm as anode catalyst in MFCs. *Electroanalysis, 23*(2), 387–394.

Logan, B. E. (2008). *Microbial fuel cells*. Hobokem: John Wiley and Sons Inc..

Logan, B. E., Hamelers, B., Rozendal, R., Schröder, U., Keller, J., Freguia, S., Aelterman, P., Verstraete, W., & Rabaey, K. (2006). MFCs: Methodology and technology. *Environmental Science & Technology, 40*(17), 5181–5192.

Logan, B., Cheng, S., Watson, V., & Estadt, G. (2007). Graphite fiber brush anodes for increased power production in air-cathode MFCs. *Environmental Science & Technology, 41*(9), 3341–3346.

Lowy, D. A., & Tender, L. M. (2008). Harvesting energy from the marine sediment-water interface: III. Kinetic activity of quinone- and antimony-based anode materials. *Journal of Power Sources, 185*(1), 70–75.

Min, B., & Angelidaki, I. (2008). Innovative MFC for electricity production from anaerobic reactors. *Journal of Power Sources, 180*(1), 641–647.

Min, B., & Logan, B. E. (2004). Continuous electricity generation from domestic wastewater and organic substrates in a flat plate MFC. *Environmental Science & Technology, 38*(21), 5809–5814.

Min, B., Kim, J., Oh, S., Regan, J. M., & Logan, B. E. (2005). Electricity generation from swine wastewater using MFCs. *Water Research, 39*(20), 4961–4968.

Min, B., Román, Ó. B., & Angelidaki, I. (2008). Importance of temperature and anodic medium composition on MFC (MFC) performance. *Biotechnology Letters, 30*(7), 1213–1218.

Min, B., Poulsen, F. W., Thygesen, A., & Angelidaki, I. (2012). Electric power generation by a submersible MFC equipped with a membrane electrode assembly. *Bioresource Technology, 118*, 412–417.

Mohan, S. V., Raghavulu, S. V., Peri, D., & Sarma, P. N. (2009). Integrated function of MFC (MFC) as bio-electrochemical treatment system associated with bioelectricity generation under higher substrate load. *Biosensors and Bioelectronics, 24*(7), 2021–2027.

Moon, J. M., Kondaveeti, S., Lee, T. H., Song, Y. C., & Min, B. (2015a). Minimum interspatial electrode spacing to optimize air-cathode MFC operation with a membrane electrode assembly. *Bioelectrochemistry, 106*(Part B), 263–267.

Moon, J. M., Kondaveeti, S., & Min, B. (2015b). Evaluation of low-cost separators for increased power generation in single chamber MFCs with membrane electrode assembly. *Fuel Cells, 15*(1), 230–238.

Oh, S., Min, B., & Logan, B. E. (2004). Cathode performance as a factor in electricity generation in MFCs. *Environmental Science & Technology, 38*(18), 4900–4904.

Pant, D., Van Bogaert, G., Diels, L., & Vanbroekhoven, K. (2010). A review of the substrates used in MFCs (MFCs) for sustainable energy production. *Bioresource Technology, 101*(6), 1533–1543.

Park, H. I., Mushtaq, U., Perello, D., Lee, I., Cho, S. K., Star, A., & Yun, M. (2007). Effective and low-cost platinum electrodes for MFCs deposited by electron beam evaporation. *Energy & Fuels, 21*(5), 2984–2990.

Qiao, Y., Li, C. M., Bao, S.-J., & Bao, Q.-L. (2007). Carbon nanotube/polyaniline composite as anode material for MFCs. *Journal of Power Sources, 170*(1), 79–84.

Richter, H., McCarthy, K., Nevin, K. P., Johnson, J. P., Rotello, V. M., & Lovley, D. R. (2008). Electricity generation by geobacter sulfurreducens attached to gold electrodes. *Langmuir, 24*(8), 4376–4379.

Rozendal, R. A., Hamelers, H. V. M., Rabaey, K., Keller, J., & Buisman, C. J. N. (2008). Towards practical implementation of bioelectrochemical wastewater treatment. *Trends in Biotechnology, 26*(8), 450–459.

Schaetzle, O., Barriere, F., & Schroder, U. (2009). An improved MFC with laccase as the oxygen reduction catalyst. *Energy & Environmental Science, 2*(1), 96–99.

Song, H.-L., Zhu, Y., & Li, J. (2015). Electron transfer mechanisms, characteristics and applications of biological cathode MFCs – A mini review. *Arabian Journal of Chemistry*. https://doi.org/10.1016/j.arabjc.2015.01.008).

Sun, J., Hu, Y., Bi, Z., & Cao, Y. (2009). Improved performance of air-cathode single-chamber MFC for wastewater treatment using microfiltration membranes and multiple sludge inoculation. *Journal of Power Sources, 187*(2), 471–479.

ter Heijne, A., Hamelers, H. V. M., Saakes, M., & Buisman, C. J. N. (2008). Performance of non-porous graphite and titanium-based anodes in MFCs. *Electrochimica Acta, 53*(18), 5697–5703.

Thorne, R., Hu, H., Schneider, K., Bombelli, P., Fisher, A., Peter, L. M., Dent, A., & Cameron, P. J. (2011). Porous ceramic anode materials for photo-MFCs. *Journal of Materials Chemistry, 21*(44), 18055–18060.

Tsai, H.-Y., Wu, C.-C., Lee, C.-Y., & Shih, E. P. (2009). MFC performance of multiwall carbon nanotubes on carbon cloth as electrodes. *Journal of Power Sources, 194*(1), 199–205.

Vologni, V., Kakarla, R., Angelidaki, I., & Min, B. (2013). Increased power generation from primary sludge by a submersible MFC and optimum operational conditions. *Bioprocess and Biosystems Engineering, 36*(5), 635–642.

Wang, K., Liu, Y., & Chen, S. (2011a). Improved microbial electrocatalysis with neutral red immobilized electrode. *Journal of Power Sources, 196*(1), 164–168.

Wang, X., Cheng, S., Zhang, X., Li, X.-Y., & Logan, B. E. (2011b). Impact of salinity on cathode catalyst performance in MFCs (MFCs). *International Journal of Hydrogen Energy, 36*(21), 13900–13906.

Wei, L., Han, H., & Shen, J. (2012). Effects of cathodic electron acceptors and potassium ferricyanide concentrations on the performance of MFC. *International Journal of Hydrogen Energy, 37*(17), 12980–12986.

Yan, Z., Wang, M., Lu, Y., Liu, R., & Zhao, J. (2014). Ethylene glycol stabilized $NaBH_4$ reduction for preparation of carbon-supported Pt–Co alloy nanoparticles used as oxygen reduction electrocatalysts for MFCs. *Journal of Solid State Electrochemistry, 18*(4), 1087–1097.

Yang, F., Ren, L., Pu, Y., & Logan, B. E. (2013). Electricity generation from fermented primary sludge using single-chamber air-cathode MFCs. *Bioresource Technology, 128*, 784–787.

You, S., Zhao, Q., Zhang, J., Jiang, J., & Zhao, S. (2006). A MFC using permanganate as the cathodic electron acceptor. *Journal of Power Sources, 162*(2), 1409–1415.

Zhang, L., Liu, C., Zhuang, L., Li, W., Zhou, S., & Zhang, J. (2009a). Manganese dioxide as an alternative cathodic catalyst to platinum in MFCs. *Biosensors and Bioelectronics, 24*(9), 2825–2829.

Zhang, Y., Min, B., Huang, L., & Angelidaki, I. (2009b). Generation of electricity and analysis of microbial communities in wheat straw biomass-powered MFCs. *Applied and Environmental Microbiology, 75*(11), 3389–3395.

Zhang, F., Saito, T., Cheng, S., Hickner, M. A., & Logan, B. E. (2010). MFC cathodes with Poly (dimethylsiloxane) diffusion layers constructed around stainless steel mesh current collectors. *Environmental Science & Technology, 44*(4), 1490–1495.

Zhang, Y., Olias, L. G., Kongjan, P., & Angelidaki, I. (2011). Submersible MFC for electricity production from sewage sludge. *Water Science and Technology, 64*(1), 50–55.

Zuo, Y., Cheng, S., Call, D., & Logan, B. E. (2007). Tubular Membrane Cathodes For Scalable power generation in MFCs. *Environmental Science & Technology, 41*(9), 3347–3353.

Chapter 11
Reactor Design for Bioelectrochemical Systems

G. Mohanakrishna, Shafeer Kalathil, and Deepak Pant

11.1 Introduction

Bioelectrochemical systems (BES) are novel hybrid systems which are designed to generate renewable energy from the low cost substrate in a sustainable way. Microbial fuel cells (MFCs) are the well studied application of BES systems that generate electricity from the wide variety of organic components and wastewaters. MFC mechanism deals with the microbial oxidation of organic molecules for the production of electrons and protons. The MFC design helps to build the electrochemical gradient on anode and cathode which leads for the bioelectricity generation. As whole reactions of MFCs happen at mild environmental and operating conditions and using waste organics as the substrate, it is defined as the sustainable and alternative option for global energy needs and attracted worldwide researchers into this research area. Apart from MFC, BES has other applications such as microbial electrolysis cells (MECs) for biohydrogen production, microbial desalinations cells (MDCs) for water desalination, and microbial electrosynthesis cells (MEC) for value added products formation. All these applications are designed to perform efficiently under mild operational conditions. Specific strains of bacteria or specifically enriched microbial consortia are acting as the biocatalyst for the oxidation and reduction of BES. Detailed function of the biocatalyst has been discussed in the other chapters of this book.

G. Mohanakrishna • D. Pant (✉)
Separation & Conversion Technologies, VITO – Flemish Institute for Technological Research, Boeretang 200, 2400 Mol, Belgium
e-mail: deepak.pant@vito.be

S. Kalathil
Division of Biological and Environmental Science & Engineering,
Water Desalination and Reuse Center, King Abdullah University of Science and Technology (KAUST), Thuwal, Saudi Arabia

D. Das (ed.), *Microbial Fuel Cell*,
https://doi.org/10.1007/978-3-319-66793-5_11

Bioelectrochemical systems mechanism involves both microbial/biological and electrochemical/physical principles. So, the design of BES is quite complex than conventional fermentation reactors. Towards optimization of the reactor design various models were tried. Even though the designs are different in various studies, the design components were found to be almost similar in all the designs. Such studies were focused to achieve improved energy recovery, high electron transfer rate, efficient energy conversion, economic reactor design for scaling up, better treatment efficiency, operational easiness, etc. The present chapter is focused to bring the knowledge related to design of BES and its components at one place. The chapter is also aimed to discuss the pros and cons of different factors that are involved in design.

11.2 Components of BES

In a BES, microbes attach to the anode surface and form a living biofilm to catalyze oxidation of organic compounds such as acetate. Hence, affinity between the anode and microbe is critical for the efficiency of the BES system such as MFCs. Also, a cathode is an integral part of the MFC where the electrons reach from the anode and are reduced to complete the reaction. Plethora of materials have already been employed as the anode and cathode materials in the MFC and some of them showed excellent performances in terms of bioelectricity production and wastewater treatment. In the present chapter we discuss current status and future perspectives of electrode materials for MFCs.

11.2.1 Anode Materials

In MFCs, anode is unique with the presence of living biofilm. The metabolically generated electrons from the biofilm are transferred to the attached anode where interfacial materials dictate the electron transfer rate (Fig. 11.1). To enhance the electron transfer process and the biofilm formation, the selection of anode material is critical. The anode material should satisfy certain features to become an ideal electrode as outlined in Fig. 11.2 (Xie et al. 2015). In addition, synthesis and fabrication of anodes should be flexible and the materials should be largely available with low cost.

11.2.1.1 Nanostructured Carbon-Based Electrodes

Carbon based electrodes have received considerable attraction as anode materials in MFCs due to their outstanding conductivity, high resistance to corrosion, and excellent biocompatibility (Kalathil and Pant 2016). Traditional carbon-based

Fig. 11.1 A schematic representation for the metabolically generated electron flow to the anode in an MFC

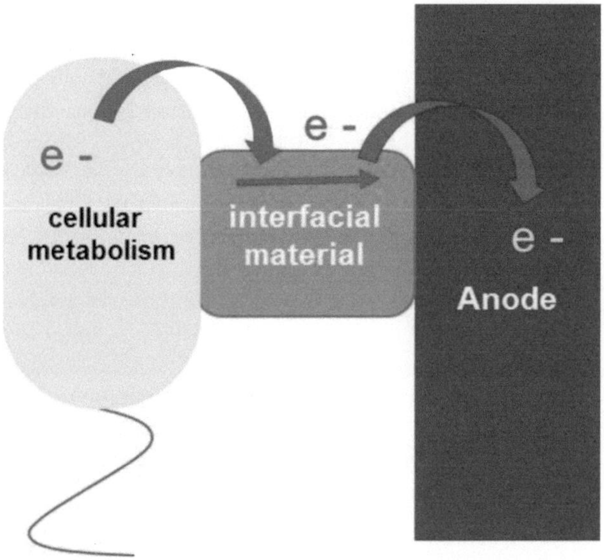

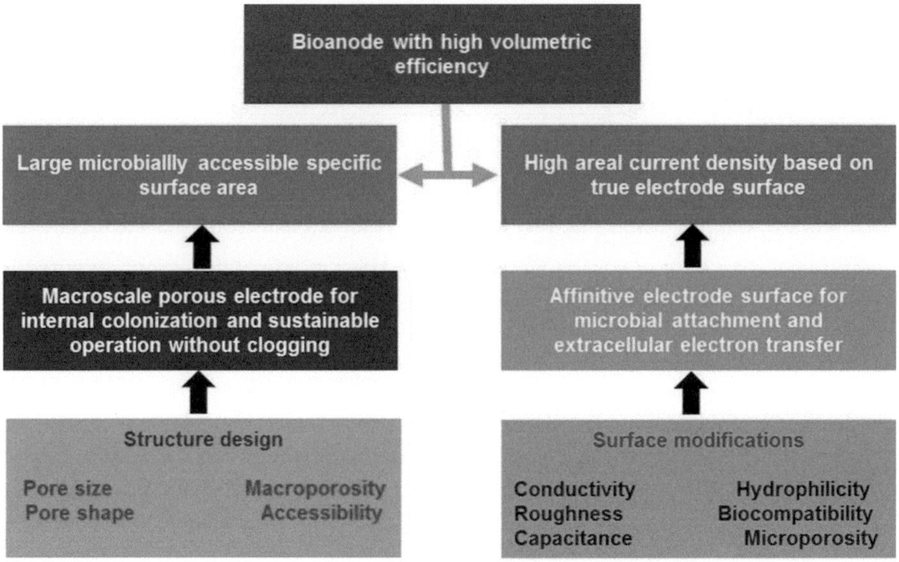

Fig. 11.2 PERT diagram for ideal characteristics of a bioanode in MFCs

electrodes such as graphite, carbon paper, carbon cloth, carbon felt and granular carbon are widely reviewed in many literature (Zhou et al. 2011). Here we focus only on nanostructured carbon materials such as carbon nanotubes, graphene, conductive polymers and metal nanoparticles.

11.2.1.2 Carbon Nanotubes

Carbon nanotube (CNT) is an attractive choice for electrode materials in MFCs due to its excellent conductivity and mechanical stability (Yazdi et al. 2016). Also, CNTs are highly biocompatible with microbes which allow biofilm growth on the CNT surface (Kalathil et al. 2013; Yazdi et al. 2016). Liu et al. (2014a) experimentally and theoretically demonstrated that CNT networks can facilitate direct extracellular electron transfer (EET) from bacterial outer membrane C-type cytochromes (OM C-Cyts) to solid electron acceptors such as electrodes. A 10-fold increase in the current production was observed with the CNT electrode as compared to a plain carbon paper with *Shewanella oneidensis* MR-1. Molecular dynamic simulation studies revealed that CNTs formed a CNT-OM C-Cyts electronic conduit to enhance the rate of EET process. The redox functional groups on the CNTs may be responsible for the formation of the electronic conduit. The study provided solid evidences on the action of CNTs in the EET process which may be helpful to fabricate CNT based electrode materials for MFCs. Erbay et al. (2015) observed that geometry of CNTs critically affects the performance of bioanodes. The study concluded that longer and loosely packed CNTs are highly suitable for the efficient EET process. Also, less amount of amorphous carbon in the CNTs facilitates microbe-electrode interactions. In another report, simple addition of CNT powders improved the performance of *Geobacter sulfurreducens* based MFC by lowering the anodic resistance and facilitating mass transfer inside the biofilm (Liang et al. 2011).

CNT hydrogel fabricated on a carbon paper improved current generation in an MFC as compared to the control (Liu et al. 2014b). The fabricated electrode provided huge surface area for the bacterial attachment and promoted direct EET due to the functional groups on CNTs. The CNT-hydrogel bioanode exhibited high stability for long term operation which is considered as one of the crucial factors for practical applications of MFCs. Many attempts have been reported for the development of micro-sized MFCs as they are highly attractive to power ultra-low-power electronics (Mink et al. 2012). A micro-sized MFC (1.25 μL) equipped with vertically grown multiwalled CNT (MWCNTs) integrated with nickel silicide (NiSi) bioanode generated a current density of 197 mA/m^2 (Mink et al. 2012). The MWCNTs provided large specific surface area for the bacterial attachment while NiSi lowered contact resistance of the modified electrode. A follow up work with a modified micro-sized MFC (75 μL) using MWCNT bioanode and air cathode performed much better than the previous work by producing a current density of 880 mA m^{-2} (Mink and Hussain 2013). However, the air cathode design allowed oxygen diffusion to the anode which deteriorated the bioanode performance. Also, the high surface area provided by the MWCNTs caused clogging of the anode with bacteria. Hence, optimization of the electrode fabrication is necessary for the long term operation of the device.

Three dimensional (3D) porous CNT bioanode showed its promising candidacy for the electrode development by producing remarkable current output as compared to traditional electrodes such as carbon cloth (Erbay et al. 2015; Xie et al. 2011). 3D structures allow the bacteria to grow at both interior and exterior of the electrode which in turn enhance the electrochemical activities of the attached biofilm. Katuri et al. (2011) fabricated 3DMWCNT/Chitosan hybrid bioanode by ice-segregation method as a bioanode for an MFC seeded with *G. sulfurreducens*. The modified bioanode showed excellent performance with a maximum current density of 24.5 A m^{-2} by providing high conductive surface area and better mass transfer for the biofilm activity. Doping of CNTs with elements is found to be a promising strategy to improve bioanode performance (Ci et al. 2012). For instance, CNTs doped with nitrogen showed better performance over bare CNTs and carbon cloth (Ci et al. 2012). The nitrogen doping increased active sites in the bioanode and marginally lowered the internal resistance caused by the electrode.

Generally, bacterial membranes are considered as negatively charged in neutral solution (Kalathil and Pant 2016). Owing to this negatively charged membranes, bacteria prefer to attach on positively charged solid surfaces. This factor inspired the researchers to create advanced bioanodes with positively charged surfaces. For example, a positively charged mesoporous-polysulfone supported single walled CNT (SWCNT) bioanode developed dense biofilm by attracting negatively charged *Shewanella* cells (Nguyen et al. 2013). The formation of dense biofilm caused enhanced current production and wastewater treatment. Also pretreatment of CNTs with acid can also create positively charged surface to enhance the charge based attraction between the anode and bacterial cells.

11.2.1.3 Graphene

Graphene is a two dimensional (2D) nanomaterial which possesses high electrical conductivity, good mechanical stability and excellent biocompatibility (Filip and Tkac 2014). Graphene also possesses higher specific surface area as compared to other carbon-based nanostructures. Owing to these properties, graphene has been largely employed in various energy devices including MFCs as an efficient electrode material (Filip and Tkac 2014). Graphene-based MFCs usually exhibit enhanced performance over activated carbon and CNT based MFCs (Yuan and He 2015). The first study of grapheme-based bioanode was reported a few years before (Zhang et al. 2011) using an *E. coli* seeded MFC. The graphene was anchored on a stainless steel mesh and utilized as the bioanode. The modified anode exhibited an enhanced current generation over the control, stainless steel mesh anode. The improved performance was mainly because of the high surface area of the anode that allowed large amount of biofilm formation. A vacuum stripped graphene scaffold bioanode showed 78 times higher current generation over the control carbon cloth in an MFC with *Pseudomonas aeruginosa* as biocatalysts (He et al. 2012). *P. aeruginosa* usually secretes exogenous mediators such as phenazine during its metabolism (Rabaey et al. 2005).

Graphene can stimulate phenazine production to accelerate mediated EET in *P. aeruginosa* (Liu et al. 2012). Moreover, the modified vacuum stripped graphene electrode showed large surface area and enhanced charge transfer rate. Similarly, a graphene deposited carbon cloth improved the performance of *P. aeruginosa* MFC due to the stimulating effect of graphene on the production of phenazine (Liu et al. 2012). Xie et al. (2012) proposed a graphene sponge decorated on a stainless steel mesh as a low cost bioanode material for the MFC. The synergic contributions of both the graphene sponge and the stainless steel in terms of electrical conductivity, and high specific surface area greatly improved the MFC performance with a maximum current density of 1.32 A m^{-2}. 3D graphene sponge fabricated bioanode using ice template method produced a maximum power density of 427 W m^{-3} which was much higher than that of a carbon felt fabricated MFC (Chen et al. 2014). The modified 3D bioanode allowed efficient mass transport for attached biofilm. Recently, a miniaturized MFC (50 μL) produced a current density of 15.51 A m^{-2} with a 3D graphene bioanode which is the highest current generation from an MFC till date (Ren et al. 2016). This outstanding current generation shows that 3D graphene can be an effective bioanode material to extend the practical application of MFCs.

11.2.1.4 Conductive Polymers

Conductive polymers show excellent conductivity, biocompatibility, high chemical stability and easiness for synthesis and processing which make them widely used electrode materials in MFCs (Kalathil and Pant 2016). Polypyrrole (PPy) is the most studied conductive polymer till date (Balint et al. 2014). PPy possesses high stability in water and reasonably high conductivity. Also, it can be easily synthesized and surface properties such as porosity can be easily tuned for various applications. Zhao et al. (2015) developed a PPy nanotube membrane bioanode by a reactive self-degraded template method. The modified bioanode produced six times higher current as compared to the control in a *Shewanella* MFC. Electrochemical impedance spectroscopy (EIS) analysis confirmed that the PPy-nanotubes greatly reduced the charge transfer resistance of the electrode that facilitated the EET process. Also, SEM images depicted coverage of biofilms on the entire PPy nanotube anode while the control carbon paper anode didn't show much biofilm formation on its surface. This indicated that the modified bioanode improved the bacterial adhesion by providing large active surface area and functional groups. A bioanode fabricated with PPy hydrogel decorated on CNTs showed enhanced current production in an MFC seeded with mixed bacterial culture by reducing internal resistance and facilitating EET (Tang et al. 2015). The porous and hydrophilic nature of the modified bioanode allowed to develop thick biofilm and unhindered mass transport for efficient metabolic activities of the attached microbes. Anthraquinone-2,6-disulphonic (AQDS) is considered as an efficient mediator in the EET process (Adachi et al. 2008). PPy deposited with AQDS acted as an excellent bioanode in *Shewanella decolorationis* S12 MFC by producing 13-fold higher current density than the control bioanode (Feng et al. 2010). The synergetic effects of PPy and AQQDS greatly improved charge transfer rate and avoided mass transfer limitations.

In addition to PPy, polyaniline (PANI) has also received considerable attention as bioanode materials in MFCs. PANI is a conducting polymer that has been widely employed in computer displays, tissue engineering, biosensors and fuel cells due to its ease of synthesis, pseudo-capacitance, stability, low cost and biocompatibility (Balint et al. 2014). Ding et al. (2012) proposed an interesting approach to control EET process by PANI-nanowire array fabricated bioanode. PANI showed multiple oxidation levels according to different applied potential. It stayed mostly in reduced state when the applied potential was −0.3 V vs. Ag/AgCl. At this potential, the EET from cells to the electrode was completely hindered due to thermodynamic barrier. At the same time, the EET from cells to electrode was enhanced at an applied potential of −0.2 V (vs. Ag/AgCl) as PANI stayed mostly in the oxidized state at this applied potential. This finding demonstrated that the EET process is highly sensitive to the applied potential and the EET can be controlled by altering external potential. A bioanode fabricated by the deposition of PANI on inorganic networks such as TiO_2 delivered high current output (1495 mW m^{-2}) in an *E. coli* MFC (Qiao et al. 2007). The main drawback of this study is that it used *E. coli* as the biocatalyst which needs mediator for the EET process and the mediator dependent MFC is not feasible for practical applications.

11.2.1.5 Metal Nanoparticles

As stated above, carbon-based materials are commonly employed as bioanode materials in MFCs. However, these materials possess a significant drawback in terms of electric conductivity which is two to three orders of magnitude below that of most metals (Baudler et al. 2015). Hence, application of metal-based bioanodes may be a suitable approach for the improved MFC. Various metal nanoparticles including noble metals (e.g. Au and Pd) and non-noble metals (e.g. Ni, Cu and Ti) have been widely employed as bioanode materials in MFCs.

Generally electrogenic bacteria are unable to form dense biofilm on the bare Au surfaces (Crittenden et al. 2006). However, deposition of Au on the surfaces of carbon-based electrodes proved to be an effective strategy to improve the performance of electrogenic bacteria such as *Shewanella oneidensis* MR-1 (Sun et al. 2010). A Au-sputtered carbon paper bioanode increased the performance of MFC inoculated with *S. oneidensis* MR-1 (Sun et al. 2010). The sputtered electrode showed improved electrochemical performances as compared to the control carbon paper which may be attributed to the enhanced MFC performance. Guo et al. (2012) constructed a bioanode by decorating Au nanoparticles on a carbon paper using layer-by-layer assembly method. The modified bioanode greatly enhanced MFC performance over the bare carbon paper. Here, Au layer on the modified electrode provided large active surface area and high conductivity for the efficient EET process. Several microbes are capable to produce metal nanoparticles when they are exposed to metal ions (Kalathil et al. 2011). A sulphate reducing bacterium,

Desulfovibrio desulfuricans formed membrane bound Pd-nano-assemblies after the exposure to Pd^{2+} ions (Wu et al. 2011). The membrane bound Pd networks facilitated electron transfer between the cells and electrode as a result of their higher electrical conductivity. Doping of bacteria with nanomaterials has shown to be an effective approach to accelerate EET processes in MFCs. For instance, doping of *Shewanella* with iron oxide/sulphide and nickel nanoparticles stimulated the EET rate as a result of enhanced bacterial electrical conductivity (Jiang et al. 2014).

It is generally believed that Cu and Ag are not biocompatible with bacteria (Grass et al. 2011). However, a recent report claimed that the antibacterial nature of these metals are not valid for electogenic bacteria (Baudler et al. 2015). Cu and Ag were utilized for the fabrication of bioanodes in MFCs. The bioanode modified with Cu and Ag produced maximum current density of 1.5 mA cm^{-2} and 1.1 mA cm^{-2}, respectively (Baudler et al. 2015). These current densities are comparable with benchmark, graphite bioanode (1 mA cm^{-2}). The similar performances of these expensive materials with carbon-based bioanodes raise a critical question that why expensive materials are employed for bioanodes as their performances are not better than the low cost carbon materials.

Titanium (Ti) has several advantages over conventional carbon and metal electrodes such as high biocompatibility, corrosion resistance and good dimensional stability (Zhou et al. 2016). However, Ti is not widely used as bioanodes in MFCs possibly due to poor electrical connection between bacterial cells and Ti. Zhou et al. (2016) proposed a possible solution to employ Ti as a bioanode by creating an oxide layer on the metal surface through heating. The oxide layer promoted the electron transfer between the cells and electrode by providing good electrocatalytic activities. Stainless steel is a material of choice for bioanode materials in MFCs as compared to conventional carbon-based materials. Stainless steel is a good conductive material with high mechanical strength and very stable at even severe experimental conditions. Also, the stainless steel is a cost-effective electrode with high resistance to corrosion that makes it as a first choice for the long-term operation. A systematic study conducted by Pocaznoi et al. (2012) has demonstrated that stainless steel can be an effective bioanode for MFCs as compared to carbon-based electrodes. The stainless steel anode produced a current density of 20.6 A m^{-2} while a graphite electrode delivered only 9.5 A m^{-2}. A major disadvantage of the stainless steel is its relatively smaller surface area as compared to traditional carbonaceous materials such as carbon cloth. To overcome this issue, a bioanode fabricated with stainless steel foam was employed in MFCs by producing a maximum current density of 80 A m^{-2} which was much higher than that of a plain stainless steel and carbon cloth (Ketep et al. 2014). Here, the foam structure provided a 3D porous structure for the efficient biofilm formation and mass transport. Several pretreatment processes have been proposed to improve the efficiency of stainless steel bioanodes. Heat treatment is a fruitful strategy to improve surface properties of the stainless steel (Guo et al. 2015).

A preheating of stainless steel at 600 °C for 5 min created an iron oxide layer on the surface that dramatically increased electrochemical interactions of the biofilm and bioanode (Guo et al. 2015). The formation of iron oxide layer protected the electrode from corrosion. The heat treated stainless electrode produced a high current density of 1.5 mA cm^{-2} by showing its promising candidacy for the low cost and scalable bioanodes for MFCs. Similarly, flame oxidation of stainless steel produced iron oxide layer on the surface which was confirmed by X-ray diffraction spectroscopy (Yamashita et al. 2016). The flame oxidized electrode exhibited enhanced current generation over the controls (untreated stainless steel and carbon cloth). Bacterial community analysis showed high abundance of *Geobacter* on the flame oxidized electrode than the controls. The stimulated *Geobacter* growth on the modified electrode could result in high current generation in the MFC. The study revealed that the bioanode materials dictate the bacterial population in the electrode attached biofilm.

11.2.2 Cathode Materials

Cathode materials are the integral parts of MFCs and they critically affect the overall performance of the system. Low durability and high cost of the cathodes hinder the practical applications of MFCs. Oxygen reduction reaction (ORR) is usually occurring at the cathode by supplying oxygen from air. ORR follows either $2e^-$ or $4e^-$ reaction pathway by producing H_2O_2 or H_2O. However, H_2O ($4e^-$) pathway is more preferable due to the large consumption of electrons. Major obstacles that prevent the cathodic performances are high overpotential and catalyst poisoning that lead to kinetic losses in ORR. Here we discuss ORR based MFC cathodes by providing their current status and future perspectives. MFC cathodes are generally categorized as chemical and biological cathodes.

11.2.2.1 Chemical Cathodes

Carbon-based materials are commonly employed as MFC cathodes due to the low cost, high stability, good biocompatibility and large surface area. However, they suffer from poor reduction activities and need to be improved greatly to establish well-performing MFCs. Deposition of Pt catalysts on the carbon materials shows highest cathodic performance. But, the high cost of Pt catalyst prevents its practical applications and hence low cost materials should be investigated for the cost-effective electrodes. Anchoring of metal oxide nanoparticles on carbon supports is a promising approach for the development of low cost cathodes. For instance, Mn_2O_3 and Fe_2O_3 cathodes exhibited good ORR performances comparable with Pt cathodes (Martin et al. 2011). Santoro et al. (2015) developed an iron-aminoantipyrine (Fe-AAPyr) cathode for ORR in a single chamber MFC. The Fe-AAPyr cathode showed better performance than that of Pt and activated carbon cathodes. A biofilm formation was detected on the Fe-AAPyr cathode after long-term operation of the MFC. Usually the biofilm formation deteriorates the performance of cathode by blocking catalytic active sites. Interestingly, the Fe-AAPyr cathode didn't exhibit any decrease in the performance even after the biofilm formation due to the long exposure to the wastewater. On the other hand, biofilm formation on the Pt cathode showed significant decrease in the ORR activity possibly due to the catalyst inactivation. These observations clearly show that the Fe-AAPyr is highly stable for long-term operation of MFCs without affecting its ORR activity under severe conditions of wastewater.

11.2.2.2 Biocathodes

Some aerobic bacteria have ability for ORR which is sometimes higher than that of conventional abiotic cathode catalysts (Kim et al. 2015). Biocathode represents a microbially catalyzed cathode which can perform reduction reactions such as ORR in MFC cathodes. It has several advantages over abiotic catalysts such as low cost, sustainability, and resistance to sulphide poisoning. Also, the biocathode has higher affinity for oxygen than Pt and usually aerobic bacteria can respire oxygen at maximum capacity with a dissolved oxygen concentration as low as 0.12 mg L^{-1} (Kim et al. 2015). Similar to bioanodes, carbon-based materials such as carbon cloth, carbon paper and graphite felt are commonly used as biocathode materials. Stainless steel is also proved to be a good candidate for biocathode formation. Mixed culture microbial consortia and pure cultures are employed to develop biocathodes (Butti et al. 2016). Zhang et al. (2012) systematically studied the role of cathode materials in the formation of MFC biocathodes by employing graphite felt, carbon paper and stainless steel mesh cathodes. The cathodes were inoculated using an anaerobic sludge. The graphite felt-biocathode showed maximum ORR activity over carbon paper and stainless steel mesh biocathodes. Increasing specific surface area for the microbial attachment is an effective way to accelerate the performance of biocathodes. Additions of graphite granules, activated carbon granules and activated carbon powder into cathode chambers enhanced MFC performance by providing high surface area for the biofilm formation that improved ORR activities of biocathodes (Tursun et al. 2016). Also, the improved biocathodes significantly reduced internal resistance of the MFC systems.

11.2.3 Membranes

MFCs usually employ membranes for separating anodes and cathodes. The membrane can prevent oxygen diffusion and substrate crossover between the chambers of MFC. However, the use of membranes adds huge cost to MFC design and hinders commercial applications. Also, they increase internal resistance and sometimes allow oxygen diffusion which adversely affects the MFC performance. Another major concern is membrane biofouling that diminishes the suitability of membrane-based MFCs for long-term applications. There are two types of membranes available for MFCs namely cation exchange membranes and anion exchange membranes.

11.2.3.1 Cation Exchange Membranes

Cation exchange membranes (CEMs) such as Nafion, Hyflon, Zirfons and Ultrex CMI 7000 are generally employed as membrane separator for MFCs due to their high proton conductivity (Leong et al. 2013). Among them, Nafion is the mostly used membrane in MFCs as it is highly proton conductive because of the negatively charged hydrophilic sulphonate group. Thinner Nafion membranes are proved to be more effective for MFCs than the thicker ones as the former contribute less ohmic resistance (Jung et al. 1998). But, the thinner membranes suffer from high permeability of oxygen and substrate crossover that lower the Coulombic efficiency of the MFC system. To avoid this issue, optimization of the membrane thickness should be achieved for the better performance of MFCs. pH drop in the anode chamber is another major concern for the Nafion-based MFC (Kim et al. 2007). At the anode, the oxidation of substrate produces protons which may cause pH drop and this pH drop is significant with Nafion membranes. It is known that bacterial respiration at the anode is inhibited by low pH (Kim et al. 2007) that deteriorates the performance of MFC. Also, the Nafion membrane allows cations transport (e.g. Na^{+}) other than protons which causes charge imbalance between the anode and cathode.

11.2.3.2 Anion Exchange Membranes

Anion exchange membranes (AEMs) usually produce higher current generation over CEMs in MFCs. The protons generated at the anode are consumed by OH^{-} ions transferred from the cathode through the AEM that can largely prevent the anodic pH drop. Due to this property, the AEM lowers ion transport resistance and the cathode resistance caused by the precipitation of transported cations (Jingmei et al. 2013).

11.3 Bioelectrochemical Cell Designs

11.3.1 Dual Chamber

Dual chamber design is the first and most common design for MFCs and MECs. The design considers an anode chamber and a cathode chamber. Both the chambers are separated by a cation exchange membrane. Anode from anode chamber and cathode from cathode chamber connects externally with resistor. Dual chamber configuration was used for different designs by changing placement of each

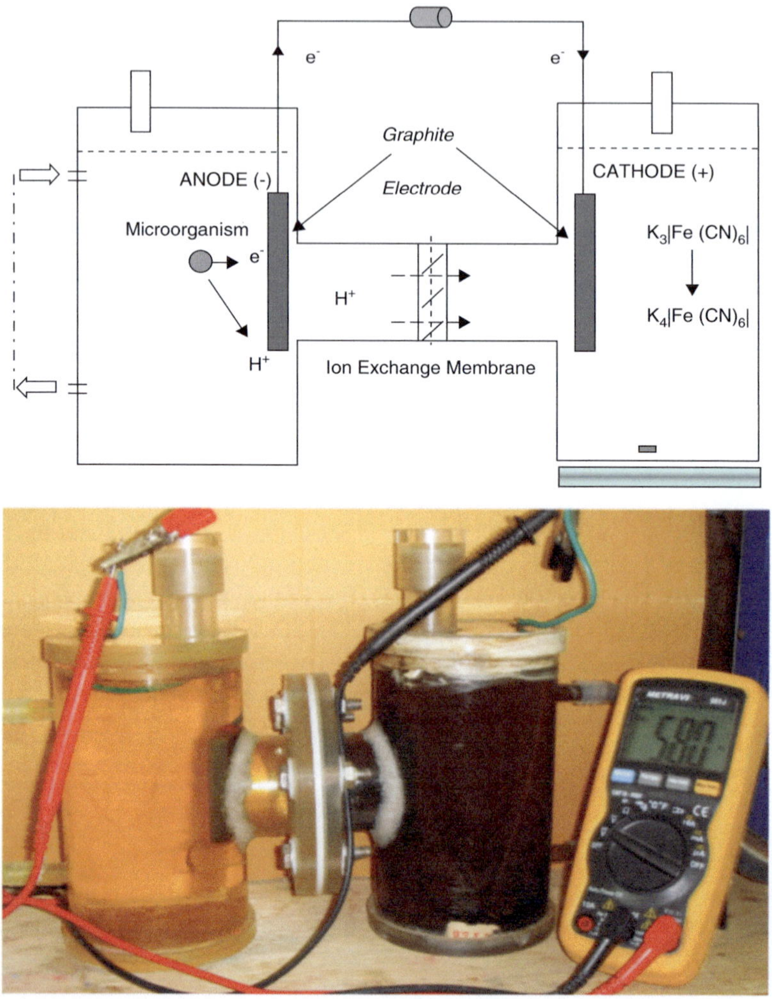

Fig. 11.3 Schematic details and photograph of H-type dual chambered MFC (Venkata Mohan et al. 2008)

chamber and using different shapes of chambers. Even though biocatalyst and the electron generation are confined to anode chamber, cathode reaction is also equally important. So, most of the studies were used with equal volumes of anode and cathode chambers. A typical H-type MFC is depicted in Fig. 11.3. Proton exchange membrane or cation exchange membranes were used as the membrane. As the design is simple and MFC performance is efficient, this design can be considered as the model for early stage researchers.

11.3.2 Single Chamber

After dual chamber configuration, single chamber design was studied. It contains only anode chamber and the cathode electrode is exposed to air. Here, atmospheric air or oxygen supplied on to the cathode surface acts as the catholyte and involves in oxygen reduction reaction (ORR). Compared to dual chamber design, this design was less efficient for electricity generation. However, the design was considered beneficial to scale up the process, where it requires less space for construction and set-up. When wastewater treatment is also major aim from the process, this design can deliver similar performance towards treatment but less bioelectricity generation. The design is also similar to the conventional anaerobic digesters for methane generation and it has the potential to convert the existing anaerobic digesters to MFCs (Fig. 11.4).

11.3.3 Stack Designs

In the course of scaling up the MFC technology stacking up the more number of MFCs into a single operating system was done. Practically, the maximum potential of single MFC (irrespective of anolyte volume) unit that can be achieved is approximately 1.0 V, which is very less and no proper application can be run with this potential except low power electronic modules and actuators has permitted the employment of MFCs in real applications as exemplified by Gastrobot (Wilkinson 2000) and EcoBots-I and -II (Ieropoulos et al. 2003; Melhuish et al. 2006). To achieve more potential and power, multiple electrodes need to be stacked to get the higher potential (Fig. 11.5). Since not much advantage is possible towards power generation with high volumes of MFCs, multiple MFCs with small volume were designed and integrated into a single operating system. First study with stacking of MFCs was reportedly done by Ieropoulos and co-workers (2003) in which approximately 50 times higher current generation was recorded than the output produced by the large MFC. The results from this study suggest that MFC scale-up may be better achieved by connecting multiple small-sized units together rather than increasing the size of an individual unit. It was also understood that the electrochemical connection and hydraulic/fluidic connections influence the MFC scaling up.

11.4 Future Perspectives and Conclusions

Electrode materials are the key components of the MFC systems and they dictate the performance of the MFC. Currently, many materials are available for the fabrication of both the anodes and cathodes. The application of improved electrodes

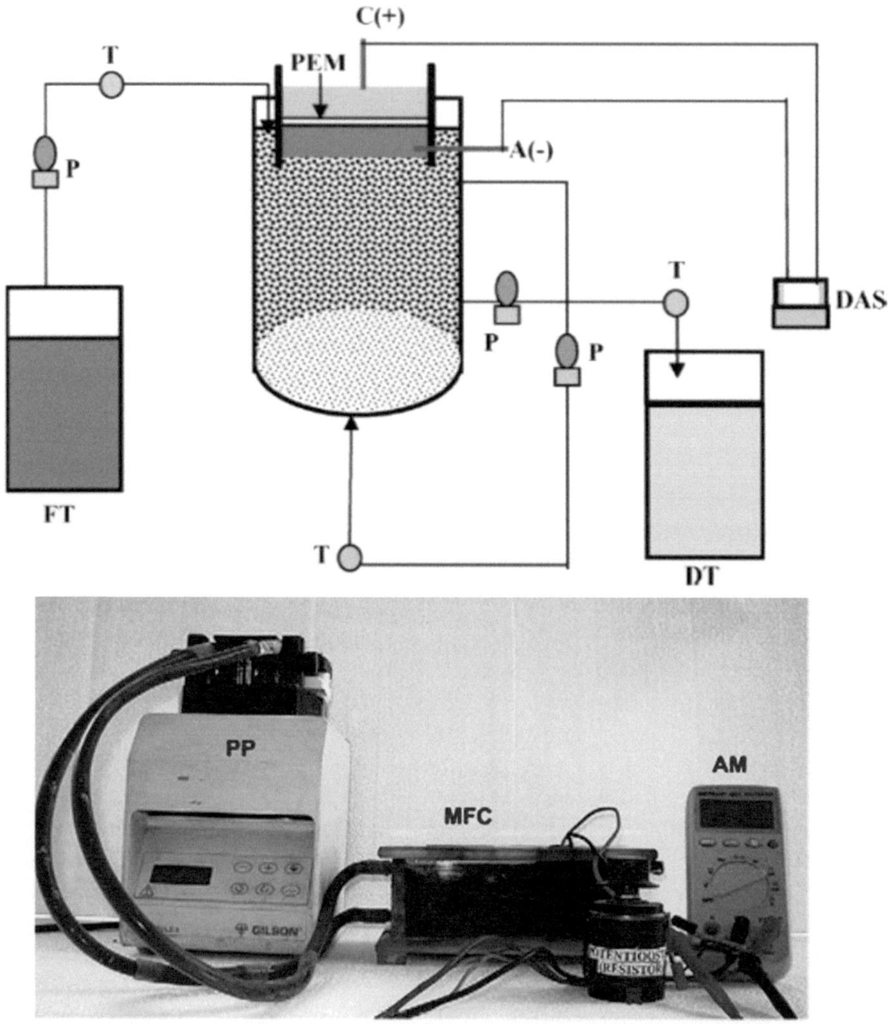

Fig. 11.4 Schematic details and photograph of single chambered MFCs used in different studies (*FT* feeding tank, *DT* decant tank, *T* pre-programmed timer, *P* peristaltic pump, *A(−)* anode, *C(+)* cathode, *PEM* proton exchange membrane, *DAS* data acquisition system, *MFC* single chambered microbial fuel cell, *PP* peristaltic pump, *AM* Multimeter, *R* variable resistor (Raghavulu et al. 2009)

significantly enhanced the MFC performance in terms of current output and wastewater treatment. However, many obstacles still need to be addressed for the practical application of these electrode materials. Even though some materials such as CNTs and graphene seem to be promising materials for the electrode development, the high cost and complicated synthesis procedures make them less attractive for the practical applications. Another major concern is the stability of the

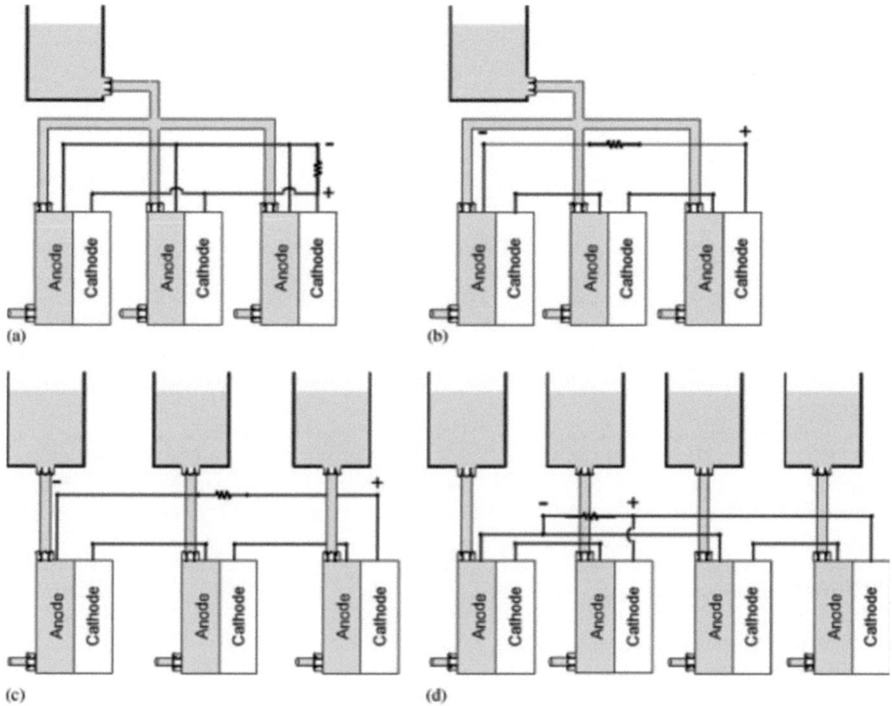

Fig. 11.5 Schematic representation of the different fluidic and electrical configurations for stacks of MFCs: (**a**) Parallel electrical connection with a common feed line; (**b**) Series electrical connection with a common feed line; (**c**) Series electrical connection but with individual feed lines; and (**d**) Series–parallel connections with individual feed lines and with an even number of MFCs (Ieropoulos et al. 2003)

electrode materials for the long-term applications. Unfortunately, most of the studies have been done for short duration with only concern on the power output by neglecting long-term stability of the materials. Hence, cost-effective and durable electrodes should be investigated to extend the MFC application in practical fields. 3D porous structures are more advisable for the electrode materials as they can allow dense biofilm formation and efficient mass transport including substrates. But, sometimes thick biofilm formation may prevent mass transfer that can create dead zones in the biofilms. Hence, a compromise on the biofilm formation and mass transport should be achieved to develop well-performing electrodes. The role of nanomaterials on the bacterial metabolism is currently largely unknown. A recent report demonstrated that CNTs can even alter bacterial respiration pathway from intracellular to extracellular (Yan et al. 2014). This observation warrants that a prior understanding on the action of nanomaterials on the bacterial metabolism is essential for the fabrication of efficient electrode materials for the MFC. The development of biocathode is a sustainable approach for the MFC cathodes. However, the current knowledge on the biocathode mechanism is very limited and further investigations may lead to efficient biocathodes.

References

Adachi, M., Shimomura, T., Komatsu, M., Yakuwaa, H., & Miyaa, A. (2008). A novel mediator-polymer-modified anode for MFCs. *Chemical Communications, 17*, 2055–2057.

Balint, R., Cassidy, N. J., & Cartmell, S. H. (2014). Conductive polymers: Towards a smart biomaterial for tissue engineering. *Acta Biomaterialia, 10*, 2341–2353.

Baudler, A., Schmidt, I., Langner, M., Greiner, A., & Schröder, U. (2015). Does it have to be carbon? Metal anodes in MFCs and related bioelectrochemical systems. *Energy and Environmental Science, 8*, 2048–2055.

Butti, S. K., Velvizhi, G., Sulonen, M. L. K., Haavisto, J. M., Koroglu, E. O., Cetinkaya, A. Y., Singh, S., Arya, D., Modestra, J. A., Krishna, K. V., Verma, A., Ozkaya, B., Lakaniemi, A.-M., Puhakka, J. A., & Mohan, S. V. (2016). Microbial electro-chemical technologies with the perspective of harnessing bioenergy: Maneuvering towards upscaling. *Renewable and Sustainable Energy Reviews, 53*, 462–476.

Chen, W., Huang, Y.-X., Li, D.-B., Yu, H.-Q., & Yan, L. (2014). Preparation of a macroporous flexible three dimensional graphene sponge using an ice template as the anode material for microbial fuel cells. *RSC Advances, 4*, 21619–21624.

Ci, S., Wen, Z., Chen, J., & He, Z. (2012). Decorating anode with bamboo-like nitrogen-doped carbon nanotubes for MFCs. *Electrochemistry Communications, 14*, 71–74.

Crittenden, S. R., Sund, C. J., & Sumner, J. J. (2006). Mediating electron transfer from bacteria. *Langmuir, 22*(23), 9473–9476.

Ding, C., Liu, H., Zhu, Y., Wan, M., & Jiang, L. (2012). Control of bacterial extracellular electron transfer by a solid-state mediator of polyaniline nanowire arrays. *Energy and Environmental Science, 5*, 8517–8522.

Erbay, C., Yang, G., Figueiredo, P. D., Sadr, R., Yu, C., & Han, A. (2015). Three-dimensional porous carbon nanotube sponges for high-performance anodes of MFCs. *Journal of Power Sources, 298*, 177–183.

Feng, C., Ma, L., Li, F., Mai, H., Lang, X., & Fan, S. (2010). A polypyrrole/anthraquinone-2,6-disulphonic disodium salt (PPy/AQDS)-modified anode to improve performance of MFCs. *Biosensors and Bioelectronics, 25*, 1516–1520.

Filip, J., & Tkac, J. (2014). Is graphene worth using in biofuel cells? *Electrochimica Acta, 136*, 340–354.

Grass, G., Rensing, C., & Solioz, M. (2011). Metallic copper as an antimicrobial surface. *Applied and Environmental Microbiology, 77*, 1541–1547.

Guo, W., Pi, Y., Song, H., Tang, W., & Sun, J. (2012). Layer-by-layer assembled gold nanoparticles modified anode and its application in MFCs. *Colloids and Surfaces A: Physicochemical and Engineering Aspects, 415*, 105–111.

Guo, K., Soeriyadi, A. H., Feng, H., Prévoteau, A., Patil, S. A., Gooding, J. J., & Rabaey, K. (2015). Heat-treated stainless steel felt as scalable anode material for bioelectrochemical systems. *Bioresource Technology, 195*, 46–50.

He, Z., Liu, J., Qiao, Y., Li, C. M., & Yang Tan, T. T. (2012). Architecture engineering of hierarchically porous chitosan/vacuum-stripped graphene scaffold as bioanode for high performance MFC. *Nano Letters, 12*, 4738–4741.

Ieropoulos, I., Greenman, J., & Melhuish, C. (2003). Imitating metabolism: Energy autonomy in biologically inspired robotics. In *Proceedings of second international symposium on imitation in animals and artifacts*, SSAISB, Aberystwyth.

Ieropoulos, I., Greenman, J., & Melhuish, C. (2008). Microbial fuel cells based on carbon veil electrodes: Stack configuration and scalability. *International Journal of Energy Research, 32* (13), 1228–1240.

Jiang, X., Hu, J., Lieber, A. M., Jackan, C. S., Biffinger, J. C., Fitzgerald, L. A., Ringeisen, B. R., & Lieber, C. M. (2014). Nanoparticle facilitated extracellular electron transfer in MFCs. *Nano Letters, 14*, 6737–6742.

Jingmei, P., An, J., Ha, P. T., Kim, T., Jang, J. K., Moon, H., & Chang, I. S. (2013). Power density enhancement of anion-exchange membrane-installed MFC under bicarbonate-buffered cathode condition. *Journal of Microbiology and Biotechnology, 23*, 36–39.

Jung, D. H., Lee, C. H., Kim, C. S., & Shin, D. R. (1998). Performance of a direct methanol polymer electrolyte fuel cell. *Journal of Power Sources, 71*, 169–173.

Kalathil, S., & Pant, D. (2016). Nanotechnology to rescue bacterial bidirectional extracellular electron transfer in bioelectrochemical systems. *RSC Advances, 6*, 30582–30597.

Kalathil, S., Lee, J., & Cho, M. H. (2011). Electrochemically active biofilm-mediated synthesis of silver nanoparticles in water. *Green Chemistry, 13*, 1482–1485.

Kalathil, S., Nguyen, V. H., Shim, J.-J., Khan, M. M., Lee, J., & Cho, M. H. (2013). Enhanced performance of a MFC using CNT/MnO_2 nanocomposite as a bioanode material. *Journal of Nanoscience and Nanotechnology, 13*, 7712–7716.

Katuri, K., Ferrer, M. L., Gutiérrez, M. C., Jiménez, R., del Monte, F., & Leech, D. (2011). Three-dimensional microchanelled electrodes in flow-through configuration for bioanode formation and current generation. *Energy & Environmental Science, 4*(10), 4201–4210.

Ketep, S. F., Bergel, A., Calmet, A., & Erable, B. (2014). Stainless steel foam increases the current produced by microbial bioanodes in bioelectrochemical systems. *Energy and Environmental Science, 7*, 1633–1637.

Kim, J. R., Cheng, S., Oh, S.-E., & Logan, B. E. (2007). Power generation using different cation, anion, and ultrafiltration membranes in MFCs. *Environmental Science and Technology, 41*, 1004–1009.

Kim, B. H., Lim, S. S., Wan Daud, W. R., Gadd, G. M., & Chang, I. S. (2015). The biocathode of microbial electrochemical systems and microbially-influenced corrosion. *Bioresource Technology, 190*, 395–401.

Leong, J. X., Wan Daud, W. R., Ghasemi, M., Liew, K. B., & Ismail, M. (2013). Ion exchange membranes as separators in MFCs for bioenergy conversion: A comprehensive review. *Renewable and Sustainable Energy Reviews, 28*, 575–587.

Liang, P., Wang, H., Xia, X., Huang, X., Mo, Y., Cao, X., & Fan, M. (2011). Carbon nanotube powders as electrode modifier to enhance the activity of anodic biofilm in MFCs. *Biosensors and Bioelectronics, 26*, 3000–3004.

Liu, J., Qiao, Y., Guo, C. X., Lim, S., Song, H., & Li, C. M. (2012). Graphene/carbon cloth anode for high-performance mediatorless MFCs. *Bioresource Technology, 114*, 275–280.

Liu, X.-W., Chen, J.-J., Huang, Y.-X., Sun, X.-F., Sheng, G.-P., Li, D.-B., Xiong, L., Zhnag, Y.-Y., Zhao, F., & Yu, H.-Q. (2014a). Experimental and theoretical demonstrations for the mechanism behind enhanced microbial electron transfer by CNT network. *Scientific Reports, 4*, 3732.

Liu, X.-W., Huang, Y.-X., Sun, X.-F., Sheng, G.-P., Zhao, F., Wang, S.-G., & Yu, H. Q. (2014b). Conductive carbon nanotube hydrogel as a bioanode for enhanced microbial electrocatalysis. *ACS Applied Materials & Interfaces, 6*, 8158–8164.

Martin, E., Tartakovsky, B., & Savadogo, O. (2011). Cathode materials evaluation in MFCs: A comparison of carbon, Mn_2O_3, Fe_2O_3 and platinum materials. *Electrochimica Acta, 58*, 58–66.

Melhuish, C., Ieropoulos, I., Greenman, J., & Horsfield, I. (2006). Energetically autonomous robots: Food for thought. *Autonomous Robots, 21*, 187–198. https://doi.org/10.1007/ s10514-006-6574-5.

Mink, J. E., & Hussain, M. M. (2013). Sustainable design of high-performance microsized MFC with carbon nanotube anode and air cathode. *ACS Nano, 7*, 6921–6927.

Mink, J. E., Rojas, J. P., Logan, B. E., & Hussain, M. M. (2012). Vertically grown multiwalled carbon nanotube anode and nickel silicide integrated high performance microsized (1.25 μL) MFC. *Nano Letters, 12*, 791–795.

Nguyen, T.-H., Yu, Y.-Y., Wang, X., Wang, J.-Y., & Song, H. (2013). A 3D mesoporous polysulfone-carbon nanotube anode for enhanced bioelectricity output in MFCs. *Chemical Communications, 49*, 10754–10756.

Pocaznoi, D., Calmet, A., Etcheverry, L., Erable, B., & Bergel, A. (2012). Stainless steel is a promising electrode material for anodes of MFCs. *Energy and Environmental Science, 5*, 9645–9652.

Qiao, Y., Bao, S.-J., Li, C. M., Cui, X.-Q., Lu, Z.-S., & Guo, J. (2007). Nanostructured polyaniline/titanium dioxide composite anode for MFCs. *ACS Nano, 2*, 113–119.

Rabaey, K., Boon, N., Hofte, M., & Verstraete, W. (2005). Microbial phenazine production enhances electron transfer in biofuel cells. *Environmental Science and Technology, 39*, 3401–3408.

Raghavulu, S. V., Mohan, S. V., Reddy, M. V., Mohanakrishna, G., & Sarma, P. N. (2009). Behavior of single chambered mediatorless microbial fuel cell (MFC) at acidophilic, neutral and alkaline microenvironments during chemical wastewater treatment. *International Journal of Hydrogen Energy, 34*(17), 7547–7554.

Ren, H., Tian, H., Gardner, C. L., Ren, T.-L., & Chae, J. (2016). A miniaturized MFC with three dimensional graphene macroporous scaffold anode demonstrating a record power density of over 10,000 W m^{-3}. *Nanoscale, 8*, 3539–3547.

Santoro, C., Serov, A., Narvaez Villarrubia, C. W., Stariha, S., Babanova, S., Artyushkova, K., Schuler, A. J., & Atanassov, P. (2015). High catalytic activity and pollutants resistivity using Fe-AAPyr cathode catalyst for MFC application. *Scientific Reports, 5*, 16596.

Sun, M., Zhang, F., Tong, Z.-H., Sheng, G.-P., Chen, Y.-Z., Zhao, Y., Chen, Y.-P., Zhou, S.-Y., Liu, G., Tian, Y.-C., & Yu, H.-Q. (2010). A gold-sputtered carbon paper as an anode for improved electricity generation from an MFC inoculated with *Shewanellaoneidensis* MR-1. *Biosensors and Bioelectronics, 26*, 338–343.

Tang, X., Li, H., Du, Z., Wang, W., & Ng, H. Y. (2015). Conductive polypyrrole hydrogels and carbon nanotubes composite as an anode for MFCs. *RSC Advances, 5*, 50968–50974.

Tursun, H., Liu, R., Li, J., Abro, R., Wang, X., Gao, Y., & Li, Y. (2016). Carbon material optimized biocathode for improving MFC performance. *Frontiers in Microbiology, 7*. https://doi.org/10.3389/fmicb.2016.00006.

Venkata Mohan, S., Saravanan, R., Raghavulu, S. V., Mohanakrishna, G., & Sarma, P. N. (2008). Bioelectricity production from wastewater treatment in dual chambered microbial fuel cell (MFC) using selectively enriched mixed microflora: Effect of catholyte. *Bioresource Technology, 99*(3), 596–603.

Wilkinson, S. (2000). *'Gastronome'—A pioneering food powered mobile robot*. Proceedings of the international conference on robotics and applications. Honolulu, Hawaii, 14–16 August 2000, IASTED-2000.

Wu, X., Zhao, F., Rahunen, N., Varcoe, J. R., Rossa, C. A., Thumser, A. E., & Slade, R. C. T. (2011). *A role for microbial palladium nanoparticles in extracellular electron transfer. Microbial Direct Electrochemistry*. Wiley Interscience.

Xie, X., Hu, L., Pasta, M., Wells, G. F., Kong, D., Criddle, C. S., & Cui, Y. (2011). Three-dimensional carbon nanotube-textile anode for high-performance MFCs. *Nano Letters, 11*, 291–296.

Xie, X., Yu, G., Liu, N., Bao, Z., Criddle, C. S., & Cui, Y. (2012). Graphene-sponges as high-performance low-cost anodes for MFCs. *Energy and Environmental Science, 5*, 6862–6866.

Xie, X., Criddle, C., & Cui, Y. (2015). Design and fabrication of bioelectrodes for microbial bioelectrochemical systems. *Energy and Environmental Science, 8*, 3418–3441.

Yamashita, T., Ishida, M., Asakawa, S., Kanamori, H., Sasaki, H., Ogino, A., Katayose, Y., Hatta, T., & Yokoyama, H. (2016). Enhanced electrical power generation using flame-oxidized stainless steel anode in MFCs and the anodic community structure. *Biotechnology for Biofuels, 9*, 62.

Yan, F.-F., Her, Y.-R., Wu, C., Cheng, Y.-Y., Li, W.-W., & Yu, H. Q. (2014). Carbon nanotubes alter the electron flow route and enhance nitrobenzene reduction by *Shewanellaoneidensis* MR-1. *Environmental Science and Technology, 1*, 128–132.

Yazdi, A. A., D'Angelo, L., Omer, N., Windiasti, G., Lu, X., & Xu, J. (2016). Carbon nanotube modification of MFC electrodes. *Biosensors and Bioelectronics*. https://doi.org/10.1016/j.bios.2016.05.033.

Yuan, H., & He, Z. (2015). Graphene-modified electrodes for enhancing the performance of MFCs. *Nanoscale, 7*, 7022–7029.

Zhang, Y., Mo, G., Li, X., Zhang, W., Zhang, J., Ye, J., Huang, X., & Yu, C. (2011). A graphene modified anode to improve the performance of MFCs. *Journal of Power Sources, 196*, 5402–5407.

Zhang, Y., Sun, J., Hu, Y., Li, S., & Xu, Q. (2012). Bio-cathode materials evaluation in MFCs: A comparison of graphite felt, carbon paper and stainless steel mesh materials. *International Journal of Hydrogen Energy, 37*, 16935–16942.

Zhao, C.-e., Wu, J., Kjelleberg, S., Chey Loo, J. S., & Zhang, Q. (2015). Employing a flexible and low-cost polypyrrole nanotube membrane as an anode to enhance current generation in MFCs. *Small, 11*, 3440–3443.

Zhou, M., Chi, M., Luo, J., He, H., & Jin, T. (2011). An overview of electrode materials in MFCs. *Journal of Power Sources, 196*, 4427–4435.

Zhou, X., Chen, X., Li, H., Xiong, J., Li, X., & Li, W. (2016). Surface oxygen-rich titanium as anode for high performance MFC. *Electrochimica Acta*. https://doi.org/10.1016/j.electacta.2016.05.103.

Chapter 12
Microfluidic Microbial Fuel Cell: On-chip Automated and Robust Method to Generate Energy

Sanket Goel

12.1 Introduction

Microbial fuel cell (MFC) is a bio-electrochemical fuel cell where microorganisms, such as bacteria and virus, are used to catalyse the redox reaction to generate energy. Due to their inherent process, MFCs lead to the production of green and clean renewable energy in a self-sustainable manner. Even though, humongous work has been carried out in MFC domain leading to the exponentially increasing scientific output over the years, there has been limitation to harness MFC as a viable, workable but cost-effective remedy to the current energy and environmental challenges due to its expensiveness, low performance and challenges to scale-up (Lee et al. 2015a, b; Wang et al. 2015).

With the ongoing rigorous discussions on harnessing the renewable energy from diverse systems, it is vital to develop new processes and devices to make the development of MFCs robust, automated, inexpensive, beyond human intervention and fast (Sinton 2014). Therefore, harnessing the unique features of micro and nanoscale in various parts of MFC, such as electrodes size and surface modification, intermediate chamber, geometry of MFC etc., have been the initial interventions to develop microscale or miniaturized MFC (Kjeang et al. 2009; Wang et al. 2011; Lee and Kjeang 2010). Such low-scales helped the MFCs to consume lower energy, provide high surface area and make them amenable to integrate into a system. Therefore, it was found to be important to continue to work to miniaturize MFCs keeping overall cost, capital cost and fabrication cost and device-simplification in mind.

S. Goel (✉)
Department of Electrical and Electronics Engineering, BITS-Pilani, Hyderabad Campus, Jawahar Nagar, Shameerpet, Hyderabad 500078, India
e-mail: sanketgoel@gmail.com

D. Das (ed.), *Microbial Fuel Cell*,
https://doi.org/10.1007/978-3-319-66793-5_12

Being a flow-driven technology, it is necessary to employ the concepts of flow in microenvironment, called as Microfluidics and such MFC are named as Microfluidic MFC (MMFC) which have shown convincing results in terms of accuracy and cost-effectiveness (Choi 2015; Yang et al. 2016b). Microfluidics, originated in 1980s, is an interdisciplinary domain where small volumes [from microlitres (10^{-6} L) to femtolitres (10^{-15} L)] are handled and manipulated in a micro-environment to realize various facets such as automation, robustness and specificity even in a multiplexed and high-throughput manner.

Harnessing such unique facets of Microfluidic MFC (MMFC) technology has received substantial consideration as it is able to provide distinctive platform capable to power mini electronic components, such as implantable pacemakers (Ren et al. 2012) and wireless network (Shantaram et al. 2005), as biosensor (Li et al. 2016a, b), fast screening of microorganisms (Mukherjee et al. 2013), toxicity detection in water (Lee et al. 2015a, b) and monitoring toxicity and environmental strains (Qian et al. 2011). With a volume of few hundreds of microlitres, in MMFC, various components, such as electrodes, membranes (if needed), fluid manipulation sub-system and other components, are integrated together to develop a fully assembled platform (Choi 2015; Yang et al. 2016b). In addition, to increase the power-density, the MMFCs can be stacked and can potentially be used to applications requiring large powers (Choi and Chae 2012).

This chapter begins with the description of the fundamentals of the microfluidic technology, both in terms of scaling and fluidics and how it is harnessed to develop MMFC with customized specifications. Subsequently, two types of MMFC, Membrane Microfluidic MFC (M^{+}MMFC) and Membraneless Microfluidic MFC (M^{-}MMFC), have been described in two separate sections, where each section in turn narrates various facets for the development of MMFC, such as fabrication, materials, dimensions, performance and power density. The chapter further describes the current challenges and future opportunities in MMFC, with the ongoing and futuristic development in both microfluidics and MFC domains. The chapter concludes while providing the roadmap of MMFCs for diverse technological applications.

12.2 Microfluidics – Basic Principles Pertaining to MFC

12.2.1 Summary of Principles

Microfluidics, originated in 1980s, is an interdisciplinary domain where small volumes (less than microlitres to nanolitres) are handled and manipulated in a micro-environment to realize various facets such as automation, robustness and specificity in a multiplexed and high-throughput manner. Except these unique highlights, these devices are prone to consume low energy, and leverage on the other ancillary upsides of micro-domain are high surface area, low cost and amenable to integrate into a system (Gravesen et al. 1993; Kirby 2013).

Microfluidics relies on three diverse physical concepts namely, Laminar flow; Surface and interfacial tension (fluid-to-surface and fluid-to-fluid); and Capillary forces (movement through a narrow passage) (Sackmann et al. 2014). A combination of such concepts has led to make these devices versatile with the aforementioned advantages. These devices, also called as lab-on-a-chip, micro-total-analysis-system (micro-TAS) or simply miniaturized-micro-devices, have been successfully harnessed and implemented in diverse applications, such as biosensing, DNA sequencing, water monitoring, defense etc. During last more than three decades, lots of work have been carried out to optimize microfluidic devices, in the whole value-chain of their development including modelling and simulation, fabrication, characterization and testing. Presently, there are several microfluidic devices available commercially after attaining rigorous clinical trials and approvals (He et al. 2016; Konwarh et al. 2016; Samiei et al. 2016).

As one of the major improvement from macro-to-microscale MFC is the miniaturized size, it is prudent to mention that MMFCs are prone to be developed by harnessing the well-proven microfabrication methods, unique properties at micro-scale, automated fluid delivery and manipulation and small response time for the on-chip (Kirby 2013). Some of the established microfabrication methods are conventional photolithography, soft photolithography, laser writing, paper-based devices and rapid prototyping (3D printing). Further, their miniaturized size leverages the unique properties at microscale, such as high surface area-to-volume (SAV) ratio, requirement to low sample, automated fluid delivery and manipulation and small response time for the on-chip reaction (Gravesen et al. 1993). It is interesting to note that microfluidic MFCs are used as a benchmarking tool to develop macroscale MFCs (Kou et al. 2016).

12.2.2 Amenability to Integration

In a conventional MFC, a membrane separates the two chambers through which hydrogen ions and other cations move freely. In MFC, the generated power is a direct function of the microbial colonization, which needs to be rigorously studied. In fact, such optimizations are obviously complicated and time consuming on macroscale. It became the prime motivation for the researchers to initiate the development of microscale MFCs by reducing the chamber volume from litres to few hundred microlitres (Kjeang et al. 2009). During this process, various integral components, such as electrodes, membrane (if needed), fluid delivery and manipulation system etc. were needed to be developed and made compatible to realize a fully integrated MMFC.

Based on the unique nature of such devices, the reactions are prone to be optimized by the flow characteristics and environment, where fluid can be precisely guided, manipulated and tracked in the range from μL (10^{-6} L) to fl (10^{-15} L)

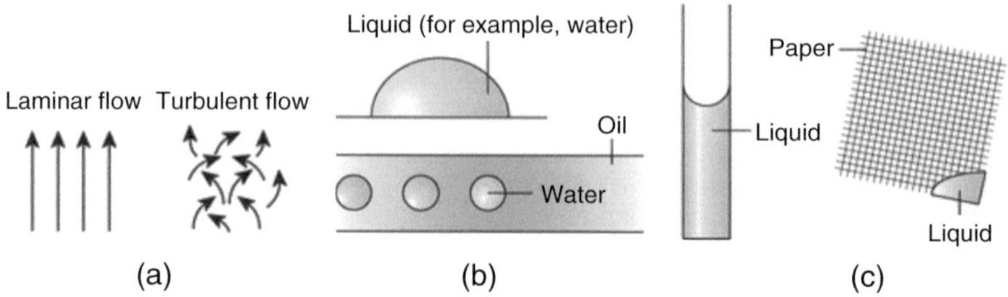

Fig. 12.1 Fundamental microfluidics concepts: (**a**) Laminar versus turbulent flow; (**b**) Surface and interfacial tension; and (**c**) Capillary forces (Reproduced from Sackmann et al. 2014 with permission from the Nature Publishing Group)

(Kirby 2013). Such devices can be easily created by using well established and proven fabrication methods and even other several compatible processes can be integrated to develop electrode and membrane (if needed). With these devices, microorganism grown can be observed in real time which leads to the monitoring of power generation (Fraiwan and Choi 2014).

12.2.3 Principle to Develop Membraneless MMFC

Except the aforementioned advantages (Sects. 12.2.1 and 12.2.2) of an MMFC device, other major benefit of such devices is the possibility to avoid the membrane leading to the development of the membraneless MMFC (M-MMFC). This is carried out by leveraging the unique properties at microscale by which the two independent laminar flow streams can be separated upto a certain length, which ensures the formation of a virtual layer separating anode and cathode. As the fluid flow at microscale is subjected to viscous forces rather than inertial forces, if the fundamental microfluid property of two laminar microfluids, kinematic viscosity, which encompasses the dynamic viscosity and density, are similar, due to diffusive mixing, a functional interface is developed until a certain distance, via controlled fluid-fluid flow (Dávila et al. 2011; Mu et al. 2006). As can be seen in Fig. 12.1, this functional interface replaces a membrane M-MMFC leading to the development of membraneless microfluidic MFC (M-MMFC).

12.3 Membraned Microfluidic MFC (M$^+$MMFC)

12.3.1 Diverse Membraned Microfluidic MFC (M$^+$MMFC)

In line with the evolution of the fabrication methods of microfluidic devices, M$^+$ MMFCs have also leveraged such fabrication methods chosen depending on

Table 12.1 Summary of the published research work in M^+MMFC

Sl. No.	Fabrication method	Cell area	Maximum power density (MPD)/ current density (MCD)	References
1	Conventional photolithography	550 mL	MPD: 2.3 nW cm^{-2}	Mu et al. (2006)
2	Bulk silicon micromachining	25 mL	MPD: 2.9 μW cm^{-2}; MCD: 214.8 μA cm^{-2}	Chen et al. (2011)
3	Conventional photolithography	144 mL	MPD: 6.5 mW cm^{-2}; MCD: 0.014 mA cm^{-2}	Dávila et al. (2011)
4	Soft lithography	15 μL	MPD: 401.2 nW cm^{-2}; MCD: 30.2 μA cm^{-2}	Siu and Chiao (2008)
5	Soft lithography	1.5 μL	MPD: 15 μW cm^{-3}; MCD: 1.3 mA cm^{-3}	Qian et al. (2009)
6	Soft lithography	5 μL	MPD: 62.5 μW cm^{-3}	Qian et al. (2011)
7	Soft lithography	0.3 μL	MCD: 2.52 μA cm^{-2} (127 μA cm^{-3})	Li et al. (2011)
8	Soft lithography		MPD: 3.2 μW cm^{-2} per well	Hou et al. (2012)
9	Soft lithography	25.4 μL	MPD: 95 mW cm^{-2}; MCD: 170 mA cm^{-2}	Choi and Chae (2013)
11	Paper based	NA	MPD: 10 μW cm^{-2}; MCD: 80 μA cm^{-2}	Fraiwan and Choi (2014)
12	Paper based	52.5 μL	25 μW cm^{-3}; 52.25 μA	Hashemi et al. (2016)
13	Laser micromachining	47 μL	MPD: 6.5 mW cm^{-2} with gold/poly microfibre (GPM)	Fraiwan et al. (2014)

applications and characteristics. We hereby review such methods for M^+MMFCs with salient specifications. Table 12.1 summarizes the main data from some of the published work in M^+MMFC.

12.3.1.1 Conventional Photolithography (Chen et al. 2011; Dávila et al. 2011; Mu et al. 2006)

The bulk silicon micromachining combined with photolithography techniques were used by Chaio et al. (Mu et al. 2006) to fabricate the electrodes for M^+MMFC. The average open circuit potential (OCP) was measured as 300–500 mV, the maximum power density was calculated as 2.3 nW cm^{-2}. In their work, Chen et al. (2011) photolithographically fabricated a 25 μL single-chamber MMFC and achieved a maximum power density of 2.9 μW cm^{-2} and a maximum current density of 214.8 μA cm^{-2}. As shown in Fig. 12.2, Dávila et al. (2011) presented a device where the electrodes were fabricated using standard photolithography and deep reactive ion etching (DRIE) on silicon wafers. Here, the electrodes have several arrays of square vertical channels, where the diffusion of the ions released during

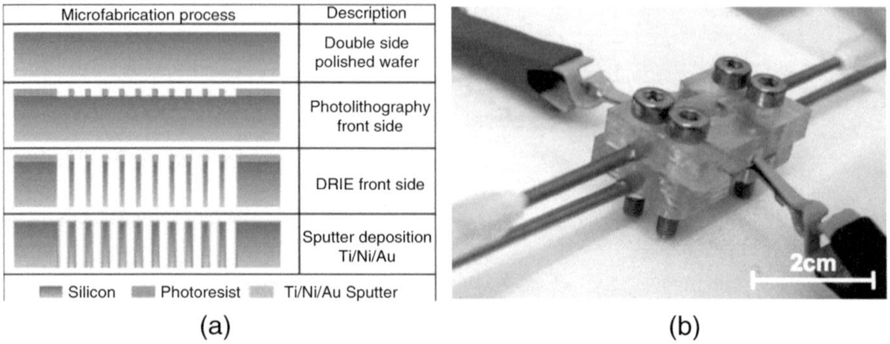

Fig. 12.2 (**a**) Steps to fabricate the device; and (**b**) Final device ready to be used as a biosensor for toxicity measurements (Reproduced from Dávila et al. 2011 with permission from the Elsevier B.V.)

oxidation towards the membrane take place. With this configuration, a maximum power density of 6.5 W cm^{-2} and maximum current densities of 0.014 mA cm^{-2} were produced. Interestingly, taking one step forward, this device was successfully used as toxicity sensor with sensing up to 0.1% v/v of formaldehyde.

12.3.1.2 Soft Lithography (Choi and Chae 2013; Li et al. 2011; Qian et al. 2009, 2011; Siu and Chiao 2008)

Polydimethylsiloxane (PDMS) based soft lithography has attained popularity due to its easy fabrication, being cost effective and flexibility to develop any kind of microfluidic device with varying properties. The standard process of soft lithography consists of realizing a master, which is the reverse replica of the final device to be developed. Thereafter, PDMS is poured, baked and released to get the final device as per the standardized protocol.

Lot of work has been carried out to use PDMS based soft lithography approach to develop membraned microfluidic MFC (M^+MMFC). A PDMS M^+MMFC with embedded micropillar structured electrodes with increased electrical output was presented by Siu and Chiao (2008). As the overall M^+MMFC is made of PDMS, the device was found to be suitable as an implantable device. The device produced a maximum OCP of 488.1 mV, power density of 401.2 nW cm^{-2} and current density of 30.2 μA cm^{-2}. In 2009, Qian et al. developed their first soft lithography based dual-chamber M^+MMFC system and performed on-chip bacterial also. With this device power density of 15 μW cm^{-3} and maximum current density of 1.3 mA cm^{-3} were achieved. The group continued their work and in 2011 presented a modified M^+MMFC device with carbon cloth electrodes, with higher power densities up to 62.5 μW cm^{-3}.

In the same year, Li et al. (2011) presented another integrated fabricated M^+ MMFC device able to generate the maximum current density of 2.542 μA cm^{-2} (equivalent to 127 μA cm^{-3}) by smartly controlling the micro-flow. Here, the

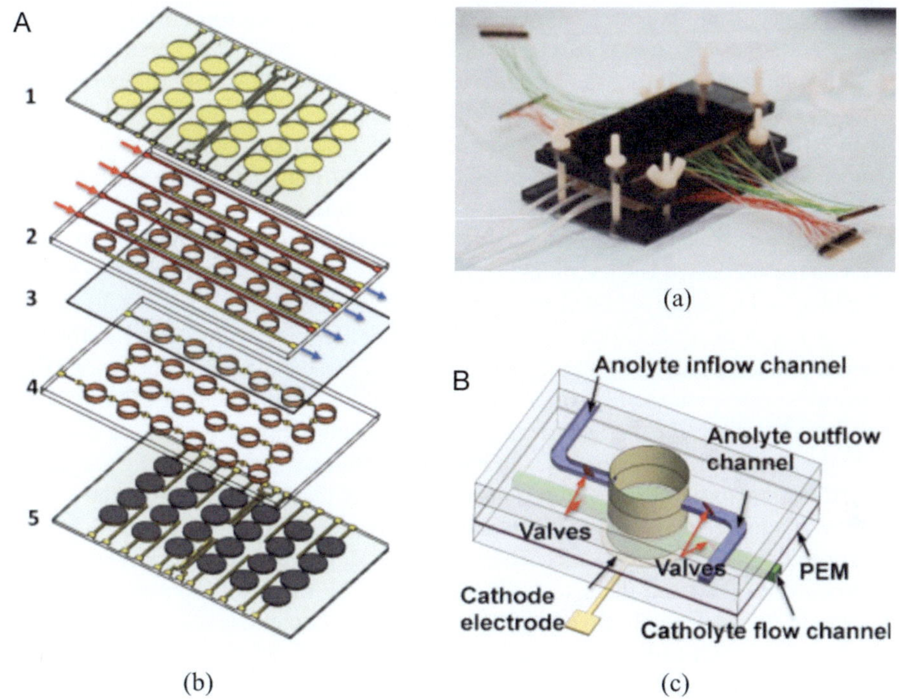

Fig. 12.3 (**a**) Schematic of 24-well M^{+}MMFC array as a high-throughput platform with various layers: (*1*) anode electrode, (*2*) anode chamber with various inflow and outflow channels, (*3*) PEM, (*4*) cathode chamber, and (*5*) cathode electrode; (**b**) Actual picture of the device; and (**c**) Schematic of a single functional M^{+}MMFC (Reproduced from Hou et al. 2012 with permission from the Royal Society of Chemistry)

microelectrodes were fabricated on a glass wafer using standard lithography technique while the microchannels were fabricated soft lithographically and later assembled. As shown in Fig. 12.3, Hou et al. (2012) developed 24-well M^{+} MMFC array device using soft lithography method where the master was fabricated by a rapid prototyping machine. The device was able to produce 3.2 $\mu W\ cm^{-2}$ per well and was shown to last for as long as 2 weeks (307 h). More recently, Choi and Chae (2013) presented M^{+}MMFC with an optimal biofilm and minimal oxygen invasion was observed into its anode chamber to generate higher power density of 95 $\mu W\ cm^{-2}$. The creation of the optical biofilm was understood by using different thicknesses of photo-definable PDMS layers.

12.3.1.3 Paper-Based Devices (Choi et al. 2015; Fraiwan and Choi 2014; Fraiwan et al. 2013; Hashemi et al. 2016)

For last more than a decade, people have started using lab filter paper to create microfluidic environment. This has been carried out by harnessing the unique ability of such filter papers to show capillary behaviour, therefore, leading to create a self-realizing and inexpensive microfluidic device with a need to develop a suitable mechanism to create boundaries for the micro-flow. Such devices have been used to fabricate M^+MMFC as well.

In 2013, Fraiwan et al. presented a microfabricated paper-based M^+MMFC with paper-based membrane with chamber where the hydrophobic walls were made of a negative photoresist. This device generated a maximum power of 5.5 μW cm^{-2} and current of 74 μA. The work has also shown that by stacking 2 M^+MMFC devices, the voltage increased by 1.9 times. Further, the same group optimized M^+MMFC (Fraiwan and Choi 2014) by stacking few such devices to run LED for 30 min. In 2015, Choi et al. (2015) developed a paper-based device with 48 wells, for high throughput and fast characterization of microbes. For the electrodes layers, PMMA substrates were patterned by micromachining and a custom-made PCB board was developed to simplify the electrical connections and measure the power produced from the 48 M^+MMFC units. The device was capable to study selected genes and to determine the ability of several mutants to generate electricity in a high throughput manner. Esquivel et al. (2014) leveraged paper-based method to develop M-MMFC where the Y-shaped paper microchannels were developed and at one end of the microchannel the membrane was positioned. An Au layer was sputtered to create the anode electrode while a carbon paper was used as a cathode electrode. With this configuration, power densities in the range of 1–5 mW cm^{-2} were measured. Lee et al. (2016) realized a paper-based, low cost, disposable and solvent-free M^-MMFC. The electrodes were fabricated with few strokes of graphite particles from a pencil. This device generated a maximum potential of 300 mV, a maximum current of 11 μA and a power of 2270 nW (0.568 μW cm^{-2}). Recently, Hashemi et al. (2016) discussed (Fig. 12.4) a self-pumped paper-based M^+MMFC which ran

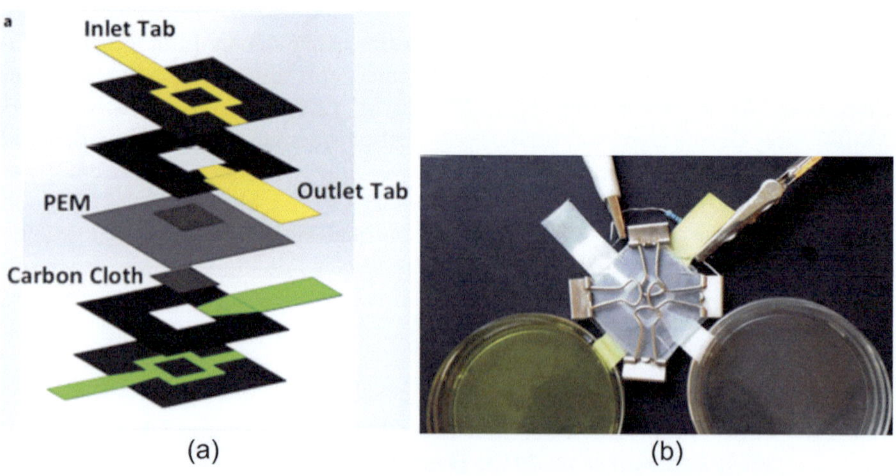

Fig. 12.4 (**a**) A schematic of the fully integrated paper-based M^+MMFC showing flows of two different samples (represented in *yellow* and *green colours*); and (**b**) A picture of the real M^+MMFC (Reproduced from Hashemi et al. 2016 with permission from the Elsevier B.V.)

for ~5 days without outside power or human interference. In a 52.5 μL chamber a power density of approximately 25 μW cm^{-3} was measured.

12.3.1.4 Laser Micromachining

Fraiwan et al. (2014) used laser micromachining to stack multiple M^{+}MMFC in order to optimize the performance, where each chamber was fabricated by patterned PMMA and plastic gasket. Here, 6 micro/nano-structured anodes in micro-sized M^{+}MMFC were compared under uniform environment. It was found that the power density of the gold/poly microfibre (GPM) and carbon nanotube (CNT) was the highest (maximum power density of 6.5 μW cm^{-2} and 4.9 μW cm^{-2} respectively).

12.3.2 Challenges in Conventional Microfluidic MFCs (M^{+}MMFC)

12.3.2.1 High Internal Resistance

In M^{+}MMFCs, the high internal resistance is observed due to the fact that the compatibility of the electrode with microorganisms was not optimized, which leads to inefficient electron transfer. Overall the high internal resistance leads to unproductive interaction between microorganism and electrode, which is detrimental.

12.3.2.2 Low Energy Density Output

The high internal resistance has severe effect on low energy density output, which may be due to low availability of microbes as well.

12.3.2.3 Oxygen Penetration

Unlike in macroscale MFCs, M^{+}MMFCs are highly susceptible to oxygen in the anodic chamber, primarily from a low overall availability of microbial communities in the microscale. This may lead to the extreme drop in anodic potential.

12.4 Membraneless Microfluidic MFC (M^-MMFC)

As can be seen from the discussion about M^+MMFC, it is clear that microfluidic environment dramatically help in not only miniaturizing the device, but also automating it in a cost-efficient manner. Of course, there are numerous challenges, such as high internal resistance and efficient scavenging of oxygen leading to low energy density. Majority of such challenges arise due to the material property, presence of the membrane and poor electron transfer rate, which are interlinked. One major way to solve such challenges is to develop membraneless microfluidic MFC (M^-MMFC), which will continue to leverage the unique properties of microfluidics but eradicate some of the limitation of the M^+MMFC. The behaviour has been explained in Sect. 12.2.3. Table 12.2 summarizes the main data from some of the published works in M^-MMFC.

Table 12.2 Summary of the published research works in M^-MMFC

Sl. no.	Fabrication method	Cell area/ volume	Maximum power density (MPD)/current density (MCD)	References
1	Soft lithography	0.3 cm^2	MPD: 2.5 mW cm^{-2}	Choban et al. (2004)
2	Soft lithography	0.5 μL	MPD: 0.65 mW cm^{-2}; MCD: 2.62 mA cm^{-2}	Sun et al. (2007)
3	Soft lithography	0.3 μL	Maximum current: 2.1 ± 0.1 μA	Li et al. (2012)
4	Soft lithography	30 μL	OCV: 246 mV	Wang and Su (2013)
5	Soft lithography	250 μL	OCV: 20 mV	Yan et al. (2014)
6	Soft lithography	80 μL	MPD: 61.8 ± 0.4 μW cm^{-2}	Ye et al. (2013)
7	Soft lithography	60 μL	MPD: 244.8 ± 3.9 μW cm^{-2}	Yang et al. (2015)
8	Paper-based	400 μL	Maximum power: 0.4 μW; maximum current: 18 μA	Veerubhotla et al. (2015)
9	Polymer micromachining	40 μL	MPD: 0.12 nW cm^{-2}; MCD: 1.9 μA cm^{-2}	Rojas et al. (2015)
10	Multi-layer bonding	3 μL	MPD: 1100 mW cm^{-3}	Li et al. (2016a, b)
11	Polymer molding	NA	MCD: 3.042 ± 2.255 nA cm^{-2}	Ahn and Schröder (2015)
12	Soft lithography	NA	MCD: 1.5 mA cm^{-2}	Li et al. (2016a, b)
13	Polymer bonding	NA	MPD: 3.8 mW cm^{-2} MCD: 32.4 mA cm^{-2}	Rathoure and Pramanik (2016)
14	Rapid prototyping (CNC, xy plotter)	100 μL	MCD: 380 mA cm^{-2}	Gurrola et al. (2016)
15	Polymeric	50 μL	MPD: 181.4 ± 135.6 μW cm^{-3}	Yang et al. (2016a)

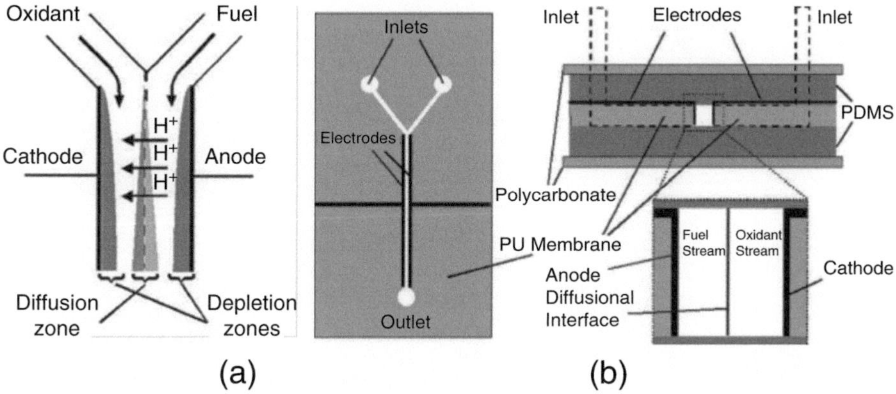

Fig. 12.5 (**a**) Schematic of the M⁻MMFC laminar flow-based fuel cell; and (**b**) Stacked assembly of M⁻MMFC (Reproduced from Choban et al. 2004 with permission from Elsevier B.V.)

12.4.1 Key Membraneless Microfluidic MFC (M⁻MMFC) and Their Comparison

Even though M⁻MMFCs have been first reported in 1990 (Dyer 1990) for conventional fuel cells, they were first realized in microfluidic environment in 2004 (Choban et al. 2004; Liu and Logan 2004) only. Choban et al. (2004) proposed a soft-lithographically fabricated microfluidic device with Y-shaped microchannel integrated with electrodes attached by sputtering and electrodeposition. The working principle and device are shown in Fig. 12.5, with which the current densities up to 8 mA cm^{-2} and power density of up to 2.5 mW cm^{-2} were achieved.

Sun et al. (2007) discussed an M⁻MMFC with 3-inlet microchannels, each for fuel, oxidant and the electrolyte solution, using the multi-stream laminar. The device was fabricated by PDMS based soft lithography and Ti-Pt electrodes were encapsulated by conventional photolithography method. Maximum power density and current density were found to be 0.65 mW cm^{-2} and 2.62 mA cm^{-2} respectively. Li et al. (2012) also presented a 0.3 μL M⁻MMFC fabricated by PDMS lithography and gold electrodes, where the current as high as 2.1 ± 0.1 μA was observed. Another M⁻MMFC, developed by Wang and Su (2013), was consisting of PDMS microchannels integrated with glass slide with sputter gold electrodes. The maximum Open Circuit Voltage (OCV) was measured as 246 mV for active microflora. The group continued their efforts to modify M⁻MMFC (Yan et al. 2014) using the anode composed of conductive microparticles generated by a T-junction microchannel. With this configuration, the maximum OCV was measured as 20 mV which lasts for more than 6 h.

Ye et al. (2013) presented an idea to realize a 80 μL M⁻MMFC producing the maximum power density of 61.8 ± 0.4 μW cm^{-2}. The device was fabricated by combining two graphite plates (acting as electrode) and two polymethylmethacrylate (PMMA) plates (microchannel). The group sustained their work

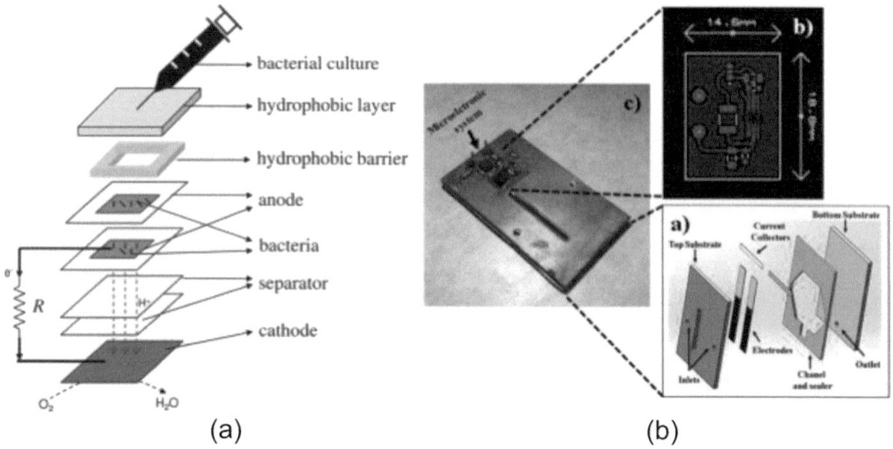

Fig. 12.6 (**a**) Schematic of the paper-based M^{-}MMFC (Reproduced from Veerubhotla et al. (2015) with permission from the Royal Society of Chemistry); and (**b**) Schematic of the M^{-}MMFC fabricated using rapid prototyping automated method (CNC, xy plotter) (Reproduced from Gurrola et al. 2016 with permission from Elsevier B.V.)

(Yang et al. 2015) and compared M^{-}MMFC with three different microchannel geometries using similar fabrication scheme (Ye et al. 2013). M^{-}MMFC with the diverging channel produced maximum power density (244.77 ± 3.89 μW cm^{-2}) because of the homogeneous biofilm and lower charge transfer resistance. As shown in Fig. 12.6, Veerubhotla et al. (2015) realized an inexpensive and disposable M^{-}MMFC fabricated on a paper with simple pencil lids. With this system, the device was shown to hold bacteria for an hour, producing an OCV of 0.4 V, a maximum current of 18 μA and maximum power of 0.4 μW.

Rojas et al. (2015) developed M^{-}MMFC by integrating gold anode (standard microfabrication method), two plexiglass gasket (to form the chamber) and carbon cloth cathode (polymer-coated on one side). The group had shown that this device produces maximum power density of 0.12 nW cm^{-2} and current density of 1.9 μA cm^{-2}. Li et al. (2016a, b) presented a vanadium M^{-}MMFC where the porous carbon paper electrodes were stacked in a multilayered fashion. The maximum power density of 1.1 W cm^{-3} were produced, which clearly manifests the importance of multi-layer electrode configuration. Ahn and Schröder (2015) discussed an M^{-}MMFC with 3-electrode system in a continuous mode and fabricated using glass and polymer with a customized polymer molding method utilizing soft-lithography. This device was used as a biosensor to detect toxicity in water as low as 0.02 mM Sodium azide with corresponding current density of 3.042 ± 2.255 nA cm^{-2}. Li et al. (Li et al. 2016a, b) presented another method to create M^{-}MMFC with 3-electrode system in using PDMS soft lithography method. Here, a gold electrode, a graphite rod and a commercial Ag/AgCl were taken as working electrode, counter electrode and as a reference electrode respectively. With this device, current density of 1.5 mA cm^{-2} was obtained and the device was used as a biosensor to detect ferric citrate as low as 0.25 mM. A self-powered air breathing M^{-}MMFC was developed by conventional polymer bonding method integrated with Pt-Ru electrodes by Rathoure and Pramanik (2016). Methanol electrooxidation was studied and the maximum power density of 3.8 mW cm^{-2} and current density of 32.4 mA cm^{-2} were measured. Gurrola et al. (2016) developed a membraneless nanofluidic device which was fabricated using rapid prototyping automated method (CNC, xy plotter) to create the M^{-}MMFC. An in-house developed test station was used to estimate the M^{-}MMFC performance which came out to be 380 mA cm^{-2}. Yang et al. (2016b) discussed to develop polymeric M^{-}MMFC with graphene based bio-cathode and bio-anode producing a maximum power density of 1.18 ± 0.14 mW cm^{-3}.

12.4.2 Salient Features of M^{-}MMFC

Based on the description of several important contributions to develop M^{-}MMFC, we can summarize following key features of M^{-}MMFC:

12.4.2.1 Membraneless

As discussed before, M^{-}MMFCs take advantage of the unique property of microfluidic environment due to which the mixing of anolyte and catholyte can be delayed as a virtual wall is formed in the form of liquid-liquid interface. This dramatically reduces the cost and other challenges occurred due to membrane such as low-chemical self-life, humidification, incorrect and inefficient fuel movement and high-ohmic resistance.

12.4.2.2 Higher Output Power Density/Current Density

As the ohmic resistance due to physical membrane is removed, all the contributory ions are able to move interface easily and contribute to the cathodic reaction. Also, the bacterial growth is better due to the modified flow conditions, the layer adjacent to the electrode surface can be quickly re-filled and the by-products can be efficiently discarded. Except being robust and integrated device, the M^{-}MMFCs use the gravity or capillary force to manipulate the flow leading to the generation of higher power density than M^{+}MMFCs.

12.4.2.3 Relatively Shorter Response Time

In general, the rate of microorganisms aggregation on the electrode surface is much higher in M^{-}MMFCs leading to fast formation of a conductive biofilm in a relatively short time. Due to the short response time the M^-MMFCs can easily be used in biosensing applications.

12.4.3 Challenges in M^-MMFC

12.4.3.1 Ensuring the Required Flow Environment

With a microfluidic environment, the key for M^-MMFC, various issues, such as microfluid-dynamics, flow-control and transport on various layers, zones and boundaries need to be properly optimized. This is more important for M^-MMFC as the virtual interface plays a crucial role to ensure proper overflow of reagents from both the regions.

12.4.3.2 Smart Integration of Various Components of M^-MMFC

Various M^-MMFC components, such as microchannel, electrodes, fluid-delivery and manipulation system, micro-pumps, micro-valves etc. need to integrate effectively, considering important parameters, such as power output, form factor, power input etc. It is critical to have the most suitable coordination between these independent units for the desirable M^-MMFC operation.

12.5 Future Opportunities

12.5.1 Electricity Generation

Being a niche fuel cell, the prime function of MMFC is visualized in terms of power feeding devices with microorganisms as feedstock. Therefore, the MMFC would play a unique role where power transmission is limited to run several low powered devices, especially in hilly and remote areas. These low powered devices include local wireless sensor network, biosensors, personalized *in vitro* diagnostic systems, toxicity and water monitoring system etc. Even though the cost to develop microfluidic chip is substantially low, the ancillary components, electrodes and other accessories add to the overall cost to operate MMFCs.

12.5.2 In Vivo *Operation*

Researchers have started testing MMFCs with bodily substances, such as glucose and urine, as feedstock. Of course, a lot more work is needed to develop a feasible and compatible system with suitable biocompatible electrodes, microfluidic device etc. Such device can be used to run implantable devices, such as pacemakers and artificial organs.

12.5.3 Input Power Requirement

Even though MMFCs are power feeding devices, to operate various constituents of MMFC, such as pumps to deliver fluid in the microchannel is necessary. Therefore, it is important to explore an MMFC with self-sustaining power requirement. The paper-based microfluidic devices, where the fluid moves under capillary force, should be one good option where the need of external pump can be avoided by various means, such as using microorganism as actuator (Kim and Breuer 2008), circulating microchannel geometry (Abaci et al. 2015), surface-tension driven flow (Xing et al. 2016), air breathing electrodes (Rathoure and Pramanik 2016) or gravity-driven flow (Mäki et al. 2014).

12.5.4 Other Applications

Except employing the MMFC in water monitoring, waste water treatment (Li et al. 2014; Papaharalabos et al. 2015), biosensing (Esquivel et al. 2014; Fraiwan et al. 2014; Li et al. Li et al. 2016a, b), biomedical diagnostics (Esquivel et al. 2014), they are strong candidates to be harnessed in other applications such as converting waste to useful products (Merrill and Logan 2009), production of low or non-alcoholic beer (Szollosi et al. 2016), hydrogen production (Li et al. 2015) etc., where the macroscale MFC have been demonstrated. Therefore, MMFC have strong potential to not be limited to their prime role to generate power, but also in a variety of applications.

12.6 Conclusion

Globally, there have been continuous debates and discussions, on various levels, to develop devices able to help with the burning issue of climate change, both in terms of being based on renewable and sustainable energy and possibly help in converting waste to useful products. Certainly, development of biofuel Fuel Cell, particularly MFC provides one such technological intervention. Like any other technology upgradation, MFCs have also been into the phases of development where several limitations have been recognized and methods and processes are proposed. One such method was to develop MFCs in microscale, called as microfluidic MFC (MMFC) where the well proven and commercially available components of microfluidic environment, such as user friendly, automated, robust, modular and cost effective, are harnessed. Another major breakthrough with MMFC is the possibility to avoid membrane by smartly optimizing the microfluidics design, which has given a unique pathway for their use in diverse applications, such as miniature electronic devices, biosensing, medical diagnostics, water monitoring and treatment, hydrogen production, toxicity detection etc. The chapter summarizes the development of MMFCs, both membraned and membraneless, where a variety of fabrication methods have been used in diverse applications. Current challenges and future opportunities of MMFCs have also been described.

References

Abaci, H. E., Gledhill, K., Guo, Z., Christiano, A. M., & Shuler, M. L. (2015). Pumpless microfluidic platform for drug testing on human skin equivalents. *Lab on a Chip, 15*(3), 882–888.

Ahn, Y., & Schröder, U. (2015). Microfabricated, continuous-flow, microbial three-electrode cell for potential toxicity detection. *BioChip Journal, 9*(1), 27–34.

Chen, Y.-P., Zhao, Y., Qiu, K.-Q., Chu, J., Lu, R., Sun, M., et al. (2011). An innovative miniature MFC fabricated using photolithography. *Biosensors and Bioelectronics, 26*(6), 2841–2846.

Choban, E. R., Markoski, L. J., Wieckowski, A., & Kenis, P. J. A. (2004). Microfluidic fuel cell based on laminar flow. *Journal of Power Sources, 128*(1), 54–60.

Choi, S. (2015). Microscale MFCs: Advances and challenges. *Biosensors and Bioelectronics, 69*, 8–25.

Choi, S., & Chae, J. (2012). An array of microliter-sized MFCs generating 100 μW of power. *Sensors and Actuators A: Physical, 177*, 10–15.

Choi, S., & Chae, J. (2013). Optimal biofilm formation and power generation in a micro-sized *MFC* (MFC). *Sensors and Actuators A: Physical, 195*, 206–212.

Choi, G., Hassett, D. J., & Choi, S. (2015). A paper-based MFC array for rapid and high-throughput screening of electricity-producing bacteria. *Analyst, 140*(12), 4277–4283.

Dávila, D., Esquivel, J. P., Sabaté, N., & Mas, J. (2011). Silicon-based microfabricated MFC toxicity sensor. *Biosensors and Bioelectronics, 26*(5), 2426–2430.

Dyer, C. K. (1990). A novel thin-film electrochemical device for energy conversion CuO. *Nature, 343*(6258), 547–548.

Esquivel, J. P., Del Campo, F. J., Gomez de la Fuente, J. L., Rojas, S., & Sabate, N. (2014). Microfluidic fuel cells on paper: Meeting the power needs of next generation lateral flow devices. *Energy & Environmental Science, 7*(5), 1744–1749.

Fraiwan, A., & Choi, S. (2014). Bacteria-powered battery on paper. *Physical Chemistry Chemical Physics, 16*(47), 26288–26293.

Fraiwan, A., Mukherjee, S., Sundermier, S., Lee, H.-S., & Choi, S. (2013). A paper-based MFC: Instant battery for disposable diagnostic devices. *Biosensors and Bioelectronics, 49*, 410–414.

Fraiwan, A., Lee, H., & Choi, S. (2014). A multianode paper-based MFC: A potential power source for disposable biosensors. *Sensors Journal, IEEE, 14*(10), 3385–3390.

Gravesen, P., Branebjerg, J., & Jensen, O. S. (1993). Microfluidics–A review. *Journal of Micromechanics and Microengineering, 3*(4), 168.

Gurrola, M. P., Ortiz-Ortega, E., Farias-Zuñiga, C., Chávez-Ramírez, A. U., Ledesma-García, J., & Arriaga, L. G. (2016). Evaluation and coupling of a membraneless nanofluidic device for low-power applications. *Journal of Power Sources, 307*, 244–250.

Hashemi, N., Lackore, J. M., Sharifi, F., Goodrich, P. J., Winchell, M. L., & Hashemi, N. (2016). A paper-based MFC operating under continuous flow condition. *Technology, 4*, 1–6.

He, Y., Wu, Y., Fu, J.-Z., Gao, Q., & Qiu, J.-J. (2016). Developments of 3D printing microfluidics and applications in chemistry and biology: A review. *Electroanalysis, 28*(8), 1658–1678.

Hou, H., Li, L., Ceylan, C. U., Haynes, A., Cope, J., Wilkinson, H. H., et al. (2012). A microfluidic MFC array that supports long-term multiplexed analyses of electricigens. *Lab on a Chip, 12*(20), 4151–4159.

Kim, M. J., & Breuer, K. S. (2008). Microfluidic pump powered by self-organizing bacteria. *Small, 4*(1), 111–118.

Kirby, B. J. (2013). *Micro- and nanoscale fluid mechanics*. Cambridge: Cambridge University Press.

Kjeang, E., Djilali, N., & Sinton, D. (2009). Microfluidic fuel cells: A review. *Journal of Power Sources, 186*(2), 353–369.

Konwarh, R., Gupta, P., & Mandal, B. B. (2016). Silk-microfluidics for advanced biotechnological applications: A progressive review. *Biotechnology Advances, 34*(5), 845–858.

Kou, S., Cheng, D., Sun, F., & Hsing, I. M. (2016). Microfluidics and microbial engineering. *Lab on a Chip, 16*(3), 432–446.

Lee, J. W., & Kjeang, E. (2010). A perspective on microfluidic biofuel cells. *Biomicrofluidics, 4*(4), 041301.

Lee, D.-J., Chang, J.-S., & Lai, J.-Y. (2015a). Microalgae–*MFC*: A mini review. *Bioresource Technology, 198*, 891–895.

Lee, H., Yang, W., Wei, X., Fraiwan, A., & Choi, S. (2015b). A microsized MFC based biosensor for fast and sensitive detection of toxic substances in water. Paper presented at the 28th IEEE International Conference on Micro Electro Mechanical Systems (MEMS), Portugal. https://doi.org/10.1109/MEMSYS.2015.7051020\t _blank10.1109/MEMSYS.2015.7051020

Lee, S. H., Ban, J. Y., Oh, C.-H., Park, H.-K., & Choi, S. (2016). A solvent-free microbial-activated air cathode battery paper platform made with pencil-traced graphite electrodes. *Scientific Reports, 6*, 28588.

Li, Z., Zhang, Y., LeDuc, P. R., & Gregory, K. B. (2011). Microbial electricity generation via microfluidic flow control. *Biotechnology and Bioengineering, 108*(9), 2061–2069.

Li, Z., Venkataraman, A., Rosenbaum, M. A., & Angenent, L. T. (2012). A laminar-flow microfluidic device for quantitative analysis of microbial electrochemical activity. *ChemSusChem, 5*(6), 1119–1123.

Li, L., Wang, G., Chen, R., Zhu, X., Wang, H., Liao, Q., et al. (2014). Optofluidics based micro-photocatalytic fuel cell for efficient wastewater treatment and electricity generation. *Lab on a Chip, 14*(17), 3368–3375.

Li, M., He, X., Zeng, Y., Chen, M., Zhang, Z., Yang, H., et al. (2015). Solar-microbial hybrid device based on oxygen-deficient niobium pentoxide anodes for sustainable hydrogen production. *Chemical Science, 6*(12), 6799–6805.

Li, L., Nikiforidis, G., Leung, M. K. H., & Daoud, W. A. (2016a). Vanadium microfluidic fuel cell with novel multi-layer flow-through porous electrodes: Model, simulations and experiments. *Applied Energy, 177*, 729–739.

Li, F., Zheng, Z., Yang, B., Zhang, X., Li, Z., & Lei, L. (2016b). A laminar-flow based microfluidic microbial three-electrode cell for biosensing. *Electrochimica Acta, 199*, 45–50.

Li, Z., Zhang, Y., LeDuc, P. R., & Gregory, K. B. (2011a). Microbial electricity generation via microfluidic flow control. *Biotechnology and Bioengineering, 108*(9), 2061–2069.

Liu, H., & Logan, B. E. (2004). Electricity generation using an air-cathode single chamber MFC in the presence and absence of a proton exchange membrane. *Environmental Science & Technology, 38*(14), 4040–4046.

Mäki, A.-J., Hemmilä, S., Hirvonen, J., Girish, N. N., Kreutzer, J., Hyttinen, J., et al. (2014). Modeling and experimental characterization of pressure drop in gravity-driven microfluidic systems. *ASME: Journal of Fluids Engineering, 137*(2), 021105.

Merrill, M. D., & Logan, B. E. (2009). Electrolyte effects on hydrogen evolution and solution resistance in microbial electrolysis cells. *Journal of Power Sources, 191*(2), 203–208.

Mu, C., Kien, B. L., & Liwei, L. (2006). Micromachined microbial and photosynthetic fuel cells. *Journal of Micromechanics and Microengineering, 16*(12), 2547.

Mukherjee, S., Su, S., Panmanee, W., Irvin, R. T., Hassett, D. J., & Choi, S. (2013). A microliter-scale MFC array for bacterial electrogenic screening. *Sensors and Actuators A: Physical, 201*, 532–537.

Papaharalabos, G., Greenman, J., Melhuish, C., & Ieropoulos, I. (2015). A novel small scale MFC design for increased electricity generation and waste water treatment. *International Journal of Hydrogen Energy, 40*(11), 4263–4268.

Qian, F., Baum, M., Gu, Q., & Morse, D. E. (2009). A 1.5 [small micro]L MFC for on-chip bioelectricity generation. *Lab on a Chip, 9*(21), 3076–3081.

Qian, F., He, Z., Thelen, M. P., & Li, Y. (2011). A microfluidic MFC fabricated by soft lithography. *Bioresource Technology, 102*(10), 5836–5840.

Rathoure, A. K., & Pramanik, H. (2016). Electrooxidation study of methanol using H_2O_2 and air as mixed oxidant at cathode in air breathing microfluidic fuel cell. *International Journal of Hydrogen Energy, 41*(34), 15287–15294.

Ren, H., Lee, H.-S., & Chae, J. (2012). Miniaturizing MFCs for potential portable power sources: Promises and challenges. *Microfluidics and Nanofluidics, 13*(3), 353–381.

Rojas, J.P., Alqarni, W., Kalantan, K., Hussain, M. M., & Mink, J. (2015). Nano-watt fueling from a micro-scale MFC using black tea waste. *International Conference on Nanotechnology (IEEE-NANO)* (pp. 955–958).

Sackmann, E. K., Fulton, A. L., & Beebe, D. J. (2014). The present and future role of microfluidics in biomedical research. *Nature, 507*(7491), 181–189.

Samiei, E., Tabrizian, M., & Hoorfar, M. (2016). A review of digital microfluidics as portable platforms for lab-on a-chip applications. *Lab on a Chip, 16*(13), 2376–2396.

Shantaram, A., Beyenal, H., Veluchamy, R. R. A., & Lewandowski, Z. (2005). Wireless sensors powered by MFCs. *Environmental Science & Technology, 39*(13), 5037–5042.

Sinton, D. (2014). Energy: The microfluidic frontier. *Lab on a Chip, 14*(17), 3127–3134.

Siu, C. P. B., & Chiao, M. (2008). A microfabricated PDMS *MFC. Journal of Microelectromechanical Systems, 17*(6), 1329–1341.

Sun, M. H., Velve Casquillas, G., Guo, S. S., Shi, J., Ji, H., Ouyang, Q., et al. (2007). Characterization of microfluidic fuel cell based on multiple laminar flow. *Microelectronic Engineering, 84*(5–8), 1182–1185.

Szollosi, A., Nguyen, Q. D., Kovacs, A. G., Fogarasi, A.-L., Kun, S., & Hegyesne-Vecseri, B. (2016). Production of low or non-alcoholic beer in MFC. *Food and Bioproducts Processing, 98*, 196–200.

Veerubhotla, R., Bandopadhyay, A., Das, D., & Chakraborty, S. (2015). Instant power generation from an air-breathing paper and pencil based bacterial bio-fuel cell. *Lab on a Chip, 15*(12), 2580–2583.

Wang, H.-Y., & Su, J.-Y. (2013). Membraneless microfluidic MFC for rapid detection of electrochemical activity of microorganism. *Bioresource Technology, 145*, 271–274.

Wang, H.-Y., Bernarda, A., Huang, C.-Y., Lee, D.-J., & Chang, J.-S. (2011). Micro-sized MFC: A mini-review. *Bioresource Technology, 102*(1), 235–243.

Wang, H., Park, J.-D., & Ren, Z. J. (2015). Practical energy harvesting for MFCs: A review. *Environmental Science & Technology, 49*(6), 3267–3277.

Xing, Y., Nourmohammadzadeh, M., Elias, J. E. M., Chan, M., Chen, Z., McGarrigle, J. J., et al. (2016). A pumpless microfluidic device driven by surface tension for pancreatic islet analysis. *Biomedical Microdevices, 18*(5), 1–9.

Yan, J., Lee, D. J., Chou, S. K., Desideri, U., Li, H., Chen, Y.-Y., et al. (2014). International conference on applied energy, ICAE2014. Fabrication of conductive microparticles as anodal electrode in microfluidic *MFC. Energy Procedia, 61*, 1565–1568.

Yang, Y., Ye, D., Li, J., Zhu, X., Liao, Q., & Zhang, B. (2015). Biofilm distribution and performance of microfluidic MFCs with different microchannel geometries. *International Journal of Hydrogen Energy, 40*(35), 11983–11988.

Yang, Y., Liu, T., Liao, Q., Ye, D., Zhu, X., Li, J., et al. (2016a). Three-dimensional nitrogen-doped Graphene Aerogel-activated carbon composite catalyst enables low-cost microfluidic MFCs with superior performance. *Journal of Materials Chemistry A*. doi: https://doi.org/10.1039/C6TA05002F.

Yang, Y., Ye, D., Li, J., Zhu, X., Liao, Q., & Zhang, B. (2016b). Microfluidic MFCs: From membrane to membrane free. *Journal of Power Sources, 324*, 113–125.

Ye, D., Yang, Y., Li, J., Zhu, X., Liao, Q., Deng, B., et al. (2013). Performance of a microfluidic MFC based on graphite electrodes. *International Journal of Hydrogen Energy, 38*(35), 15710–15715.

Chapter 13
Diagnostic Tools for the Assessment of MFC

Jhansi L. Varanasi, Ramya Veerubhotla, and Debabrata Das

13.1 Introduction

Microbial fuel cells (MFCs) are renewable sustainable technologies that can convert biodegradable wastes directly into electricity. This conversion is usually brought about by electrogenic bacteria that can degrade the organic/inorganic substrates via their metabolisms and transfer the released electrons to a solid electron acceptor (the electrodes). A typical MFC comprises different components such as an anode, a cathode, a separator, electrolyte and electrical circuits. Each of these components have pivotal role in depicting the overall performance of MFCs. At present, the power outputs of MFCs are too low which restricts its practical applicability. To overcome the bottlenecks, a better understanding of all the components and their limitations is required.

Over the years, several advances have been made in developing innovative analytical techniques to evaluate the MFCs' performance. MFC as a whole is an integrated technology comprising various interrelated disciplines such as electrochemistry, material science, microbiology, molecular biology etc. Hence, an in-depth knowledge of all the diagnostic techniques and the associated theories is essential to select the appropriate tool for the diagnosis. Usually, most of the techniques that are traditionally used for chemical fuel cells such as polarization, voltammetry etc. are also employed for analysing MFCs. However, since the operational conditions and the performances vary significantly for both these systems, it is a prerequisite to take into account these differences while performing the analysis or interpreting the data.

J.L. Varanasi • R. Veerubhotla • D. Das (✉)
Department of Biotechnology, Indian Institute of Technology Kharagpur,
Kharagpur 721302, India
e-mail: ddas.iitkgp@gmail.com

D. Das (ed.), *Microbial Fuel Cell*,
https://doi.org/10.1007/978-3-319-66793-5_13

In this chapter, emphasis is made on the various diagnostic tools that are used to analyse performance of MFCs with respect to each component i.e. the microbial biofilms, electrode configurations, nature of electrolyte, material characteristics etc. In the following section, a brief background on the different performance indicators like current, voltage, power density etc. that are used for comparison of MFCs in literature is provided so that the readers get acquainted with the terminologies and later each analytical technique is described in details.

13.2 Reporting Data Using Typical Performance Indicators

13.2.1 Open Circuit Voltage (OCV)

OCV is the emf of the cell that is measured when both anode and the cathode leads of a fuel cell are not connected to each other (Logan et al. 2006). It is the maximum voltage that is obtained from a cell at infinite resistance as the OCV value does not take the voltage losses into account. It is expected that the OCV should ideally be equal to the emf of the cell. However, in practical, the losses dominate resulting in lowering its value drastically.

13.2.2 Half-Cell Potential

The half-cell potential is the electrode potentials for individual electrodes in the system against a known reference electrode. This is an important parameter that is helpful in understanding the contribution of each electrode to the overall emf of the cell (Logan et al. 2006). A reference electrode is the electrode whose electrode potential is stable and constant in a definite known set of potential window at a particular pH. Typical examples of the reference electrodes used are Ag/AgCl standard electrodes, Standard Hydrogen Electrode (SHE) and Normal Hydrogen Electrode (NHE). Understanding the role and the contribution of individual electrodes to the net voltage generation is crucial to improve the associated components and hence the overall performance.

13.2.3 Current Density

Current generated by an MFC is calculated using the Ohm's law ($V = IR$, where V is the operating voltage, I is the current generated and R is the external resistance), current density is denoted either as current per surface area of the electrode (A cm^{-2}) or current per unit volume of the anode chamber (A cm^{-3}).

$$I_d = \frac{I}{A_{\text{ele}}} \tag{13.1}$$

$$I_d = \frac{t}{Vol} \tag{13.2}$$

where I_d is the current density, I is the current obtained in amperes, A_{ele} is the surface area of the electrodes and Vol is the volume of the anode chamber (Sun et al. 2016). Current density not only provides fundamental insights into the electron transfer rate of the cell but also becomes the preliminary parameter to access the overall performance.

13.2.4 Power Density

The amount of power generated by the electrochemical reactions that occur in an MFC is represented as $P = V.I$, where V is the operating voltage (V) of the cell and I is the current (A) at a specific resistance. It is desirable to obtain a stable and consistent power output from an MFC to support practical real-world applications. However, MFCs experience severe fluctuations in the power output due to the changes in metabolic activity of the microbes, decrease in the activity of the catalysts, membrane fouling, accumulation of the cell debris etc. (Sun et al. 2016). Researchers adopt a more reliable parameter called power density (P_d) to compare various cells with different configurations or volumes. Depending on the parameter being considered in the cell, power density is obtained by normalizing the total power obtained either by the surface area of the electrode or the volume of specific chamber or the cell itself as follows.

$$P_d \frac{V \times I}{A_{\text{ele}}} \tag{13.3}$$

$$I_d = \frac{V \times I}{Vol} \tag{13.4}$$

13.2.5 Columbic Efficiency

It is the total ratio of coulombs that can be recovered in the form of current to the total coulombs that are inherently contained in the substrate. Depending on the mode of operation, the net columbic efficiency is expressed.

13.2.5.1 Batch or Fed-Batch Mode of Operation

The total coulombs recovered are a function of output current integrated over a time interval t. The energy contained in the substrate is given as $F\ b\ \Delta M$, where F is the Faraday's constant, b is the number of electrons per mole of substrate and ΔM is the molecular mass of the substrate. Since chemical oxygen demand (COD) is used as indirect measure of the organic content, and every mole of oxygen requires four electrons for the oxidation ($b = 4$), the columbic efficiency is expressed as follows:

$$CE = \frac{M \int_{\mathrm{C}}^{\mathrm{t}} Idt}{Fb(Vol)\Delta COD} \tag{13.5}$$

13.2.5.2 Continuous Mode of Operation

For MFCs operated under continuous mode of operation with a steady state current I, the overall columbic efficiency is estimated as follows:

$$CE = \frac{MI}{Fbk\Delta COD} \tag{13.6}$$

In the above equation, k corresponds to the influent flow rate of wastewater or feed, and ΔCOD is the difference between the initial and final COD content. The primary goal of all the MFCs employing wastewaters is to attain high columbic efficiency in order to ensure electrogenesis. Certain approaches such as suppression of methanogenesis in the MFC system, improved catalysis and use of appropriate current collector in the reactor improve the columbic efficiency.

13.2.6 *Energy Efficiency*

Analogous to the columbic efficiency, the ratio of the total energy generated from the system in terms to electric output to the total energy contained in the substrate is given by its energy efficiency. The total energy recovery is formulated as follows:

$$EE = \frac{\int_0^{\mathrm{t}} VIdt}{\Delta Hm} \tag{13.7}$$

EE corresponds to the energy efficiency, DH is the heat of combustion of the substrate and m is the mass of substrate added. The amount of the energy recovered depends upon the nature and biodegradability of the substrate (Logan et al. 2006). Normalized energy recovery is another important parameter expressed as the EE normalized per unit volume of the wastewater or ΔCOD of the wastewater (Sun et al. 2016). It is expressed as follows:

$$NER = \frac{EE}{Vol} \tag{13.8}$$

$$NER = \frac{EE}{\Delta COD} \tag{13.9}$$

13.3 Performance Evaluation via Electro-chemical Tools

13.3.1 Polarization

The polarization study is crucial for analysis of the cell as well as independent components as it provides valuable information to characterize fuel cells. Basically, polarization represents the I-V characteristics of the cell. It can be done either for the whole cell or independently for anode and cathode employing a reference electrode. Experimentally, it is performed in two ways: (1) By performing as the linear-sweep voltammetry technique at low scan rates using a potentiostat; and (2) Using a resistance box where a different set of external resistances can be varied at convenience (Logan et al. 2006). The user manually varies the resistances and records the stable operating voltage using a multimeter or a voltage monitoring device. Using Ohm's law, the current density is calculated for different resistances and is plotted against the current density. For reproducibility, it is recommended to vary the resistances in both increasing and decreasing order. The choice of the resistances to be varied primarily depends on the volume of the reactor and the internal resistance of the cell. A typical polarization curve can be depicted as in Fig. 13.1.

The polarization graph of a fuel cell can be divided into three zones (Fig. 13.1a):

1. Zone (1) where there is a sharp decline in the operating voltage following OCV. This is the zone where losses called activation losses are dominant. As the name suggests, these losses arise due to the activation energy required to initiate the oxidation and reduction reactions at various components of the fuel cell.
2. Zone (2) corresponds to a steady decline in the voltage with an increase in the current. These losses can be attributed to ohmic losses which arise mainly due to the materials involved in the system. Resistance offered by electrodes, membranes, electrolyte used, interconnects, as well as the current collectors contribute to the total Ohmic losses of the cell.
3. Zone (3) corresponds to the region at high currents where the voltage readings fall drastically due to losses incurred by concentration polarization. They mainly constitute the improper mass transfer at various locations leading to drop in the potential. Possible cases are substrate limitation for the bacteria, lower diffusion

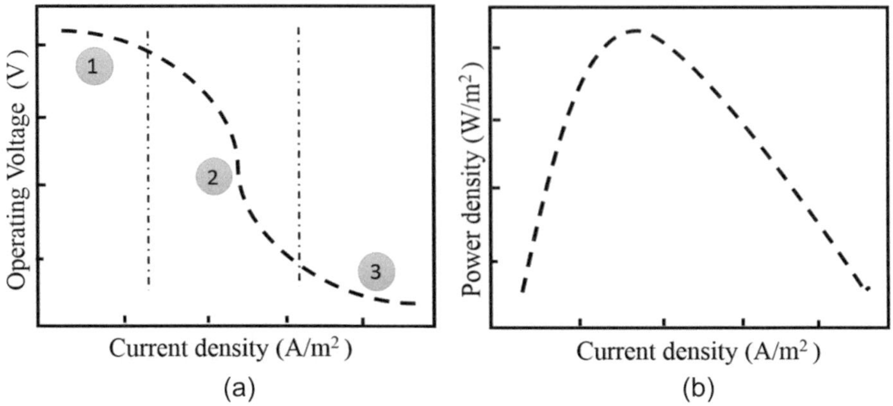

Fig. 13.1 (**a**) Polarization plot of an MFC. The zones (*1*), (*2*) and (*3*) in the figure correspond to the zones where activation, ohmic and concentration polarization losses are dominant respectively; (**b**) Power density curve of MFC

rates of the reaction species etc. Hence, the polarization losses provide information on various losses that dominate in a cell indicating the appropriate improvement needed in the system.

Further, power plot is a derived plot from the polarization graph. The power graph is a representation of power density against the current density. Typical power density plots for an MFC are depicted in Fig. 13.1b. The power slowly increases to reach a maximum point called Maximum Power Point (MPP) and falls beyond this point to reach short circuit condition or the zero current values. The point where the maximum attainable power is reached in a cell can be usually considered as R_{int} provided the power graph is a perfect semi-circle. Thus the power graph can indicate the maximum power that a specific system can generate besides a preliminary estimate of the internal resistance.

13.3.2 Current Interruption (CI)

CI is used to compute the internal resistance of MFCs. In this method, an MFC is operated at a fixed external resistance (closed circuit conditions) till a stable current is observed over a period of time. Once a stable current is achieved, the current flow is rapidly interrupted by disconnecting the circuit (open circuit conditions) and the voltage transients are observed (Fig. 13.2).

The internal resistance can be calculated as (13.10)

$$R_{int} = [(V_2 - V_1)R_{ext}]/I \tag{13.10}$$

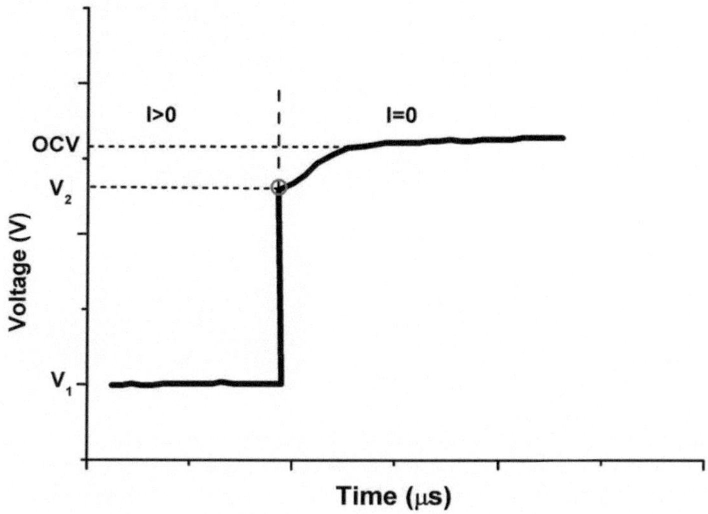

Fig. 13.2 Voltage waveform during CI technique where V_1 is the closed circuit voltage, V_2 is the steep increase in voltage during current interruption and OCV is the open circuit voltage

where R_{int} = internal resistance (Ω), V_1 = closed circuit voltage (V), V_2 = voltage after steep increase (V), R_{ext} = external resistance (Ω) and I = closed circuit current.

The resistance obtained using this technique corresponds only to Ohmic overpotential as it is an instantaneous process as compared to other potential losses (activation and concentration overpotentials) which have longer relaxation times (Zhao et al. 2009). The main advantage of this technique is that a single data point is generated which could be easily interpreted from the V-I curve. However, the major disadvantage is that the imposed interruption cause significant perturbations only for a short duration of time (few milli-seconds) and thus any delay in the data collection can misinterpret the results.

13.3.3 Voltammetry Techniques

Voltammetry includes basic electroanalytical techniques that are used to analyse the reactivity of an analyte in an electrochemical half reaction. In this method, the applied potential (E_{app}) of the electrode in contact with the analyte is varied and the resulting current (I) is monitored. The obtained I-E_{app} curve is known as voltammogram. The principal instrument used for conducting voltammetry experiments is called a potentiostat comprising three electrode systems viz. the working, the counter and the reference electrodes (Fig. 13.3). The working electrode (which is in contact with the analyte) is maintained at the constant desired potential to

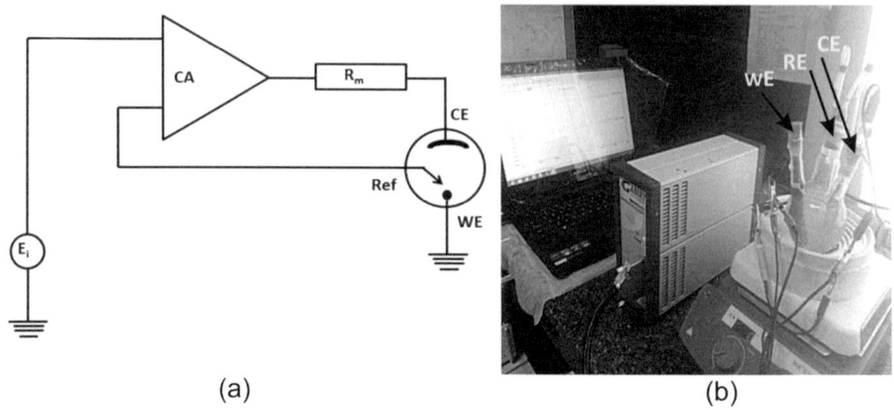

Fig. 13.3 (**a**) Schematic of working principle of a potentiostat comprising three electrodes viz. working, reference and counter: a control amplifier which maintains the voltage between the reference and working electrodes close to the input voltage E_{i} and the resistance R_{m} across which the current drop is measured. (**b**) A working potentiostat with an electrochemical cell used in the laboratory

facilitate electron transfer to and from the analyte of interest. The reference electrode has a constant potential and is used to maintain a constant potential at the working electrode. The most commonly used reference electrodes used in laboratories include saturated calomel electrode (SCE; +0.25 V vs. SHE) or silver/silver chloride electrodes (Ag/AgCl; +0.210 V vs. SHE). The counter electrode is used to complete the circuit by passing the current required to control the potential at the working electrode. For two-electrode configurations, the reference and the counter are connected to the same electrode.

For MFC applications, the classical voltammetry techniques can be used to deduce the electron transfer mechanisms between the electrogenic bacteria and the electrodes. Over the past few years, several researchers have applied voltammetry techniques for characterization of biofilms and to evaluate the kinetics of bioelectrochemical reactions (Marsili et al. 2008). Depending upon the study of interest, different voltammetry techniques have been employed for analysing MFCs as described below.

13.3.3.1 Linear Sweep Voltammetry (LSV)

In this technique, the potential of the working electrode is varied linearly as a function of time and the current response is recorded. The voltage is varied between an upper and lower limit at a constant scan rate $\left(v = \frac{dE}{dt}\right)$ (Fig. 13.4). Any peaks observed during the potential scan correspond to an oxidized or a reduced species.

The characteristics of the LSV depend upon several factors including the rate of the electron transfer and the reactivity of the intermediate/mediator species. It is generally used as an alternative polarization technique for MFC applications. It can

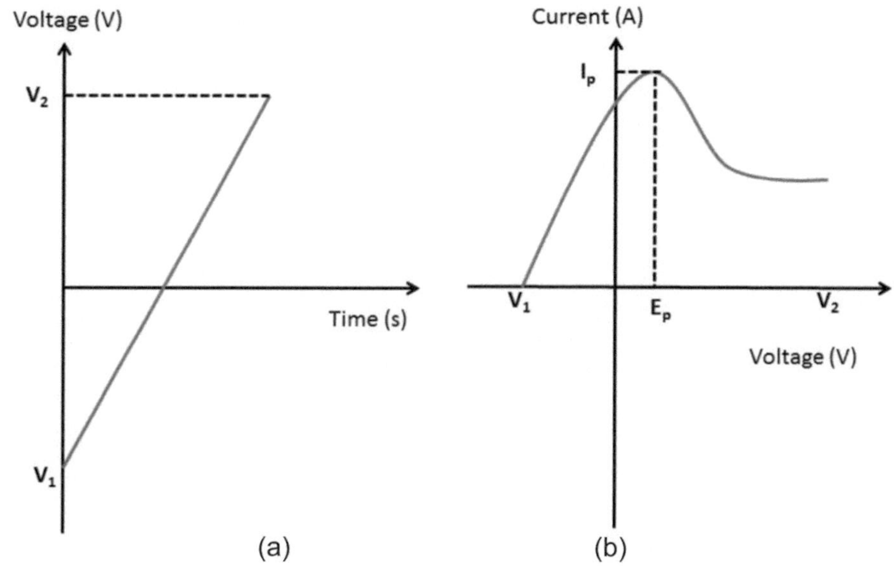

Fig. 13.4 (**a**) Voltage vs. time plot during an LSV scan from the *lower limit* V_1 to *upper limit* V_2. (**b**) A typical linear sweep voltammogram with E_p (peak potential) and I_p (peak current) for a single electrode transfer reaction

also be used to identify unknown species and to quantify the concentration of analyte in the solution. LSV is useful for deducing the electron transfer mechanisms involved in an irreversible electrochemical reaction. For example, Cheng et al. (2009) used LSV to examine the conversion of electrical current to methane via electromethanogenesis which is an irreversible reaction. However, this technique can only be used at very low scan rates (1–5 mV s^{-1}) as at high scan rates the capacitive currents increase drastically which cannot be compensated electronically.

13.3.3.2 Cyclic Voltammetry (CV)

CV is a powerful tool devoted for estimating the nature of electron transfer mechanisms, formal potentials and reversibility of electroactive species, mass transfer influence and kinetic behaviour of microbe-electrode interactions (Harnisch and Freguia 2012). It is similar to LSV analysis in which the potential is scanned between an upper and lower limit. However once the voltage reaches the upper limit, the scan is ramped reversibly and the voltage is swept back to the lower limit (Fig. 13.5).

In the plot of a CV curve, both anodic and cathodic peaks are observed for a reversible redox reaction while only a single peak is observed for an irreversible redox system. Thus the reaction mechanisms can be elucidated by observing the CV

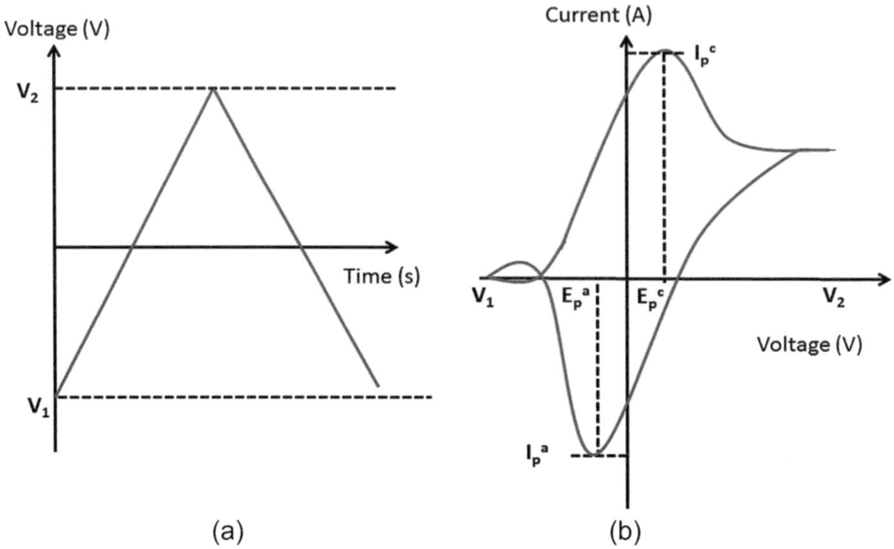

Fig. 13.5 (**a**) Voltage vs. time plot during a CV scan from the *lower limit* V_1 to *upper limit* V_2 and again back to the *lower limit* V_1. (**b**) A typical cyclic voltammogram with $E_p{}^a$ (anodic peak potential), $E_p{}^c$ (cathodic peak potential), $I_p{}^a$ (anodic peak current) and $I_p{}^a$ (cathodic peak current) for a reversible single electrode transfer reaction

plots. The formal potentials for particular redox specie can be estimated via CV using the following equation:

$$E_f = \frac{E_{pa} + E_{pc}}{2} \tag{13.11}$$

where E_{pa} is anodic peak potential and E_{pc} is cathodic peak potential.

The obtained formal potentials from CV can be compared to the standard biological redox potentials (E^o) and thus the redox entity involved in the bioelectrochemical reaction can be determined. Since the biological reactions occurring in the MFCs are far more complex than the typical electrochemical redox reactions, more than one peak can be obtained during CV scan corresponding to different redox entities (Marsili et al. 2008). There is also a chance that the bioelectrochemical reaction is followed by a chemical reaction which leads to the absence of a reversible peak in the CV scan. In such cases it becomes difficult to elucidate the limiting reaction step in the process. Thus, CVs should be carefully scrutinized when analysing biological reactions in MFCs. Due to the poor sensitivity of this technique, CV is seldom used for quantitative analysis of the redox species (Fricke et al. 2008).

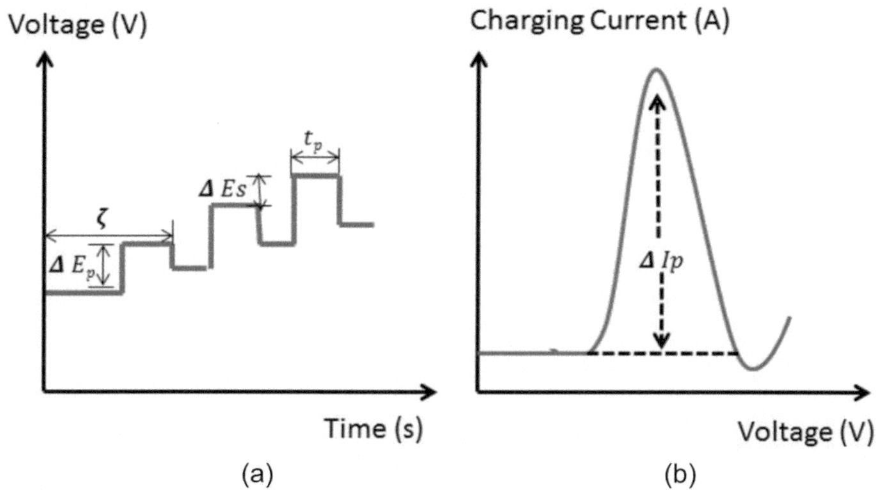

Fig. 13.6 (**a**) Voltage vs. time plot. (**b**) Voltammogram for DPV scan with τ cycle time; ΔE_p fixed pulse potential; ΔE_s fixed change in potential per cycle and t_p pulse time

13.3.3.3 Differential Pulse Voltammetry (DPV)

In this technique, the potential is scanned with a series of pulses such that each pulse is fixed at small amplitude. Current is monitored at the beginning and the end of each pulse and this differential reading of current generates a peak-shaped voltammogram (Fig. 13.6). The basis of this technique is the difference in the rate of the decay of charging and Faradaic current when a potential pulse is applied. Since the Faradic current is a function of time, its rate of decay is much slower than the charging current and thus the current monitored after each pulse solely consist of Faradic current. This technique is very sensitive and can be used for qualitative and quantitative estimation of redox species.

DPV technique is very sensitive and can be used for qualitative and quantitative estimation of redox species. This technique is used to compare the concentration of the redox active molecule (membrane bound FAD molecule) playing active role in the exocellular electron transfer in MFC at different time intervals (Sharma et al. 2013). By combining this technique with spectrophotometric or chromatographic techniques can bring about the broader details of the nature of the electroactive species.

13.3.3.4 Chronoamperometry (CA)

It is a sub class of voltammetry technique in which a step potential is applied and the resulting current is monitored over a period of time (Fig. 13.7). The most useful equation in chronoamperometry is the Cottrell equation (Eq. 13.12), which

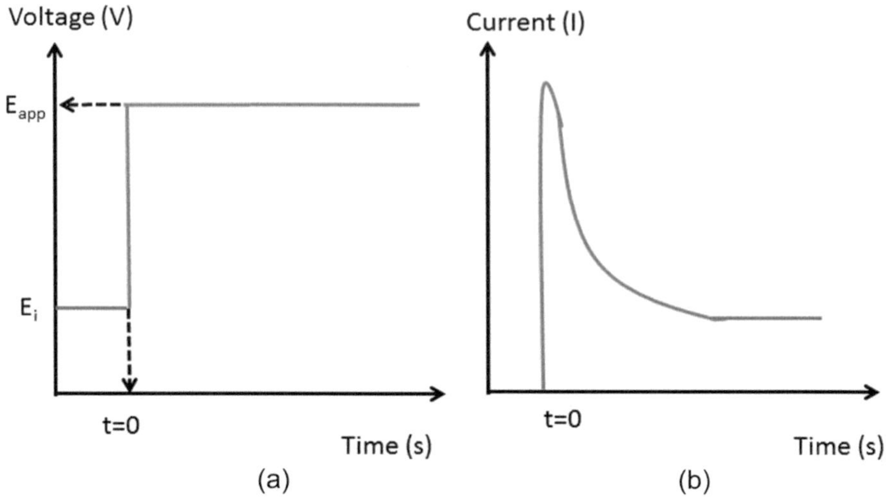

Fig. 13.7 (**a**) The potential-time profile during chronoamperometry, E_i is initial value and E_{app} is the applied potential. (**b**) The corresponding response of the current due to change in the potential

describes the exact form of current-time dependence with respect to the electrode at a constant potential.

$$i_t = \frac{nFAC_0 D_0^{1/2}}{n^{1/2} t^{1/2}} \tag{13.12}$$

where n = stoichiometric number of electrons involved in the reaction; F = Faraday's constant (96,485 C/equivalent), A = electrode area (cm^2), C_0 = concentration of electroactive species (mol/cm^3), and D_0 = diffusion constant for electroactive species (cm^2/s).

With a known electrode area, measurement of either n or D_0 for an electroactive species is easily accomplished.

In MFC, CA technique has been used for the identification of electrogenic species in the natural environment (Parot et al. 2008), formation of electroactive biofilms on electrode surfaces (Cercado-Quezada et al. 2010), to distinguish between the capacitive and Faradaic currents (Khilari et al. 2015) etc. With the discovery of current consuming electrotrophs, the scope of CA techniques has widened in the detection of direct conversion of current to useful products such as alcohols.

13.3.4 Butler-Volmer Analysis and Tafel Plots

The fundamentals of electron transfer kinetics and the associated overpotentials are described by the Butler–Volmer equation (Eq. 13.13). It assumes that the electrode kinetics is only governed by the charge transfer between the analyte and the electrode while the Ohmic and concentration overpotentials are negligible:

$$j = Ai_0\left\{e^{\frac{\mathrm{an}\mathrm{F}\eta}{\mathrm{RT}}} - e^{\frac{(1-a)nF\eta}{\mathrm{RT}}}\right\} \tag{13.13}$$

where j is the current density, A is the electrode active surface area, i_o is the exchange current density, a is the charge transfer barrier (symmetry coefficient), n is number of electrons involved in the electrode reaction, and h is the charge transfer overpotential.

The exchange current density i_o is the maximum current that could be extracted at negligible overpotential at equilibrium (i.e. cathodic current = anodic current). For an electrochemical system, higher the $i_{o,}$ the faster will be the reaction and vice versa. The Butler–Volmer equation has two limiting conditions viz. high overpotential condition where the equation simplifies to yield Tafel equation (Eq. 13.14) and low overpotential region where the equation simplifies to yield polarization resistance (Eq. 13.15).

$$\ln i = i_v \pm \frac{\alpha nF\eta}{RT} \tag{13.14}$$

$$i = i_0\frac{nF}{RT}(\eta - \boldsymbol{E}) \tag{13.15}$$

where i represents current (A) and h is overpotential (V) and (anF/RT) is the Tafel slope and E is the equilibrium potential.

Tafel slopes are important parameters that describe the charge transfer capabilities of electrogenic bacteria. Lower the Tafel slopes, higher will be the electron transfer to the electrode surface. Estimation of Tafel slopes also helps in deducing the electron transfer coefficient (α) which is also a crucial parameter that describes the symmetry between the forward and reverse reactions. Thus, knowledge of these fundamental parameters can provide valuable insights over the bacterial-electrode electron transfer mechanisms in MFCs.

13.3.5 Electrochemical Impedance Spectroscopy (EIS) Analysis

EIS is an important electrochemical tool used by the researchers primarily to examine the internal resistance of the cells. As mentioned earlier, there are three types of major polarization losses that incur in an MFC. Not only can EIS estimate the value of these losses at various operating conditions but can also provide information on the individual contribution of each loss (He and Mansfeld 2009). Owing to its simplicity, non-destructive property, ability to deduce diverse and valuable insights about the cells, it is being widely employed by various researchers to study MFC performance. EIS has been employed not only to analyse various losses that arise from operating the system but also to generate and analyse various novel electrodes, the biofilm properties and even to compare various components of the system. The uniqueness of this method relies on the fact that the cell under consideration is subjected to a known AC (alternating current) current unlike other techniques which make use of DC (Sekar and Ramasamy 2013).

A potentiostat is the basic instrument required to perform this study. Depending on what is being analysed, the technique can be performed either in a two-electrode mode or a three-electrode mode. In the two-electrode mode, the anode/cathode (basically the reaction centre or the component that needs to be analysed) becomes the working electrode while the other becomes counter electrode. Similar connections are made in the three-electrode mode where the reference electrode of the cell is also connected to the potentiostat which serves as the reference. A small electrical AC perturbance is given to the system and the resulting impedance as well as phase shift are monitored with time. The data generated from an EIS plot basically consists of two major plots namely: (1) Bode plot and (2) Nyquist plot. Typical Nyquist and Bode plots are represented in Fig. 13.8.

When an EIS scan is performed, a small AC sinusoidal signal is applied on the system within a definite frequency range, the results of which are represented as Nyquist plot and Bode plot. Nyquist plot is a simple representation of real part of impedance on the X-axis and the complex part of the impedance on Y-axis with each point corresponding to the impedance value at a definite frequency. The graph mainly comprises a semi-circle and sometimes a straight line following the semi-circle may also be observed. The diameter of the semi-circle is the charge transfer resistance (R_p) or the polarization resistance of the cell while the straight line is an indication of capacitance of the cell. Bode plot provides further information on the frequency, phase angles and impedance. However, in practical situations, the impedance plots are a complex combination of capacitances and resistances and they vary from system to system.

13.4 Material Characterization Methods

13.4.1 Scanning Electron Microscopy (SEM)

Scanning electron microscopy (SEM) is a powerful technique capable of examining the surface structure of the materials samples and is often used in MFCs for electrode and biofilm characterization. In this the scanning electron microscope

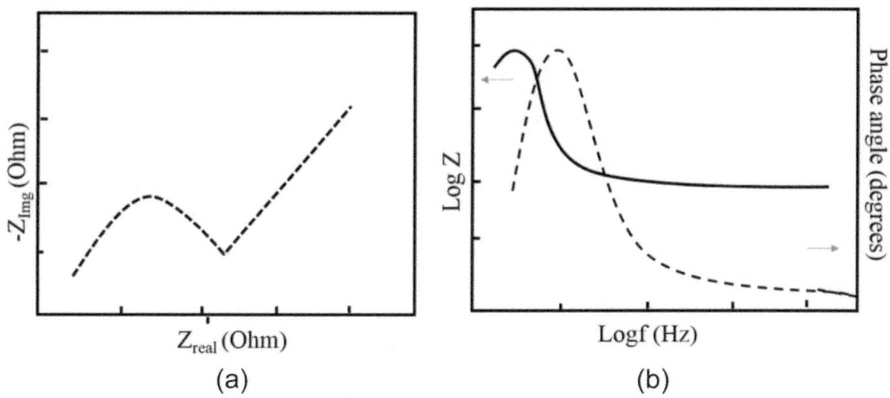

Fig. 13.8 (**a**) Nyquist plot. (**b**) Bode plot of an MFC system

scans a focussed electron beam over a surface to create an image. The electrons in the beam interact with the sample, producing various signals that can be used to obtain information about the surface topography and composition. The energy X-ray spectroscopy (EDX) is another useful analytical tool integrated with SEM. EDX is used to determine the elemental composition of surface films. EDS data are typically collected from an area in the specimen.

13.4.2 Transmission Electron Microscopy (TEM)

TEM forms an image by accelerating a beam of electrons that pass through the specimen. In TEM, electrons are accelerated to 100 KeV or higher (up to 1 MeV), projected onto a thin specimen (less than 200 nm) by means of the condenser lens system, and penetrate the sample thickness either undeflected or deflected. The greatest advantages that TEM offers are the high magnification ranging from 50 to 106 and its ability to provide both image and diffraction information from a single sample.

In MFCs, TEM analysis is usually performed to quantitatively measure the size, distribution, and morphology of the nanoparticles that are used as catalysts on the electrode surfaces. Apart from material characteristics, TEM also provides valuable information on the spatial arrangement and cellular ultra-structure of the biofilm. Thus, internal cross-sectional detail of the individual microorganisms and the overall biofilm can be visualized using TEM. The main disadvantages of electron microscopy being the cost and the preparation time required for visualization of the samples.

13.4.3 X-Ray Diffraction (XRD)

XRD is a rapid analytical technique primarily used for phase identification of a crystalline material and can provide information on unit cell dimensions. X-ray diffraction is based on constructive interference of monochromatic X-rays and a crystalline sample. These X-rays are generated by a cathode ray tube, filtered to produce monochromatic radiation, collimated to concentrate, and directed toward the sample. The interaction of the incident rays with the sample produces constructive interference (and a diffracted ray) when conditions satisfy Bragg's Law

$$n\lambda = 2d \sin\theta \tag{13.16}$$

where n is a positive integer, λ is the diffracted wavelength, d is the distance between atomic layers in a crystal and θ is the angle of incidence.

This law relates the wavelength of electromagnetic radiation to the diffraction angle and the lattice spacing in a crystalline sample. These diffracted X-rays are then detected, processed and counted. By scanning the sample through a range of 2θ angles, all possible diffraction directions of the lattice should be attained due to the random orientation of the powdered material.

13.4.4 BET Surface Area Measurements

Brunauer-Emmett-Teller (BET) surface area analysis provides precise specific surface area evaluation of materials by nitrogen multilayer adsorption measured as a function of relative pressure using a fully automated analyser. The technique encompasses external area and pore area evaluations to determine the total specific surface area in $m^2\ g^{-1}$ yielding important information in studying the effects of surface porosity and particle size in many applications. For MFCs in particular, BET analysis is used for determining electrode specific surface area along with the pore volume and area distributions that characterise porosity.

13.4.5 Other Methods

Apart from the techniques mentioned above, several other methods are employed to analyse the membrane and electrode properties including atomic force microscopy (AFM) which facilitates the observation of three-dimensional surface structures of the nanomaterial under study including the biological structures (tissues, cells, biomolecules, atomic absorption spectroscopy (AAS) which provides insights on the elemental composition of the material under study. Fourier transform infrared spectroscopy (FTIR) is used to identify or quantify the material under study based on the absorption and transmittance of infrared radiations, Raman spectroscopy which provides information about molecular vibrations that can be used for sample identification and quantitation etc. Depending upon the nature and complexity of the material under study, any of the aforementioned techniques can be used for the analysis purpose.

13.5 Techniques for Microbial Community Analysis

13.5.1 DGGE

Denaturing gradient gel electrophoresis (DGGE) is a molecular fingerprinting method that separates polymerase chain reaction (PCR)-generated DNA products. The polymerase chain reaction of environmental DNA can generate templates of differing DNA sequence that represent many of the dominant microbial organisms. However, since PCR products from a given reaction are of similar size (bp), conventional separation by agarose gel electrophoresis results only in a single DNA band that is largely non-descriptive. DGGE can overcome this limitation by separating PCR products based on sequence differences that results in differential denaturing characteristics of the DNA. During DGGE, PCR products encounter increasingly higher concentrations of chemical denaturant as they migrate through a polyacrylamide gel. Upon reaching a threshold denaturant concentration, the weaker melting domains of the double-stranded PCR product will begin to denature at which time migration slows dramatically. Differing sequences of DNA (from different bacteria) will denature at different denaturant concentrations resulting in a pattern of bands. Each band theoretically representing a different bacterial population present in the community. Once generated, fingerprints can be uploaded into databases in which fingerprint similarity can be assessed to determine microbial structural differences between environments or among treatments.

13.5.2 ARDRA

Amplified rDNA (ribosomal DNA) Restriction Analysis (ARDRA) is the extension of the technique of RFLP (restriction fragment length polymorphism) to the gene encoding the small (16s) ribosomal subunit of bacteria. The technique involves an enzymatic amplification using primers directed at the conserved regions at the ends of the 16s gene, followed by digestion using tetracutter restriction enzymes. The pattern obtained is said to be representative of the species analysed. Patterns obtained from several restriction enzymes can be used to phylogenetically characterize cultured isolates and 16s genes obtained through cloning from community DNA.

Based on the simple formula for the frequency of random occurrence of a restriction site, a 4-bp sequence occurs once every 256 bp. At least three restriction enzymes must be used for the analysis to overcome the probability of certain restriction enzymes to yield similar patterns for not so unrelated organisms. The amplicon to be analysed must preferably correspond to a size of greater than 1000 bp, purely for the sake of encountering a greater possibility of the restriction site. The amplicons must be preferably purified before digestion.

13.5.3 *Pyrosequencing*

It is a DNA synthesis technique which uses the principle of "synthesis by sequence" i.e. each of the four nucleosides are fed sequentially to the DNA template by the DNA polymerase and after the successful incorporation of any base, a pyrophosphate is released which is detected in an enzyme cascade that emits light. The general principle of pyrosequencing can be described as:

$$(\text{Oligonucleotide})_n + \text{nucleotide} \xrightarrow{\text{Polymerase}} (\text{Oligonucleotide})_{n+1} + \text{PPi} \quad (13.17)$$

$$\text{PPi} + \text{APS} \xrightarrow{\text{ATP sulfurylaze}} \text{ATP} + \text{Sulfate} \quad (13.18)$$

$$\text{ATP} + \text{luciferin} + O_2 \xrightarrow{\text{luciferase}} \text{AMP} + \text{PPi} + \text{oxyluciferin} + \text{Light} \quad (13.19)$$

In the first step, a polymerase catalyses the incorporation of nucleotides to a nucleic acid chain, next after each nucleotide is successfully incorporated a pyrophosphate (PPi) is released which is incorporated into ATP by ATP sulfurylase. Once luciferin is added in the presence of luciferase enzyme, the ATP is degraded to AMP and light is generated. This light is detected by the sensors and a peak is observed (Pyrogram). The height of each peak corresponds to the number of nucleotides incorporated.

Pyrosequencing has been used extensively for microbial genotyping applications. Since, the sequences are synthesized using new copies of DNA template, unambiguous results are obtained using this technique. Unlike hybridization techniques, pyrosequencing allows the identification of a large number of species using a single conserved primer. Consequently, DNA extracted from multiple microbe species can be sequenced in the same pyrosequencing run.

13.5.4 *Other Molecular Techniques*

Other molecular techniques used for identification and characterization of microbial biofilms and communities include CLSM (confocal laser scanning microscopy), flow cytometry, fluorescent in-situ hybridization (FISH), single-strand conformation polymorphism (SSCP), restriction fragment length polymorphism (RFLP), etc. For detailed description of these techniques, readers are advised to refer Chap. 9.

13.6 Waste and Wastewater Analysis

Standard laboratory analytical methods or test procedures that are prescribed by APHA (American Public Health Association) for examining waste and wastewater are used in MFCs to analyse amount of waste degraded by the microbes. These methods make use of various diagnostic tools which include measurement of hardness (mg L^{-1} as $CaCO_3$) by spectrophotometer, volatile residue by Muffle Furnace, inorganic anions by ion chromatography, alkalinity by colorimetric method, total Kjeldahl nitrogen, chemical oxygen demand (COD), biochemical oxygen demand (BOD) etc. The working principle and theories for these techniques can be found elsewhere.

13.7 Conclusions

This chapter summarizes various techniques utilized in the literature to study and analyse the performance of the MFC system as a whole as well as individual components of the same. The fundamentals of the techniques have been presented in terms of material analysis, microbe community structure, electrochemical analysis, wastewater analysis and the performance evaluation techniques of the MFCs. There is a need for the development of novel electrochemical tools to further probe the fundamental electrochemical processes that occur in MFCs.

References

Cercado-Quezada, B., Delia, M. L., & Bergel, A. (2010). Testing various food-industry wastes for electricity production in MFC. *Bioresource Technology, 101*, 2748–2754.

Cheng, S., Xing, D., Call, D. F., & Logan, B. E. (2009). Direct biological conversion of electrical current into methane by Electromethanogenesis. *Environmental Science & Technology, 43*, 3953–3958.

Fricke, K., Harnisch, F., & Schröder, U. (2008). On the use of cyclic voltammetry for the study of anodic electron transfer in MFCs. *Energy & Environmental Sciences, 1*, 144.

Harnisch, F., & Freguia, S. (2012). A basic tutorial on cyclic voltammetry for the investigation of electroactive microbial biofilms. *Chemistry, an Asian Journal, 7*, 466–475.

He, Z., & Mansfeld, F. (2009). Exploring the use of electrochemical impedance spectroscopy (EIS) in MFC studies. *Energy & Environmental Science, 2*, 215–219.

Khilari, S., Pandit, S., Varanasi, J. L., Das, D., & Pradhan, D. (2015). Bifunctional manganese ferrite/polyaniline hybrid as electrode material for enhanced energy recovery in MFC. *ACS Applied Materials & Interfaces, 7*, 20657–20666.

Logan, B. E., Hamelers, B., Rozendal, R., Schröder, U., Keller, J., Freguia, S., Aelterman, P., Verstraete, W., & Rabaey, K. (2006). MFCs: Methodology and technology. *Environmental Science & Technology, 40*, 5181–5192.

Marsili, E., Rollefson, J. B., Baron, D. B., Hozalski, R. M., & Bond, D. R. (2008). Microbial biofilm voltammetry: Direct electrochemical characterization of catalytic electrode-attached biofilms. *Applied and Environmental Microbiology, 74*, 7329–7337.

Parot, S., Délia, M.-L., & Bergel, A. (2008). Forming electrochemically active biofilms from garden compost under chronoamperometry. *Bioresource Technology, 99*, 4809–4816.

Sekar, N., & Ramasamy, R. P. (2013). Electrochemical impedance spectroscopy for MFC characterization. *Journal of Microbial and Biochemical Technology, S6*, 1–14.

Sharma, M., Jain, P., Varanasi, J. L., Lal, B., Rodríguez, J., Lema, J. M., & Sarma, P. M. (2013). Enhanced performance of sulfate reducing bacteria based biocathode using stainless steel mesh on activated carbon fabric electrode. *Bioresource Technology, 150*, 172–180.

Sun, M., Zhai, L.-F., Li, W.-W., & Yu, H.-Q. (2016). Harvest and utilization of chemical energy in wastes by MFCs. *Chemical Society Reviews, 45*, 2847–2870.

Zhao, F., Slade, R. C. T., & Varcoe, J. R. (2009). Techniques for the study and development of MFCs: An electrochemical perspective. *Chemical Society Reviews, 38*, 1926–1939.

Chapter 14
Modelling of Reaction and Transport in Microbial Fuel Cells

Ramya Veerubhotla, Sajal Kanti Dutta, and Saikat Chakraborty

14.1 Introduction

Understanding the fundamental processes underlying the microbial fuel cells (MFCs) can provide valuable insights in recognizing the key limiting factors, the scope of improvement of the system which in turn helps in the scaling-up the process. The science behind an MFC is complex and it involves a subtle interplay of various fields such as microbiology, physics and electrochemistry (Zhang and Halme 1995). A comprehensive understanding of various parameters involved in the process is essential for the improvement of power generation and to explore further applications. Modelling the system prior to experimentation can provide various perspectives and alternatives saving time and money. The physics of the process can be understood using quantitative predictions using modelling. It also provides valuable information about the dynamics of a process and thus important in reactor design and scale-up. Consequently, efficient monitoring of the process as well as precise control may be achieved through modelling (Marcus et al. 2007). Multi-scale modelling is crucial for a well-defined understanding of the process at both micro and macro scales.

The present chapter reviews the existing models of MFCs, with an emphasis on models pertaining to anode and the microbes, which distinguish the process from chemical fuel cells. Most of the previous reports show models on the specific components and their associated transport modelling owing to the complexity of

R. Veerubhotla
Department of Biotechnology, Indian Institute of Technology Kharagpur,
Kharagpur 721302, India

S.K. Dutta • S. Chakraborty (✉)
Department of Chemical Engineering, Indian Institute of Technology Kharagpur,
Kharagpur 721302, India
e-mail: saikat@che.iitkgp.ernet.in

D. Das (ed.), *Microbial Fuel Cell*,
https://doi.org/10.1007/978-3-319-66793-5_14

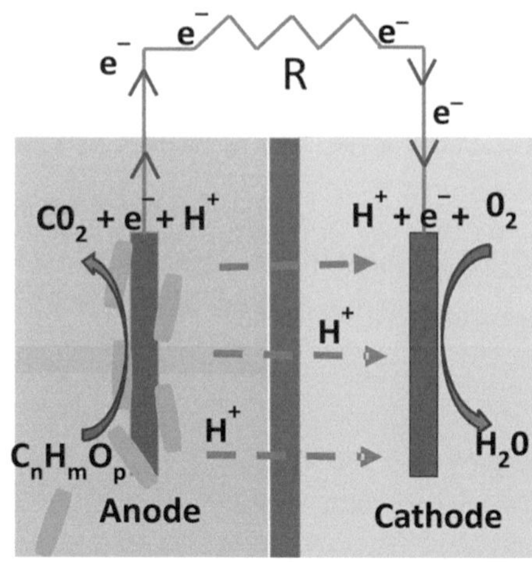

Fig. 14.1 Physical categorization showing the individual parts along with the reactions involved in the process

the process. Hence, we define the individual components of the MFC as illustrated in Fig. 14.1 to elucidate and identify different reactions that simultaneously occur in the system. For the sake of convenience, we categorize the existing models on MFCs based on the physical categorization as well as on the interaction between the biocatalysts and the electrodes (Fig. 14.2).

14.2 Principle of an MFC

As described earlier, MFC is a diverse system which involves mainly the following components – anode, anolyte, bacteria, separator, cathode and catholyte. At anode, the bacteria, acting as a biocatalyst, oxidizes the substrate(s) (Logan et al. 2006) leading to the formation of degradation products, electrons and protons (Fig. 14.1). In reality, since wastewater is used in MFCs, it contains a wide variety of substrates and their mixtures and the products depend on the type of the substrate (Lovley 2006).

A general stoichiometric equation of the oxidation reaction can be written as:

$$C_nH_mO_p + (2n - p)\,H_2O \rightarrow nCO_2 + (m + 4n - 2p)\,H^+ + (m + 4n - 2p)\,e^- \tag{14.1}$$

The bacteria involved in the MFCs, called exoelectrogens, have a capability to utilize anode as the final electron acceptor in anaerobic conditions. The details of the electron transfer mechanisms are described elsewhere. Usually, mixed cultures are used in MFCs as they help in recovering maximum energy from wastewater. The main purpose of the separator is to physically separate the anode and cathode chambers, while allowing an efficient and desired ion exchange across them. Depending on the type of membrane (anion exchange, cation exchange or proton

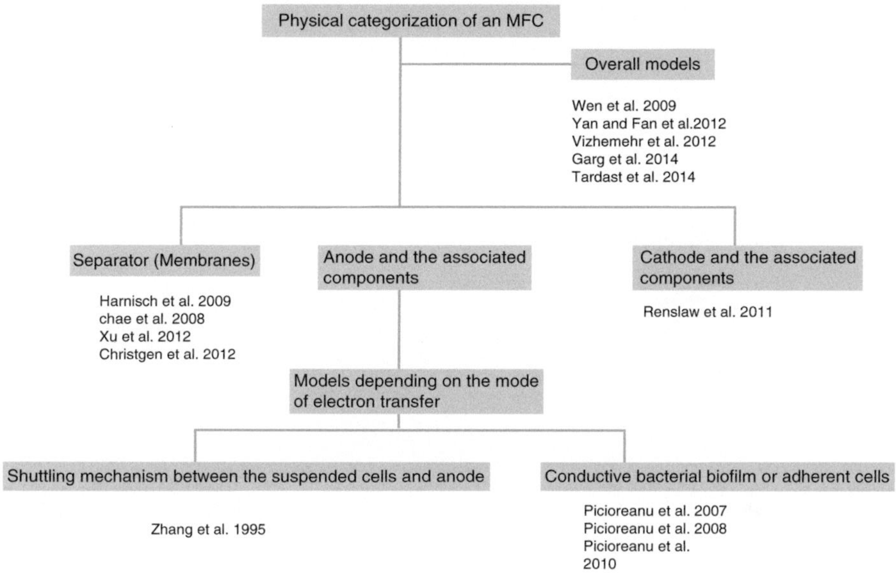

Fig. 14.2 Classification of models based on the physical categorization

exchange), ions (anions, cations or protons) typically move across the chambers to facilitate the cathode reaction. At cathode, reduction occurs resulting in utilization of electrons that flow through the circuit. Following are the major reactions that happen at cathode depending on whether oxygen or ferrycyanide is used as catholyte.

$$O_2 + H^+ + e^- \rightarrow H_2O \tag{14.2}$$

$$\mathrm{Fe(III)} + e^- \rightarrow \mathrm{Fe(II)} \tag{14.3}$$

14.3 Classification of the Models

The modelling approaches of the MFC are diverse. We further segregate the existing models of the literature into heterogeneous and homogeneous models depending on the phases being considered by the models (Fig. 14.3). This segregation approach would further help the readers to pick up the model to be applied depending upon the level of complexity. In reality, MFC involves multi-phase heterogeneous reactions. However, for the sake of simplicity, several modelling approaches use a single phase model and represent a heterogeneous system through pseudo-homogenous models. We show through our phase-based categorization in this review that while hetero models can capture the physics of the system at a more detailed level, they are difficult to simulate. Homogenous models, on the other

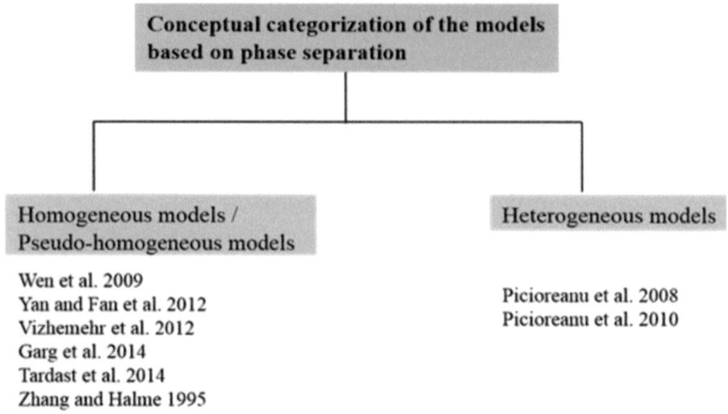

Fig. 14.3 Classification of models based on the phase categorization

hand, lump the multi-phase physics of the system into a single phase thus leading to a loss of information, but are much easier to compute. Pseudo-homogenous models are used for engineering purposes and render them amenable to scale-up. They are particularly useful when large numbers of species are involved.

The following section briefly introduces various models and discusses some of the important models in details.

14.4 Overall Models

Various researchers attempted to model the entire MFC as a system instead of individual components and most of their work included prediction of the maximum current, quantifying the voltage losses and optimizing the power output. Wang et al. (2008) modelled a continuous MFC system operated on brewery wastewater. The model quantifies various types of losses in an MFC system including losses due to reaction kinetics (η_{act}), ohmic polarization (η_{ohmic}) and mass transport (η_{conc}) and introduces a new concept called the 'parasitic loss due to gross leakage' (j_{leak}) given by the following equations where V corresponds to the real output voltage, E_{thermo} is the thermodynamically predicted voltage, and j is the current density.

$$V = E_{thermo} - \eta_{act} - \eta_{ohmic} - \eta_{conc} \tag{14.4}$$

$$j_{gross} = j + j_{leak} \tag{14.5}$$

Yan and Fan (2013) extended the application of a PID controller combined with a fuzzy logic by incorporating it into an MFC system. These techniques enable operation of the system with a constant voltage output minimizing the disturbances. Further, predictive model by Stratford et al. interrelates the diversity of the microbial species in the biofilm as well as the bulk liquid of the anode chamber using

Shannon and Simpson index. Their results suggest that a strong correlation exists between the relative abundance of species in a microbial community and the diversity of anode bacteria (Shannon index) (Stratford et al. 2014) combined with the DNA content has a strong impact on the power output of the MFC. Some more advanced and predictive models include application of Artifical Intelligence methods (Tardast et al. 2014; Garg et al. 2014). However, the methods discussed do not give detailed insights into the underlying mechanisms and are only helpful for quantitative predictions. They are also highly specific and are further limited by operating conditions, reactor configuration and bacterial community.

14.5 Models Pertaining to Anode-Bacterial Interactions

The following section discusses the models comprising the interaction between anode and the microbes. These models assume that anode is the limiting factor in the production of electricity. We classify such models involving the bacterial species and the anode into two different types based on the mode of the level of complexity of their representation:

1. Models involving the suspended bacteria and a planar electrode - Shuttling mechanism
2. Models involving the bacteria adhered to the electrode and biofilms (formation of conductive matrix) – DET mechanism.

This type of models essentially involves molecules called shuttles or mediators which are released by the bacteria into the solution. These molecules shuttle the electrons from cell to the electrode by a series of oxidation and reduction reactions. This mechanism is well-established and the power output can be potentially increased by adding certain mediators that extract the electrons produced by the bacterial metabolism. Zhang and Halme (1995) attempted to model an MFC involving quinone as the mediator. The model assumes that the proton transport through the membrane and the oxygen reduction reaction at cathode are fast as compared to the biochemical reactions. A sequential reaction scheme was adopted with the following steps which forms the basis of this diffusion-based model. The first step is the transfer of this molecule to the interior of the cell followed by reduction and subsequent export to the bulk liquid (Fig. 14.4). The following equations are defined by Zhang and Halme (1995) through the model. The rate of substrate $[S]$/electron donor degradation with time t, is given as Eq. 14.6 in relation to the bacterial concentration $[X]$ where k_1 is the reaction rate constant.

$$\frac{d[S]}{dt} = k_1 \frac{[S][X]}{[S] + k_s} \tag{14.6}$$

With an assumption that the mediator is neither utilized nor dissociated, the kinetics involving mediator are given in a similar fashion as enzyme kinetics where

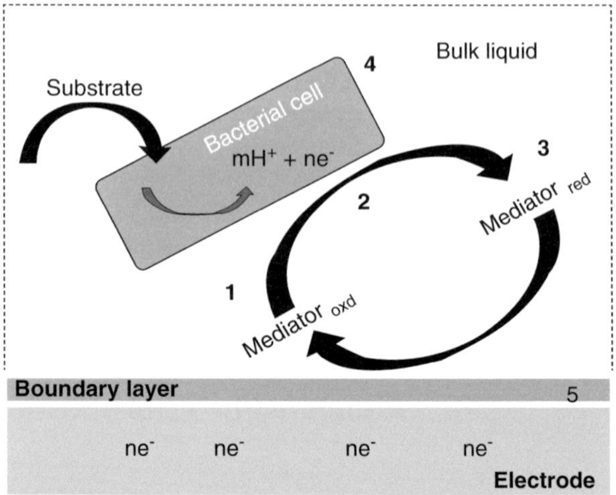

Fig. 14.4 Electron transport to the anode through shuttling mechanism

$[HNQ]$ and $[HNQH_2]$ are concentrations of 2-hydroxy-1,4-naphthoquinone and its reducing ones, respectively. $[M_{\text{red}}]$ and $[M_{\text{oxd}}]$ are the reducing and oxidizing intermediates, k_1, k_2, m, k_3 and k_{-2} are reaction rate constants in (mmol/g h), (1/mM h), (mmol/g h), (1/h) and (l/h), while k_d and k_s are dissociation and substrate limiting constant in (l/h) and (mM).

$$\frac{d[M_{\text{red}}]}{dt} = m\frac{[S][X]}{[S]+k_s} - k_2[HNQ][M_{\text{red}}] - k_d[M_{\text{red}}] + k_{-2}[HNQH_2] \tag{14.7}$$

$$\frac{d[HNQ]}{dt} = -k_2[HNQ][M_{\text{red}}] - k_3[HNQH_2] + k_{-2}[HNQH_2] \tag{14.8}$$

The model links up the current generation in an MFC with the mediator and substrate concentrations by combining the biochemical reactions and the electrical circuit through the following equation

$$I = nFk_3[HNQH_2] = nFk_3\{[HNQ^*] - [HNQ]\} \tag{14.9}$$

where e, n and F are electron, electron number and Faraday's constant respectively and I is the current density (mA). Although the net electromotive force E of the system is a resultant of various factors including oxygen concentration $[O_2]$, proton concentration $[H^+]$ on both the anode and cathode chambers as well as the equivalents of the mediator, certain assumptions were made and the final E value was mathematically represented as follows

$$E = E^* + \left(\frac{RT}{2F}\right)\ln\left(\frac{[HNQH_2]}{[HNQ]}\right) \tag{14.10}$$

E^* is a constant for certain MFC system and T represents the temperature (K).

The model could successfully predict the effect of the mediator concentration on the output. An increasing mediator concentration caused an immediate increase in the current output as compared to the increase in substrate concentration. The model is also helpful to predict the adequate concentration of mediator that should be added to a system. This model is one of the most preliminary works related to the modelling of MFCs before the mechanisms such as conduction through the nanowires were not discovered. Further, Picioreanu et al. (2010) proposed an elaborate and extended model for suspended biomass with an added mediator with an emphasis on mass balances of various chemical substances involved in the cell. Their design consists of glucose as the substrate and ferro/ferri-cyanide as the chemical catholyte. The model also accounts for the background current generated by the cells in the substrate limiting condition owing to their endogenous substrate storage. Another added advantage of this model was that it could be applied to other bio-electrochemical systems along with MFCs. The model adopts a spatial compartment division of anode into bulk liquid and the mass transfer boundary across anode where homogenous conditions exist in the bulk liquid (Fig. 14.4). Reaction scheme is as per the model including Stoichiometric coefficient Y_s. It depends on the electron content of that substrate.

$$Y_sS + M_{\text{oxd}} \rightarrow M_{\text{red}} + \text{products} \tag{14.11}$$

$$M_{\text{red}} \rightleftarrows M_{\text{oxd}} + nH^+ + ne^- \tag{14.12}$$

Unlike the previous model, the biological reaction rate is given by double Monod equation as follows.

$$r_1 = k_1[X]\frac{[s]}{k_s + [S]}\frac{[M_{\text{oxd}}]}{k_{M_{\text{oxd}}} + [M_{\text{oxd}}]} \tag{14.13}$$

The model suggests that a higher mass transfer rate of the mediator improves the power output. Furthermore, the increase in the biomass also lead to higher power outputs as they facilitated a quick and more reduced mediator available for the oxidation. The model also shows that a better mixing at the anode compartment lead to a thinner boundary layer and higher current in the system. Monod half-saturation constant also significantly affected the current generation.

14.5.1 Background Current and Modelling of Endogenous Metabolism

When MFCs are operated in fed-batch mode, non-zero current readings are usually observed when the substrate exhausts. This can result due to three different possibilities arising from the endogenous metabolism. Either the part of biomass is consumed resulting in mediator oxidation or internal reserves of bacteria can

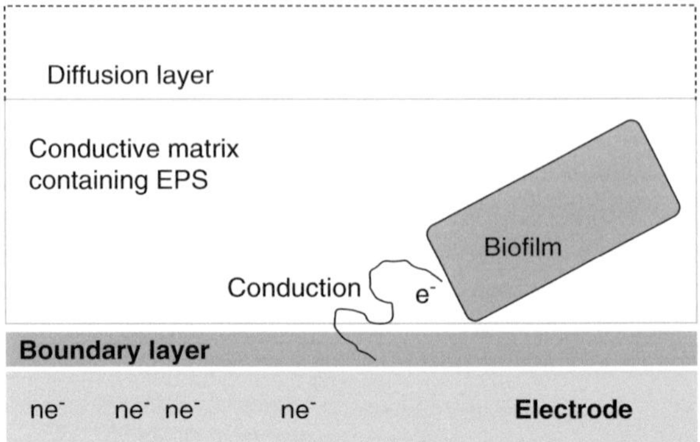

Fig. 14.5 Electron transport via conduction through bacterial nanowires

participate in the metabolism or it might also happen that a slower degradation of the intermediates obtained from the utilization of a primary substrate can cause subsequent mediator reduction and thus current generation (Figs. 14.5 and 14.6). Different models were proposed to understand the inherent mechanisms (Picioreanu et al. 2010). In the first case, a part of the biomass is consumed to produce electricity utilizing the mediator as described in Eqs. 14.14 and 14.15.

$$Y_{\mathrm{endo}}X + M_{\mathrm{oxd}} \rightarrow M_{\mathrm{red}} + \text{products} \tag{14.14}$$

$$r_2 = k_2[X]\frac{[M_{\mathrm{oxd}}]}{k_{\mathrm{M_{oxd}}} + [M_{\mathrm{oxd}}]} \tag{14.15}$$

The simulation results of these equations indicated that a higher rate of this biomass utilization lead to increase background currents and a gradual decline in the peak currents on further addition of substrates. The second mechanism which involves the storage of the substrate in the form of certain polymeric compounds followed by their degradation are given by Eqs. 14.16 and 14.17, respectively.

$$Y_{\mathrm{S}}S = M_{\mathrm{oxd}} \rightarrow M_{\mathrm{red}} + \text{products} + Y_{\mathrm{p}}polyS \tag{14.16}$$

$$Y_{\mathrm{p}}polyS = M_{\mathrm{oxd}} \rightarrow M_{\mathrm{red}} + \text{products} \tag{14.17}$$

Assuming that the oxidation of the polymeric substances also follows a similar stoichiometry and if the concentration of such substances in a cell is given by $[polyS]$,

$$r_2 = k_2[X]\frac{[polyS]}{K_{\mathrm{s}} + [polyS]}\frac{[M_{\mathrm{oxd}}]}{k_{\mathrm{M_{oxd}}} + [M_{\mathrm{oxd}}]} \tag{14.18}$$

It was observed by their group that the complex nature of these substances results in a slower degradation rates.

This section will cover the models comprising a stationary anode and a conductive biofilm. Following the electron conduction by the shuttles, researchers also

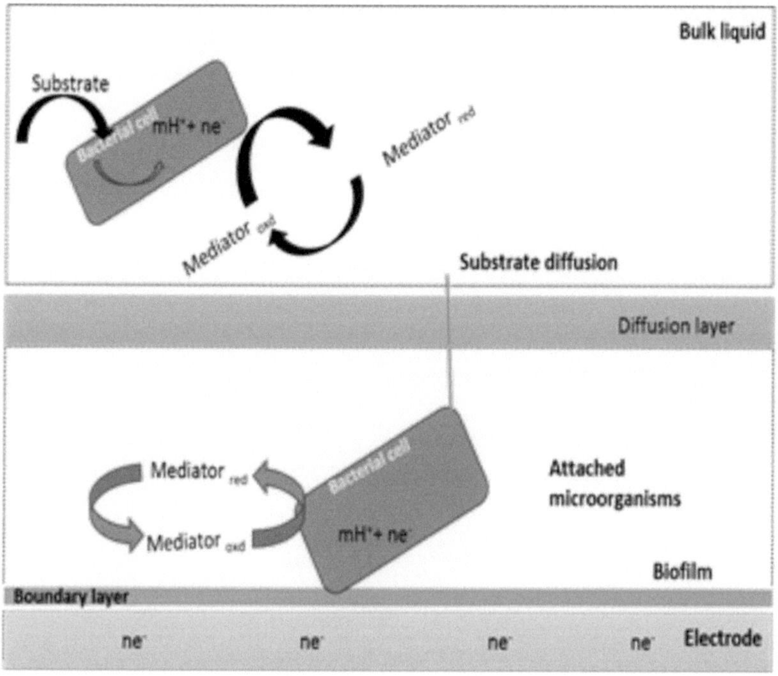

Fig. 14.6 Electron transport via shuttling and adherent microbes

noticed a significant electricity production even in the absence of the shuttles. This led to the proposal of two more theories for the electron conduction from the cell surface to the electrode. Notable works describing this mechanism is by Marcus et al. (2007). They proposed the first dynamic, one-dimensional model using a Nernst M onod equation. Their model shows the relation key parameters such as biofilm growth, its conductivity, electron donor (substrate) utilization and subsequently electricity generation in the MFCs. Their results demonstrate that a mass transfer of the substrate becomes dominant in biofilms with higher conductivity. One of the important hypothesis proposed by them is that high active biomass responsible for the electricity generation is promoted by an environment with high shear. The model is based on the assumption that substrate concentration, biomass per unit area of anode and an increase in the electric potential are directly proportional to the current generated. The governing equations proposed of the model can be divided into four major parts as described below.

Part 1: Rate of substrate utilization and respiration

$$\frac{d[S]}{dt} = k_1 \frac{[S]}{k_{\mathrm{S}} + [S]} \frac{S_{\mathrm{a}}}{K_{\mathrm{S_a}} + S_{\mathrm{a}}} \tag{14.19}$$

where S_a corresponds to electron acceptor (EA) concentration, K_{Sa} = half-max-rate *EA* concentration, $k_1 = q_{max}\phi_a$ (ϕ_a = volumetric fraction of active biomass (dimensionless) and q_{max} being the maximum specific rate of substrate utilization). Since the electron acceptor in a physical sense is the anode, whose concentration cannot be defined, the potential of the anodic EA (E_{anode}) is related to the concentration of *EA* using the Nernst Equation as follows:

$$E_{anode} = E_A^0 - \frac{RT}{nF} \ln\left(\frac{S_a^0}{S_a}\right) \quad (14.20)$$

where $S_a{}^o$ = standard anodic-EA concentration, $E_A{}^o$ = standard reduction potential for the anodic EA whereas $E_{k_a} = E_A^0 - \frac{RT}{F} \ln\left(\frac{S_a^0}{K_{S_a}}\right) E_{K_a}$ being the anodic acceptor potential for the half maximum-rate.

Part 2: Steady state electron donor mass balance

$$0 = D_{S,f}\frac{\partial^2}{\partial z^2} - X_{f.a}q \quad (14.21)$$

Following are the boundary conditions used by their model to solve the above equation.

(a) There exists no flux at the anode surface

$$0 = d_{ED,f}\frac{\partial S_d}{\partial z}\bigg|_{z=0}$$

(b) There exists continuity of flux at the interface of external diffusion layer with the outer surface of the biofilm.

$$D_{ED,I}\frac{\partial S_d}{\partial z}\bigg|_{z=L_f} = D_{ED,f}\frac{\partial S_d}{\partial z}\bigg|_{z=L_f} = \left(\frac{D_{ED,I}}{L}\right)\left(S_{d,bulk} - S_{d,Surface}\right) \quad (14.22)$$

Part 3: Steady state electron balance and the application of Ohm's law

This equation signifies the total electron balance of the biofilm anode in which the first term corresponds to the change in electron flux, the second term corresponds to conversion factor associated with the substrate to electron conversion and the third to the endogenous respiration of the microbes where the terms g_1 and g_2 are assigned to electron equivalence of ED and the active biomass (based on an empirical formula for microbial cells, $C_nH_aN_bO_c$).

$$0 = \frac{\partial j}{\partial z} + \frac{F\gamma_1}{T} f_e^0 X_{\mathrm{f,a}} q + \frac{F\gamma_2}{T} X_{\mathrm{f,a}} r_{\mathrm{res}} \quad (14.23)$$

A fixed potential at the anode ($\eta|_{z=0} = V_{\mathrm{anode}}$) and no electron flux condition at the diffusion layer of the biofilm $\left(\frac{\partial \eta}{\partial z}\Big|_{z=L_f} = \frac{j|_{z=L_f}}{\kappa_{\mathrm{bio}}} = 0\right)$ are used as boundary condition to solve the equation.

The parameter T in the equation corresponds to the time conversion factor (86,400 s day^{-1}) and fe^0 fraction of electrons from the ED used for energy generation to support synthesis (dimensionless). One of the most fundamental assumption that defines this model is the assumption that biofilm acts as a porous conductive matrix. Then according to Ohm's law,

$$0 = \kappa_{\mathrm{bio}} \frac{\partial \eta}{\partial z} + j \quad (14.24)$$

Part 4: Biomass balance

Finally, a mass balance on the biomass is represented as follows which indicates that the summation of both active (ϕ_a) and inert mass (ϕ_i) of the biomass is unity. The active biomass contributes to electricity generation and the inert biomass comprises of the extracellular polymeric matrix, nanowires and other conductive materials.

$$1 = \phi_a + \phi_i \quad (14.25)$$

This model is the basis of further models developed by Picioreanu et al. (2008) (that combines the anaerobic digestion and the electrogenesis) and Merkey and Chopp (2012) (study of the anode geometry on the performance).

14.6 Models Pertaining to Membrane/Separator

Harnisch et al. (2009) modelled the ion transfer across the membrane in bio-electrochemical systems using a simplified one dimensional model. Their set-up comprised a double chambered MFC with platinum mesh electrodes separated by a proton exchange membrane, Nafion 117. The model assumes anode as the source and the cathode as the sink with identical rates of reaction at both anode and cathode. The entire system is divided into three major components: A, A^1 – which represent the anode and cathode chambers, B, B^1 – representing the diffusion layers adhered to the membrane and C – representing the Nafion 117 membrane. If c_i is the concentration of the ion species diffusing across the membrane, charge transport can be written as follows where S is the source term (both biochemical and other external reactions are also considered) and flux density is written as mmol s^{-1} cm^{-2} and is given by the continuity equation.

$$\frac{\partial c_{\mathrm{i}}(x)}{\partial t} = -\frac{\partial}{\partial x} f_{\mathrm{i}}(x) + S \tag{14.26}$$

The ion flux density is described by the Nernst Planck equation as follows:

$$f_{\mathrm{i}}(x) = -D_{\mathrm{i,m}} \frac{\partial c_{\mathrm{i}}(x)}{\partial x} + \frac{z_i F}{RT} D_{\mathrm{i,m}} c_{\mathrm{i}}(x) \Phi(x) \tag{14.27}$$

As the equation indicates, the first term corresponds to diffusion while the second one is due to the migration of the ion species. $D_{i,m}$ represents the diffusion coefficient of an ion species i in the cell m. The term $\Phi(x)$ is the electric field given by the equation below:

$$\Phi(x) = \frac{1}{\varepsilon_{\mathrm{m}} \varepsilon_{\mathrm{a}}} \int p(x') dx' \tag{14.28}$$

$$\rho(x) = F \Sigma z_{\mathrm{i}} c_{\mathrm{i}}(x) \tag{14.29}$$

$\rho(x)$ is the charge density, ε_0 the dielectric constant of vacuum and ε_{m} represents the relative dielectric coefficient for the different compartments. Different zones were distinguished in each compartment of the system (Fig. 14.7) and a 1-D finite volume method was used to model the system. Depending on the zone, grids were formed in such a way the model depicts reality for accuracy. The equations are simplified and solved using a central difference scheme for discretization.

The simulation results predict that the concentration of the positively charged ions is low at the interface of anode-membrane (region B) as compared to the bulk liquid as well as cathode interface. This unequal distribution of ions results in the development of a potential drop across the membrane which varies with the concentration of ions in both the chambers. The model concludes that low concentrated electrolyte solutions suffer from problems with high membrane resistance as compared to highly concentrated electrolyte solutions. However, the problem of pH splitting is more pronounced in the former.

The effect of membrane fouling on the overall performance of system is modelled and experimentally validated by Xu et al. (2012). Since microbes can adhere to the membrane surface as well, a thick biofilm formation on the membrane combined by the extracellular polymeric matrix secretion can subsequently block the movement of ions across the surface and reduce the performance. The current profiles were plotted with time for both raw and fouled membrane systems and modelled using the following equation.

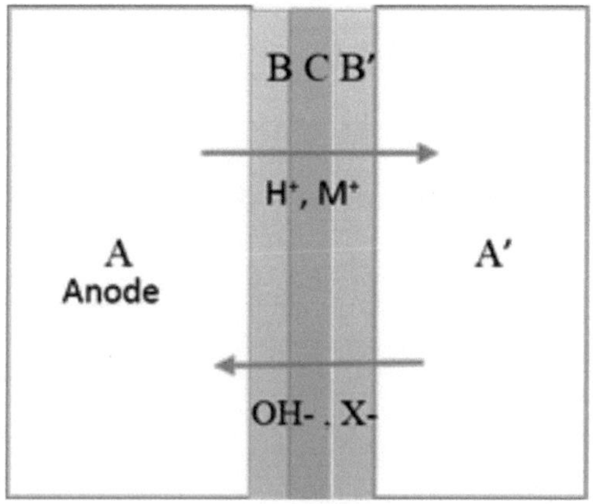

Fig. 14.7 Transport of ions across the separator in MFC

$$i = \frac{dQ}{dt} = \frac{\sum Na^+, K^+, NH_4^+, Ca^+, Mg^{2+} V_2 F Z_c dc_{2,t}}{dt} \quad (14.30)$$

where Q is net positive charge in the form of cation species in MFC which move through the PEM, V_2 is the cathode chamber liquid volume, Z_c is the valence of the cation species, $c_{2,t}$ is the concentration of cations in the cathode chamber and $\frac{dc_{2,t}}{dt}$ is found using the diffusion-migration equation proposed by Harnisch et al. (2009). Other details and know-how of important parameters influencing the membrane performance such as oxygen transfer coefficient, ion conductivity, membrane permeability and diffusion coefficients are discussed by several researchers (Christgen et al. 2015; Chae et al. 2008).

14.7 Models Pertaining to Oxygen Reduction Reaction (ORR) Kinetics at Cathode

In an MFC, the dissolved oxygen present in the electrolyte gets reduced at the cathode by combining with the diffused protons from the anode and the electrons obtained through the closed circuit, to form water.

$$O_2 + 4H^+ + 4e^- \rightleftarrows 2H_2O \quad (14.31)$$

The oxygen reduction may be a rate limiting step of the whole process based on the reaction activation energy. The kinetics of the oxygen reduction will depend on cathode size, dissolved oxygen concentration, and temperature. Renslow et al. (2011) developed a kinetic model for a sediment MFC. For a non-isothermal system, the equilibrium constant of solubility [K_{eq,O_2}] for dissolved oxygen concentration [C_{O_2}] at the cathode chamber is represented by

$$C_{O_2} = \frac{po_2}{K^0_{eq.O_2} e^{\frac{\Delta H^0}{R}\left(\frac{1}{T^0} - \frac{1}{T}\right)}} \tag{14.32}$$

where p_{O_2} is the partial pressure of oxygen, $K^0{}_{eqD2}$ is the Henry's constant at standard conditions, ΔH^0 is the standard enthalpy change and, T^0 is the reference temperature.

Further, the reduction potential E_{ORR} and the equilibrium constant of the ORR, K_{ORR}, in the cathode chamber are related through the following equation

$$K_{ORR} = e^{\frac{nFE_{ORR}}{RT}} \tag{14.33}$$

And finally, the electron potential of the ORR is function of temperature which is as follows:

$$E_{ORR} = \frac{E^0_{ORR}}{T^0} T + \frac{\Delta H^0_{ORR}}{nFT^0} \left(T - T^0\right) \tag{14.34}$$

$$\text{where } k_{f_{ORR}} = k^0_{ORR} e^{-\frac{nF\alpha}{RT}(e - E + E_{loss})} \tag{14.35}$$

E_{loss} includes both ohmic loss and potential loss, k_{fORR} is the rate constant of the forward reaction and $k^0{}_{ORR}$ is the standard heterogenous rate constant.

Finally, the total current density in the MFC is given by

$$j = -nFC_{O_2} k_{f_{ORR}} \tag{14.36}$$

Thus, the temperature and dissolved oxygen concentration in the cathode chamber determine current density and hence the overall efficiency of MFC.

14.8 Conclusion

This chapter presents a complete classification of the existing MFC models from the literature based on the two-tier characterization. At the first level, the models are characterized based on their spatial location of the component of the MFC. At the next level, a further characterization based on the phase is performed. The phase categorization segregates the models based on the level of complexity into homogeneous and heterogeneous systems. The homogeneous/pseudo-homogeneous models are easy to solve, simulate but lumps information of various phases into one while heterogeneous models are computationally intensive but they preserve the detailed information about the system. It is for the individual researcher to decide whether to pick pseudo-homogenous models against heterogeneous ones based on the level of details one wants to explore.

References

Chae, K., Choi, M., Ajayi, F., & Park, W. (2008). Mass transport through a proton exchange membrane (nafion) in microbial fuel cells. *Energy & Fuels, 22*, 169–176.

Christgen, B., Scott, K., Dolfing, J., Head, I. M., & Curtis, T. P. (2015). An evaluation of the performance and economics of membranes and separators in single chamber MFCs treating domestic wastewater. *PloS One, 10*, e0136108.

Garg, A., Vijayaraghavan, V., Mahapatra, S. S., Tai, K., & Wong, C. H. (2014). Performance evaluation of MFC by artificial intelligence methods. *Expert Systems with Applications, 41*, 1389–1399.

Harnisch, F., Warmbier, R., Schneider, R., & Schröder, U. (2009). Modeling the ion transfer and polarization of ion exchange membranes in bioelectrochemical systems. *Bioelectrochemistry, 75*, 136–141.

Logan, B. E., Hamelers, B., Rozendal, R., Schroder, U., Keller, J., Freguia, S., Aelterman, P., Verstraete, W., & Rabaey, K. (2006). MFCs: Methodology and technology. *Environmental Science & Technology, 40*(17), 5181–5192.

Lovley, D. R. (2006). Bug juice: Harvesting electricity with microorganisms. *Nature Reviews. Microbiology, 4*, 497–508.

Marcus, A. K., Torres, C. I., & Rittmann, B. E. (2007). Conduction-based modeling of the biofilm anode of an MFC. *Biotechnology and Bioengineering, 98*, 1171–1182.

Merkey, B. V., & Chopp, D. L. (2012). The performance of an MFC depends strongly on anode geometry: A multidimensional modeling study. *Bulletin of Mathematical Biology, 74*, 834–857.

Picioreanu, C., Katuri, K. P., Head, I. M., Van Loosdrecht, M. C. M., & Scott, K. (2008). Mathematical model for MFCs with anodic biofilms and anaerobic digestion. *Water Science and Technology, 57*(7), 965–971.

Picioreanu, C., Katuri, K. P., Van Loosdrecht, M. C. M., Head, I. M., & Scott, K. (2010). Modelling MFCs with suspended cells and added electron transfer mediator. *Journal of Applied Electrochemistry, 40*, 151–162.

Renslow, R., Donovan, C., Shim, M., Babauta, J. T., Nannapaneni, S., Schenk, J., & Beyenal, H. (2011). Oxygen reduction kinetics on graphite cathodes in sediment MFCs. *Physical Chemistry Chemical Physics, 13*, 21573–21584.

Stratford, J. P., Beecroft, N. J., Slade, R. C. T., Greening, A., & Avignone-Rossa, C. (2014). Anodic microbial community diversity as a predictor of the power output of MFCs. *Bioresource Technology, 156*, 84–91.

Tardast, A., Rahimnejad, M., Najafpour, G., Ghoreyshi, A., Premier, G. C., Bakeri, G., & Oh, S. E. (2014). Use of artificial neural network for the prediction of bioelectricity production in a membraneless MFC. *Fuel, 117*, 697–703.

Wang, X., Feng, Y. J., & Lee, H. (2008). Electricity production from beer brewery wastewater using single chamber MFC. *Water Science and Technology, 57*, 1117–1121.

Xu, J., Sheng, G.-P., Luo, H.-W., Li, W.-W., Wang, L.-F., & Yu, H.-Q. (2012). Fouling of proton exchange membrane (PEM) deteriorates the performance of MFC. *Water Research, 46*, 1817–1824.

Yan, M., & Fan, L. (2013). Constant voltage output in two-chamber MFC under fuzzy pid control. *International Journal of Electrochemical Science, 8*, 3321–3332.

Zhang, X. C., & Halme, A. (1995). Modelling of a MFC process. *Biotechnology Letters, 17*, 809–814.

Chapter 15
Bioremediation and Power Generation from Organic Wastes Using Microbial Fuel Cell

Jhansi L. Varanasi and Debabrata Das

15.1 Introduction

Worldwide energy consumption has increased drastically over the years due to the increasing population and economic growth. Modern energy services, electricity in particular, are a key enabler of economic and social development of a country. Rapid industrialization has led to the accelerated use of fossil fuels limiting their accessibility and thus causing difficulty to exploit these sources in the future. Thus, to meet the energy demands of the growing population, research has been focussed for the development of clean and green alternative technologies for energy generation. Bioelectrochemical systems (BES) represent such green technologies that utilize biocatalysts for bioenergy generation using wastes and wastewaters as feedstock (Pant et al. 2012a; Rabaey et al. 2009; Sleutels et al. 2012). BESs not only lead to the sustainable renewable energy generation but also help in reducing the costs incurred in waste treatment systems. Over the last couple of years, many possible applications for BESs have been emerged with respect to the oxidation and/or reduction of organic matter at anode and cathode respectively. Among the BESs, most widely studied are the microbial fuel cells (MFCs) that convert chemical energy to electrical energy through microbial oxidation of biodegradable organic matters (Oh et al. 2010; Wang and Ren 2013; Zhang 2012).

Different configurations of MFCs have been employed for simultaneous bioremediation and wastewater treatment along with electricity generation. The unique microbial species used in these systems are known as electroactive bacteria (EAB) that have the ability to donate/accept electrons in the surroundings through their cell surfaces. Several different EABs have been reported for their ability to treat specific

J.L. Varanasi (✉) • D. Das
Department of Biotechnology, Indian Institute of Technology Kharagpur,
Kharagpur 721302, India
e-mail: v.jhansi15@gmail.com

D. Das (ed.), *Microbial Fuel Cell*,
https://doi.org/10.1007/978-3-319-66793-5_15

pollutant present in the organic wastes. In this chapter, emphasis is laid on the basic principles of MFCs for electricity generation by EABs through degradation of organic matter. Certain specific examples of MFCs for bioremediation and wastewater treatment applications have been described and the current challenges and limitations of MFC technology along with the future directions have been briefly discussed.

15.2 Basic Principles of Power Generation from Organic Wastes in MFC

The basic principles of power generation in MFC have been illustrated in detail in Chap. 2. In general, a typical MFC mainly comprises anode and cathode chambers which are separated by an ion exchange membrane (Fig. 15.1). Usually the bacterial degradation of organic matter occurs at anode leading to the generation of protons, electrons and CO_2. The protons diffuse through the ion-selective membrane creating a potential difference between the anode and cathode. This in turn causes the electrons to flow and traverse through the external circuit to the cathode where they recombine with the protons and the terminal electron acceptor (usually O_2) to form water and electricity. The specific electrode potential depends upon the separate half-cell reactions occurring at the anode and the cathode. These electrode potentials can be deduced via Nernst equation similar to an electrochemical cell (Logan et al. 2006).

In general, the anodic and cathodic reactions of an MFC with acetate as electron donor and O_2 as the terminal electron acceptor can be given as:

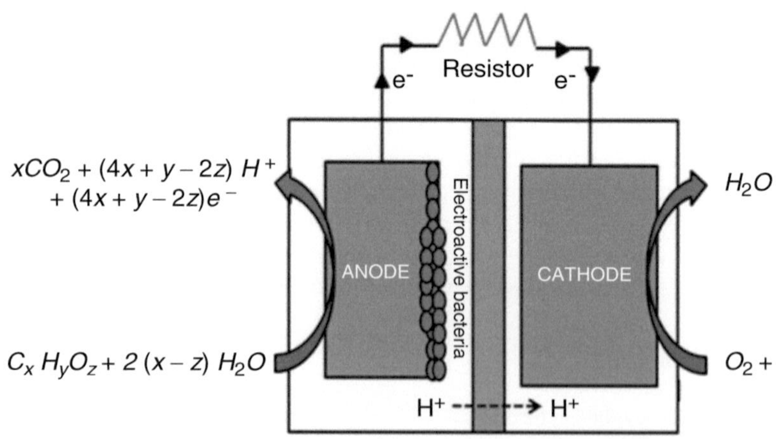

Fig. 15.1 Schematic of a typical MFC with an anode and cathode separated through a selective membrane

$$\text{Anode}: CH_3COOH + 2H_2O \rightarrow 2CO_2 + 8H^+ + 8e^- (E_{an} = -0.296\ V) \quad (15.1)$$

$$\text{Cathode}: 2O_2 + 8H^+ + 8e^{-1} \rightarrow H_2O\ (E_{cat} = 0.805\ V) \quad (15.2)$$

Thus, a cell voltage of $E_{cell} = 1.1$ V ($E_{cat} - E_{an}$) can be obtained from an MFC using acetate as substrate and O_2 as electron acceptor.

Apart from O_2, certain bacteria can utilize other inorganic compounds as electron acceptors such as nitrate, sulphate, manganese etc. in anaerobic conditions. By utilizing such microbes at cathode, remediation of inorganic pollutant compounds can be achieved. The bioremediation of specific pollutants using MFC technology will be described in details in subsequent sections.

15.3 Electrode Mechanisms

Depending upon the type of pollutant and the electroactive bacteria (EAB), both anode and cathode can be used for bioremediation purpose. Different mechanisms prevail at the electrode surfaces that lead to oxidation/reduction of the organic substrates. As described in Chap. 5, several extracellular electron transfer mechanisms of EABs have been elucidated that directly or indirectly interact with the electrode surfaces to transfer or uptake electrons obtained from the organic pollutant. Depending upon the redox reaction involved, the EABs can either be electrode oxidizing bacteria (at anode) and electrode reducing bacteria (at cathode). Apart from microbial metabolic reactions, partial chemical and electrochemical reactions also prevail in the anode and cathode chambers that influence the pollutant removal and wastewater treatment processes (Venkata Mohan and Srikanth 2011).

15.3.1 Reactions at Anode

Oxidation of substrates at anode can be microbialy catalysed as well as chemically or electrochemically induced. These reactions at anode and the substrate degradation are dependent upon the cathodic reduction processes and the terminal electron acceptor. Oxygen is the most widely used electron acceptor at cathode due to its high electronegativity. At anode, the substrates (pollutant/wastewater) are oxidized by microbial electron transfer reactions which lead to the production of intermediate reductant contaminants or convert to CO_2, protons and electrons. In addition, due to the presence of strong oxidizing agents a potential difference is created between anode and cathode which in turn increases the redox potential of the system and induces chemical or electrochemical reactions. These chemical/electrochemical reactions can cause partial oxidation of the substrate leading to the formation of intermediate reductant compounds. The intermediate contaminants/compounds can further be microbialy oxidized or can act as mediators for electron

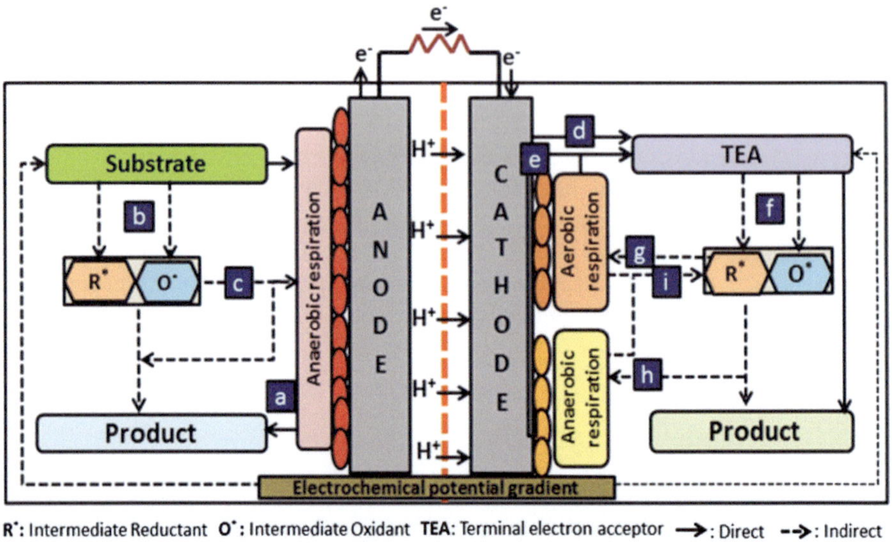

Fig. 15.2 Schematic illustration of the possible bio-electrochemical reactions occurring at anode and biocathode during MFC operation: (*a*) Microbial oxidation of substrate; (*b*) Formation of intermediate redox compounds by chemical/electrochemical/microbial induced reactions; (*c*) Anaerobic oxidation/reduction of intermediate redox compounds by microbes; (*d*) Electrochemical reduction; (*e*) Microbial reduction using aerobic or anaerobic biocathode; (*f*) Formation of intermediate redox compounds by microbial/chemical/electrochemical induced reactions; (*g*) Microbial oxidation/reduction of intermediate redox compounds by aerobic biocathode; (*h*) Microbial oxidation/reduction of intermediate redox compounds by anaerobic biocathode; and (*i*) Formation of intermediate redox compounds by microbial metabolism and further reduction using chemical/electrochemical reactions

transfer between the bacteria and anode surface (Venkata Mohan and Srikanth 2011). Apart from microbial and induced chemical/electrochemical reactions, electrode sorption also has shown to play an important role for the conversion and removal of trace organic pollutants (Wang et al. 2015a, b). Usually, due to the presence of different types of contaminants present in the wastewater, a combination of different mechanisms as described above can occur simultaneously. A general schematic of possible reactions occurring at the anode and cathode of MFC is shown in Fig. 15.2.

15.3.2 Reactions at Cathode

The electrons obtained by the substrate oxidation at anode flow via the external circuit to cathode where the reduction reaction takes place in the presence of terminal electron acceptors (TEA). Depending upon the type of MFC system and its application, TEA can be varied at the cathode. As described previously, oxygen is the most commonly used TEA as it is highly electronegative, abundant in nature and sustainable due to its reduction product being water (Dopson et al. 2016). It is observed that in the absence of O_2 environmental contaminants such as nitrate, hydrocarbons, azo dyes etc. can be utilized as potential electron acceptors in MFC systems thereby resulting in effective remediation of such pollutant containing waste streams at cathode (Pant et al. 2012).

Similar to anode, depending upon the microenvironment, different reactions are possible at cathode (Fig. 15.2). The microenvironment at cathode can be chosen based on the type and nature of the pollutant to be treated. Generally, in the absence of oxygen, the reduction reaction is catalysed by microbial metabolic reactions which can be either aerobic or anaerobic in nature. Like in anode, the prevailing chemical/electrochemical reactions can also lead to partial reduction of the pollutant resulting in an intermediate oxidant formation. This intermediate oxidant again can further be reduced microbially or can induce certain chemical/electrochemical reaction to generate product of interest. The energy output and treatment efficiency for this process, however, varies with the type of reaction occurring at cathode.

15.4 MFC Configurations

Several designs and configurations of MFCs have been developed for simultaneous wastewater treatment, pollutant remediation and bioelectricity generation. In general, they can be broadly classified into double chambered, single chambered, U-tube, upflow and stack MFCs (Fig. 15.3). The type of MFC used depends upon the purpose of its application. The most primitive MFC design is a double chambered MFC in which the anode and cathode chambers are separated by a cation or proton exchange membrane. These systems have high internal resistance and complex design and, therefore, scale up of double chambered MFCs is challenging.

To minimize the complexity and cost of the process, single chambered air-cathode MFCs were developed that could obtain high volumetric power densities (Logan et al. 2006). In this configuration, a single chamber (anode) is attached to the membrane cathode assembly (MCA) such that one side of the cathode is bonded to the membrane while the other side is air-facing. This design negates the use of external air supply to the cathode and thus proves to be cost effective.

For bulk-scale wastewater treatment and simultaneous electricity generation, upflow and U-tube MFCs are considered to be a promising configuration (Deng et al. 2010). This configuration is a hybrid of upflow anaerobic sludge blanket (UASB) and MFC that combines the advantages of both the systems. In this design, the substrate is fed continuously to the reactor from the bottom of the anodic chamber such that an up-flow hydraulic pattern is created. Due to this, continuous mixing of the anolyte is ensured and thus the use of a mechanical agitator can be avoided. Upflow MFCs have shown high power outputs as compared to the single chambered.

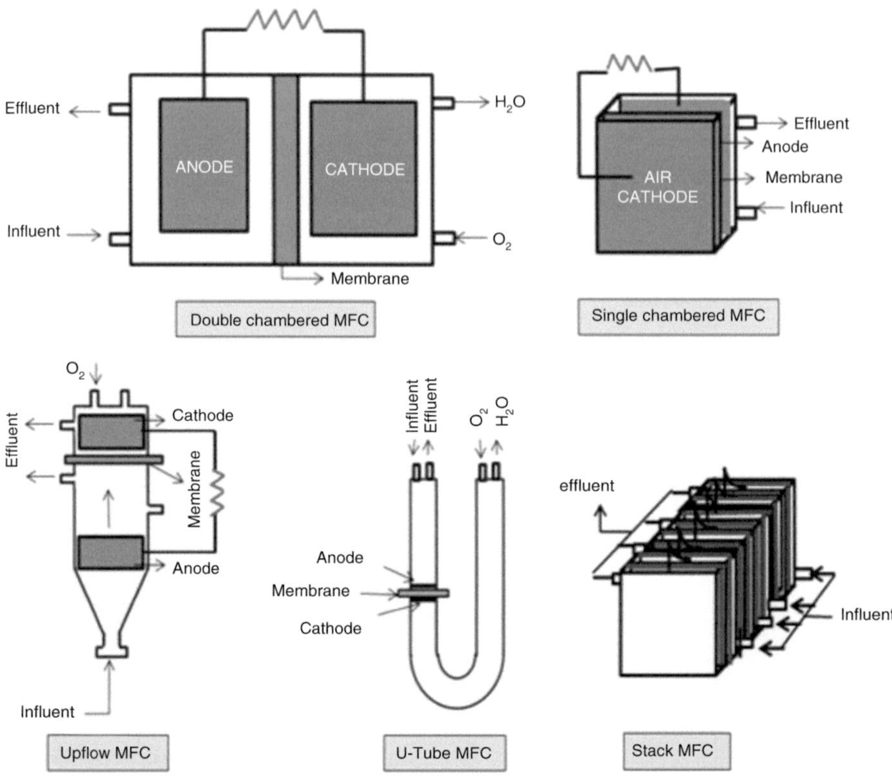

Fig. 15.3 Different configurations of MFCs used for bioremediation, power generation and wastewater treatment

A scalable configuration of MFCs which has been used for practical application is the Stack MFCs. In this configuration, several MFCs are interconnected either in series or in parallel. Since current is an extensive property with respect to the surface area of electrodes, stack MFCs are considered to be the most appropriate configuration for obtaining high voltage and current outputs along with higher wastewater treatment efficiencies (Aelterman et al. 2006; Pasupuleti et al. 2015).

15.5 Microbial Remediation Using MFC-Based Technologies

Bioremediation is the process of consumption or degradation of environmental pollutants by the use of naturally occurring or deliberately introduced microorganisms to clean a polluted site. Though bioremediation process is considered to be an efficient, cost effective and environmentally friendly technology, the major challenge for this process is the lack of contact between the pollutant and the microbes and the slow kinetics (Wang et al. 2015b). MFC-based technologies emerge as alternative bioremediation processes by utilizing EABs to oxidize/reduce pollutant at a specific site. Sometimes the pollutants themselves can act as mediators in electron transfer and can be treated in the process. Over the past few decades, several studies related to the removal of specific environmental pollutants such as azo dyes, polycyclic hydrocarbons and its derivatives, heavy metals, radioactive compounds etc. using MFC-based technologies have been explored (Table 15.1). These studies reflect the practical feasibility of MFC-based technologies for real-time removal of environmental pollutants at contaminated sites in a sustainable and economical manner.

15.5.1 MFC-Assisted Biodegradation of Azo Dyes

Azo dyes are aromatic compounds comprising one or more azo groups (−N=N−) and are the most widely used synthetic dyes used in commercial applications (Solanki et al. 2013). These compounds when degraded result in production of mutagenic or carcinogenic degraded products and if released into the environment can pose serious threat to human health and the natural environment (Chen 2015). These dyes are water soluble and highly stable in nature and at present harsh physicochemical methods (coagulation, flocculation, adsorption, membrane filtration etc.) are used for their removal from industrial effluents (Pandey et al. 2007). These physicochemical methods are energy and cost intensive and they often lead to the production of secondary waste streams that need further treatment and/or disposal. Biological processes for azo dye degradation have also been studied extensively using enzymes or whole cells (aerobic/anaerobic) and prove to be effective alternatives to the physicochemical methods of decolourization of wastewater. Nevertheless, the huge cost of enzymes, product inhibition, incomplete degradation and slow kinetics are the major challenges for the application of these processes at commercial scale. Enzymatic decolourization is now widely used for the decolourization of dye wastewater. However, this method is also facing several problems such as cost of enzymes, enzyme stability and product inhibition.

Recently, MFCs have been employed for the application of azo dye treatment by utilizing both the anode and cathode chambers (Chen 2015; Solanki et al. 2013). In anode, anaerobic degradation of azo dye occurs via co-metabolism i.e. in the presence of another organic compound (carbon source) which acts as a primary or co-substrate. The EABs utilize primary substrate as electron donor and a portion of electron released are used to generate electricity while the other portion is utilized for azo dye reduction thereby competing with anode for electrons. Apart from co-metabolism, direct anaerobic degradation of azo dyes has also been reported in MFCs in the absence of other organic compounds (Solanki et al. 2013). In such systems dye decolourization occurs via breakdown of the azo bond

Table 15.1 Bioremediation of specific pollutants using MFC technology

Pollutant	MFC configuration	Redox reaction	Electron donor	Electron acceptor	P_d ($W\ m^{-3}$)	CE (%)	RE (%)	References
Phenol	Double chamber	Anodic oxidation	Phenol	Oxygen	9.1	1.5	95.5	Luo et al. (2009)
Pyridine	Double chamber	Anodic oxidation	Pyridine	Ferricyanide	1.7	5	95	Zhang et al. (2009a, b)
p-nitrophenol	Single chamber	Anodic oxidation	p-nitrophenol	Oxygen	n.a.	*	80	Yuan et al. (2010)
Petroleum sludge	Single chamber	Anodic oxidation	Petroleum sludge	Oxygen	0.54	*	41.8	Mohan and Chandrasekhar (2011)
Nitrate	Double chamber	Cathodic reduction	Domestic wastewater	Nitrate	1.68	66	49	Fang et al. (2011)
Petroleum hydrocarbons	U-tube	Anodic oxidation	Petroleum hydrocarbon	Oxygen	n.a.	*	79	Wang et al. (2012)
Estriol	Single chamber	Anodic oxidation	Estriol	Oxygen	0.83	*	50.2	Kiran Kumar et al. (2012)
Ethynyl-estradiol			Ethynyl-estradiol		0.99		44.6	
Benzene	Double chamber	Anodic oxidation	Benzene	Oxygen	*	*	69	Rakoczy et al. (2013)
Sulphide			Sulphide				59	
Oxalate	Double chamber	Anodic oxidation	Oxalate	*	*	33.9	100	Bonmatí et al. (2013)
Sulphate	Double chamber	Anodic oxidation	Sulphate	Ferricyanide	0.5	*	77.9	Lee et al. (2014)

Toluene	Double chamber	Anodic oxidation	Toluene	Ferricyanide	0.1	11.6	96	Wu et al. (2014)
Methyl orange	Double chamber	Anodic oxidation	Methyl orange	Ferricyanide	2.08	7.5	75.1	Guo et al. (2014)
Tetrathionate	Double chamber	Anodic oxidation	Tetrathionate	Ferric ion	1.3	4.9	66	Sulonen et al. (2015)
Aniline	Double chamber	Anodic oxidation	Aniline	Ferricyanide	1.71	5.1	91.2	Cheng et al. (2015)
Ethanolamine	Single chamber	Anodic oxidation	Ethanolamine	Oxygen	10	20	53.4	Shin et al. (2015)
Sulfamethoxazole	Double chamber	Anodic oxidation	Sulfamethoxazole	Ferricyanide	*	*	50	Wang et al. (2015b)
Ethanolamine	Double chamber	Anodic oxidation	Ethanolamine	Nitrate	8.41	28.5	80.9	An et al. (2015)
Nitrate		Cathodic reduction				52.4	58.6	An et al. (2015)
Ethylene glycol	Single chamber	Anodic oxidation	Ethylene glycol	Oxygen	0.13	*	92	Hosseinpour et al. (2016)

P_d Power density, *CE* Columbic efficiency, *RE* Removal efficiency
*Data not reported

at the anode while complete degradation of intermediates occurs at cathode. Anaerobic degradation of various dyes such as congo red (Cao et al. 2010), active brilliant red (Sun et al. 2009), acid orange (Mu et al. 2009) etc. have been reported at anode with 75–90% removal efficiencies (Solanki et al. 2013).

Azo dyes can also be degraded in the cathode chamber by receiving electrons from the cathode electrode. Such reactions are already well established for electrochemical cells in which the chromophoric linkage of azo dyes is reduced to degradable colourless amines. Similar mechanisms can be employed in MFCs systems in which the pollutants with high redox potentials such as nitro-aromatic compounds, metal ions like manganese (VII), chromium (VI) or uranium (VI) etc. can be used and treated. The feasibility of utilizing different toxic azo dyes as electron acceptors at cathode was demonstrated by Liu et al. Several dyes such as methyl orange, orange I, orange II etc. were studied for the degradation in cathode (Liu et al. 2009). The mechanisms of dye degradation in cathode are similar to the anaerobic anodic degradation. However, additional protons and electrons are transferred to the cathode via the membrane and the external circuit respectively which can also be utilized for the degradation of dye.

The performance of an MFC for azo dye degradation depends on several factors such as type and structure of dye used, its concentration, operating pH, wastewater quality, external resistance used etc. (Solanki et al. 2013). These factors not only influence the degradation process but also affect power generation capacity of MFC system as a whole.

15.5.2 Bioremediation of Hydrocarbons and Their Derivatives

Organic compounds comprising hydrogen and carbon are known as hydrocarbons while their derivatives have a functional group in place of hydrogen atom. Major hydrocarbon pollutants include BTEX (benzene, toluene, ethylbenzene and xylenes), PAHs (polycyclic aromatic hydrocarbons) and TPHs (total petroleum hydrocarbons) which impose serious health and environmental concerns and thus require to be eliminated. BTEXs are usually found in petroleum derivatives such as petrol (gasoline) and have harmful effects on the central nervous system of humans. They also lead to the contamination of soil and groundwater that are near to the petroleum and/or natural gas production sites. The amount of BTEX at a site is used to assess the relative risk or seriousness of contamination at that particular site. PAHs consist of two or more fused benzene rings and/or pentacyclic molecules arranged in various structural configurations. Due to their low water solubility they can persist in the environment for longer duration. Though they are found in ubiquitous environments, they are most prevalent contaminants in soils (Sherafatmand and Ng 2015). TPHs are mixtures of hydrocarbons that are found in crude oil and can contaminate the environment. They mainly comprise hexane, benzene, toluene, xylenes, naphthalene, fluorine, gasoline constituents, mineral oils etc. Like BTEXs, some TPH compounds can affect the central nervous system or can cause serious effects on the blood, immune system, lungs, skin and eyes. Some TPH compounds have also been shown to affect reproduction and the developing foetus in animals. At present, all the hydrocarbon pollutants (BTEXs, PAHs and TPHs) and their derivatives are degraded using different bioremediation techniques such as biosparging, biostimulation, bioaugmentation etc. However, these techniques have several disadvantages such as low kinetics, low contact between the microbe and the pollutant, competition for survival between the new and the already present microbes etc. (Wang et al. 2015a).

Several studies have suggested MFCs to be an alternate remediation technology that could help mitigate the problems associated with the existing bioremediation techniques (Morris and Jin 2008; Sherafatmand and Ng 2015; Wang et al. 2012). MFCs can couple the hydrocarbon degradation to energy production (in the form of electricity) by employing electrogenic bacteria that could utilize hydrocarbons as electron donors at anode. For this purpose, major studies have reported the use of sediment MFC (SMFC) that utilizes indigenous microbes present in the soil/sediments to remove organic compounds. SMFCs typically consist of an anode buried inside the soil at the site of interest and a cathode in the top of the soil exposed to air (De Schamphelaire et al. 2008). Sherafatmand and Ng reported bioremediation of PAHs in water originated from soil with consistent power generations of 6.02 ± 0.34 and 3.63 ± 0.37 mW/m^2 by the aerobic and anaerobic SMFC respectively. The bioremediation capabilities of 41.7%, 31.4% and 36.2% removal of naphthalene, acenaphthene and phenanthrene, respectively in the aerobic environment and 76.9%, 52.5% and 36.8%, respectively in the anaerobic environment were achieved (Sherafatmand and Ng 2015). Wang et al. reported the use of U-tube MFCs for enhanced degradation of TPHs. They reported the degradation rates to be enhanced by 120% with simultaneous 125 ± 7 C of charge output (0.85 ± 0.05 mW/m^2 in the tested period (25 days). These studies suggest that utilizing MFCs for hydrocarbon remediation is the most successful technology nearing commercialization with several pilot and field studies as compared to other applications (Wang et al. 2015a).

15.5.3 Removal of Heavy Metals

Heavy metals are group of metals and metalloids (such as Cd, Cr, Cu, Ni, As, Pb, Zn etc.) that have atomic density of greater than 4000 kg/m^3. They are used extensively in industrial, medical and household applications. However, when exposed in the environment they can pose various health and environmental concerns since they are not biodegradable and can accumulate in living tissues causing serious diseases and disorders (Wang et al. 2015a, b). Due to the high market value of these metals, studies have been focused on recovery of these metals rather than their degradation. Numerous approaches have been studied for the development of cheaper and more effective technologies for heavy metal recovery from contaminated wastewater such as adsorption, membrane separation (including ultrafiltration (UF), nanofiltration (NF) and reverse osmosis (RO), electrotreatments (such as electrodialysis), photocatalytic processes etc. (Barakat 2011). These conventional methods are energy intensive and become ineffective if metals concentrations are below 1–100 mg/L (Barakat 2011).

MFCs appear to be effective in recovering heavy metals from wastewater (Mathuriya and Yakhmi 2014). Metal ions can be reduced and deposited by bacteria by utilizing both anode and cathode chambers of MFC. In principle, metal pollutants can be recovered at anode by the oxidative action of microbes while they can serve as alternative electron acceptors at cathode in place of oxygen. Nitrate, trichloroethene, perchlorate etc. have been demonstrated as effective electron acceptors in the MFC cathode chamber (Mathuriya and Yakhmi 2014). Different mechanisms have been elucidated for simultaneous wastewater treatment and heavy metal recovery in MFCs (Wang et al. 2015a, b) which include:

1. Metal with a redox potential higher than the MFC anode potential can be spontaneously reduced e.g. Au(III), Ag(I), Cu(II) etc.
2. Metals with lower redox potentials can be recovered by applying an external power supply (microbial electrolysis cell) to force the electrons to travel from the anode to the abiotic cathode e.g. Ni(II), Cd(II), Zn(II) etc.
3. Microbial reduction of metal oxides on the cathode with or without using an external potential.

The recovery of metallic species in MFCs has several benefits as compared to other conventional methods such as eliminating the need of external energy input for the treatment process, recovery of metals present in lower concentrations etc. Nevertheless, metals with low redox potentials still require stringent operating conditions and external input to drive the reaction in MFCs and thus extensive research is required to optimize and enhance the process efficiency.

15.5.4 Other Pollutants

Apart from the applications described above, MFCs have also been employed to treat several other pollutants such as chlorinated organic compounds, perchlorate reduction, sulfide removal, trace organic compounds etc. (Pant et al. 2012; Wang et al. 2015a, b). These pollutants can be effectively removed in MFCs since they provide a unique environment where both oxidation and reduction reactions can occur simultaneously along with the different microbial reactions as described in Fig. 15.1. Though most of these studies were conducted in lab scale, successful demonstration of MFCs for environmental remediation in pilot scale have also been reported (Wang et al. 2015a, b). Apart from treatment of one specific pollutant at a time, different contaminants that co-exist in soil, sediment or groundwater can also be removed using MFC technology though more studies are required to understand the mechanisms of treatment under such conditions.

15.6 Organic Wastes and Wastewater as Potential Feedstocks for MFCs

Over the past few decades, numerous studies have reported the application of MFCs to treat wastes and wastewaters with simultaneous electricity generation (Pant et al. 2012c). In these systems, the organic content is degraded with the help of electrogenic bacteria which convert the chemical energy of the organic waste/wastewater directly into electricity. Removal efficiency as high as 95% have been reported so far. MFCs prove to be attractive technologies for the waste/wastewater treatment as they reduce the input energy requirement as compared to the conventional wastewater treatment technologies. They also produce less sludge during the treatment making it environmentally friendly process. Several solid wastes and wastewaters like food wastes, cattle manure, domestic, industrial, and agricultural wastewaters have been studied in MFCs for bioelectricity generation which are elucidated in subsequent sections.

15.6.1 Solid Residual Wastes

Solid residual wastes usually cannot be reused, recycled or composted and need stringent disposal technologies such as landfill and incineration. A strategy to remove these materials and products from the waste stream is the use of MFC-based technologies. The primary goal is to obtain high organic removal rather than achieving higher power outputs. However, to make the MFC system self-sufficient, the exploitation of MFC for simultaneous power generation and waste treatment is necessary. The solid waste residues are majorly composed of complex molecules such as cellulose and hemicellulose which can be actively utilized for bioelectricity generation. Due to the different operational conditions, reactor configurations, types of electrodes, membranes and microorganisms involved, it is difficult to compare the performances of MFCs. However, a rough approximation can be made to evaluate the performance in terms of volumetric power densities and removal efficiencies. Under different operational and experimental conditions of MFCs, the power densities achieved with different solid residual wastes such as corn stover, cattle manure, food and vegetable waste etc. range 2–100 W/m^3 with COD removal efficiencies ranging from 40 to 90% (Table 15.2). These studies suggest that the energy-generating capacities of MFCs vary significantly, depending on the composition, strength and solution chemistry of wastes.

Table 15.2 List of different solid residual wastes used as feedstocks for power generation in MFCs

Organic waste	MFC configuration	Inoculum source	Electron donor	Electron acceptor	P_d (W m^{-3})	CE (%)	RE (%)	References
Corn stover	Single chamber	Municipal wastewater	Corn stover	Oxygen	14.89	3.6	42	Wang et al. (2009)
Wheat straw hydrolysate	Double chamber	Clarifier wastewater	Wheat straw	Ferricyanide	20.66	37.1	*	Zhang et al. (2009b)
Furfural	Double chamber	Anaerobic and aerobic sludge	Furfural	Ferricyanide	103	10.2	95	Luo et al. (2010)
	Single chamber			Oxygen	18	30.3	68	
Vegetable waste	Single chamber	UASB sludge	Vegetable waste	Oxygen	2.72	*	62.8	Venkata Mohan et al. (2010)
Lignocellulosic biomass	Single chamber	Wastewater plant	Lignocellulosic biomass	Oxygen	*	21	83	Catal et al. (2011)
Cattle dung	Double chamber	Biogas slurry	Cattle dung	Potassium permanganate	0.22	2.79	73.9	Zhao et al. (2012)
Food waste leachate	Double chamber	Anaerobic sludge	Food waste leachate	Oxygen	9.95	63.4	85.4	Li et al. (2013)
Cheese whey	Upflow	Digested sludge	Cheese whey	Phosphate buffer	1.3	3.9	60	Kelly and He (2014)
Rice straw hydrolysate	Single chamber	Anaerobic sludge	Rice straw hydrolysate	Oxygen	12.58	17.9	72	Wang et al. (2014)
Municipal solid waste	Upflow	Granular sludge	Municipal solid waste	Oxygen	0.71	4.9	52.8	Karluvalı et al. (2015)
Potato waste	Double chamber	Anaerobic consortia	Potato waste	Oxygen	0.39	*	56	Du and Li (2015)
Composite food waste	Single chamber	Food waste slurry	Composite food waste	Oxygen	118	*	*	Bardone and Bravi (2016)
Fruit and vegetable waste	Single chamber	Compost	Fruit and vegetable waste	Oxygen	71.43	*	40.3	Md Khudzari et al. (2016)
Algal biomass	Double chamber	Returned activated sludge	Algal biomass	Phosphate buffer	*	36.6	90	Mohammadi Khalfbadam et al. (2016)

Pd Power density, *CE* Columbic efficiency, *RE* Removal efficiency
*Data not reported

Complex substrates present in wastes require higher energy to break down as compared to simple substrates and thus in turn yield lower energy outputs as compared to pure substrates. Certain complex substrates like lignocellulosic biomass can be detrimental to electrogenic bacteria and may require the use of pretreatment strategies prior to substrate utilization by MFCs. Different pretreatment methods such as mechanical, thermal, chemical, biological and the combination of these have been reported in literature for the hydrolysis of complex substrates into simple sugars or low-molecular weight compounds (Ariunbaatar et al. 2014). These hydrolysis products are ideal substrates to support bioelectricity generation in MFC systems.

15.6.2 Organic Wastewater

Wastewater treatment at present utilize aerobic and anaerobic biologic treatment technologies which include activated sludge, trickling filters, sequencing batch reactors (SBR), upflow anaerobic sludge blankets (UASB), anaerobic filters, constructed wetlands, or a combination of these. These technologies provide sufficient effluent quality. They are usually energy and cost intensive. MFCs on the other hand could be used for generating energy along with wastewater treatment and thus can offset the operational costs of wastewater treatment plants (Pant et al. 2012c). Apart from reducing the overall energy consumption in the treatment process, MFCs produce much less secondary sludge making the process environmental friendly. In fact while performing energy balance analysis for MFC systems, it was observed that theoretically the energy generated by MFC process is much higher as compared to the energy consumption (Kelly and He 2014). However, practically the energy recoveries from the MFCs have been much lower due to the prevailing internal resistance of the systems. The presence of alternate electron acceptors such as nitrate, nitrite, sulphate etc. can impair energy recovery from wastewater.

Several wastewater streams originating from different sources such as distillery waste, brewery waste, food processing waste, palm oil mill effluent etc. which are readily available and are rich in organic content have been used in MFCs for power generation (Table 15.3). Most of these studies have been conducted using mixed cultures so as to avoid stringent aseptic conditions and to their ability to utilize wide variety of substrates. Usually (Pant et al. 2012c) MFCs have shown to be effective in a COD range of 3–5 g/L. The absence of microbial growth inhibiting agents in these wastewaters adds up to an additional advantage. Since the COD concentration of the wastewater originating from industries or agriculture is much higher, integrated treatment systems coupling MFCs with other wastewater treatment technologies such as dark fermentation or anaerobic digestion processes etc. have also been employed to treat wastewaters and enhance the overall energy recovery (Pandit et al. 2014; Varanasi et al. 2015). In the integration process, the complex wastewater is first converted to volatile fatty acids (VFA) via the acidogenic pathway and

Table 15.3 An overview of different wastewaters used as feedstocks for power generation in MFCs

Organic waste	MFC configuration	Inoculum	Electron donor	Electron acceptor	P_d (W m^{-3})	CE (%)	RE (%)	References
Beer brewery wastewater	Single chamber	Anaerobic mixed consortia	Brewery wastewater	Oxygen	9.52	19.7	43	Wen et al. (2009)
Rice mill wastewater	Double chamber	Anaerobic sludge	Rice mill wastewater	Permanganate	2.31	21.2	96.5	Behera et al. (2010)
Distillery wastewater	Upflow	Enriched anaerobic consortia	Distillery wastewater	Oxygen	1.74	*	72.8	Mohanakrishna et al. (2010)
Palm oil mill effluent	Double chamber	Palm oil mill effluent sludge	Palm oil mill effluent	Oxygen	62.2	32	23	Jong et al. (2011)
Alcohol distillery wastewater	Upflow	Anaerobic digestion sludge	Alcohol distillery wastewater	Oxygen	0.02	*	90	Huang et al. (2011a)
Pharmaceutical wastewater	Single chamber	Enriched anaerobic consortia	Pharmaceutical wastewater	Oxygen	2.69	*	85	Velvizhi and Mohan (2012)
Domestic wastewater	Single chamber	*Shewanella oneidensis MR-1*	Domestic wastewater	Oxygen	1.01	0.71	26	Nimje et al. (2012)
Fermented food waste	Single chamber	Mixed culture	Mixed volatile fatty acids	Oxygen	0.09	46	95	Pant et al. (2013)
Molasses wastewater	Single chamber	Mixed culture	Molasses wastewater	Oxygen	5.06	*	59	Sevda et al. (2013)
Fermented cane molasses	Single chamber	Anaerobic mixed consortia	Mixed volatile fatty acids	Oxygen	3.02	3.6	80.2	Pandit et al. (2014)
Fermented cellobiose	Double chamber	Enriched anaerobic consortia	Mixed volatile fatty acids	Ferricyanide	1.23	12.2	75	Varanasi et al. (2015)
Palm oil mill effluent	Double chamber	Anaerobic sludge	Palm oil mill effluent	Potassium permanganate	0.32	24	70	Baranitharan et al. (2015)
Municipal wastewater	Stack	Primary sludge	Municipal wastewater	Oxygen	44.1	24.8	24	Hiegemann et al. (2016)
Seafood processing wastewater	Upflow	Waste activated sludge	Seafood processing wastewater	Oxygen	2.21	23.73	83	Jayashree et al. (2016)

Pd Power density, *CE* Columbic efficiency, *RE* Removal efficiency
*Data not reported

these VFAs in turn are utilized in MFCs by the EAB. The integrated MFC technologies have proved to be better treatment systems achieving removal efficiencies of 70–90% with the overall energy recoveries 30–40% (Chookaew et al. 2014; Pasupuleti et al. 2015; Varanasi et al. 2015; Wang et al. 2011).

15.7 Challenges

Several attempts have been made in the development of various MFC technologies for enhanced energy recovery and simultaneous waste/wastewater treatment. Their practical real-field applications have been limited due to the associated operational and economic challenges. Although many pilot scale studies have been performed using real-time wastewater, the outputs obtained is far behind from those obtained with the bench studies under similar operational conditions (Du et al. 2007). Several operational factors such as limited membrane transport, ohmic losses, activation losses, unstable voltage for long duration of time, columbic losses etc. limit the performance of MFCs during large scale operation (Logan 2010). Though stacks cells provide appreciable outputs and stability, they are limited by the voltage reversal arising at high current densities. Growth of excessive unwanted biomass and biofouling of membranes can also severely affect the long-term performance of pilot-scale MFCs. To utilize MFCs in real-world applications such as environmental remediation and wastewater treatment, more flexible reactor configurations will be required that can adapt to the physico-chemical environment to which they are constructed. Use of expensive electrodes and membranes materials, their pretreatment methods, installation and operational costs, use of extra current collectors and precious metal catalysts etc. contribute to the economic constraints for the large scale production of these systems (Pant et al. 2012c). These constraints can be overcome by utilizing cheaper electrodes and treatment strategies. Use of aeration in cathode chamber also leads to the increasing costs of MFCs for wastewater treatment and utilizing biocathodes or single chambered air-cathode MFCs could be a possible solution for such systems. Implementing biocathodes not only reduces the costs but can also lead to the production of value added compounds (Huang et al. 2011b). Developing membrane-less MFCs is another strategy that could be used to further reduce the overall costs and improve the performance and treatment efficiencies of MFCs. However, in such a scenario, the distance between the electrodes might increase and thus in-turn the internal resistance of the systems may increase. Further research should be made to compare the performances of membrane and membraneless systems and if needed cheaper membrane materials or membrane cathode assemblies should be used which can further improve the performance of the system. Though at present the outputs of MFCs are far behind than the theoretical values, the ongoing research to tackle the above mentioned challenges can lead to successful commercialization of these technologies.

15.8 Conclusions and Future Prospects

Direct waste to energy conversion by employing MFC-based technologies appears to be the most promising solution to tackle the global energy and wastewater management related issues. The present chapter discusses in brief the recent advances made with respect to environmental remediation as well as waste/wastewater treatment by employing MFCs. Several studies suggest that MFCs achieve as high as 90% removal efficiencies though the energy recoveries are poor. To make this process energy efficient, considerable attention must be given to the complex reactor configurations, type of electrodes, membranes and the external circuit components which collectively affect the internal resistance of the system. To avoid strict aseptic conditions, enrichment of electrogen-rich consortium is desirable. For achieving high energy recoveries from MFCs, pre-treatment of wastes and/or integration with existing wastewater treatment technologies like fermentation, anaerobic digestion, activated sludge process etc. appear to be more realistic, cost-efficient and feasible. With the recent developments of novel cost effective materials and cell components, superior performance is expected from MFCs that could expand their applicability for real-field applications. Wide applications of MFCs have emerged using biocathodes which include bioproduct development and its recovery along with power generation. It is anticipated that with these upcoming improvements and the few pilot-scale studies, the commercialization of MFCs is the next step for a sustainable and economical bioenergy production.

References

Aelterman, P., Rabaey, K., & Verstraete, W. (2006). Continuous electricity generation at high voltages and currents using stacked MFCs. *Environmental Science & Technology, 40*, 3388–3394.

An, B.-M., Song, Y.-H., Shin, J.-W., & Park, J.-Y. (2015). Two-chamber MFC to simultaneously remove ethanolamine and nitrate. *Desalination and Water Treatment, 57*(17), 7866–7873.

Ariunbaatar, J., Panico, A., Esposito, G., Pirozzi, F., & Lens, P. N. L. (2014). Pretreatment methods to enhance anaerobic digestion of organic solid waste. *Applied Energy, 123*, 143–156.

Barakat, M. A. (2011). New trends in removing heavy metals from industrial wastewater. *Arabian Journal of Chemistry, 4*, 361–377.

Baranitharan, E., Khan, M. R., Prasad, D. M. R., Teo, W. F. A., Tan, G. Y. A., & Jose, R. (2015). Effect of biofilm formation on the performance of MFC for the treatment of palm oil mill effluent. *Bioprocess and Biosystems Engineering, 38*, 15–24.

Behera, M., Jana, P. S., More, T. T., & Ghangrekar, M. M. (2010). Bioelectrochemistry Rice mill wastewater treatment in MFCs fabricated using proton exchange membrane and earthen pot at different pH. *Bioelectrochemistry, 79*, 228–233.

Bonmatí, A., Sotres, A., Mu, Y., Rozendal, R., & Rabaey, K. (2013). Oxalate degradation in a bioelectrochemical system: Reactor performance and microbial community characterization. *Bioresource Technology, 143*, 147–153.

Cao, Y., Hu, Y., Sun, J., & Hou, B. (2010). Explore various co-substrates for simultaneous electricity generation and Congo red degradation in air-cathode single-chamber MFC. *Bioelectrochemistry, 79*, 71–76.

Catal, T., Fan, Y., Li, K., Bermek, H., & Liu, H. (2011). Utilization of mixed monosaccharides for power generation in MFCs. *Journal of Chemical Technology and Biotechnology, 86*, 570–574.

Chen, B. (2015). Exploring MFC-assisted bioremediation of textile dyes. *Energy Conversion Energy Reviews, 1*(1), 11–18.

Cheng, H.-Y., Liang, B., Mu, Y., Cui, M.-H., Li, K., Wu, W.-M., & Wang, A.-J. (2015). Stimulation of oxygen to bioanode for energy recovery from recalcitrant organic matter aniline in MFCs (MFCs). *Water Research, 81*, 72–83.

Chookaew, T., Prasertsan, P., & Ren, Z. J. (2014). Two-stage conversion of crude glycerol to energy using dark fermentation linked with MFC or microbial electrolysis cell. *New Biotechnology, 31*, 179–184.

De Schamphelaire, L., Rabaey, K., Boeckx, P., Boon, N., & Verstraete, W. (2008). Outlook for benefits of sediment MFCs with two bio-electrodes. *Microbial Biotechnology, 1*, 446–462.

Deng, Q., Li, X., Zuo, J., Ling, A., & Logan, B. E. (2010). Power generation using an activated carbon fiber felt cathode in an upflow MFC. *Journal of Power Sources, 195*, 1130–1135.

Dopson, M., Ni, G., & Sleutels, T. H. (2016). Possibilities for extremophilic microorganisms in microbial electrochemical systems. *FEMS Microbiology Reviews, 40*, 164–181.

Du, H., & Li, F. (2015). Size effects of potato waste on its treatment by MFC. *Environmental Technology, 37*(10), 1305–1313.

Du, Z., Li, H., & Gu, T. (2007). A state-of-the-art review on MFCs: A promising technology for wastewater treatment and bioenergy. *Biotechnology Advances, 25*, 464–482.

Fang, C., Min, B., & Angelidaki, I. (2011). Nitrate as an oxidant in the cathode chamber of a MFC for both power generation and nutrient removal purposes. *Applied Biochemistry and Biotechnology, 164*(4), 464–474.

Frattini, D., Falcucci, G., Minutillo, M., Ferone, C., Cioffi, R., & Jannelli, E. (2016). On the effect of different configurations in air-cathode MFCs fed by composite food waste for energy harvesting. *Chemical Engineering Transactions*, 49, 85–90 ISBN 978-88-95608-40-2.

Guo, W., Feng, J., Song, H., & Sun, J. (2014). Simultaneous bioelectricity generation and decolorization of methyl orange in a two-chambered MFC and bacterial diversity. *Environmental Science and Pollution Research, 21*, 11531–11540.

Hiegemann, H., Herzer, D., Nettmann, E., Lübken, M., Schulte, P., Schmelz, K.-G., Gredigk-Hoffmann, S., & Wichern, M. (2016). An integrated 45L pilot MFC system at a full-scale wastewater treatment plant. *Bioresource Technology, 218*, 115–122.

Hosseinpour, M., Asadi, M., Eliato, T. R., Vossoughi, M., & Alemzadeh, I. (2016). Ethylene glycol biodegradation in MFC. *Energy Sources, Part A: Recovery, Utilization, and Environmental Effects, 38*(8), 1096–1102.

Huang, J., Yang, P., Guo, Y., & Zhang, K. (2011a). Electricity generation during wastewater treatment: An approach using an AFB-MFC for alcohol distillery wastewater. *Desalination, 276*, 373–378.

Huang, L., Regan, J. M., & Quan, X. (2011b). Electron transfer mechanisms, new applications, and performance of biocathode MFCs. *Bioresource Technology, 102*, 316–323.

Jayashree, C., Tamilarasan, K., Rajkumar, M., Arulazhagan, P., Yogalakshmi, K. N., Srikanth, M., & Banu, J. R. (2016). Treatment of seafood processing wastewater using upflow MFC for power generation and identification of bacterial community in anodic biofilm. *Journal of Environmental Management, 180*, 351–358.

Jong, B. C., Liew, P. W. Y., Juri, M. L., Kim, B. H., Mohd Dzomir, A. Z., Leo, K. W., & Awang, M. R. (2011). Performance and microbial diversity of palm oil mill effluent MFC. *Letters in Applied Microbiology, 53*(6), 660–667.

Karluvalı, A., Köroğlu, E. O., Manav, N., Çetinkaya, A. Y., & Özkaya, B. (2015). Electricity generation from organic fraction of municipal solid wastes in tubular MFC. *Separation and Purification Technology, 156*, 502–511.

Kelly, P. T., & He, Z. (2014). Understanding the application niche of MFCs in a cheese wastewater treatment process. *Bioresource Technology, 157*, 154–160.

Kiran Kumar, A., Venkateswar Reddy, M., Chandrasekhar, K., Srikanth, S., & Venkata Mohan, S. (2012). Endocrine disruptive estrogens role in electron transfer: Bio-electrochemical remediation with microbial mediated electrogenesis. *Bioresource Technology, 104*, 547–556.

Lee, D.-J., Liu, X., & Weng, H.-L. (2014). Sulfate and organic carbon removal by MFC with sulfate-reducing bacteria and sulfide-oxidising bacteria anodic biofilm. *Bioresource Technology, 156*, 14–19.

Li, X. M., Cheng, K. Y., & Wong, J. W. C. (2013). Bioelectricity production from food waste leachate using MFCs: Effect of NaCl and pH. *Bioresource Technology, 149*, 452–458.

Liu, L., Li, F., Feng, C., & Li, X. (2009). MFC with an azo-dye-feeding cathode. *Applied Microbiology and Biotechnology, 85*, 175–183.

Logan, B. E. (2010). Scaling up MFCs and other bioelectrochemical systems. *Applied Microbiology and Biotechnology, 85*, 1665–1671.

Logan, B. E., Hamelers, B., Rozendal, R., Schröder, U., Keller, J., Freguia, S., Aelterman, P., Verstraete, W., & Rabaey, K. (2006). MFCs: Methodology and technology. *Environmental Science & Technology, 40*, 5181–5192.

Luo, H., Liu, G., Zhang, R., & Jin, S. (2009). Phenol degradation in MFCs. *Chemical Engineering Journal, 147*, 259–264.

Luo, Y., Liu, G., Zhang, R., & Zhang, C. (2010). Power generation from furfural using the MFC. *Journal of Power Sources, 195*(1), 190–194.

Mathuriya, A. S., & Yakhmi, J. V. (2014). MFCs to recover heavy metals. *Environmental Chemistry Letters, 12*, 483–494.

Md Khudzari, J., Tartakovsky, B., & Raghavan, G. S. V. (2016). Effect of C/N ratio and salinity on power generation in compost MFCs. *Waste Management, 48*, 135–142.

Mohammadi Khalfbadam, H., Cheng, K. Y., Sarukkalige, R., Kaksonen, A. H., Kayaalp, A. S., & Ginige, M. P. (2016). A bio-anodic filter facilitated entrapment, decomposition and in situ oxidation of algal biomass in wastewater effluent. *Bioresource Technology, 216*, 529–536.

Mohan, S. V., & Chandrasekhar, K. (2011). Self-induced bio-potential and graphite electron accepting conditions enhances petroleum sludge degradation in bio-electrochemical system with simultaneous power generation. *Bioresource Technology, 102*, 9532–9541.

Mohanakrishna, G., Venkata Mohan, S., & Sarma, P. N. (2010). Bio-electrochemical treatment of distillery wastewater in MFC facilitating decolorization and desalination along with power generation. *Journal of Hazardous Materials, 177*, 487–494.

Morris, J. M., & Jin, S. (2008). Feasibility of using MFC technology for bioremediation of hydrocarbons in groundwater. *Journal of Environmental Science and Health. Part A, Toxic/Hazardous Substances & Environmental Engineering, 43*, 18–23.

Mu, Y., Rabaey, K., Rozendal, R. A., Yuan, Z., & Keller, J. (2009). Decolorization of Azo dyes in bioelectrochemical systems. *Environmental Science & Technology, 43*, 5137–5143.

Nimje, V. R., Chen, C.-Y., Chen, H.-R., Chen, C.-C., Huang, Y. M., Tseng, M.-J., Cheng, K.-C., & Chang, Y.-F. (2012). Comparative bioelectricity production from various wastewaters in MFCs using mixed cultures and a pure strain of Shewanella oneidensis. *Bioresource Technology, 104*, 315–323.

Oh, S. T., Kim, J. R., Premier, G. C., Lee, T. H., Kim, C., & Sloan, W. T. (2010). Sustainable wastewater treatment: How might MFCs contribute. *Biotechnology Advances, 28*, 871–881.

Pandey, A., Singh, P., & Iyengar, L. (2007). Bacterial decolorization and degradation of azo dyes. *International Biodeterior and Biodegradation, 59*, 73–84.

Pandit, S., Balachandar, G., & Das, D. (2014). Improved energy recovery from dark fermented cane molasses using MFCs. *Frontiers of Chemical Science and Engineering, 8*, 43–54.

Pant, D., Singh, A., Van Bogaert, G., Irving Olsen, S., Singh Nigam, P., Diels, L., & Vanbroekhoven, K. (2012). Bioelectrochemical systems (BES) for sustainable energy production and product recovery from organic wastes and industrial wastewaters. *RSC Advances, 2*, 1248–1263.

Pant, D., Arslan, D., Van Bogaert, G., Gallego, Y. A., De Wever, H., Diels, L., & Vanbroekhoven, K. (2013). Integrated conversion of food waste diluted with sewage into volatile fatty acids through fermentation and electricity through a fuel cell. *Environmental Technology, 34*, 1935–1945.

Pasupuleti, S. B., Srikanth, S., Venkata Mohan, S., & Pant, D. (2015). Continuous mode operation of MFC stack with dual gas diffusion cathode design for the treatment of dark fermentation effluent. *International Journal of Hydrogen Energy, 40*, 12424–12435.

Rabaey, K., Angenent, L., & Schroder, U. (2009). *Bioelectrochemical systems: From extracellular electron transfer to biotechnological application*. London: IWA Publishing.

Rakoczy, J., Feisthauer, S., Wasmund, K., Bombach, P., Neu, T. R., Vogt, C., & Richnow, H. H. (2013). Benzene and sulfide removal from groundwater treated in a MFC. *Biotechnology and Bioengineering, 110*, 3104–3113.

Sevda, S., Dominguez-Benetton, X., Vanbroekhoven, K., De Wever, H., Sreekrishnan, T. R., & Pant, D. (2013). High strength wastewater treatment accompanied by power generation using air cathode MFC. *Applied Energy, 105*, 194–206.

Sherafatmand, M., & Ng, H. Y. (2015). Using sediment MFCs (SMFCs) for bioremediation of polycyclic aromatic hydrocarbons (PAHs). *Bioresource Technology, 195*, 122–130.

Shin, J.-W., Seo, S.-J., Maitlo, H. A., & Park, J.-Y. (2015). The enhancement of ammonium removal from ethanolamine wastewater using air-cathode MFCs coupled to ferric reduction. *Bioresource Technology, 190*, 466–473.

Sleutels, T. H. J. a., Ter Heijne, A., Buisman, C. J. N., & Hamelers, H. V. M. (2012). Bioelectrochemical systems: An outlook for practical applications. *ChemSusChem, 5*, 1012–1019.

Solanki, K., Subramanian, S., & Basu, S. (2013). MFCs for azo dye treatment with electricity generation: A review. *Bioresource Technology, 131*, 564–571.

Sulonen, M. L. K., Kokko, M. E., Lakaniemi, A.-M., & Puhakka, J. A. (2015). Electricity generation from tetrathionate in MFCs by acidophiles. *Journal of Hazardous Materials, 284*, 182–189.

Sun, J., Hu, Y., Bi, Z., & Cao, Y. (2009). Simultaneous decolorization of azo dye and bioelectricity generation using a microfiltration membrane air-cathode single-chamber MFC. *Bioresource Technology, 100*, 3185–3192.

Varanasi, J. L., Roy, S., Pandit, S., & Das, D. (2015). Improvement of energy recovery from cellobiose by thermophillic dark fermentative hydrogen production followed by MFC. *International Journal of Hydrogen Energy, 40*, 8311–8321.

Velvizhi, G., & Mohan, S. V. (2012). Electrogenic activity and electron losses under increasing organic load of recalcitrant pharmaceutical wastewater. *International Journal of Hydrogen Energy, 37*, 5969–5978.

Venkata Mohan, S., & Srikanth, S. (2011). Enhanced wastewater treatment efficiency through microbially catalyzed oxidation and reduction: Synergistic effect of biocathode microenvironment. *Bioresource Technology, 102*, 10210–10220.

Venkata Mohan, S., Mohanakrishna, G., & Sarma, P. N. (2010). Composite vegetable waste as renewable resource for bioelectricity generation through non-catalyzed open-air cathode MFC. *Bioresource Technology, 101*, 970–976.

Wang, H., & Ren, Z. J. (2013). A comprehensive review of microbial electrochemical systems as a platform technology. *Biotechnology Advances, 31*, 1796–1807.

Wang, X., Feng, Y., Wang, H., Qu, Y., Yu, Y., Ren, N., Li, N., Wang, E., Lee, H., & Logan, B. E. (2009). Bioaugmentation for electricity generation from corn Stover biomass using MFCs. *Environmental Science & Technology, 43*(15), 6088–6093.

Wang, A., Sun, D., Cao, G., Wang, H., Ren, N., Wu, W.-M., & Logan, B. E. (2011). Integrated hydrogen production process from cellulose by combining dark fermentation, MFCs, and a microbial electrolysis cell. *Bioresource Technology, 102*, 4137–4143.

Wang, X., Cai, Z., Zhou, Q., Zhang, Z., & Chen, C. (2012). Bioelectrochemical stimulation of petroleum hydrocarbon degradation in saline soil using U-tube MFCs. *Biotechnology and Bioengineering, 109*, 426–433.

Wang, Z., Lee, T., Lim, B., Choi, C., & Park, J. (2014). Microbial community structures differentiated in a single-chamber air-cathode MFC fueled with rice straw hydrolysate. *Biotechnology for Biofuels, 7*, 9.

Wang, H., Luo, H., Fallgren, P. H., Jin, S., & Ren, Z. J. (2015a). Bioelectrochemical system platform for sustainable environmental remediation and energy generation. *Biotechnology Advances, 33*, 317–334.

Wang, L., Wu, Y., Zheng, Y., Liu, L., Zhao, F., et al. (2015b). Efficient degradation of sulfamethoxazole and the response of microbial communities in MFCs. *RSC Advances, 5*, 56430–56437.

Wen, Q., Wu, Y., Cao, D., Zhao, L., & Sun, Q. (2009). Electricity generation and modeling of MFC from continuous beer brewery wastewater. *Bioresource Technology, 100*, 4171–4175.

Wu, C.-H., I, Y.-P., Chiu, Y.-H., & Lin, C.-W. (2014). Enhancement of power generation by toluene biodegradation in a MFC in the presence of pyocyanin. *Journal of the Taiwan Institute of Chemical Engineers, 45*, 2319–2324.

Yuan, S.-J., Sheng, G.-P., Li, W.-W., Lin, Z.-Q., Zeng, R. J., Tong, Z.-H., & Yu, H.-Q. (2010). Degradation of organic pollutants in a Photoelectrocatalytic system enhanced by a MFC. *Environmental Science & Technology, 44*(14), 5575–5580.

Zhang, Y. (2012). *Energy recovery from waste streams with MFC-based technologies*. Doctoral dissertation.

Zhang, C., Li, M., Liu, G., Luo, H., & Zhang, R. (2009a). Pyridine degradation in the MFCs. *Journal of Hazardous Materials, 172*, 465–471.

Zhang, Y., Min, B., Huang, L., & Angelidaki, I. (2009b). Generation of electricity and analysis of microbial communities in wheat straw biomass-powered MFCs. *Applied and Environmental Microbiology, 75*, 3389–3395.

Zhao, G., Ma, F., Wei, L., Chua, H., Chang, C.-C., & Zhang, X.-J. (2012). Electricity generation from cattle dung using MFC technology during anaerobic acidogenesis and the development of microbial populations. *Waste Management, 32*, 1651–1658.

Chapter 16
Removal and Recovery of Metals by Using Bio-electrochemical System

A. Evren Tugtaş and Bariş Çalli

16.1 Introduction

One of the potential uses of bio-electrochemical systems (BES) is the removal and recovery of metals (Nancharaiah et al. 2015). Every day, significant amounts of metals are released to the environment via anthropogenic activities such as mining, smelting, metal refining, fossil fuel combustion, waste and sewage sludge incineration, electroplating discharges, disposal of electronic waste and photographic supplies as well as manufacturing of printed circuit board, semiconductor, paint, stainless steel, pigments, photoelectric cells, glass, pesticides etc. Conventionally, metals are removed by precipitation as insoluble salts, ion exchange, adsorption, biosorption and microbial reduction from metal contaminated wastewater (Elouear et al. 2009; Fu and Wang 2011; Choi and Hu 2013; Tao et al. 2011b, 2012; Wang and Ren 2014; Lim et al. 2015; Zhang et al. 2016). The conventional methods are usually impractical at low concentrations and cost-inefficient because of high operational costs, energy consumption and production of excessive amounts of hazardous wastes (Ntagia et al. 2016). The barriers in removing and recovering of metals from waste streams may be overcome by using BES.

16.2 Principles of Bioelectrochemical Systems (BESs)

BES reactors generally consist of an anode, a cathode and a separator. Biodegradable materials are oxidized in the anode chamber and the electrons generated as a result of oxidation are transferred to the cathode. In the cathode compartment these

---

A. Evren Tugtaş • B. Çalli (✉)
Department of Environmental Engineering, Marmara University, 34722 Kadikoy, Istanbul, Turkey
e-mail: baris.calli@marmara.edu.tr

D. Das (ed.), *Microbial Fuel Cell*,
https://doi.org/10.1007/978-3-319-66793-5_16

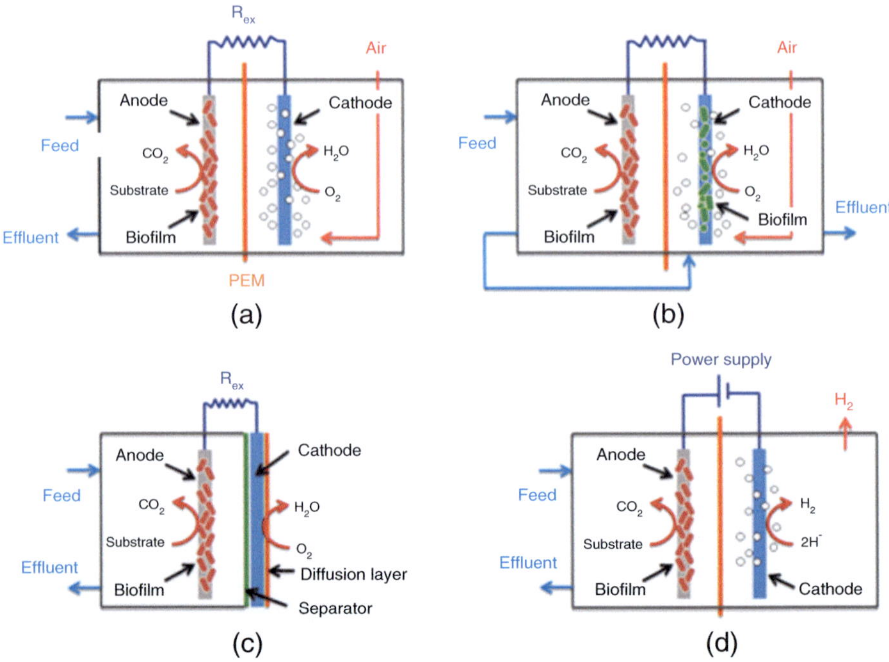

Fig. 16.1 Different BES configurations: (**a**) Two-chamber MFC; (**b**) Biocathode MFC; (**c**) Single chamber air cathode MFC; and (**d**) Two-chamber MEC for hydrogen production

electrons can be used for the direct electricity generation, reduction of chemicals or organic matter (Logan et al. 2006; Lovley 2006; Nancharaiah et al. 2016; Rabaey and Verstraete 2005; Cavdar et al. 2011; Tugtas et al. 2011, 2013). These systems have been studied intensively in terms of configuration, electrochemistry, microbiology and potential application areas and as a result microbial fuel cells (MFCs), microbial electrolysis cells (MECs) and microbial desalination cells (MDCs) have been identified as the representative of BES systems (Kelly and He 2014). In MFCs, current produced by BESs are directly harvested and used as electricity (Arends and Verstraete 2012). In MFCs, electrochemically active bacteria gain electrons by oxidizing the organic matter and transfer electrons to the surface of the anode electrode made of an inert and conductive material (e.g. carbon). The electrons are then transferred through an external resistor to the other electrode used as cathode (Fig. 16.1a–c). As a result, chemical energy in the organic matter is converted to electricity. On the surface of the cathode electrode, an oxidizing agent is used as electron acceptor, the electrons, and the protons that pass through a proton exchange membrane from anode to cathode compartment combine and form water molecules; thus the circuit is completed (He and Angenent 2006; Logan et al. 2006).

MECs (Fig. 16.1d) are the extension of MFCs; in addition to biologically produced current, some extra energy can be supplied to enhance the chemical reactions in the cathode (Logan et al. 2006). If no external electron acceptor is supplied to the cathode chamber, electrons that arrive at the cathode can combine with the protons to generate value added products such as hydrogen, ethanol, methane, hydrogen peroxide etc. (Rozendal et al. 2008). In MECs, the electron transfer from anode to cathode does not occur spontaneously and thus an extra external energy is needed in addition to that generated by microorganisms, to drive the process (Rozendal et al. 2008).

BESs are generally configured as conventional two-chamber MFCs, bio-cathode MFCs, single chamber air cathode MFCs and two-chamber MECs (Fig. 16.1). The MFCs are operated in batch, semi-continuous, or continuous mode as other bio-reactors. The electrochemically active bacteria grow on the bio-cathode makes it a cheap and efficient alternative to chemical cathodes (Fig. 16.1b). In single chamber air cathode MFCs, cathode chamber is removed and cathode electrode is exposed directly to the air (Fig. 16.1c). Instead of producing electricity, MFCs in the form of MECs may also be used to produce hydrogen (H_2) and some valuable chemicals from soluble organic compounds. In MECs the potential generated at the anode is augmented with an additional voltage to generate H_2 at the cathode (Liu et al. 2005) (Fig. 16.1d). Figure 16.1 shows the schematics of these four BES configurations.

16.3 Metals in the Environment

Metals are ubiquitous in the biosphere and essential for the industry and infrastructure; however, industrial revolution resulted in redistribution of metals causing accumulation in terrestrial and aquatic environments (Gadd 2010). Metals have been categorized as light, toxic, heavy, semi-metal (metalloids) and trace depending on several chemical and physical properties (Sparks 2005). Thirteen metals and metalloids (Ag, As, Be, Cd, Cr, Cu, Hg, Ni, Pb, Sb, Se, Tl and Zn), which can be derived from both natural and anthropogenic sources, are considered as priority pollutants (Sparks 2005). Hydrological processes are the main mechanism of metal transport and deposition in natural environments (Foster and Charlesworth 1996). Atmospheric deposition of metals originating from natural or anthropogenic sources is the major mechanism for metal input to soils and plants (Sparks 2005). In terrestrial systems, soils are the sink for metals, whereas in aquatic systems, sediments are the receiver of the metals (Sparks 2005). Natural waters are being increasingly polluted with metals worldwide, which possess adverse effects on biota and human health (Gadd 2010; Nancharaiah et al. 2016). Metals are non-biodegradable, most of them are toxic or carcinogenic and tend to accumulate in fats and tissues of living organisms (Fu and Wang 2011). Therefore, in order to decrease metal pollution, strict regulations along with appropriate treatment technologies are required worldwide.

Traditional metal removal technologies include: chemical precipitation, ion exchange, adsorption, membrane filtration, coagulation and flocculation, flotation and electrochemical treatment (Fu and Wang 2011; Wang and Ren 2014).

In addition to traditional metal removal methods, bio-electrochemical systems have been utilized for metal removal and recovery (Lu et al. 2015; Wang and Ren 2014). In the following sections, the recent developments and progress in metal removal and recovery by using BES technology are discussed.

16.4 Bio-electrochemical Metal Removal and Recovery

16.4.1 Arsenic

Arsenic is a ubiquitous element ranking 20th in abundance in earth's crust and it naturally occurs in over 200 different mineral forms (Mandal and Suzuki 2002). Arsenic is a silver-grey brittle crystalline with atomic weight of 74.9 g mol^{-1} (Mohan and Pittman 2007). It is in the form of arsenious acid, arsenic acid, arsenites, methylarsenic acid, dimethylarsinic acid, arsenates, and arsine in the environment (Mohan and Pittman 2007). Arsenic can be found in most rocks at concentrations of 0.5–2.5 mg kg^{-1} and can be concentrated in some reducing marine sediments up to 3000 mg kg^{-1} (Mandal and Suzuki 2002). Arsenic in natural waters is a worldwide problem and many countries are suffering from contamination of groundwater resources with arsenic (Mohan and Pittman 2007). Adverse health effects of arsenic can be listed as pulmonary, cardiovascular, gastrointestinal, hematological, hepatic, renal, neurological, developmental, reproductive, immunologic, genotoxic, mutagenic and carcinogenic (Mandal and Suzuki 2002). Drinking water standard for arsenic has been adopted as 10 ppb (0.01 mg L^{-1}) by WHO and US-EPA (Mohan and Pittman 2007). Physical and chemical treatment technologies have been used to remove arsenic from water. In addition, BES has been employed to remove arsenic.

Arsenite removal was successfully achieved using MFC and zero valent iron (MFC-ZVI) hybrid system with removal efficiencies greater than 96% within 2 h (Table 16.1) (Xue et al. 2013). As a result of low voltage produced by MFC, more oxidants were driven from H_2O_2 leading effective oxidation of arsenite to arsenate which further improved the removal (Xue et al. 2013). In another study, where single stage MFC was used with carbon fibre felt anode and Pt enriched carbon paper cathode, arsenite was completely converted to arsenate within 7 days of operation (Table 16.1) (Li et al. 2016b).

16.4.2 Cadmium (Cd)

Cadmium contamination in soils and crops is mainly due to application of cadmium containing fertilizers and sewage sludge to land, mining activities and atmospheric deposition of airborne cadmium (Jarup and Akesson 2009). Cadmium is an

Table 16.1 Removal and recovery of arsenic (As) via BES

BES type	Electrode materials (A: anode, C: cathode)	Membrane	Anode inoculum	Electron donor	Electron donor removal efficiency	Metal salt used	Metal removal efficiency	Maximum power density	References
MFC-ZVI hybrid system	A: High purity iron C: Carbon felt	None	Anaerobic sludge from a WWTP	Glucose, 0.8 g L^{-1}	–	As_2O_3, 300 μg L^{-1} As(III)	> 96% (pH 8, experiment duration: 2 h)	477 mW m^{-2}	Xue et al. (2013)
Air cathode MFC	A: Carbon fibre felt C: Plain carbon paper (0.5 mg cm^{-2} Pt)	None	Anaerobic sludge	Glucose, 0.75 g L^{-1}	84% TOC removal	$NaAsO_2$, 200 μg L^{-1} As(III)	100% converted to As(V), experiment duration: 7 days	752.6 ± 17 mW m^{-2}	Li et al. (2016b)

extremely toxic element known to cause kidney damage, bone disease, cancer and increased mortality. Physical and chemical treatment technologies such as precipitation, coagulation and flocculation, ion exchange and electrochemical technologies have been used to remove cadmium from water. BES has been successfully applied to remove cadmium from water.

There are several studies in the literature investigating removal and recovery of cadmium by bioelectrochemical systems (Table 16.2). Choi et al. (2014) investigated a two chambered MEC with carbon brush anode and carbon cloth cathode electrodes where chromium was oxidized in the anode chamber and cadmium was reduced in the cathode chamber. Cadmium removal greater than 89% was observed with 2 W m^{-2} power applied. In addition, 22.5 W m^{-2} power was generated via Cr and Cd removing two chambered MEC configuration, which was approximately 11 times higher than the supplied energy (Table 16.2) (Choi et al. 2014). In another study, Cd removal was assessed using graphite fibre brush as anode and stainless steel mesh as cathode electrode and 50–67% Cd(II) removal was observed within 24 h (Table 16.2) (Colantonio and Kim 2016). In another study, 0.5 V voltage in a MEC using graphite felt was used as an anode electrode and titanium sheet (TSh), nickel foam (NF) or carbon cloth (CC) was used as cathode materials where cadmium reduction of 41.9%, 39% and 46.6% were observed, respectively (Wang et al. 2016a).

16.4.3 Chromium (Cr)

Chromium (VI) is a priority toxic chemical and it is used in industries such as electroplating, lumber and wood processing, leather tanning, dye/pigment manufacturing, cement, fungicide production (Huang et al. 2010; Tandukar et al. 2009). Chromium (VI) is a known mutagen, carcinogen and teratogen and its accumulation in the environment causes serious threat to human and other living organisms (Tandukar et al. 2009). Chromium (VI) also exerts stress on soil microbiota and enters the food chain; therefore, contaminated soils need to be remediated (Wang et al. 2016a). Chromium (VI) is highly water soluble and mobile; therefore, it needs to be converted to less toxic and less soluble Cr (III) (Wang et al. 2016b).

Removal or recovery of chromium using MFCs has been extensively studied in the literature (Table 16.3). In order to remove chromium mainly two chambered MFCs were utilized. Plain carbon felt, graphite plate, graphite brush, graphite fibre and stainless steel block were generally used as the anode electrodes, whereas plain graphite, graphite plate, graphite granules, or graphite blocks were used as the cathode materials (Table 16.3). Power generation up to 6.4 W m^{-3} was obtained and the highest hexavalent chromium efficiency of 100% was reached in the literature studies via MFCs (Huang et al. 2011; Li et al. 2008a; Tandukar et al. 2009; Wang et al. 2016b). Literature studies revealed that the complete hexavalent chromium reduction to Cr(III) can be achieved via two chambered MFCs and the

Table 16.2 Removal and recovery of cadmium (Cd) via BES

BES type	Electrode materials (A: anode, C: cathode)	Membrane	Anode inoculum	Cathode inoculum	Electron donor	Metal salt at cathode	Metal removal efficiency	Voltage/ power generated or applied	References
tMEC	A: Carbon brush C: Carbon cloth	1st cell: Cation exchange membrane 2nd cell: Anion exchange membrane	Anaerobic sludge	Abiotic	1 g L^{-1} sodium acetate + Cr (VI) reducing MFC	$CdSO_4$ (50–200 ppm Cd(II)	>89%	Generated: 22.5 W m^{-2} applied: 2 W m^{-2}	Choi et al. (2014)
Single chamber MEC	A: Graphite fibre brush C: Stainless steel mesh	None	An MEC used for WAS digestion	–	0.5 mg L^{-1} sodium acetate	$CdCl_2$, 12.26 mg L^{-1} Cd(II) in anode	50–67% Cd(II), 24 h	0.4–1.0 V cathode potential of −1.0 V vs. Ag/AgCl.	Colantonio and Kim (2016)
tMEC	A: Graphite felt C: Titanium sheet (TSh) nickel foam (NF) carbon cloth (CC)	CEM	Suspended bacteria from acetate-fed MFCs reactor	Abiotic	Sodium acetate, 1 g L^{-1}	Cd(II), 50 mg L^{-1}	TSh: 5.27 mg L^{-1} h^{-1} (41.9%) NF: 4.96 mg L^{-1} h^{-1} (39.4%) CC: 5.86 mg L^{-1} h^{-1} (46.6%)	0.5 V	Wang et al. (2016a)

Table 16.3 Removal and recovery of chromium (Cr) via BES

BES type	Electrode materials (A: Anode, C: Cathode)	Membrane	Anode inoculum	Cathode inoculum	Electron donor removal efficiency	Electron donor	Metal salt at cathode	Metal removal efficiency	Voltage/ power generated or applied	References
tMFC	A: Plain carbon felt C: Plain graphite	Proton exchange membrane	Anaerobic digester of a WWTP	Abiotic	–	1 g L^{-1} sodium acetate	$K_2Cr_2O_7$; 204 ppm Cr^{6+}	99.5% of Cr^{6+} and 66.2% of total chromium (pH 2, 35 °C, 25 h)	1600 mW/m^2 (204 ppm)	Li et al. (2008a)
tMFC	A: Graphite plate C: Graphite plate	Proton exchange membrane	–	Denitrifiying and anaerobic mixed cultures	–	Sodium acetate	$K_2Cr_2O_7$; 63 mg L^{-1} Cr^{6+}	100% Cr^{6+} (pH 7.2–7.6, 22–24 °C, 5 h)	55.5 mW m^{-2} (63 mg L^{-1})	Tandukar et al. (2009)
tMFC	A: Graphite plate C: Graphite plate submerged in granular graphite	Proton exchange membrane	–	Abiotic	–	1.6 g L^{-1} acetate	$K_2Cr_2O_7$; 39.2 mg L^{-1} Cr^{6+}	2.4 ± 0.2 mg g^{-1} VSS h^{-1}	2.4 ± 0.1 W m^{-3}	Huang et al. (2010)
tMFC	Porous carbon felts as anode and cathode	Cation exchange membrane	MFC effluent	Soil	–	Sodium acetate	Cr^{6+} contaminated soil	50% to 99.1%	–	Wang et al. (2016b)
Tubular tMFC	A: Graphite granules C: Graphite brush		MFC effluent	Primary clarifier effluent		1 g L^{-1} acetate	$K_2Cr_2O_7$; 20 mg L^{-1} of Cr(VI)	>98%	6.4 W m^{-3}	Huang et al. (2011)

produced Cr(III) can be removed via settling and removal of biomass from the cathode. However, the major issue of the process is the separation of adsorbed and non-adsorbed Cr (III) from the biomass which limits the practical application (Tandukar et al. 2009).

16.4.4 Cobalt (Co)

Lithium cobalt oxide ($LiCoO_2$) is extensively utilized in lithium-ion batteries (Huang et al. 2013, 2014a). Annual lithium-ion battery waste is estimated as 200–500 million tons containing 5–15% Co by weight and 2–7% Li by weight (Xu et al. 2008). Recovery of cobalt is highly desirable as it is a rare, precious and expensive metal.

Traditional methods such as hydrometallurgical, and bioleaching can be applied to recover cobalt; however, considering the economic and environmental disadvantages of the traditional methods, MFCs became an interest in cobalt recovery (Huang et al. 2014a; Jiang et al. 2014; Wang et al. 2015a). In cases where MEC systems were used to recover cobalt, voltage ranging from 0.2 V to 0.7 V was applied and greater than 75% recovery was achieved when graphite was used as an anode material and carbon, graphite, titanium sheet, nickel foam, woven mesh or carbon cloth were used as a cathode material (Table 16.4) (Huang et al. 2014a, b; Jiang et al. 2014; Wang et al. 2015b). Cobalt recovery percentages greater than 66% was achieved via MFC systems resulted in power generation ranging from 258 to 1500 mW m^{-3} (Table 16.4) (Huang et al. 2013, 2015).

16.4.5 Copper (Cu)

Copper is a crucial element for microorganisms, plants and animals at very low concentrations. It has a vital role in the synthesis of several enzymes. Nevertheless, it becomes potentially toxic to living organisms at elevated concentrations (Bilal et al. 2013). Copper can be found in waste streams generated from mining and smelting, printed circuit board, semiconductor and paint manufacturing, metallurgical, electroplating, wire drawing and copper polishing industries. Conventionally, it is removed by precipitation as insoluble salt or metallic copper, ion exchange and adsorption from wastewater (Nacharaiah et al. 2015). Copper removed from waste streams may be recovered and used in different applications (Ter Heijne et al. 2010).

Wu et al. (2016) reported that a two-chamber MFC which was operated in multiple batch cycles and had stainless steel woven mesh cathode fed with 50 mg L^{-1} of Cu (II) exhibited a maximum power density of 6.5 W m^{-3} (26.9 A m^{-3}) and a copper removal efficiency of 99.7 $\pm$ 0.4% (Table 16.5). Their results have shown

Table 16.4 Removal and recovery of cobalt (Co) via BES

BES type	Electrode materials (A: Anode, C: Cathode)	Membrane	Anode inoculum	Cathode inoculum	Electron donor	Electron donor removal efficiency	Metal salt at cathode	Metal removal efficiency	Voltage/ power generated or applied	References
tMEC	A: Graphite felt C: Carbon rod	Cation exchange membrane	Wastewater from a primary sedimentation tank	–	Sodium acetate, 1 g L^{-1}	0.8 g co g^{-1} COD	$CoCl_2$, 50 mg L^{-1} Co(II)	Co (II) reduction 7 mg L^{-1} h^{-1} pH 6	0.3 V from cobalt leaching MFC	Huang et al. (2014a)
tMFC	A: Graphite felt C: Graphite felt	Cation exchange membrane	Wastewater from a primary sedimentation tank	–	Sodium acetate, 0.38 g COD L^{-1}	CE, 42.7 ± 2.2% at pH 1.0	$LiCoO_2$ particles; solid/ liquid ratios of 50–1000 mg L^{-1}	99.1%, 50 mg L^{-1}, 48 h, pH 1	258 ± 9 mW m^{-3}	Huang et al. (2013)
tMEC	A: Graphite fibre C: Porous graphite felt	Cation exchange membrane	Pre-acclimated MFC	Anaerobic digester of a WWTP receiving domestic and metalworking wastewaters	Sodium acetate (12.2 mM)	0.266 ± 0.001 mol Co mol^{-1} COD; 0.113 mol CH_4 mol^{-1} COD; 0.103 ± 0.003 mol acetate mol^{-1} COD	Co (II) (0.34 mM)	88.1% of Co (II)	0.2 V	Huang et al. (2014b)
tMEC	A: Graphite brush C: Graphite felt	Cation exchange membrane	Pre-acclimated MFC	Abiotic	Acetate, 1 g L^{-1}	0.81 mol Co mol^{-1} COD	847 μM Co(II)	92.2% in 6 h	0.3–0.5 V	Jiang et al. (2014)
tMFC	A: Graphite fibre C: Porous graphite felt	Cation exchange membrane	Wastewater from a primary sedimentation tank	Aerobic and anaerobic sludge, sediment and metalworking wastewater	Acetate, 1 g L^{-1}	0.24 ± 0.00 mol Co $(OH)_2$ mol^{-1} COD	$CoCl_2$; Co(II), 0.339 mM	0.079 ± 0.001 mmol L^{-1} h^{-1}	1500 mW m^{-3}	Huang et al. (2015)

that the initial deposition of Cu on the cathode is critical for efficient and continuous Cu(II) removal and power generation in the long-run.

Ntagia et al. (2016) showed the microbial hydrogen oxidation on a non-catalyzed graphite anode coupled with cathodic copper reduction in an MFC for the first time to simultaneously recover copper and produce power. They reported a maximum power density of 0.25 W m^{-2} (0.48 A m^{-2}), but no copper removal efficiency (Table 16.5). In another study, ter Heijne et al. (2010) obtained a maximum power density of 0.43 W m^{-2} (1.7 A m^{-2}) for acetate oxidation in the anode coupled to cathodic copper reduction with the same MFC configuration used by Ntagia et al. (2016). Likewise, Tao et al. (2011a) used glucose as organic electron donor for copper electrodeposition via cathodic reduction in a two-chamber MFC. They obtained a maximum power density of 339 mW m^{-3} and achieved Cu(II) removal efficiencies above 99% at 200 mg L^{-1} initial concentrations (Table 16.5).

In another study, a two-chamber membraneless MFC was successfully used for Cu(II) removal at cathode by using 5 g L^{-1} glucose as electron donor in the anode (Tao et al. 2011b). A maximum power density of 47 mW m^{-3} and 93 $\pm$ 0.1% Cu (II) removal efficiency was obtained at 200 mg L^{-1} initial Cu(II) concentration in 144 h (Table 16.5). Elemental Cu and brownish-red colour Cu_2O crystals were observed on the cathode as an indication of Cu deposition.

16.4.6 Mercury (Hg)

Mercury is one of the rarest elements on earth. It arises naturally through weathering of rocks, volcanic eruptions and deep-sea vents or from anthropogenic activities such as burning of fossil fuels, incineration of mercury-containing wastes (batteries, fluorescent light bulbs, etc.), use of fungicides containing mercury and catalysts. In water, it exists as soluble salts of chloride, sulphide or organic acids. Mercury which has already precipitated on the ground may evaporate back to the atmosphere. It is a neurotoxin and generally enters into the body from the environment, via the consumption of shellfish and fish from mercury-contaminated water bodies as well as through exposure to emissions from power plants and incinerators burning mercury-containing fuels and manufacturing processes. Exposure to mercury may damage the brain and kidneys and particularly children are considered under risk (Berlin et al. 2007).

There are different chemical and biological methods used to reduce soluble Hg (II) to insoluble Hg(0). The insoluble Hg(0) may then be adsorbed and removed from water. A diverse group of microorganisms are capable of reducing the soluble Hg(II) to insoluble elemental mercury (Nancharaiah et al. 2015). Because of its high standard potential, Hg(II) is a potential electron acceptor to be used in the cathode of an MFC (Wang et al. 2011). Hg(II) can be precipitated with chloride at low pH values.

Table 16.5 Removal and recovery of copper (Cu) via BES

BES type	Electrode materials (A: Anode, C: Cathode)	Membrane	Anode inoculum	Cathode inoculum	Electron donor	Electron donor removal efficiency	Metal salt at cathode	Metal removal efficiency	Voltage/power output or applied	References
tMFC	A: Graphite plate C: Graphite foil pressed on titanium plate	Bipolar membrane	Mixed microbial culture from another MFC	Abiotic	Acetate, 20 mM	CE = 43%	$CuCl_2$, 1 g/L cu(II), pH 3	>99.98%	0.8 W m^{-2} (3.2 A m^{-2})	Ter Heijne et al. (2010)
tMFC	A: Graphite felt C: Graphite plate	Proton exchange membrane	Sludge from an anaerobic digester	Abiotic	Glucose, 5 g g L^{-1}	CE = 3.65%	$CuSO_4$, 200 mg L^{-1} of Cu(II)	>99%, 144 h, pH 7	26 mW m^{-3}	Tao et al. (2011a)
tMFC	A: Graphite plate C: Graphite plate	Membrane-less	Sludge from an anaerobic digester	Abiotic	Glucose, 5 g g L^{-1}	CE = 0.78%	$CuSO_4$, 200 mg L^{-1} of Cu(II)	93 ± 0.1%, 144 h	47 mW m^{-3}	Tao et al. (2011b)
tMFC	A: Graphite felt C: Stainless steel woven mesh	Cation exchange membrane	Mixed microbial culture from another MFCs	Abiotic	Acetate, 1 g L^{-1}	CE > 100% because of cathode corrosion	$CuCl_2$ solution (50 mg L^{-1} of Cu (II))	99.7 ± 0.4%	6.5 W m^{-3} (26.9 A m^{-3})	Wu et al. (2016)
tMEC	A: Graphite foil C: Graphite plate	Anion exchange membrane	Mixed microbial culture from an active MFC	Abiotic	3, 10 and 30 mL min^{-1} of H_2 gas	Na	$CuCl_2$ solution (1 g L^{-1} of Cu(II) at pH 4)	Na	0.48 A m^{-2} (0.25 W m^{-2})	Ntagia et al. (2016)

Wang et al. (2011) have operated an MFC by feeding the cathode chamber with Hg(II) as electron acceptor and investigated the bioelectrochemical removal of mercury from water. They obtained a maximum power density of 433.1 mW m^{-2} (1.44 A m^{-2}) and a cathodic Hg(II) removal efficiency of above 98% at pH 2 and concluded that lower the pH and higher the initial Hg (II) concentration result in a higher maximum power density (Table 16.6). The removed Hg(II) deposited as Hg (0) on the cathode surface and as Hg_2Cl_2 precipitate at the bottom of cathode chamber.

16.4.7 Gold (Au)

Gold is a rare, precious metal usually found in elemental form on earth in rocks, vein and alluvial deposits. After being extracted and refined, it is mainly used in jewellery and also in electronics due to its excellent corrosion resistance and high electrical conductivity (Spitzer and Bertazzoli 2004). If elemental gold is released into the environment it does not result in bioaccumulation or any other ecological problems because it is insoluble and its biodegradation is expected to be very poor. Electronic wastes and electroplating solutions release significant amounts of gold into the environment (Choi and Hu 2013). Therefore, currently there is a growing interest on recovery of precious metals from electronic wastes and electroplating waste streams. Gold is typically recovered from waste streams and leachate of scrap by chemical precipitation. However, chemical precipitation methods are quite inefficient at low concentrations (Nancharaiah et al. 2015). The other methods are electrochemical recovery, biosorption and microbial reduction (Choi and Hu 2013).

Choi and Hu (2013) have tested the MFC technology as a cost-effective alternative for recovery of gold from tetrachloroaurate which is used as electron acceptor at cathode. They recovered 99.89 of the Au(III) at an initial concentration of 200 mg L^{-1} and obtained 6.58 W m^{-2} maximum power density by using 2000 mg L^{-1} Au(III) and 12.2 mM acetate as catholyte and anolyte, respectively. The details of this study are given in Table 16.7.

16.4.8 Nickel (Ni)

Nickel is a hard and ductile metal used in nickel alloys, nickel-cadmium batteries, coins, industrial plumbing, machinery parts, stainless steel, spark plugs, electroplating and catalysts. Generally, very low concentration of nickel is found in soil, water and air as oxides and sulphides. Nickel chloride, sulphate and nitrate are the soluble and bioavailable nickel salts. The major sources of nickel emissions are released from fossil fuel combustion, mining, nickel metal refining, electroplating discharges, manufacturing facilities and sewage sludge incineration (ATSDR 1997).

Table 16.6 Removal and recovery of mercury (Hg) via BES

BES type	Electrode materials (A: Anode, C: Cathode)	Membrane	Anode inoculum	Cathode inoculum	Electron donor	Electron donor removal efficiency	Metal salt at cathode	Metal removal efficiency	Voltage/ power output or applied	References
tMFC	A: Graphite felt C: Carbon paper	Anion exchange membrane	From a waste-water treatment plant	Abiotic	0.82 g L^{-1} sodium acetate	CE = 1.6–4%	$HgCl_2$; 100 mg L^{-1} Hg(II)	98.2–99.5% at pH 2	433.1 mW m^{-2} (1.44 A m^{-2})	Wang et al. (2011)

Table 16.7 Removal and recovery of gold (Au) via BES

BES type	Electrode materials (A: Anode, C: Cathode)	Membrane	Anode inoculum	Cathode inoculum	Electron donor	Electron donor removal efficiency	Metal salt at cathode	Metal removal efficiency	Voltage/power output or applied	References
tMFC	A: Carbon brush C: Carbon cloth	Cation exchange membrane	From a waste-water treatment plant	Abiotic	1 g L^{-1} sodium acetate	CE = 57%	$AuCl_4^-$; 200 and 2000 mg L^{-1}	99,89% (200 mg L^{-1}, 25 h)	6.58 W/m^{-2} (2000 mg L^{-1})	Choi and Hu (2013)

Table 16.8 Removal and recovery of nickel (Ni) via BES

BES type	Electrode materials (A: Anode, C: Cathode)	Membrane	Anode inoculum	Cathode inoculum	Electron donor	Electron donor removal efficiency	Metal salt at cathode	Metal removal efficiency	Voltage/ power output or applied	References
tMEC	A: Carbon felt C: Stainless steel mesh	Bipolar membrane	From a mature MFC	Abiotic	Sodium acetate, 1 g L^{-1}	CE = 35 ± 3.6%- 68 ± 3.4%	$NiSO_4$ solution; 50–1000 mg L^{-1} of Ni(II)	99 ± 0.6% at 50 mg L^{-1} 33 ± 4.2% at 1000 mg L^{-1}	0.9 V	Qin et al. (2012)
tMEC	A: Carbon felt C: Carbon cloth coated with platinum catalyst	Bipolar membrane	Effluent of a mature MFC	Abiotic	Sodium acetate, 1 g L^{-1}	Na	$NiSO_4$ solution; 500 mg L^{-1} of Ni (II); pH = 3	96.9 ± 3.1% at 500 mg L^{-1}	0.7 V	Luo et al. (2015)
tMEC	A: Carbon fibre brush C: Copper sheet	Bipolar membrane	Anaerobic granular sludge supernatant	Abiotic	Acetate, 0.5 g L^{-1}	CE > 90%	$NiSO_4$ solution; 1 g L^{-1} of Ni (II);	40.7%; pH 3	0.5 V	Cai et al. (2016)

Although it is an essential nutrient for some microorganisms and plants, exposure of high concentration of nickel may cause a variety of adverse effects on living organisms (Klein and Costa 2007). Conventionally, nickel is removed from waste streams with coagulation/flocculation, chemical precipitation, adsorption and ion exchange. There are also very recent studies about its bioelectrochemical removal and recovery (Cai et al. 2016).

Qin et al. (2012) investigated the removal of Ni(II) using a two-chamber MEC and compared the Ni(II) removal performance to the results obtained with an identical electrolysis cell (EC) and an MFC. They reported that with 0.9 V applied voltage the Ni(II) removal efficiency of MEC was 99 $\pm$ 0.6% at 50 mg L^{-1} initial concentration and was almost three times higher than those obtained with EC and MFC. When the initial concentration increased to 1000 mg L^{-1}, Ni(II) removal efficiency of MEC decreased to 33 $\pm$ 4.2%, while the amount of Ni(II) removed increased consistently with the initial concentration (Table 16.8). In another study, Luo et al. (2015) studied the selective recovery of Cu(II) and Ni(II) ions by using a fed-batch MEC following an MFC. Almost all of the Cu(II) ions and a small fraction of Ni(II) were removed at the cathode of MFC while the rest of the Ni (II) was recovered on the cathode of the subsequent MEC with an applied voltage of 0.7 V (Table 16.8). The very small amount of Ni(II) removed at the cathode of MFC was certainly due to adsorption, because the cathode potential of MFC was not adequate for the reduction of Ni(II). Likewise, Cai et al. (2016) studied the cathodic Ni(II) reduction at initial pH 3 in MECs by testing different cathode materials. The copper sheet cathode achieved the highest Ni(II) removal efficiency of 40.7% with 0.5 V applied voltage and 0.61 kWh kg^{-1} energy consumption (Table 16.8). Ni (II) was mainly removed with electrochemical reduction on the copper sheet cathode while a lesser amount with chemical precipitation.

16.4.9 Selenium (Se)

On the earth, selenium is found in water bodies, soils, rocks and the atmosphere. Through a biogeochemical cycle, it is released from sources rich in selenium such as organic-rich black shales, phosphatic rocks and coals. Pollution of the environment with selenium may occur naturally from weathering of seleniferous rocks and soils. Anthropogenic selenium contamination is caused by mining, agricultural drainage, combustion of Se containing coal, refinement of metals and manufacturing of electronics, stainless steel, pigments, semi-conductors, glass, photoelectric cells and pesticides (Nancharaiah and Lens 2015b). Release of significant amounts of selenium from agricultural and industrial activities into the environment may lead to accumulation of Se in aquatic organisms and cause acute and chronic toxicities (Catal et al. 2009).

Selenium (Se) is a critical element because of its intensive use in high-tech electronics and being an essential trace element for living organisms. However, above a certain concentration it becomes a potential toxicant. In order to remove and recover selenium from polluted water bodies, bacterial biomineralization based applications is a promising alternative to conventional precipitation with concurrent adsorption (Nancharaiah and Lens 2015a). Having various oxidation states (−II, 0, +VI and +VI) in chemically different inorganic and organic and physically solid, liquid and gas forms make the biogeochemical selenium cycle quite complex. Generally, the oxyanions of selenium—selenite and selenate—are soluble and bioavailable under aerobic conditions. However, elemental selenium and metal selenides have limited solubilities and thus are immobilized under normal conditions (Nancharaiah and Lens 2015b). Selenite is more toxic than selenate. Therefore, bioaccumulation is more serious in water bodies contaminated with selenite (Catal et al. 2009). Stringent standards are set out to regulate the discharge of selenium in order to minimize its bioaccumulation and enrichment in food chain. For instance, the US EPA recommends a discharge limit of 5 μg L^{-1} for Se (Nancharaiah et al. 2015).

As an innovative approach, Catal et al. (2009) used air cathode MFC and evaluated the simultaneous electricity generation and selenite removal using carbon sources like acetate and glucose. They obtained removal efficiency above 88% at initial Se(IV) concentrations of 5–75 mg L^{-1} in acetate fed MFCs. When instead of acetate, glucose was used as substrate, removal efficiencies above 99% were achieved up to 200 mg L^{-1} of Se(IV). The maximum power output was 2.9 W m^{-2} and attained at 25 mg L^{-1} Se(IV) with acetate as substrate (Table 16.9). The bright red deposit observed as a result of elemental Se generation through the reduction of selenite showed that the mixed bacterial culture enriched for power generation in air cathode MFC can use selenite as electron acceptor as well as oxygen.

16.4.10 Silver (Ag)

Silver is a rare metal and occurs naturally as mineral deposits together with other metals. Anthropogenic silver emissions result from manufacture and disposal of some electrical and photographic supplies, coal combustion, smelting operations and cloud seeding. The biogeochemical silver cycling consists of natural and anthropogenic releases to the soil, water and atmosphere, wet and dry deposition, long-distance transportation of fine particulate matters in the atmosphere and sorption to sediments and soils. In nature, silver mainly occurs in the form of sulphide or exists together with sulphides of lead, copper, iron and gold, which are basically insoluble (Howe and Dobson 2002). The elemental silver is not soluble in water and within the silver salts silver nitrate ($AgNO_3$) is the only soluble one. Silver nitrate is widely used in photography, electroplating, ceramics and ink manufacturing industry. Silver ions in an aqueous solution can be recovered by ion exchange, extraction, biosorption and electrochemical reduction (Lim et al. 2015).

Table 16.9 Removal and recovery of selenium (Se) via BES

BES type	Electrode materials (A: Anode, C: Cathode)	Membrane	Anode inoculum	Electron donor	Electron donor removal efficiency	Metal salt at cathode	Metal removal efficiency	Voltage/power output or applied	References
Air cathode MFC	A: Carbon cloth C: Wet proof carbon cloth	Membrane-less	Mixed bacterial culture from a wastewater treatment plant	Sodium acetate, 2 g L^{-1} or glucose 1.2 g L^{-1}	CE = 19% with acetate at 50 mg L^{-1} of Se(IV); CE = 38% with glucose at 150 mg L^{-1} of Se(IV)	Sodium selenite; 0–400 mg L^{-1} Se (IV)	99%; 48 h; pH = 7 (acetate fed MFC)	2.9 W m^{-2} at 25 mg L^{-1} Se (IV)	Catal et al. (2009)

In the environment, high concentrations of silver may be determined close to the sewage discharge points, electroplating plants, areas seeded with silver iodide and mining sites. Generally, silver ions are very toxic to microorganisms. However, severe silver inhibitions are not experienced on microbial activity in sewage treatment plants. Because, since silver forms complexes rapidly with and adsorbed by other compounds, its bioavailability is reduced significantly (Howe and Dobson 2002). On the other hand, dissolved silver may easily bio-accumulate in living organisms.

Lim et al. (2015) studied the recovery of silver from a synthetic wastewater by using two-chamber MFC as an efficient and cost-effective technology. They achieved about 98% removal efficiency after 10 h at 1000 mg L^{-1} initial Ag (I) concentrations. The maximum power density attained was 1.93 W m^{-2} (4.25 A m^{-2}) at 2000 mg L^{-1} (Table 16.10). The results of SEM and EDS analyses verified that the shiny metal particles on the surface of graphite felt cathode electrode were metallic silver (Lim et al. 2015). In another study, Wang et al. (2013) investigated the recovery of silver from ammonia chelated silver alkaline wastewater coupled with power production in a two-chamber MFC. They achieved 99.9% Ag(I) recovery and 83% COD removal efficiencies simultaneously (Table 16.10). The maximum power output of the system was 317 mW m^{-2}. The deposit on the surface of graphite cathode was analyzed with XRD and identified as elemental silver (Wang et al. 2013). Likewise, Tao et al. (2012) showed the feasibility of metallic silver recovery from a synthetic photographic wastewater and achieved above 95% Ag(I) removal and 109 mW m^{-2} power density in a dual-chamber MFC (Table 16.10). The results of XRD analyses revealed that the deposits on the graphite cathode were metallic silver with purity above 91%.

16.4.11 Vanadium (V)

Vanadium is an abundant element occurring naturally in various minerals, phosphate rocks, iron ores and crude oil and it is released to atmosphere as a result of volcano emissions and continental dusts (Zwolak 2014). Vanadium is steel grey, corrosion resistant and most common valences are +3, +4 and +5 (Barceloux 1999). Humans are generally exposed to vanadium as a result of food consumption as vanadyl or vanadate and it is absorbed from gastrointestinal tract and transferred to tissues, liver, spleen, kidneys, testicles and bones (Zwolak 2014). Vanadium is not carcinogenic and only weakly mutagenic; generally health effects include upper respiratory tract irritation (Barceloux 1999).

Vanadium behaves as an effective electron acceptor in the cathode chambers of MFCs and removal of vanadium along with power generation is possible (Table 16.11) (Hao et al. 2016; Li et al. 2016a; Zhang et al. 2015). Single and two chambered MFC studies revealed that greater than 400 mW m^{-2} has been reached with metal removal efficiencies ranging between 25 and 76% when carbon is used as the anode and cathode material (Table 16.11). In addition to vanadium

Table 16.10 Removal and recovery of silver (Ag) via BES

BES Type	Electrode materials (A: Anode, C: Cathode)	Membrane	Anode inoculum	Cathode inoculum	Electron donor	Electron donor removal efficiency	Metal salt at cathode	Metal removal efficiency	Voltage/power output or applied	References
tMFC	A: Graphite plate C: Graphite felt	Proton exchange membrane	Mixed culture from a lab BES	Abiotic	Sodium acetate, 1.28 g L^{-1}	93.6–97.8% COD removal	$AgNO_3$, 1 mM of Ag(I)	95% in 36 h, at pH 4	109 mW m^{-2}	Tao et al. (2012)
tMFC	A: Carbon cloth C: Graphite	Bipolar membrane	From a fruit waste treating anaerobic reactor	Abiotic	Sodium acetate, 1.6 g L^{-1}	83% COD removal	5 mM Ag_2SO_4	99.9% ag (I), 21 h, pH 9.2	317 mW m^{-2}	Wang et al. (2013)
tMFC	A: Graphite felt C: Graphite felt	Anion exchange membrane	From an anaerobic digester	Abiotic	Sodium acetate, 0.82 g L^{-1}	NA	$AgNO_3$, 50–4000 ppm Ag(I)	92.3–98.2%	1.93 W m^{-2} (4.25 A/m^{-2}) at 2000 mg L^{-1}	Lim et al. (2015)

Table 16.11 Removal and recovery of vanadium (V) via BES

BES type	Electrode materials (A: Anode, C: Cathode)	Membrane	Anode inoculum	Cathode inoculum	Electron donor	Electron donor removal efficiency	Metal salt at cathode	Metal removal efficiency	Voltage/ power output or applied	References
tMFC	A: Carbon fibre felt C: Carbon fibre felt	Proton exchange membrane	Anaerobic granular sludge from an UASB reactor	Abiotic	Glucose, 0.75 g L^{-1}	NA	$NaVO_3$; 75 mg L^{-1} V(V)	76.8 ± 2.9%	418.6 ± 11.3 mW m^{-2}	Zhang et al. (2015)
sMFC	A: Carbon fibre felt C: Plain carbon paper	Membrane-less	Anaerobic granular sludge from an UASB reactor	–	Glucose, 0.8 g L^{-1}	NA	$NaVO_3.2H_2O$; 75 mg L^{-1} V (V)	77.6% (12 h)	589.1 mW m^{-2}	Hao et al. (2016)
tMFC	A: Carbon fibre felt C: Carbon fibre felt	Proton exchange membrane	Anaerobic sludge	Abiotic	Glucose, 0.812 g L^{-1} and sodium sulphide, 100 mg S L^{-1}	NA	$NaVO_3.2H_2O$; 150 mg L^{-1} V (V)	NA	553.9 mW m^{-2}	Li et al. (2016a)

removal through MFC, vanadium can also be used to enhance power outputs of MFCs due to its excellent redox characteristics (Li et al. 2016a).

16.5 Conclusions

BES is considered a promising, cost-efficient and sustainable technology to remove and recover metals from wastewater and leachate. The drawbacks such as high operational cost and energy consumption, inefficiency at low concentrations and excessive hazardous waste production of conventional chemical precipitation, ion exchange, adsorption, biosorption and microbial reduction processes may be overcome with BES technology. In the literature, there are many studies demonstrating the cathodic reduction (in some cases oxidation) of the ions of arsenic, cadmium, chromium, cobalt, copper, gold, mercury, nickel, selenium, silver and vanadium in combination with the oxidation of an organic substance in the anode of a BES. Parameters affecting the efficiency of cathodic metal removal and recovery are initial metal ion concentration, pH, anode potential or applied voltage, internal resistance and conductivity of electrolytes. The biosorption and precipitation are the other mechanisms frequently contributing to cathodic metal reduction in BESs. Future studies have to focus on selective sequential recovery of metals from mixed metal solutions and real waste streams.

References

Arends, J. B. A., & Verstraete, W. (2012). 100 years of microbial electricity production: Three concepts for the future. *Microbial Biotechnology, 5*(3), 333–346.

ATSDR—Agency for Toxic Substances and Disease Registry. (1997). *Toxicological profile for nickel (update)*. Altanta: Public Health Service, U.S. Department of Health and Human Services.

Barceloux, D. G. (1999). Vanadium. *Journal of Toxicology-Clinical Toxicology, 37*(2), 265–278.

Berlin, M., Rudolfs, K. Z., & Fowler, B. A. (2007). Mercury. In G. F. Nordberg, B. A. Fowler, M. Nordberg, & L. T. Friberg (Eds.), *Handbook on toxicology of metals* (3rd ed., pp. 743–758). Amsterdam: Elsevier Science Publishers B.V.

Bilal, M., Shah, J. A., Ashfaq, T., Gardazi, S. M. H., Tahir, A. A., Pervez, A., Haroon, H., & Mahmood, Q. (2013). Waste biomass adsorbents for copper removal from industrial wastewater–A review. *Journal of Hazardous Materials, 263*, 322–333.

Cai, W.-F., Geng, D.-L., & Wang, Y.-H. (2016). Assessment of cathode materials for Ni (II) reduction in microbial electrolysis cells. *RSC Advances, 6*, 31732–31738.

Catal, T., Bermek, H., & Liu, H. (2009). Removal of selenite from wastewater using microbial fuel cells. *Biotechnology Letters, 31*, 1211–1216.

Cavdar, P., Yilmaz, E., Tugtas, A. E., & Calli, B. (2011). Acidogenic fermentation of municipal solid waste and its application to bio-electricity production via microbial fuel cells (MFCs). *Water Science and Technology, 64*(4), 789–795.
Choi, C., & Hu, N. (2013). The modeling of gold recovery from tetrachloroaurate wastewater using a microbial fuel cell. *Bioresource Technology, 133*, 589–598.
Choi, C., Hu, N., & Lim, B. (2014). Cadmium recovery by coupling double microbial fuel cells. *Bioresource Technology, 170*, 361–369.
Colantonio, N., & Kim, Y. (2016). Cadmium (II) removal mechanisms in microbial electrolysis cells. *Journal of Hazardous Materials, 311*, 134–141.
Elouear, Z., Bouzid, J., & Boujelben, N. (2009). Removal of nickel and cadmium from aqueous solutions by sewage sludge ash: Study in single and binary systems. *Environmental Technology, 30*(6), 561–570.
Foster, I. D. L., & Charlesworth, S. M. (1996). Heavy metals in the hydrological cycle: Trends and explanation. *Hydrological Processes, 10*(2), 227–261.
Fu, F., & Wang, Q. (2011). Removal of heavy metal ions from wastewaters: A review. *Journal of Environmental Management, 92*(3), 407–418.
Gadd, G. M. (2010). Metals, minerals and microbes: Geomicrobiology and bioremediation. *Microbiology-Sgm, 156*, 609–643.
Hao, L., Zhang, B., Cheng, M., & Feng, C. (2016). Effects of various organic carbon sources on simultaneous V(V) reduction and bioelectricity generation in single chamber microbial fuel cells. *Bioresource Technology, 201*, 105–110.
He, Z., & Angenent, L. T. (2006). Application of bacterial biocathodes in microbial fuel cells. *Electroanalysis, 18*(19–20), 2009–2015.
Howe, P. D., & Dobson, S. (2002). *Silver and silver compounds: Environmental aspects* (Concise international chemical assessment document 44). Stuttgart: World Health Organization.
Huang, L. P., Chen, J. W., Quan, X., & Yang, F. L. (2010). Enhancement of hexavalent chromium reduction and electricity production from a biocathode microbial fuel cell. *Bioprocess and Biosystems Engineering, 33*(8), 937–945.
Huang, L. P., Chai, X. L., Chen, G. H., & Logan, B. E. (2011). Effect of set potential on hexavalent chromium reduction and electricity generation from biocathode microbial fuel cells. *Environmental Science and Technology, 45*(11), 5025–5031.
Huang, L., Li, T., Liu, C., Quan, X., Chen, L., Wang, A., & Chen, G. (2013). Synergetic interactions improve cobalt leaching from lithium cobalt oxide in microbial fuel cells. *Bioresource Technology, 128*, 539–546.
Huang, L., Jiang, L., Wang, Q., Quan, X., Yang, J., & Chen, L. (2014a). Cobalt recovery with simultaneous methane and acetate production in biocathode microbial electrolysis cells. *Chemical Engineering Journal, 253*, 281–290.
Huang, L., Yao, B., Wu, D., & Quan, X. (2014b). Complete cobalt recovery from lithium cobalt oxide in self-driven microbial fuel cell – Microbial electrolysis cell systems. *Journal of Power Sources, 259*, 54–64.
Huang, L., Liu, Y., Yu, L., Quan, X., & Chen, G. (2015). A new clean approach for production of cobalt dihydroxide from aqueous Co(II) using oxygen-reducing biocathode microbial fuel cells. *Journal of Cleaner Production, 86*, 441–446.
Jarup, L., & Akesson, A. (2009). Current status of cadmium as an environmental health problem. *Toxicology and Applied Pharmacology, 238*(3), 201–208.
Jiang, L., Huang, L., & Sun, Y. (2014). Recovery of flakey cobalt from aqueous Co(II) with simultaneous hydrogen production in microbial electrolysis cells. *International Journal of Hydrogen Energy, 39*(2), 654–663.
Kelly, P. T., & He, Z. (2014). Nutrients removal and recovery in bioelectrochemical systems: A review. *Bioresource Technology, 153*, 351–360.

Klein, C., & Costa, M. (2007). Nickel. In G. F. Nordberg, B. A. Fowler, M. Nordberg, & L. T. Friberg (Eds.), *Handbook on toxicology of metals* (3rd ed., pp. 743–758). Amsterdam: Elsevier Science Publishers B.V.

Li, Z., Zhang, X., & Lei, L. (2008). Electricity production during the treatment of real electroplating wastewater containing Cr^{6+} using microbial fuel cell. *Process Biochemistry, 43*(12), 1352–1358.

Li, Y. L., Zhang, B. G., Cheng, M., Li, Y. L., Hao, L. T., & Guo, H. M. (2016a). Spontaneous arsenic (III) oxidation with bioelectricity generation in single-chamber microbial fuel cells. *Journal of Hazardous Materials, 306*, 8–12.

Li, J., Zhang, B., Song, Q., & Borthwick, A. G. L. (2016b). Enhanced bioelectricity generation of double-chamber air-cathode catalyst free microbial fuel cells with the addition of non-consumptive vanadium (v). *RSC Advances, 6*(39), 32940–32946.

Lim, B. S., Lu, H., Choi, C., & Liu, Z. X. (2015). Recovery of silver metal and electric power generation using a microbial fuel cell. *Desalination and Water Treatment, 54*(13), 3675–3681.

Liu, H., Grot, S., & Logan, B. E. (2005). Electrochemically assisted microbial production of hydrogen from acetate. *Environmental Science and Technology, 39*(11), 4317–4320.

Logan, B. E., Hamelers, B., Rozendal, R. A., Schrorder, U., Keller, J., Freguia, S., Aelterman, P., Verstraete, W., & Rabaey, K. (2006). Microbial fuel cells: Methodology and technology. *Environmental Science and Technology, 40*(17), 5181–5192.

Lovley, D. R. (2006). Microbial fuel cells: Novel microbial physiologies and engineering approaches. *Current Opinion in Biotechnology, 17*(3), 327–332.

Lu, Z. H., Chang, D. M., Ma, J. X., Huang, G. T., Cai, L. K., & Zhang, L. H. (2015). Behavior of metal ions in bioelectrochemical systems: A review. *Journal of Power Sources, 275*, 243–260.

Luo, H., Qin, B., Liu, G., Zhang, R., Tang, Y., & Hou, Y. (2015). Selective recovery of Cu^{2+} and Ni^{2+} from wastewater using bioelectrochemical system. *Frontiers of Environmental Science & Engineering, 9*, 522–527.

Mandal, B. K., & Suzuki, K. T. (2002). Arsenic round the world: A review. *Talanta, 58*(1), 201–235.

Mohan, D., & Pittman, C. U. (2007). Arsenic removal from water/wastewater using adsorbents – A critical review. *Journal of Hazardous Materials, 142*(1–2), 1–53.

Nancharaiah, Y. V., & Lens, P. N. L. (2015a). Selenium biomineralization for biotechnological applications. *Trends in Biotechnology, 33*(6), 323–330.

Nancharaiah, Y. V., & Lens, P. N. L. (2015b). Ecology and biotechnology of selenium-respiring bacteria. *Microbiology and Molecular Biology Reviews, 79*(1), 61–80.

Nancharaiah, Y. V., Mohan, S. V., & Lens, P. N. L. (2015). Metals removal and recovery in bioelectrochemical systems: A review. *Bioresource Technology, 195*, 102–114.

Nancharaiah, Y. V., Mohan, S. V., & Lens, P. N. L. (2016). Biological and bioelectrochemical recovery of critical and scarce metals. *Trends in Biotechnology, 34*(2), 137–155.

Ntagia, E., Rodenas, P., ter Heijne, A., Buisman, C. J. N., & Sleutels, T. H. J. A. (2016). Hydrogen as electron donor for copper removal in bioelectrochemical systems. *International Journal of Hydrogen Energy, 41*, 5758–5764.

Qin, B., Luo, H., Liu, G., Zhang, R., Chen, S., Hou, Y., & Luo, Y. (2012). Nickel ion removal from wastewater using the microbial electrolysis cell. *Bioresource Technology, 121*, 458–461.

Rabaey, K., & Verstraete, W. (2005). Microbial fuel cells: Novel biotechnology for energy generation. *Trends in Biotechnology, 23*(6), 291–298.

Rozendal, R. A., Hamelers, H. V. M., Rabaey, K., Keller, J., & Buisman, C. J. N. (2008). Towards practical implementation of bioelectrochemical wastewater treatment. *Trends in Biotechnology, 26*(8), 450–459.

Sparks, D. L. (2005). Toxic metals in the environment: The role of surfaces. *Elements, 1*(4), 193–197.

Spitzer, M., & Bertazzoli, R. (2004). Selective electrochemical recovery of gold and silver from cyanide aqueous effluents using titanium andvitreous carbon cathodes. *Hydrometallurgy, 74*, 233–242.

Tandukar, M., Huber, S. I., Onodera, T., & Pavlostathis, S. G. (2009). Biological chromium (VI) reduction in the cathode of a microbial fuel cell. *Environmental Science and Technology, 43*(21), 8159–8165.

Tao, H.-C., Li, W., Liang, M., Xu, N., Ni, J.-R., & Wu, W.-M. (2011a). A membrane-free baffled microbial fuel cell for cathodic reduction of Cu(II) with electricity generation. *Bioresource Technology, 102*(7), 4774–4778.

Tao, H.-C., Liang, M., Li, W., Zhang, L.-J., Ni, J.-R., & Wu, W.-M. (2011b). Removal of copper from aqueous solution by electrodeposition in cathode chamber of microbial fuel cell. *Journal of Hazardous Materials, 189*(1–2), 186–192.

Tao, H.-C., Gao, Z.-Y., Ding, H., Xu, N., & Wu, W.-M. (2012). Recovery of silver from silver(I)-containing solutions in bioelectrochemical reactors. *Bioresource Technology, 111*, 92–97.

Ter Heijne, A., Liu, F., van der Weijden, R., Weijma, J., Buisman, C. J. N., & Hamelers, H. V. M. (2010). Copper recovery combined with electricity production in a microbial fuel cell. *Environmental Science & Technology, 44*(11), 4376–4381.

Tugtas, A. E., Cavdar, P., & Calli, B. (2011). Continuous flow membraneless air cathode microbial fuel cell with spunbonded olefin diffusion layer. *Bioresource Technology, 102*, 10425–10430.

Tugtas, A. E., Cavdar, P., & Calli, B. (2013). Bio-electrochemical post-treatment of anaerobically treated landfill leachate. *Bioresource Technology, 128*, 266–272.

Wang, H., & Ren, Z. J. (2014). Bioelectrochemical metal recovery from wastewater: A review. *Water Research, 66*, 219–232.

Wang, Z., Lim, B., & Choi, C. (2011). Removal of Hg^{2+} as an electron acceptor coupled with power generation using a microbial fuel cell. *Bioresource Technology, 102*, 6304–6307.

Wang, Y.-H., Wang, B.-S., Pan, B., Chen, Q.-Y., & Yan, W. (2013). Electricity production from a bio-electrochemical cell for silver recovery in alkaline media. *Applied Energy, 112*, 1337–1341.

Wang, Q., Huang, L. P., Yu, H. T., Quan, X., Li, Y. X., Fan, G. F., & Li, L. (2015a). Assessment of five different cathode materials for Co(II) reduction with simultaneous hydrogen evolution in microbial electrolysis cells. *International Journal of Hydrogen Energy, 40*(1), 184–196.

Wang, H., Luo, H., Fallgren, P. H., Jin, S., & Ren, Z. J. (2015b). Bioelectrochemical system platform for sustainable environmental remediation and energy generation. *Biotechnology Advances, 33*(3–4), 317–334.

Wang, C., Deng, H., & Zhao, F. (2016a). The remediation of chromium (VI)-contaminated soils using microbial fuel cells. *Soil and Sediment Contamination, 25*(1), 1–12.

Wang, Q., Huang, L., Pan, Y., Zhou, P., Quan, X., Logan, B. E., & Chen, H. (2016b). Cooperative cathode electrode and in situ deposited copper for subsequent enhanced Cd(II) removal and hydrogen evolution in bioelectrochemical systems. *Bioresource Technology, 200*, 565–571.

Wu, D., Huang, L., Quan, X., & Puma, G. L. (2016). Electricity generation and bivalent copper reduction as a function of operation time and cathode electrode material in microbial fuel cells. *Journal of Power Sources, 307*, 705–714.

Xu, J. Q., Thomas, H. R., Francis, R. W., Lum, K. R., Wang, J. W., & Liang, B. (2008). A review of processes and technologies for the recycling of lithium-ion secondary batteries. *Journal of Power Sources, 177*(2), 512–527.

Xue, A., Shen, Z. Z., Zhao, B., & Zhao, H. Z. (2013). Arsenite removal from aqueous solution by a microbial fuel cell-zerovalent iron hybrid process. *Journal of Hazardous Materials, 261*, 621–627.

Zhang, B., Tian, C., Liu, Y., Hao, L., Liu, Y., Feng, C., Liu, Y., & Wang, Z. (2015). Simultaneous microbial and electrochemical reductions of vanadium (V) with bioelectricity generation in microbial fuel cells. *Bioresource Technology, 179*, 91–97.

Zhang, D., Wang, J., Zhao, J., Cai, Y., & Lin, Q. (2016). Comparative study of nickel removal from synthetic wastewater by a sulfate-reducing bacteria filter and a zero valent iron-sulfate-reducing bacteria filter. *Geomicrobiology Journal, 33*(3–4), 318–324.

Zwolak, I. (2014). Vanadium carcinogenic, immunotoxic and neurotoxic effects: A review of in vitro studies. *Toxicology Mechanisms and Methods, 24*(1), 1–12.

Chapter 17
Sediment Microbial Fuel Cell and Constructed Wetland Assisted with It: Challenges and Future Prospects

Md. T. Noori, M.M. Ghangrekar, and C.K. Mukherjee

17.1 Introduction

In recent years, the research work focus in energy sector has been shifted towards the renewable energy due to continuous depletion of conventional energy sources. On the other hand, exponentially increasing pollution in water reserves has stimulated phenomenal debates among researchers, pollution control agencies, and stakeholders in search of sustainable solution to remediate it. Sediment microbial fuel cell (SMFC) is one of the most promising approaches to address these two highly recognized problems together (Sajana et al. 2013b). In addition, SMFCs can offer distinctive opportunity to understand the flow of energy through electrochemically active bacteria, energy collection efficiency from natural systems, and the role of SMFCs for power generation and *in situ* bioremediation in the natural environment (Sajana et al. 2013a). SMFCs comprise two electrically conductive electrodes as anode and cathode placed 5–10 cm beneath the free surface of sediment and free water surface, respectively (Fig. 17.1a). Chemical energy associated with organic matter present in the sediment and water gets converted to electron and proton during oxidation catalyzed by microorganisms, working as biocatalyst on anode surface. Sediment permits the flow of protons from anode to cathode side serving as proton permeable natural medium. The anode collects extracellular electrons and transfer them to the cathode through an external circuit. On cathode, oxygen or other chemical oxidant (like nitrate) serve as terminal

Md.T. Noori • C.K. Mukherjee
Department of Agricultural and Food Engineering, Indian Institute of Technology Kharagpur, Kharagpur 721302, India

M.M. Ghangrekar (✉)
Department of Civil Engineering, Indian Institute of Technology Kharagpur, Kharagpur 721302, India
e-mail: ghangrekar@civil.iitkgp.ernet.in

D. Das (ed.), *Microbial Fuel Cell*,
https://doi.org/10.1007/978-3-319-66793-5_17

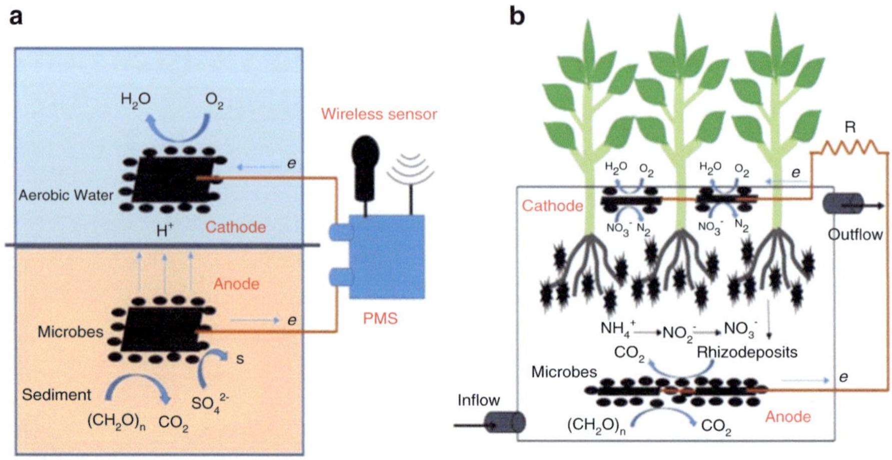

Fig. 17.1 Schematic diagram of (**a**) Sediment microbial fuel cell; and (**b**) Constructed wetland microbial fuel cell

electro acceptor (TEA), which combines with electron and proton and produce water or other reduced product (Rismani-Yazdi et al. 2008). In addition, anions and cations can be used for charge balanced in the SMFCs based on their concentration in the fluid (Kim et al. 2007). Natural phenomenon of redox charge gradient have been used for development of SMFCs. Table 17.1 shows the brief summary of half-cell equations (anodic and cathodic) which can take place on anode and cathode during bioconversion of organic matter to electricity.

In last decade, application of various types of SMFCs in different environment have been demonstrated for wastewater treatment (Fang et al. 2013), bioremediation of aquaculture sediment (Sajana et al. 2013b), and powering remote sensors (Ewing et al. 2014). All these have been shown to be of great interest of research in order to seek sustainable solution to mitigate pollution threat and power recovery. However, the lacuna of the SMFCs lie in poor power production and recovery of electrons from substrate (coulombic efficiency) due to deprived electrode kinetics. The performance of SMFCs has been remedied by various modifications in the SMFC in recent times, rendering it as an alternative for aquatic sediment bioremediation and source of bioenergy that has found its niche.

The constructed wetland (CW) and microbial fuel cell (MFC) are two different biological systems which are capable of degrading organic matter in distinct way. CWs depend upon ecological functions similar to natural wetland and are largely based on plant interactions, but it is still unknown that which plant population can enhance the treatment performance of CWs. However, some researchers manifested the relation between plant root canopy and density and functional performance of microbial population on the treatment performance of CWs. However the relationship lacks adequate scientific evidence (Hammer 1989; Reed et al. 1995). On the other hand, an MFC provides controllable option for wastewater treatment and power recovery (Tiwari et al. 2016). Kinetics of anode and cathode can be enhanced by manipulation of

Table 17.1 Anodic and cathodic half-cell reactions in MFCs and corresponding standard potentials (E^0) vs. Standard Hydrogen Electrode (SHE)

Anodic reactions	E^0 (V)	Cathodic reactions	E^0 (V)
$C_6H_{12}O_6 + 6\ H_2O \rightarrow 6CO_2 + 24H^+ + 24e^-$	−0.43	$O_2 + 4H^+ + 4e^- \rightarrow 2H_2O$	1.23
$H_2 \rightarrow 2H^+ + 2e$	−0.4	$MnO_2(s) + 4H^+ + 2e \rightarrow Mn^{2+} + 2H_2O$	1.23
$CH_3COO^- + 2H_2O \rightarrow 2CO_2 + 7H^+ + 8e^-$	−0.28	$MnO_4^- + 8H^+ + 5e \rightarrow Mn^{2+} + 4H_2O$	1.5
$H_2S \rightarrow S^o + 2H^+ + 2e$	−0.28	$Fe(CN)_6^{3-} + e \rightarrow Fe(CN)_6^{4-}$	0.361
$H_2S + 4H_2O \rightarrow SO_4^{2-} + 6H^+ + 8e$	−0.22	$Fe^{3+} + e \rightarrow Fe^{2+}$	0.77
$CH_4 + 2H_2O \rightarrow CO_2 + 8H^+ + 8e$	−0.24	Fumarate + H^+ + e→Succinate	0.03
$NADH \rightarrow NAD^+ + H^+ + 2e$	−0.32	$2NO_3^- + 12H^+ + 10e \rightarrow N_2 + 6H_2O$	0.74
		$NO_3^- + 2H^+ + 2e \rightarrow NO_2^- + H_2O$	0.433

Source: He and Angenent (2006)

microbial consortia using physical and chemical treatment (Rajesh et al. 2014; Tiwari and Ghangrekar 2015); whereas using oxygen reduction catalyst the cathode kinetics can be enhanced (Noori et al. 2016a). The similarity of substrate degradation characteristics of both biological system using microorganisms led to the concept of combination of CWs and MFCs, which resulted in a most promising approach of CW-MFC for wastewater treatment and renewable energy tapping. The very first report on performance of CW-MFC was documented by Yadav et al. (2012) which was used to treat synthetic wastewater containing azo dye. CW-MFCs or SMFCs possess narrow difference in terms of system architecture. The difference of these two systems could be pointed out based on their feed uptake mechanism by electricigens. SMFCs allow electricigens to take feed from the rhizodeposits, exudates and secondary metabolites of aquatic animals, whereas rhizodeposits and wastewater serves as substrate medium for electricigens in CW-MFC (Strik et al. 2008). It is reported that the root of living plants can increase the substrate to the electricigens, which may result in as much as 18-times higher power as compared to the fresh water SMFCs (Timmers et al. 2012).

This chapter addresses the SMFC and CW-MFC to develop better understanding of the parameters influencing performance of these. A brief summary of previous research has also been included which deals with the application of SMFCs as power source for operating wire-less sensors. With limited researches performed in the past, the aspects of CW-MFCs are discussed for summarizing the current trends, application potential and future research needs to improve the performance.

17.2 Fundamentals of SMFCs and CW-MFCs

SMFCs have added advantage for field application because they require less attention for operation and maintenance, can power remote sensors (Gong et al. 2011) and also can provide *in situ* remediation of aquaculture ponds for maintaining healthy aquatic environment (Sajana et al. 2013a). A SMFC can be easily fabricated and installed by inserting an anode in sediment up to a depth of 5–10 cm from sediment-water interface in which degradation of organic matter and collection of electrons occurs. A cathode should be placed just below the air-water interface, on which reduction of TEA (mostly O_2) occurs by combining electrons and protons. The anode and cathode should be connected through corrosion resistant conductive materials (such as copper and aluminum) across an external load as shown in Fig. 17.1a.

There are three possible mechanisms reported for electron transfer from microorganisms to anode: (1) Direct contact of *c*-type cytochromes; (2) Nanowires (conductive pili); and (3) Redox mediators or electron shuttle (Sajana et al. 2013a). The losses during oxidation of organic matter to generate electron and proton and subsequent reduction in cathode are considered as bottlenecks of these systems. These losses can be listed as thermodynamic loss, activation loss, ohmic loss and concentration loss and can be seen during polarization study. Many studies in past few years have successfully identified the parameters affecting the performance of SMFCs and CW-MFCs. These parameters include the electrode material (Dumas et al. 2007), distance between the electrodes and pH (Sajana et al. 2013b), temperature (Liu et al. 2005), dissolved oxygen (DO) near the cathode (Saravanan et al. 2010), organic matter in the sediment (Sajana et al. 2014) etc. In successive sections, brief description on the parameters affecting the performance of SMFCs and CW-MFCs is presented.

Though the concept of CW-MFC is new for wastewater treatment and simultaneous recovery of bio-electricity, these two distinctive systems, CWs and MFCs, have been explored widely for wastewater treatment. CW-MFCs are subclass of SMFCs and between them the feeding mechanism to the electricigens are possibly the main distinction. CWs possess anaerobic and aerobic strata throughout their soil depth and the water column (Yadav et al. 2012); hence a CW-MFC can be developed by embedding an anode in the deep layer of soil and a cathode on the water column (soil surface) or in rhizosphere (Fig. 17.1b). The incorporation of plants in SMFC creates system similar to the CW-MFC. Therefore, in some literatures both the technologies were placed in the same category (Xu et al. 2015).

17.3 Factors Affecting the Performance of SMFCs and CW-MFCs

17.3.1 Electrode Materials

Power generation in SMFCs is truly based on the characteristics of electrode materials and it was found limited by kinetics of anode and cathode. As for example, an anode should possess biocompatibility for bacterial cell adhesion, it should be highly conductive and super hydrophilic (Wu et al. 2015); whereas a better cathode material with excellent catalytic activity and high electronic conductivity can enhance oxygen reduction reaction (ORR) (Noori et al. 2016b). Numerous kinds of electrode materials such as stainless steel wire mesh (Song et al. 2011), graphite plates (Mohan et al. 2009) and carbon cloth (Des Jarlais et al. 2013) have been investigated in SMFCs. The electrode material used should be corrosion resistant, since it has to survive in highly exhaustive environment. Due to chances of high corrosion, the SS wire mesh has limited application in marine environment over carbon based electrode materials. Graphite granules with graphite rod as anode demonstrated ever highest power density of 380 mW m^{-2} till date (Nielsen et al. 2007). Song et al. (2012) reported power density of 75 mW m^{-2} using activated carbon felt as anode in fresh water SMFC.

Like SMFCs, the electrode materials affect the performance of CW-MFC. Due to high corrosion potential of iron-based electrodes in water logged medium (soil and sediments) they are susceptible to corrosion and cannot be used for prolonged period. Therefore, carbon based electrodes are always preferred for such applications since they can offer long term sustainability, high electrical conductivity and non-oxidative in nature and moreover they can facilitate large surface area for microbial attachment for biomass growth. Dordio and Carvalho (2013) reported enhanced COD removal by biosorption process from CWs using carbon granules. Granular activated carbon (GAC) as biocathode material for CW-MFC was found to be the most suitable with a power density of 55.05 mW m^{-2} as compared to the other tested materials, for instance carbon cloth (28.9 mW m^{-2}) and stainless steel (1.76 mW m^{-2}) (Liu et al. 2014). The enhanced performance using GAC in CW-MFC was attributed to its higher surface area to support ORR and the rational utilization of capillary action. Furthermore, the size of carbon granules was also observed to have influence on the performance of SMFCs, with smaller size of graphite granules between 0.25 mm and 0.5 mm, the current density was found to be 77.7 mA m^{-2} as compared to the lower value of 37.9 mA m^{-2} with large granule size between 1 and 5 mm (Arends et al. 2012).

In spite of electrode materials, the shape and architecture also have depicted profound effect on the performance of SMFCs. A better geometry of electrode with high surface area can facilitate better substrate diffusion resulting in high redox kinetics, thereby, enhancing the power recovery from SMFCs. Various shape of carbon-based electrode materials have been tested to evaluate its effect on the performance of SMFCs. Graphite rod anode in SMFC containing acetate enriched sediment recovered a power density of 19.57 mW m^{-2} as compared to the lower value of 8.72 mW m^{-2} with graphite disk anode (Sacco et al. 2012). Li et al. (2009) demonstrated that the SMFC with solid column graphite anode could be a better anode as compared to the graphite disk anode material due to enhanced surface area. The power density obtained from SMFC using graphite column anode was found to be 20.2 mW m^{-2}, which was 1.35-times higher than the SMFC using graphite disk anode (14.9 mW m^{-2}). Higher power density with graphite column anode can be attributed to the less diffusion hindrance of substrate due to large surface area.

17.3.2 Electrode Spacing and External Resistance

Distance between the electrodes can regulate internal resistance of SMFCs by regulating ohmic overpotential loss. Loss in potential energy experienced by MFCs during movement of proton and electron via natural voltage gradient between anode and cathode causes ohmic overpotential loss (Singh et al. 2010). The ohmic overpotential losses are proportional to the current and behaves linearly as current increases (Rismani-Yazdi et al. 2008). It can be calculated by measuring the gradient of linear portion of voltage vs. current curve. A reduced spacing in electrodes in MFCs can decrease ohmic overpotential loss by reducing proton transfer energy from anode to cathode (Krishnaraj and Jong Sung 2015). Current density obtained from the SMFC was observed to be a function of electrode distance; as the distance between anode and cathode was increased from 12 cm to 100 cm, the current density decreased from 11.5 A m^{-2} to 2.11 A m^{-2} (Hong et al. 2009). Sajana et al. (2013b) reported similarly on the reduction of power density of 3.1 mW m^{-2} with an electrode spacing of 100 cm as compared to the electrode spacing of 50 cm (4.29 mW m^{-2}). However, the chemical oxygen demand (COD) and total nitrogen (TN) removal efficiency was noticed higher in SMFC with 100 cm electrode spacing as compared to the 50 cm.

According to the Ohm's law, $I = V/R_{ex}$ (where I is current generation, V is voltage and R_{ex} is external resistance), the sustainable I from an MFC and SMFC is a function of R_{ex}. The effect of R_{ex} on the performance of MFC has been demonstrated in earlier studies (Del Campo et al. 2014; Pinto et al. 2011). Increase in R_{ex} from 100 Ω to 1000 Ω drove positive effect on voltage generation from SMFC with substantial increase in operating voltage from 20 mV to 550–600 mV (Song et al. 2010a). Power density obtained from SMFC with varied R_{ex} was found to be enhanced from 0.064 mW m^{-2} when SMFC was operated with R_{ex} of 10 Ω to 0.413 mW m^{-2} (at R_{ex} of 100 Ω) and to 2.4 mW m^{-2} at R_{ex} of 1000 Ω (Hong et al. 2009). Song et al. (2010a) observed similar trend of power production of 0.0, 0.73, 1.66, 2.81 and 3.15 mW m^{-2} corresponding to the applied R_{ex} of 0, 100, 400, 800 and 1000 Ω in fresh water SMFC. This could be attributed to the fact that as the external resistance approached close to internal resistance the power production increases, and the internal resistance of SMFCs is generally higher. However, among all R_{ex} tested, higher organic matter removal efficiency of 29% at external resistance of 100 Ω was obtained in SMFC, whereas lower organic matter removal efficiency of 10.3% was obtained at R_{ex} of 1000 Ω.

17.3.3 Effect of Catalysts and Mediators

Role of catalysts in MFCs had been well documented and most of the results showed significantly higher power output as compared to the SMFCs provided without catalyzed cathode or anode. As for example, results have proven that the complex substrates such as cellulose and molasses can be effectively used in MFCs as substrate in presence of *Clostridium* biocatalysts (Niessen et al. 2005). Moreover, it has been observed that the sufficient availability of H_2 in anode could increase the methanogenic activity (Conrad 2002) and reduces the performance of SMFCs. Anode coated with platinum-poly (3,4-ethylenedioxythiophene) (PtPEDOT) bilayer composite biocatalysts can oxidize H_2 at anode, which decrease the methanogenic activity on anode of SMFC and improves the performance (Rosenbaum et al. 2005). An anode modified with mediators such as anthraquinone-1, 6-disulphonic acid (AQDS) and 1, 4-napthoquinonone (NQ) had enhanced the power density of SMFCs. A fivefold higher power density of 98 mW m^{-2} in SMFC was obtained using AQDS modified graphite plate as compared to the SMFC using plane graphite plate without any modification (Reimers et al. 2001).

Oxygen is the most feasible and sustainable TEA for the application of SMFCs due to high reduction potential and abundant availability in pond and marine environment in dissolved form. However, slow-moving ORR and high overpotential losses had been a bottleneck to achieve considerable power from the SMFCs (Noori et al. 2016c). Therefore, the cathode reduction kinetics need to speed-up using suitable catalysts. Platinum (Pt) catalyzed cathodes demonstrated promising results when used in MFCs and SMFCs due to reduced activation energy barrier to accomplish ORR. He et al. (2007) reported power density of 49 mW m^{-2} using Pt catalyzed carbon cloth cathode in SMFC. A platinum (Pt) modified carbon felt cathode could produce 207 mW m^{-2} from marine SMFC (Mathis et al. 2008). Though Pt catalyzed cathode delivered attractive results, its high cost and acute poisoning due to presence of H_2S could be a challenging task to implement in SMFCs, especially in marine environment. Hence, low-cost iron-cobalt based catalyst was developed to replace Pt. Cathode mounted on carbon paper with iron doped tetramethoxyphenyl porphyrin (Fe-CoTMPP) catalyst noted almost 300-times higher power density of 62 mW m^{-2} as compared to plain carbon paper (0.2 mW.m^{-2}) (Scott et al. 2008).

Natural water bodies containing diverse microorganism population are capable of performing catalytic activity for ORR (He and Angenent 2006). Later, this distinctive property of microorganisms shaped the opportunity of biocathode development. Hasvold et al. (1997) observed enhance ORR due to formation of biofilm on cathode which reveals that the cathode biofilm can function as biocatalyst. Application of biocathodes in SMFCs can be advantageous for several reasons. First, the cost of construction and operation of SMFCs may be lowered. Second, metal catalysts or artificial electron mediators could be poisoned by pollutants present in natural water. Third, microorganisms can function as catalysts to assist the electron transfer. Maximum power density of 1 W m^{-3} was observed using floating foam box reinforced carbon cloth biocathode in marine SMFC (Wang et al. 2012). Algal biocathode has been seen to produce oxygen in cathode, which could be an added benefit to overcome oxygen depletion in cathode (Mohan et al. 2014). Berk and Canfield (1964) reported maximum open circuit potential of 0.96 V with short circuit current of 750 mA m^{-2} using blue-green marine algae in the cathode of MFC. An algae assisted cathode could produce maximum power density of 21 mW m^{-2} and could be further enhanced to 38 mW m^{-2} using carbon nanotube coated cathode (Wang et al. 2014). Due to chances of acute poisoning of metal-based catalyst in aquatic environment, thermodynamic overpotential loss occurs, thereby, reducing the power output. Hence, the use of biocathodes has been advocated as sustainable solution for SMFCs.

17.3.4 Effect of pH, Dissolved Oxygen and Temperature

The pH of water and sediment plays an important role in the performance of MFCs and SMFCs, a mid-range alkaline pH range between 7 and 8 was suggested to obtain high current (He et al. 2008). Under alkaline range (pH 9), the biofilm attached to the anode was found to be more electrochemically active as compared to the acidic pH 5. At pH <6, reduction in power generation was also reported (Behera and Ghangrekar 2009). However, acidophilic pH around 6 or less may impart positive affect on the metabolism of microorganisms, which results in releasing additional electrons and protons (Mohan et al. 2009). In a different study with acidic pH <3, the SMFC demonstrated sustainable power density and current density of 0.3 W m^{-2} and 3.5 A m^{-2}, respectively (García-Muñoz et al. 2011). Sajana et al. (2013b) reported slight reduction in COD removal efficiency when pH of feed was increased from 6.5 (79%) to 8.5 (77%). Moreover, at pH 8.5 SMFC produced higher power density of 4.29 mW m^{-2} as compared to the power density of 3.5 mW m^{-2} obtained at pH of 6.5. The effect of pH on the performance of SMFCs is still confusing and no clear concluding remarks can be drawn from the previous experiments possibly due to the dynamic behaviour of biological system. However, a better performance can be expected in the pH ranging between 6 and 9.

The dissolved oxygen (DO) concentration at the cathode reaction interface greatly affects power recovery of SMFC and should be high enough to maintain the ORR (An et al. 2011). The DO concentration in natural aquatic environment fluctuates by microbial activity due to presence of organic matter (Zhang et al. 2009) as well as with fluctuations in temperature (Manasrah et al. 2006). For example, growth of microorganisms on cathode consume oxygen during respiration and if the re-oxygenation rate is lower than consumption rate, the water becomes oxygen depleted (Nguyen et al. 2006). An et al. (2011) developed bi-functional anti-microbial and catalytic cathode using silver nanoparticle (Ag-NPs) to overcome the problem of oxygen depletion. Results showed that after getting stable OCV of 0.67 V on 9th day in SMFC using plain graphite cathode the OCV was observed to be declining during consecutive days of operation until 50 d due to microbial growth of biomass (9.69 g of cell protein/g of electrode). As compared to the plain graphite cathode, less microbial growth in Ag-NPs treated cathode (5.3 g of cell protein/g of electrode) prevented to deplete DO concentration, which resulted in maintaining consistent cell voltage. Furthermore, when the DO concentration was increased from 3 to 7 mg/l, the current density obtained from SMFC was enhanced from 23 mA m^{-2} to 25.5 mA m^{-2} (Hong et al. 2009).

The performance of SMFC is found to be greatly influenced by temperature due to uneven fluctuation in DO and effect on microbial activity. Though at low temperature, the DO concentration would be high in water but most of the anaerobic microorganisms to be developed on anode show their activity in the mid temperature range of 20–25 °C excluding *Geobacteraceae,* which can grow at 4 °C. The current density was found to be increased from 15.6 mA m^{-2} to 52.6 mA m^{-2} when the operating temperature was increased from 10 to 35 °C (Hong et al. 2009). Schamphelaire et al. (2008) also observed reduction in power density from 231 mW m^{-2} to 157 mW m^{-2} when the temperature was decreased from 20 °C to 13.2 °C in rice field soil SMFC. Renslow et al. (2011) observed that the performance of freshwater SMFC decreased linearly with decrease in the temperature. The decrease temperature can reduce the microbial activity, resulting in high electrode resistance and less power recovery. Huang et al. (2012) reported disrupted anode kinetics due to seasonal change in environmental temperature.

17.3.5 Plants

The plant interactions regulate the ecological function of CWs similar to the natural wetland. However, it is still unclear that which plant types can enhance the performance of CWs. The relation between plants root canopy and density as a function of microbial population and its effect on the treatment performance of CWs was attempted to establish, but unfortunately the relationship has not resulted in sound evidence (Reed et al. 1995). Perhaps to use the locally available plants in study area of CW-MFCs would be a better solution (Xu et al. 2015). It would be of great interest to understand the effect of density and canopy of plants on the microbial growth and on the electrode kinetics. Plants can influence the distribution of electron donor/ acceptor by increasing the oxygen concentration due to their physical effect on water flow. Plants also have number of other functions such as creating surface area from bacterial attachment and biofilm formation, supplying carbon to the microorganism, up taking of some contaminants etc. Inclusion of plant roots of *Ipomoea aquatica* at the cathode was reported to improve the power generation of CW-MFCs by 142% as compared to the unplanted and rhizosphere-anode CW-MFCs (Liu et al. 2013). As shown in Table 17.2, *Ipomoea aquatica* and *Phragmites austrails* are the two major species of plant which have been majorly investigated in the CW-MFCs for phytoremediation of wastewater.

Table 17.2 Performance comparison of CW-MFCs under different substrate flow pattern and aquatic plant

Wastewater	Flow pattern	Plant	COD removal (%)	CE (%)	Power density ($mW\ m^{-3a}$)	References
Swine	Vertical	*Phragmites australis*	76.5	0.1–0.6	42	Zhao et al. (2013)
Synthetic	Batch mode	*Canna indica*	75	75	0.05–0.06 $mW\ m^{-2}$	Yadav et al. (2012)
Synthetic	Vertical subsurface	*Ipomoea aquatica*	95	2.8–3.9	44.63 $mW\ m^{-2}$	Liu et al. (2014)
Synthetic	Vertical	*Ipomoea aquatica*	94.8	0.39–1.29	12.42 $mW\ m^{-2}$	Liu et al. (2013)
Swine	Vertical	*Phragmites australis*	80		268	Doherty et al. (2015b)
Synthetic	Vertical	*Ipomoea aquatica*	85.7	0.5	852	Fang et al. (2015)
Synthetic	Vertical	*Ipomoea aquatica*	85.7	0.58–1.71	5.62 $mW.m^{-2}$	Fang et al. (2013)
Municipal	Horizontal	*Phragmites australis*	81.6	–	< 1	Corbella et al. (2014)
Synthetic	Horizontal	*Phragmites australis*	95	0.3–0.5	94	Villasenor et al. (2013)

[a]Otherwise stated; *CE* coulombic efficiency

17.3.6 Operating Conditions

For developing the natural redox gradient, which is an obligatory parameter for producing current from bio-electrochemical systems, most of the CW-MFCs were operated under up-flow regime of feeding (Fang et al. 2015; Liu et al. 2013). This type of feeding arrangement can minimize the DO concentration at anode and ensure higher substrate availability while maximizing DO at cathode. However, the up-flow regime to maintain natural redox gradient results in large electrode distance and subsequently contributes higher ohmic resistance to the system (Doherty et al. 2015a). For example, the internal resistance of 500 Ω obtained from CW-MFC (Doherty et al. 2015a) was found higher than 33 Ω for a multi-electrode MFC with separator electrode assembly (Ahn and Logan 2012).

Doherty et al. (2015b) proposed a design to minimize electrode separation and enhancing the power recovery by 70% wherein anode and cathode were separated with glass wool and combined flow, up-flow at anode and down-flow at cathode, was adopted simultaneously. However, long term operation of this design of CW-MFC resulted in clogging problem for the plants roots, hence not allowing them to penetrate the wetland soil subsurface (Doherty et al. 2015b). The performance of CW-MFC was enhanced by using bentonite layer as a separator and recirculating the flow of substrate from bottom to top of the wetland. However, the electricity recovery was compromised at higher organic loading because the anode was not capable to fully oxidize the organics (Villasenor et al. 2013).

17.4 Electricity Generation as a Function of Wastewater Treatment

Constructed wetlands are being considered as low-cost solution for wastewater treatment from past few decades (Hammer 1989). Lots of research have been conducted to enhance the performance of CWs by improvising different design, including different species of plants, manipulating soil characteristics and integrating other biological system such as MFCs. The very first report in integrated CW and MFC system demonstrated 75% COD removal efficiency (Yadav et al. 2012). CW-MFC planted with *Ipomoea aquatic*a demonstrated slightly higher COD removal efficiency of 94.8% than that obtained from unplanted CW-MFC (92.1%) (Liu et al. 2013). Unlike the COD removal, a substantial difference in total nitrogen efficiency was observed in planted CW-MFC (90.8%) and unplanted CW-MFC (54.4%) possibly due to assimilation of nitrogen in plants. Furthermore, planted CW-MFC showed enhanced power density of 12.42 mW m^{-2} as compared to the unplanted CW-MFC (5.13 mW m^{-2}).

Furthermore, CW-MFCs were also found to be capable of removing specific compounds such as azo dye from wastewater. Fang et al. (2013) obtained 91.1% removal efficiency of azo dye active brilliant red X-3 (ABRX3) and power density of 0.3 W m^{-3} when a CW-MFC was operated under 3 d HRT. Discolouration efficiency was also found to be affected by the operation modes such as close circuit and open circuit. As for example, in the same study, a 15% higher discolouration was observed when CW-MFC was operated under current generation mode (close circuit mode) than that of open circuit mode. CW-MFC treating high strength synthetic wastewater containing 500 mg/l of methylene blue dye demonstrated 93.1% discolouration rate with power density of 15.73 mW m^{-2} after 48 h of contact time (Yadav et al. 2012). Anode acts as an insoluble terminal electron acceptor while promoting the degradation of dye thereby increasing the metabolic rate of anaerobic microorganism and enhancing the substrate consumption, which eventually facilitates more electrons to accelerate discolouration rate from wastewater.

Though the inclusion of MFCs in CWs improves the COD removal efficiency (Doherty and Zhao 2015), only 0.05% to 3.9% of COD removal could be converted into electricity (Doherty et al. 2015a). Most of the researchers have reported low coulombic efficiency (CE) in SMFC, up to 3.9% (Table 17.2), suggesting that very little amount of electricity would be possible to convert from degradation of bulk organic compounds.

17.5 Scaling Up of SMFCs and Operating Wireless Sensors

SMFC is a promising alternative renewable energy source which can generate electricity for powering remote sensors, requires low maintenance and can provide alternate wastewater treatment option at low cost. Scaling-up of this technology is quite difficult with a specific configuration. However, researchers have claimed that Watt-level of power density could be obtained from MFCs and SMFCs. For example, Song et al. (2010b) demonstrated an MFC with energy generating capacity of 100 W m^{-3}, whereas a 30 ml MFC could generate a power density of 4.3 W m^{-2} (Fan et al. 2012). However, these normalized power densities were estimated based on the results obtained from laboratory scale MFCs. In the initial stage of development of SMFCs, it was expected that the power output from a SMFC would improve proportionally with increase in electrode size, but practically the power density does not depend on the surface area of the current limiting electrode (Ewing et al. 2014). A study revealed that for enhancing the power density from MFCs up to two-fold, the electrode surface area should be increased by 100-times (Dewan et al. 2008). This way of enhancing power from SMFCs or MFC for real time application does not seem feasible solution at all, since a huge electrode area would be problematic to bury in sediment in remote location. Moreover, it would be implacable to install such a huge SMFC to operate a single remote sensor. However, providing a power management system coupled with charge pumps and supercapacitors may be a feasible solution (Gong et al. 2011; Tang et al. 2015).

Ewing et al. (2014) developed a strategy to operate 2.5 W remote sensor using power obtained from MFC by intermittent harvesting and storing in supercapacitors. Multiple small-size electrodes with parallel connection rather than using a big single electrode may be a good solution for getting applicable power to operate wire-less sensors (Ewing et al. 2014). SMFCs fabricated with four anodes of 0.36 m^2 surface area (0.09 m^2 each) connected in parallel provided the power of 2.3 mW vs. 0.64 mW, where the latter was obtained from the SMFC using a single anode with surface area of 0.36 m^2. This power obtained was used to operate a wire-less temperature sensor using customized power management system (PMS). A 18 mW metrological buoy has been set-up by Naval Research Laboratory, USA (NRL, USA) powered by benthic attended generators (BUGs) for remotely monitoring air-temperature, water-temperature, pressure and relative humidity (Tender et al. 2008). To remotely monitor environmental parameters and military tactical surveillance via wire-less sensors are the foremost promising applications of SMFCs. As far as SMFC is concerned, a wire-less sensor cannot be operated with power generated from SMFC due to inconsistent and low output voltage. Therefore, a PMS was developed to store sufficient energy in supercapacitors for intermittent use and to boost the voltage using DC-DC convertor up to the requirement of sensors (in most of the cases 5 V). Bandyopadhyay et al. (2013) propelled an underwater 25 W bio-robot vehicle for 165 s at a time using power recovered from SMFC. Furthermore, movement of fish and other aquatic life has been monitored using ultrasonic sensor powered by SMFC (Donovan et al. 2013). From the above discussion, it can be concluded that the SMFC could be a promising renewable source of energy, but certain controllable parameters such as electrode materials, electrode spacing, shape, external resistance etc. need further attention to improve performance. Moreover, to operate wire-less sensors, an optimized PMS could provide a long-term solution. Unfortunately, to the best of our knowledge, there is no research available which had used CW-MFC as renewable power source to operate wire-less sensors.

17.6 Conclusion

Various issues, challenges and opportunities of SMFCs and CW-MFCs have been reviewed. In the present situation, the pollution threats in fresh water bodies and depleting conventional sources of energy are the two main brainstorming concerns across the globe. Development of SMFCs is expected to provide solution to these problems. In near future SMFCs might take niche of many available treatment technologies to offer sustainable solution to sediment and water remediation and energy harvesting. However, the challenges related to fabrication, installation and performance optimization are still under development stage. Corrosion-free carbon-based materials such as carbon cloth, carbon/graphite felt, graphite plates/disc/column etc. should be used in SMFCs or CW-MFCs due to their appreciable performance and prolonged stability in exhaustive environment. The electrodes offer large surface area for biofilm development/to accomplish higher ORR. However, this is found to have noticeable effect on the wastewater treatment and electricity recovery. To obtain high electrical current, comprehensive strategy on the operating parameters including optimized electrode spacing, external resistance, plant type, pH, temperature and DO need to be explored while maintaining natural redox gradient, substrate availability and required condition to trigger ORR in the system. Finally, an efficient PMS would provide opportunity to utilize the power generated by SMFCs for wire-less sensor operation for tactical surveillance, metrological monitoring etc.

References

Ahn, Y., & Logan, B. E. (2012). A multi-electrode continuous flow microbial fuel cell with separator electrode assembly design. *Applied Microbiology and Biotechnology, 93*, 2241–2248.

An, J., Jeon, H., Lee, J., & Chang, I. S. (2011). Bifunctional silver nanoparticle cathode in microbial fuel cells for microbial growth inhibition with comparable oxygen reduction reaction activity. *Environmental Science & Technology, 45*, 5441–5446.

Arends, J. B., Blondeel, E., Tennison, S. R., Boon, N., & Verstraete, W. (2012). Suitability of granular carbon as an anode material for sediment microbial fuel cells. *Journal of Soils and Sediments, 12*, 1197–1206.

Bandyopadhyay, P. R., Thivierge, D. P., McNeilly, F. M., & Fredette, A. (2013). An electronic circuit for trickle charge harvesting from littoral microbial fuel cells. *IEEE Journal of Oceanic Engineering, 38*, 32–42.

Behera, M., & Ghangrekar, M. M. (2009). Performance of microbial fuel cell in response to change in sludge loading rate at different anodic feed pH. *Bioresource Technology, 100*, 5114–5121.

Berk, R. S., & Canfield, J. H. (1964). Bioelectrochemical energy conversion. *Applied Microbiology, 12*, 10–12.

Conrad, R. (2002). Control of microbial methane production in wetland rice fields. *Nutrient Cycling in Agroecosystems, 64*, 59–69.

Corbella, C., Garfí, M., & Puigagut, J. (2014). Vertical redox profiles in treatment wetlands as function of hydraulic regime and macrophytes presence: Surveying the optimal scenario for microbial fuel cell implementation. *Science of the Total Environment, 470*, 754–758.

Del Campo, A. G., Canizares, P., Lobato, J., Rodrigo, M., & Morales, F. F. (2014). Effects of external resistance on microbial fuel cell's performance. In *Environment, energy and climate change II* (pp. 175–197). Cham: Springer.

Des Jarlais, D. C., Feelemyer, J. P., Modi, S. N., Abdul-Quader, A., & Hagan, H. (2013). High coverage needle/syringe programs for people who inject drugs in low and middle income countries: A systematic review. *BMC Public Health, 13*, 1.

Dewan, A., Beyenal, H., & Lewandowski, Z. (2008). Scaling up microbial fuel cells. *Environmental Science & Technology, 42*, 7643–7648.

Doherty, L., & Zhao, Y. (2015). Operating a two-stage microbial fuel cell–constructed wetland for fuller wastewater treatment and more efficient electricity generation. *Water Science and Technology, 72*, 421–428.

Doherty, L., Zhao, Y., Zhao, X., Hu, Y., Hao, X., Xu, L., & Liu, R. (2015a). A review of a recently emerged technology: Constructed wetland–microbial fuel cells. *Water Research, 85*, 38–45.

Doherty, L., Zhao, Y., Zhao, X., & Wang, W. (2015b). Nutrient and organics removal from swine slurry with simultaneous electricity generation in an alum sludge-based constructed wetland incorporating microbial fuel cell technology. *Chemical Engineering Journal, 266*, 74–81.

Donovan, C., Dewan, A., Heo, D., Lewandowski, Z., & Beyenal, H. (2013). Sediment microbial fuel cell powering a submersible ultrasonic receiver: New approach to remote monitoring. *Journal of Power Sources, 233*, 79–85.
Dordio, A. V., & Carvalho, A. J. P. (2013). Organic xenobiotics removal in constructed wetlands, with emphasis on the importance of the support matrix. *Journal of Hazardous Materials, 252*, 272–292.
Dumas, C., Mollica, A., Féron, D., Basséguy, R., Etcheverry, L., & Bergel, A. (2007). Marine microbial fuel cell: Use of stainless steel electrodes as anode and cathode materials. *Electrochimica Acta, 53*, 468–473.
Ewing, T., Ha, P. T., Babauta, J. T., Tang, N. T., Heo, D., & Beyenal, H. (2014). Scale-up of sediment microbial fuel cells. *Journal of Power Sources, 272*, 311–319.
Fan, Y., Han, S.-K., & Liu, H. (2012). Improved performance of CEA microbial fuel cells with increased reactor size. *Energy & Environmental Science, 5*, 8273–8280.
Fang, Z., Song, H.-L., Cang, N., & Li, X.-N. (2013). Performance of microbial fuel cell coupled constructed wetland system for decolorization of azo dye and bioelectricity generation. *Bioresource Technology, 144*, 165–171.
Fang, Z., Song, H.-L., Cang, N., & Li, X.-N. (2015). Electricity production from azo dye wastewater using a microbial fuel cell coupled constructed wetland operating under different operating conditions. *Biosensors and Bioelectronics, 68*, 135–141.
García-Muñoz, J., Fernández, V. M., De Lacey, A. L., Malki, M., & Amils, R. (2011). Electricity generation by microorganisms in the sediment-water interface of an extreme acidic microcosm. *International Microbiology, 14*, 73–81.
Gong, Y., Radachowsky, S. E., Wolf, M., Nielsen, M. E., Girguis, P. R., & Reimers, C. E. (2011). Benthic microbial fuel cell as direct power source for an acoustic modem and seawater oxygen/temperature sensor system. *Environmental Science & Technology, 45*, 5047–5053.
Hammer, D. A. (1989). *Constructed wetlands for wastewater treatment: Municipal, industrial and agricultural*. Boca Raton: CRC Press.
Hasvold, Ø., Henriksen, H., Melv, E., Citi, G., Johansen, B. Ø., Kjønigsen, T., & Galetti, R. (1997). Sea-water battery for subsea control systems. *Journal of Power Sources, 65*, 253–261.
He, Z., & Angenent, L. T. (2006). Application of bacterial biocathodes in microbial fuel cells. *Electroanalysis, 18*, 2009–2015.
He, Z., Shao, H., & Angenent, L. T. (2007). Increased power production from a sediment microbial fuel cell with a rotating cathode. *Biosensors and Bioelectronics, 22*, 3252–3255.
He, Z., Huang, Y., Manohar, A. K., & Mansfeld, F. (2008). Effect of electrolyte pH on the rate of the anodic and cathodic reactions in an air-cathode microbial fuel cell. *Bioelectrochemistry, 74*, 78–82.
Hong, S. W., Chang, I. S., Choi, Y. S., & Chung, T. H. (2009). Experimental evaluation of influential factors for electricity harvesting from sediment using microbial fuel cell. *Bioresource Technology, 100*, 3029–3035.
Huang, Y., He, Z., Kan, J., Manohar, A. K., Nealson, K. H., & Mansfeld, F. (2012). Electricity generation from a floating microbial fuel cell. *Bioresource Technology, 114*, 308–313.
Kim, J. R., Cheng, S., Oh, S.-E., & Logan, B. E. (2007). Power generation using different cation, anion, and ultrafiltration membranes in microbial fuel cells. *Environmental Science & Technology, 41*, 1004–1009.
Krishnaraj, R. N., & Jong Sung, Y. (2015). *Bioenergy: Opportunities and challenges*. Boca Raton: CRC Press, Taylor and Francis group.
Li, J. H., Fu, Y. B., Liu, J., Li, A. L. & Ma, D. D. (2009). Effect of electrode shape on power and internal resistance in benthic microbial fuel cell material on marine sediment. *Advanced Materials Research*. Trans Tech Publ, pp. 2195–2198.
Liu, H., Cheng, S., & Logan, B. E. (2005). Power generation in fed-batch microbial fuel cells as a function of ionic strength, temperature, and reactor configuration. *Environmental Science & Technology, 39*, 5488–5493.

Liu, S., Song, H., Li, X., & Yang, F. (2013). Power generation enhancement by utilizing plant photosynthate in microbial fuel cell coupled constructed wetland system. *International Journal of Photoenergy, 2013*, 1–10.

Liu, S., Song, H., Wei, S., Yang, F., & Li, X. (2014). Bio-cathode materials evaluation and configuration optimization for power output of vertical subsurface flow constructed wetland–microbial fuel cell systems. *Bioresource Technology, 166*, 575–583.

Manasrah, R., Raheed, M., & Badran, M. I. (2006). Relationships between water temperature, nutrients and dissolved oxygen in the northern Gulf of Aqaba, Red Sea. *Oceanologia, 48*, 237–253.

Mathis, B., Marshall, C., Milliken, C., Makkar, R., Creager, S. E., & May, H. (2008). Electricity generation by thermophilic microorganisms from marine sediment. *Applied Microbiology and Biotechnology, 78*, 147–155.

Mohan, S. V., Srikanth, S., Raghuvulu, S. V., Mohanakrishna, G., Kumar, A. K., & Sarma, P. (2009). Evaluation of the potential of various aquatic eco-systems in harnessing bioelectricity through benthic fuel cell: Effect of electrode assembly and water characteristics. *Bioresource Technology, 100*, 2240–2246.

Mohan, S. V., Srikanth, S., Chiranjeevi, P., Arora, S., & Chandra, R. (2014). Algal biocathode for in situ terminal electron acceptor (TEA) production: Synergetic association of bacteria–microalgae metabolism for the functioning of biofuel cell. *Bioresource Technology, 166*, 566–574.

Nguyen, T. N., Woodward, R. T., Matlock, M. D., Denzer, A., & Selman, M. (2006). *A guide to market-based approaches to water quality*. Washington, DC: World Resource Institute.

Nielsen, M. E., Reimers, C. E., & Stecher, H. A. (2007). Enhanced power from chambered benthic microbial fuel cells. *Environmental Science & Technology, 41*, 7895–7900.

Niessen, J., Schröder, U., Harnisch, F., & Scholz, F. (2005). Gaining electricity from in situ oxidation of hydrogen produced by fermentative cellulose degradation. *Letters in Applied Microbiology, 41*, 286–290.

Noori, M. T., Ghangrekar, M. M., Mitra, A., & Mukherjee, C. (2016a). Enhanced power generation in microbial fuel cell using MnO_2-catalyzed cathode treating fish market wastewater. In *Proceedings of the first international conference on recent advances in bioenergy research* (pp. 285–294). New Delhi: Springer.

Noori, M. T., Ghangrekar, M. M., & Mukherjee, C. (2016b). V_2O_5 microflower decorated cathode for enhancing power generation in air-cathode microbial fuel cell treating fish market wastewater. *International Journal of Hydrogen Energy, 41*, 3638–3645.

Noori, M. T., Jain, S. C., Ghangrekar, M. M., & Mukherjee, C. K. (2016c). Biofouling inhibition and enhancing performance of microbial fuel cell using silver nano-particles as fungicide and cathode catalyst. *Bioresource Technology, 220*, 183–189.

Pinto, R., Srinivasan, B., Guiot, S., & Tartakovsky, B. (2011). The effect of real-time external resistance optimization on microbial fuel cell performance. *Water Research, 45*, 1571–1578.

Rajesh, P., Noori, M. T., & Ghangrekar, M. M. (2014). Controlling methanogenesis and improving power production of microbial fuel cell by lauric acid dosing. *Water Science and Technology, 70*, 1363–1369.

Reed, S. C., Crites, R. W., & Middlebrooks, E. J. (1995). *Natural systems for waste management and treatment*. New York: McGraw-Hill.

Reimers, C. E., Tender, L. M., Fertig, S., & Wang, W. (2001). Harvesting energy from the marine sediment-water interface. *Environmental Science & Technology, 35*, 192–195.

Renslow, R., Donovan, C., Shim, M., Babauta, J., Nannapaneni, S., Schenk, J., & Beyenal, H. (2011). Oxygen reduction kinetics on graphite cathodes in sediment microbial fuel cells. *Physical Chemistry Chemical Physics, 13*, 21573–21584.

Rismani-Yazdi, H., Carver, S. M., Christy, A. D., & Tuovinen, O. H. (2008). Cathodic limitations in microbial fuel cells: An overview. *Journal of Power Sources, 180*, 683–694.

Rosenbaum, M., Schröder, U., & Scholz, F. (2005). In situ electrooxidation of photobiological hydrogen in a photobioelectrochemical fuel cell based on *Rhodobacter sphaeroides*. *Environmental Science & Technology, 39*, 6328–6333.

Sacco, N. J., Figuerola, E. L., Pataccini, G., Bonetto, M. C., Erijman, L., & Cortón, E. (2012). Performance of planar and cylindrical carbon electrodes at sedimentary microbial fuel cells. *Bioresource Technology, 126*, 328–335.

Sajana, T., Ghangrekar, M. M., & Mitra, A. (2013a). Application of sediment microbial fuel cell for in situ reclamation of aquaculture pond water quality. *Aquacultural Engineering, 57*, 101–107.

Sajana, T., Ghangrekar, M. M., & Mitra, A. (2013b). Effect of pH and distance between electrodes on the performance of a sediment microbial fuel cell. *Water Science and Technology, 68*, 537–543.

Sajana, T., Ghangrekar, M. M., & Mitra, A. (2014). Effect of presence of cellulose in the freshwater sediment on the performance of sediment microbial fuel cell. *Bioresource Technology, 155*, 84–90.

Saravanan, R., Arun, A., Venkatamohan, S., & Kandavelu, T. (2010). Membraneless dairy wastewater-sediment interface for bioelectricity generation employing Sediment Microbial Fuel Cell (SMFC). *African Journal of Microbiology Research, 4*, 2640–2646.

Schamphelaire, L. D., Bossche, L. V. D., Dang, H. S., Höfte, M., Boon, N., Rabaey, K., & Verstraete, W. (2008). Microbial fuel cells generating electricity from rhizodeposits of rice plants. *Environmental Science & Technology, 42*, 3053–3058.

Scott, K., Cotlarciuc, I., Hall, D., Lakeman, J., & Browning, D. (2008). Power from marine sediment fuel cells: The influence of anode material. *Journal of Applied Electrochemistry, 38*, 1313–1319.

Singh, D., Pratap, D., Baranwal, Y., Kumar, B., & Chaudhary, R. (2010). Microbial fuel cells: A green technology for power generation. *Annals of. Biological Research, 1*, 128–138.

Song, T. S., Yan, Z. S., Zhao, Z. W., & Jiang, H. L. (2010a). Removal of organic matter in freshwater sediment by microbial fuel cells at various external resistances. *Journal of Chemical Technology and Biotechnology, 85*, 1489–1493.

Song, Y.-C., Yoo, K. S., & Lee, S. K. (2010b). Surface floating, air cathode, microbial fuel cell with horizontal flow for continuous power production from wastewater. *Journal of Power Sources, 195*, 6478–6482.

Song, T.-S., Yan, Z.-S., Zhao, Z.-W., & Jiang, H.-L. (2011). Construction and operation of freshwater sediment microbial fuel cell for electricity generation. *Bioprocess and Biosystems Engineering, 34*, 621–627.

Song, T. S., Tan, W. M., Wu, X. Y., & Zhou, C. C. (2012). Effect of graphite felt and activated carbon fiber felt on performance of freshwater sediment microbial fuel cell. *Journal of Chemical Technology and Biotechnology, 87*, 1436–1440.

Strik, D. P., Snel, J. F., & Buisman, C. J. (2008). Green electricity production with living plants and bacteria in a fuel cell. *International Journal of Energy Research, 32*, 870–876.

Tang, N., Hong, W., Ewing, T., Beyenal, H., Kim, J.-H., & Heo, D. (2015). A self-sustainable power management system for reliable power scaling up of sediment microbial fuel cells. *IEEE Transactions on Power Electronics, 30*, 4626–4632.

Tender, L. M., Gray, S. A., Groveman, E., Lowy, D. A., Kauffman, P., Melhado, J., Tyce, R. C., Flynn, D., Petrecca, R., & Dobarro, J. (2008). The first demonstration of a microbial fuel cell as a viable power supply: Powering a meteorological buoy. *Journal of Power Sources, 179*, 571–575.

Timmers, R. A., Rothballer, M., Strik, D. P., Engel, M., Schulz, S., Schloter, M., Hartmann, A., Hamelers, B., & Buisman, C. (2012). Microbial community structure elucidates performance of Glyceria maxima plant microbial fuel cell. *Applied Microbiology and Biotechnology, 94*, 537–548.

Tiwari, B. R., & Ghangrekar, M. M. (2015). Enhancing Electrogenesis by pretreatment of mixed anaerobic sludge to be used as inoculum in microbial fuel cells. *Energy & Fuels, 29*, 3518–3524.

Tiwari, B. R., Noori, M. T., & Ghangrekar, M. M. (2016). A novel low cost polyvinyl alcohol-Nafion-borosilicate membrane separator for microbial fuel cell. *Materials Chemistry and Physics, 182*, 86–93.

Villasenor, J., Capilla, P., Rodrigo, M., Canizares, P., & Fernandez, F. (2013). Operation of a horizontal subsurface flow constructed wetland–microbial fuel cell treating wastewater under different organic loading rates. *Water Research, 47*, 6731–6738.

Wang, A., Cheng, H., Ren, N., Cui, D., Lin, N., & Wu, W. (2012). Sediment microbial fuel cell with floating biocathode for organic removal and energy recovery. *Frontiers of Environmental Science & Engineering, 6*, 569–574.

Wang, D.-B., Song, T.-S., Guo, T., Zeng, Q., & Xie, J. (2014). Electricity generation from sediment microbial fuel cells with algae-assisted cathodes. *International Journal of Hydrogen Energy, 39*, 13224–13230.

Wu, X. Y., Tong, F., Song, T. S., Gao, X. Y., Xie, J. J., Zhou, C. C., Zhang, L. X., & Wei, P. (2015). Effect of zeolite-coated anode on the performance of microbial fuel cells. *Journal of Chemical Technology and Biotechnology, 90*, 87–92.

Xu, B., Ge, Z., & He, Z. (2015). Sediment microbial fuel cells for wastewater treatment: Challenges and opportunities. *Environmental Science: Water Research & Technology, 1*, 279–284.

Yadav, A. K., Dash, P., Mohanty, A., Abbassi, R., & Mishra, B. K. (2012). Performance assessment of innovative constructed wetland-microbial fuel cell for electricity production and dye removal. *Ecological Engineering, 47*, 126–131.

Zhang, Q., Li, Z., Zeng, G., Li, J., Fang, Y., Yuan, Q., Wang, Y., & Ye, F. (2009). Assessment of surface water quality using multivariate statistical techniques in red soil hilly region: A case study of Xiangjiang watershed, China. *Environmental Monitoring and Assessment, 152*, 123–131.

Zhao, Y., Collum, S., Phelan, M., Goodbody, T., Doherty, L., & Hu, Y. (2013). Preliminary investigation of constructed wetland incorporating microbial fuel cell: Batch and continuous flow trials. *Chemical Engineering Journal, 229*, 364–370.

Chapter 18
Fundamentals of Microbial Desalination Cell

Soumya Pandit, Shruti Sarode, and Debabrata Das

18.1 Introduction

Microbial desalination cells (MDCs) are bioelectrochemical tools which exploit organic matter in wastewater to use as an energy source for desalinating salt water. These cells desalinate water by expending the electric potential gradient established by exoelectrogenic bacteria to drive ion transport through a series of ion exchange membranes (Kim and Logan 2013). This device has the potential to solve the world's freshwater crisis. As such, a significant increase in the number of installed desalination capacities were noticed since the 1980s (Greenlee et al. 2009). However, conventional technology of desalination consumes high amounts of energy which is a matter of concern as the stocks of fossil fuels are rapidly depleting. Theoretically, if we consider a thermodynamically reversible process at 50% water recovery, the minimum energy required for desalination of typical sea water is approximately 1 kWh m^{-3}. The most efficient systems of seawater desalination using reverse osmosis have achieved an energy requirement of only 1.8–2.2 kWh m^{-3}. However, considering the energy needs for pre-treatment, pumping, etc., the overall consumption stands at about 3–4 kWh m^{-3}, which is not efficient at all (Liu et al. 2011). MDCs have emerged as a solution to this problem because they accomplish environment friendly wastewater treatment and at the same time they drastically diminish the energy expenditure for desalination (Cao et al. 2009).

S. Pandit (✉) • S. Sarode • D. Das
Department of Biotechnology, Bioprocess Engineering Laboratory, Indian Institute of Technology Kharagpur, Kharagpur 721302, India
e-mail: sounip@gmail.com

D. Das (ed.), *Microbial Fuel Cell*,
https://doi.org/10.1007/978-3-319-66793-5_18

18.2 Ion Exchange Membrane (IEM) Based MDC

18.2.1 Reactor Design

The initial design of MDC was a three-chambered device consisting of the anode, the middle desalination chamber and the cathode. In this model, the anode chamber is fed with the wastewater which contains organic matter. The bacteria exploit this organic matter as substrate to grow as a biofilm on the anode. These bacteria oxidize the substrates by acting on them. The electrons released in the process are collected by the anode and are transferred via an external circuit to the cathode where these electrons combine with oxygen and protons to form water. The middle chamber has salt water containing various cations and anions. It is bounded by an anion exchange membrane (AEM) on the anode side and a cation exchange membrane (CEM) on the cathode side (Cao et al. 2009). The electrode oxidation and reduction reactions generate a 1–1.5 V electric potential gradient or electric field between the electrodes (open circuit voltage), which attracts the cations and anions in the salt water towards the anode and the cathode respectively, thus desalinating the water in the middle chamber (Fig. 18.1).

This separation of ions can be accelerated and increased by using multiple membrane pairs between the electrodes, resembling to the equipment used in case of electro dialysis (ED) desalination systems (Chen et al. 2011). Here, the middle

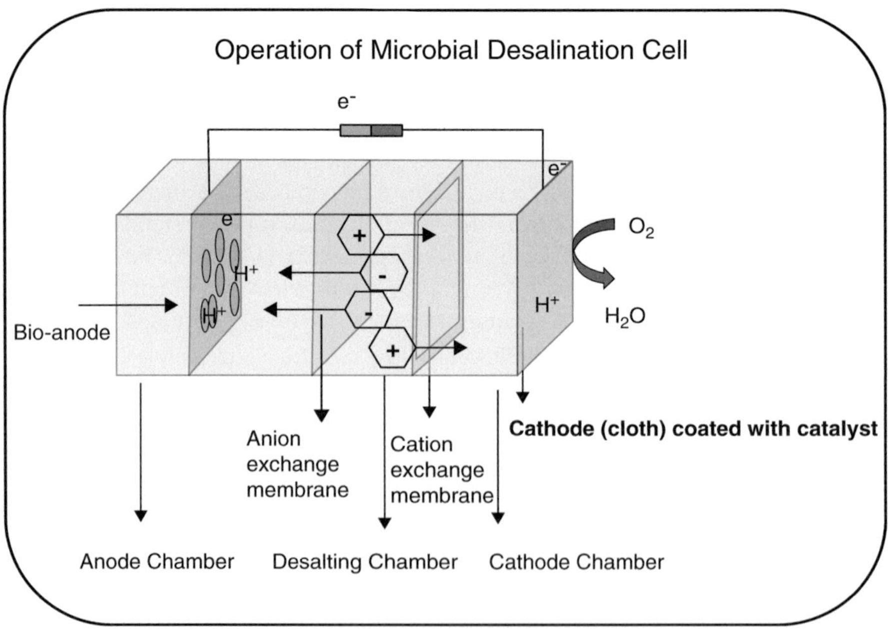

Fig. 18.1 Schematic of a prototype 3-chamber MDC

chamber consists of a stack of alternating AEMs and CEMs which forms repeating pairs of desalting (dilute) and concentrating (concentrate) cells. The principle on which it relies is that the number of monovalent ion pairs separated from the salt water per electron transferred is equal to the number of repeated cell pairs (or membrane pairs) in the stack. But the internal resistance of the MDC is also directly proportional to the number of membrane pairs in the stack i.e. more the number of membranes more is the resistance of the MDC. As in an MDC, the voltage used for desalination is limited to that produced solely by the electrode reactions (since no external voltage is applied), the voltage per cell pair decreases with increase in the number of cell pairs. Hence, a balance is necessary between the two limiting conditions such that maximum number of ions can be separated and at the same time a sufficient amount of potential difference is generated (Jacobson et al. 2011a). To enhance the performance by minimizing the internal resistance and simultaneously allowing larger number of cell pairs, the strategy adopted should be to keep the inter-membrane distance as low as possible. Summary of different MDC configurations is depicted in Table 18.1.

Air Cathode Microbial Desalination Cell A typical air cathode MDC is a 3-chambered cell consisting of anode chamber, desalting chamber and cathode chamber. But unlike a conventional MDC, it uses atmospheric oxygen as an electron source (Fig. 18.2). According to studies, the salinity of water was reduced by 63% in a single cycle by MDC using carbon cloth electrode with platinum as catalyst (Torres 2012). Alternatively, activated carbon and cobalt tetramethoxyphenyl-porphyrin can also be used so that cobalt tetra-methoxyphenylporphyrin acts as a catalyst while the increase in surface area is carried out by activated carbon which further improves the performance of MDC. Using atmospheric oxygen as the electron acceptor is also advantageous in terms of environmental sustainability since oxygen has negligible toxic effects compared to other chemicals. One of the disadvantages of an air cathode MDC is that the redox kinetics in ambient conditions is slow as compared to that using other catholytes. This would necessitate the need for more expensive catalytic materials e.g. platinum, to reduce the activation over potential for oxygen reduction (Kim and Logan 2013). Another concern associated with air cathodes is the high energy required to maintain the dissolved oxygen concentration in the cathode. There are three ways to minimize the impacts of these problems. The first is by either exposing the MDC to the atmosphere, another is using passive methods to achieve oxygen transfer in the cathodes or the use of high surface area carbon substrates to achieve acceptable levels of oxygen reduction without the need for expensive platinum catalysts (Rahimnejad et al. 2015).

The use of biocathodes as catalysts in MDCs is becoming increasingly popular due to self-regenerating and sustainability (Kokabian and Gude 2015). The biocathodes allow the electroactive bacteria in the cathode biofilm to act as catalysts and promote the oxidation-reduction reactions, which in turn enhances water desalination (Kokabian and Gude 2013). Moreover, more potential at the anode is experienced if the biofilm growth is denser, therefore, producing more

Table 18.1 Summary of different MDC configurations

MDC configurations	Key features	Advantages
Air cathode MDC	1. Oxygen used as a terminal electron acceptor	1. High reduction potential
	2. Approximately 63 percent salt reduction	2. Self-generating and sustainable
	3. Catalyzes reduction reactions through use of microbes	3. Enhanced water desalination
Biocathode MDC	Catalyzes reduction reactions through use of microbes	1. Self-generating and sustainable
		2. Enhanced water desalination
		3. Reduced start-up time
Stack structure MDC	Alternating AEMs and CEMs.	1. Improves charge transfer efficiency
	Total desalination rate approximately 1.4 times that of a typical MDC unit	2. Increased separation of ion pairs from saltwater
		3. Increased energy recovery
Recirculation MDC	Catholyte and anolyte solutions sequentially re-circulated through cell	1. Increased power density
		2. Increased desalination efficiency
Microbial electrolysis desalination and chemical-production cell	Formed by introducing an acid-production chamber and a bipolar membrane in an MDC	1. Reduced pH fluctuations
		2. Increased desalination rate
Capacitive MDC	1. Incorporates the concept of capacitive deionization	1. Reduced salt contamination in anode and cathode chambers
	2. Double-layer capacitor formed on the surface electrodes	2. Resolving ion migration problems
	3. Salts should be continually removed from electrodes	
Upflow MDC	Tubular reactor containing two compartments separated by IEMs	1. Efficient fluid mixing within the chambers
		2. Easier to scale up
		3. Increased power density
		4. Improved desalination efficiency
Osmotic MDC	1. AEM replaced with FO membranes	1. Improves overall desalination performance
	2. Dilute saltwater by increasing water flux	2. Enhances organic matter removal from wastewater
	3. Potassium ferricyanide usually used as catalyst	3. Cost of FO membrane lower than cost of AEM

(continued)

Table 18.1 (continued)

MDC configurations	Key features	Advantages
Bipolar membrane MDC	1. Anion and cation selective layers laminated together to make BPM	1. High ion selectivity
	2. BPM placed next to the anode chamber, making a four-chamber MDC	2. Low water splitting voltage drop and electrical resistance
		3. Long-life duration of BPM
		4. Increased desalination efficiency
		5. Maintaining desired pH in anode chamber
Decoupled MDC	1. Anode and cathode units placed directly in salt solution	1. Easier to control and vary the liquid volume ratios
	2. Stainless steel mesh wrapped with carbon cloth used as an electrode	2. Ease in repairing or replacement of any damaged parts
		3. Easy to scale up

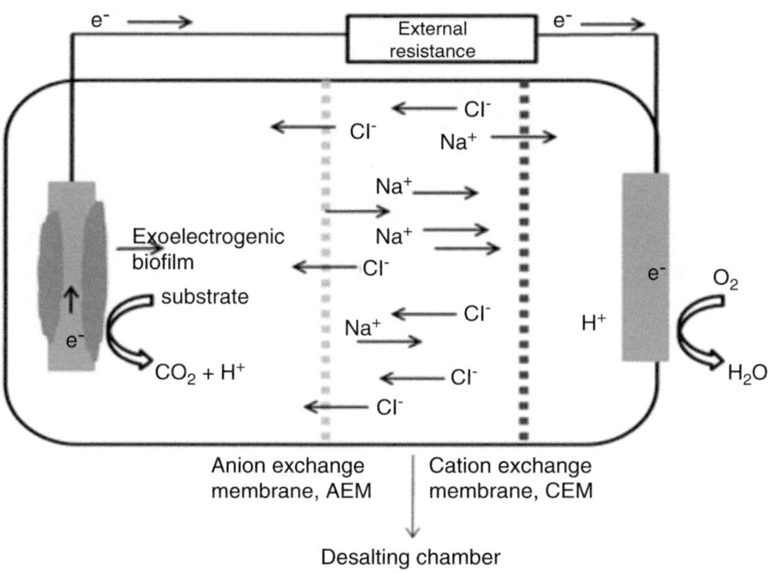

Fig. 18.2 Schematic of a typical air cathode MDC

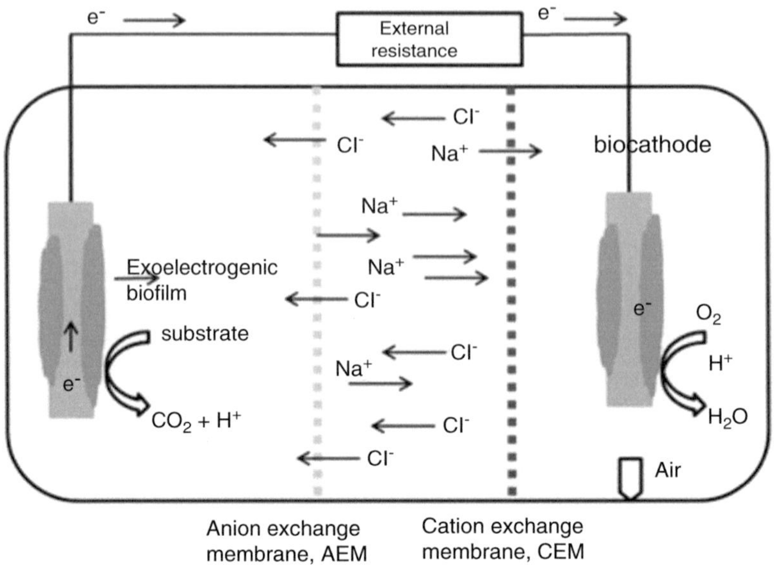

Fig. 18.3 Schematic of biocathode MDC

power (Fig. 18.3). If the biocathode is optimized, it may also help reducing the startup time for the MDC and, therefore, improve the cell performance. Wen et al. (2012) reported the use of biocathode MDC, in which the biocathode was aerobic, containing carbon felt and bacterial catalysts.

18.2.2 *Junction Potential and Water Transport*

Although the electric potential gradient is the major factor responsible for separating ions in an MDC, there are certain other factors like IEM junction potential and water transport across IEM which also affect desalination rate. In three chambered systems, these factors act as additional driving forces for desalination but in the case of stacked multiple cell pair systems, they generally reduce the desalination efficiency.

In the three chambered MDCs, the middle chamber containing salt water is usually hypertonic with respect to its neighbouring anodic and cathodic chambers. Typically, sea water in the middle chamber has dissolved solid concentration of 30–40 g/L while the wastewater in the anodic chamber has almost 0.5–0.9 g/L of dissolved solids. Due to this steep potential gradient, salt ions in the middle chamber are driven to the neighbouring anodic and cathodic chambers. The junction potential ($\Delta\Phi_{jct}$) quantifies this driving force between sea water (sw) and waste water (ww) and is given by the Eq. 18.1 (Bard and Faulkner 2001):

$$\left|\Delta\Phi_{jct}\right| = \left(\frac{RT}{F}\right)\left|\sum_{i}\left(\frac{t_i}{z_i}\right)\ln\left(\frac{a_{i,sw}}{a_{i,ww}}\right)\right| \quad (18.1)$$

where R is gas constant, T – absolute temperature, F – Faraday's constant, z – ionic charge, a – activity of the ionic species I and t is transport number (fractional contribution of ionic flux to current density for the given IEM). For an ideally permselective CEM in a NaCl solution, $t_{Na+} = 1$ and $t_{Cl-} = 0$. Taking the activity ratio ($a_{i,sw}/a_{i,ww}$) as the ratio of the concentrations of dissolved solids in sea water and domestic wastewater (i.e., 35 g L^{-1}/0.5 g L^{-1} = 70), the sum of junction potentials in a three chambered MDC is about 0.2 V (at 25 °C and considering 95% permselective IEMs). This voltage is only around 30% of the voltage generated by the electrode reactions. It is observed that most of the decrease in conduction in the desalination chamber is due to the concentration gradient as mentioned before (Kim and Logan 2011). This concentration gradient also induces osmotic inflow of water into the middle chamber, thus diluting it and aiding in desalination in the three chambered system. According to the assumptions taken earlier, the osmotic pressure typically turns out to be 14.4 atm for seawater-wastewater couple.

$$\Pi = \Delta cRT \quad (18.2)$$

As the IEMs cannot withstand such high osmotic pressure, water must be allowed to enter the middle chamber from the adjacent chambers and release the pressure during MDC operation. This dilutes the seawater in the middle chamber and helps in desalination in the three chambered MDC (Jacobson et al. 2011b).

Electro osmosis is another driving force for water transport across the IEM in which water transport occurs because of ionic movement through the nano-scale pores of the IEM. The ions are in the hydrated form and thus have water molecules surrounding them in all directions. When such ions move through the IEM pores, they drag the water molecules along with them and hence cause efflux of water from the middle chamber resulting in the concentration of the salt water. However, it has been observed that osmotic water transport generally dominates the electro osmosis phenomenon and thus the net volume of water in the middle chamber increases. Electro osmotic water transport is directly proportional to the current density through the IEMs and osmotic water transport is directly proportional to the hydraulic residence time (Zhang and He 2012).

The osmotic water transport and junction potentials prove beneficial for water desalination in case of three chambered MDCs because wastewater usually contains lesser concentration of dissolved solids than seawater. But it is not true in case of stacked MDCs with multiple membrane pairs because the concentration of ions in the concentrate cells are higher than that in the dilute cells (due to desalination) and this adversely affects the desalination process. For instance, in a 5-cell paired stack system, the junction potential losses consumed up to 18% of the total potential difference between the microbial anode and the air-breathing cathodes generated due to the electrode reactions. In stack systems, the direction of water transport due to osmosis and electro osmosis is the same i.e. from dilute cells to concentrate cells, resulting in the concentration of the dilute cells by up to 30% in a 5-cell stack (Kim and Logan 2011).

18.3 MDC Performance

18.3.1 Salinity Removal

In three chambered MDCs, the salinity removal can be above 90% from 30 to 35 g L^{-1} NaCl solutions having conductivities similar to that of seawater. However, it is found that very high salinity removals require large volumes of non-salty anolyte and catholyte, about 55–133 times the volume of seawater per batch. This high requirement can be reduced with the use of stacked MDCs. 98% salinity removal from 35 g L^{-1} NaCl has been achieved using 5-cell pair stacked MDC with a reduced anolyte volume (13 times the desalinated water). For 40–60% salinity removal, the required wastewater is only 2–3 times the desalinated water. This indicates that MDCs are more effective when used for partial desalination of seawater. MDCs can also be used to desalinate brackish water. 99% desalination of 10 g L^{-1} NaCl has been achieved with anolyte volume 14 times that of desalinated water. For 86% salinity removal of same salt water sample, anolyte required was six times that of desalinated water (Luo et al. 2012). This implies that volume of wastewater required is also dependent on the initial concentration of salts (or salinity) of the seawater sample to be desalinated.

18.3.2 Maximum Current vs. Maximum Power

The operating conditions for an MDC are chosen in such a way that they represent a trade-off between maximizing current density (hence the rate of treatment) and the power density. Higher the current amount, higher is the rate of desalination. This is a preferable operation condition from the perspective of minimizing reactor hydraulic retention time and maximizing rate of treatment. Maximum reported current densities range from 0.7 to 8.4 A m^{-2}, which is most likely, achieved using low internal resistances. However, sometimes low performance might be observed even at low resistances due to unfavourable anode potential, which affects current generation by exoelectrogenic bacteria in the MDC. At low external resistances, power densities can rapidly reduce and current densities can double back to lower values in a phenomenon termed as Type D power overshoot (Watson and

Logan 2011). However, this phenomenon can be overcome over time through proper acclimation of the reactors to low values of resistance.

Another strategy for MDC operation is to maximize electrical power production. With oxygen reduction reaction at cathode, the highest generated power reported for MDCs is 31 W m^{-3} while with ferricyanide catholyte and normalized anolyte volume; the maximum power reported is 65 W m^{-3}. If electrical power costs are low, it would be more economical to try to achieve higher current densities rather than power densities (Jacobson et al. 2011b). The experiments with MDCs must be designed to address the trade-off in these operation conditions. The power vs current relationship for MDCs forms a semi-circular plot. Under this situation, maximum power operation will induce half of the maximum possible current, decreasing the rate of desalination. However, the electrical energy harvested from the MDCs can be utilized in downstream desalination processes including reverse osmosis (Jacobson et al. 2011b).

18.3.3 Current Efficiency

The current efficiency is defined the amount of ions separated per electron transferred at the electrodes. It is shown in Eq. 18.3 (McGovern et al. 2014):

$$\eta_{\mathrm{i}} = \frac{(FzV\,\Delta c)}{\left(N_{\mathrm{CP}} \int idt\right)} \tag{18.3}$$

where Δc is reduction in the concentration of saltwater, V – volume desalinated, N_{CP} – number of cell pairs and i is current. The wide variations in current efficiencies in MDCs suggest that ionic separation is greatly influenced by osmotic water transport into the middle chamber, thus diluting saltwater and improving current efficiency (Chen et al. 2011). This current efficiency is an important factor which can be used to estimate the integrity of the IEMs used. Current efficiencies below 90% indicate IEM integrity failure or huge current losses along feed channels and those above 100% indicate high amount of water transport through the IEMs.

18.3.4 Coulombic Efficiency

Coulombic efficiency (CE) is the ratio of the total number of electrons transferred from the anode to the maximum number of electrons which can be generated due to the bacterial oxidation reactions of the substrate at the anode. It is shown in Eq. 18.4 (Logan 2008):

$$CE = \frac{\left(M_{O_2} \int idt\right)}{\left(n_e F V_{an} \Delta COD\right)} \tag{18.4}$$

where V_{an} is volume of the anode chamber, ΔCOD – change in chemical oxygen demand of the wastewater during MDC operation, M_{O2} – molecular mass of O_2 and n_e is required number of electrons to reduce oxygen to water. The coulombic efficiency thus indicates what fraction of the substrate was utilized by the bacteria for current production. High coulombic efficiency can be achieved by well-isolating the anode chamber from oxygen by the use of multiple IEMs (Veerman et al. 2008).

18.3.5 COD Removal

Most of the MDC experiments have been performed by using easily biodegradable substrates like acetate or xylose which have given high values of COD removal (>70%). But when typical domestic wastewater is used, the COD removal percentage is observed to be much less (~54%) (Luo et al. 2012). This confirms that bacteria like *Geobacter sulfurreducens* and *Pelobacter propionicus* have excellent capability to degrade wastewater but the COD removal is not sufficient for safe discharge of the sewage.

18.3.6 Effects of Electrolyte pH

In the type of reactors in which anodic and cathodic chambers are different, pH imbalance may occur because protons are generated at the anode and hydroxyl ions are generated at the cathode. In single chambered MFCs, overall pH is neutralized due to mixing within the reactor and a bulk neutral pH is maintained. In an MDC, however, the major part of the ionic flux through the IEM pores, which is necessary to balance charge, is due to salt ions like Na^+ and Cl^-. Thus protons and hydroxyl ions are unable to migrate away from their respective chambers and can create severe pH imbalances. The pH imbalance at the anode can prove catastrophic for the exoelectrogenic bacteria, thus diminishing current production. The sensitivity of exoelectrogenic bacteria to anolyte acidification is one of the limitations of MDC applications (Kim and Logan 2011).

At the cathode, in absence of a buffer, the pH can increase up to 12. For the cathodic reactions, a unit increase in pH can result in 59 mV reduction in cell voltage (at 25 °C). Hence at pH 12, the generated cell voltage would be 0.295 V less than that generated at pH 7. An efficient way to solve this problem of pH imbalance is to recirculate the anolyte and catholyte solutions continuously. Recirculation helps stabilize the anolyte pH at around 6 even with low concentration of buffer

(25 mM phosphate). However, the coulombic efficiency drastically reduces with the recirculation because oxygen present in the cathodic chamber enabled the aerobes present there to degrade part of the substrate in a wasteful process. Hence the recirculation is needed to be optimized so as to minimize substrate losses while still achieving the aim of pH balance (Kim and Logan 2013).

18.3.7 Salinity Effects on Exoelectrogenic Bacteria

During the desalination process, chloride ions are rapidly transported through the AEM into the anodic chamber which increases their concentration there. It has been observed that Cl^- concentrations of up to 300 mM do not adversely affect power generation. However, at 500 mM Cl^- concentration, the maximum power output is reduced by 12%. Increasing KCl concentrations increased power generations but only up to 300 mM. This implies that exoelectrogenic bacterial activity is affected at very high concentrations of anions produced while using high content of organic substrates in the anolyte (Lefebvre et al. 2012). Again, the type of bacterium used also determines the effect of high salt concentrations. The activity of *Pelobacter propionicus* decreased and that of *Geobacter sulfurreducens* increased with increasing NaCl concentration in the anolyte, thus suggesting that a given salinity level may favour a particular type of microbe only (Oh and Logan 2006).

18.3.8 Cathode Reactions: O_2 Reduction vs. H_2 Evolution

At the cathode, three separate reactions have been used for MDCs viz. the oxygen reduction reaction (ORR), the hydrogen evolution reaction (HER) and the ferricyanide reduction reaction (Harnisch and Schröder 2010). The ferricyanide reduction is not a sustainable reaction for MDCs, even though its use gives better power production as compared to oxygen. With use of acetate as the substrate in the anolyte, ORR can give up to 1.1 V of potential difference in the open circuit condition. However, the voltage available for desalination is only 0.5–0.6 V due to electrode overpotential losses. HER is not a thermodynamically favourable reaction for acetate-based MDCs and hence an external voltage of 0.11 V is needed at neutral pH to make it occur. However, the recovery of H_2 at the cathode compensates for the invested external energy (Chen et al. 2012a). The advantage of using HER in MDC is that the current densities can be significantly increased by applying higher potentials.

18.4 Types of Microbial Desalination Cells (MDCs)

18.4.1 Osmotic MDCs

It is a type of MDC in which the AEM separating the anodic chamber and the middle chamber is replaced with a forward osmosis (FO) membrane (Fig. 18.4). It is a special type of membrane which maximizes osmotic water transport from the anolyte to the saltwater, thus diluting it and enhancing the desalination process (Zhang and He 2012). The osmotic water transport occurs due to huge difference in the osmolality of the saltwater and the wastewater. However, the introduction of the FO membrane destroys the system's ability of selectively separating the ions and hence the overall rate of ionic separation from the middle chamber and the current efficiencies reduce drastically. Further research is required in this field to improve upon the lacunae in the process (Zhang and He 2013).

18.4.2 Bipolar Membrane MDCs

It is a type of MFC which not only desalinates water but also produces acid (HCl) and base (NaOH) solutions simultaneously. In such MDCs, a bipolar membrane is positioned next to the anode chamber, creating four chambers in all (Fig. 18.5).

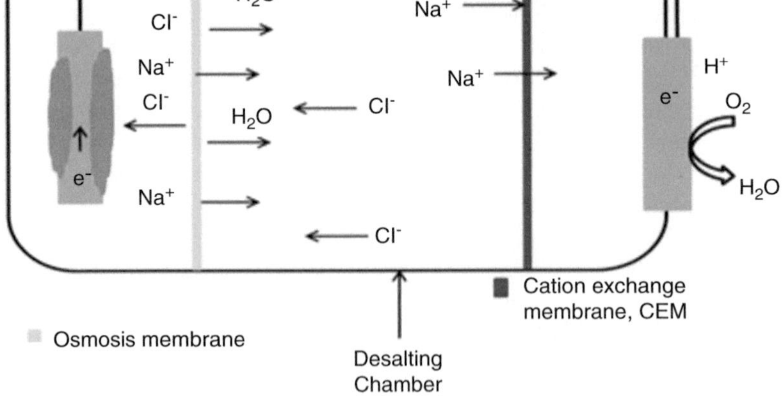

Fig. 18.4 Schematic of Osmotic MDC having osmotic membrane close to bioanode chamber

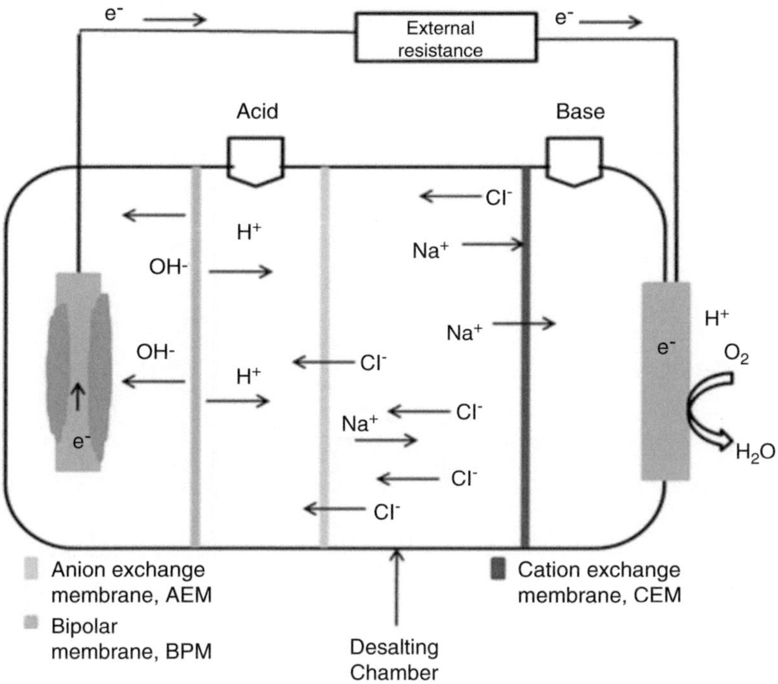

Fig. 18.5 Schematic of a bipolar MDC having bipolar membrane close to anode chamber

A bipolar membrane comprises a CEM and an AEM laminated together to form a single membrane. With the application of a sufficiently high electrical potential difference, water is split into protons and hydroxyl ions at the interface of the laminated IEMs in such a way that hydroxyl ions are liberated towards the anode chamber and protons towards the other side where it combines with the incoming chloride ions from the middle chamber to form hydrochloric acid. On the cathode side, the hydroxyl ions produced combine with the incoming sodium ions from the middle chamber and form sodium hydroxide. These reactions play a major role in avoiding the fall of pH in the anode chamber, thus tackling a crucial limitation in MDCs (Forrestal et al. 2012a). Only thing to be dealt with in a bipolar membrane MDC is the limitation posed by the requirement of application of external voltage to operate the bipolar membrane. Theoretically, the minimum voltage required to split water is 0.83 V (at 1 M concentration and 25 °C) but in practice, the required voltage exceeds 1.2 V. Hence, we can see that the potential difference generated by the MDC alone is not sufficient to meet the requirement for bipolar membrane operation and external voltage of about 1 V is needed to be applied to make it work.

The water splitting at the bipolar membrane plays a key role in the desalination of water too. The extent of desalination is also more because of the maintenance of anolyte pH. Moreover, sufficient external power is required to ensure quality of the produced chemicals. If the costs of the chemicals produced and desalination are

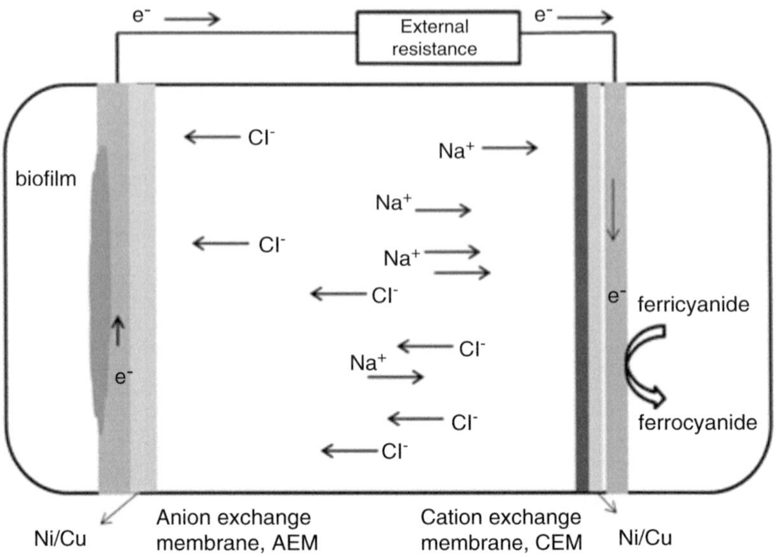

Fig. 18.6 Schematic of a capacitive MDC having Ni/Cu electrode sandwiched between ion exchange membrane and electrode

high enough, the expenses of the external energy supplement can be recovered (Chen et al. 2012b).

18.4.3 Capacitive Microbial Desalination Cell

One of the flaws in conventional MDCs is an increase in the salt concentration of the anolyte and catholyte because of the concentration of the salts which are removed from the saline water. This would result in the prevention of the reuse of anolyte and catholyte in the desalination cell and, therefore, have to be replaced frequently. Furthermore, this ion accumulation may cause concerns for water reuse, where the total dissolved solids (TDS) is regulated. To avoid this problem, a capacitive MDC (cMDC) scheme (Fig. 18.6) has been developed which works on the principle of salt removal by incorporating the concept of capacitive deionization into the system (Forrestal et al. 2012a). Using a double-layer capacitor formed on the high surface electrodes, the ions can be adsorbed when a saline solution flows between the two charged electrodes. A double-layer capacitor on the surface electrodes adsorbs the ions from the saline water, and when the potential gradient has been removed, the ions are allowed to flow back into the liquid. This way, the saltwater is deionized by "electrochemical salt adsorption" on the electrodes without the anode and cathode chambers being contaminated by the salt (Forrestal et al. 2012b).

18.5 Challenges and Perspective

18.5.1 Control of pH

The decrease in the pH of the anolyte is a serious issue because low pH adversely affects the activity of the exoelectrogenic bacteria and the current generated falls rapidly. One of the techniques to counter this, as discussed before, is the recirculation of the anolyte and catholyte. A problem faced here is the reduction of coulombic efficiency due to wasteful aerobic oxidation of substrate at the cathode. However, if we use HER instead of ORR, we can have air-sealed cathodes and so no loss of coulombic efficiency would take place. Another precaution to keep in mind would be to ensure minimal ionic flux through the recirculation tube lines because this ionic flux reduces the current flowing through the IEMs and hence lowers the current efficiency. This can be done by using periodic recirculation, so that no ions flow through the tubing in between pumping periods (Chen et al. 2012a, b).

Another strategy adopted is to blend the catholyte having a high pH with the wastewater flowing into the anode chamber so that the protons released at the anode reactions would neutralize the equal number of hydroxyl ions generated by the cathode reactions and thus maintain the neutral pH. The catholyte volume to be blended with the anolyte should be kept small so as to minimize dilution of organic matter concentration in the wastewater feed (Lakshminarayanaiah 1969).

Use of acidophilic bacteria as the exoelectrogenic microbes in the anode has also been tested and has evolved as an approach to avoid loss of current in the anode with proton production. These bacteria thrive well in acidic conditions and hence low pH in the anode would not affect their activity (Ping and He 2013). For instance, *Acidiphilium cryptum* produces open circuit voltages of about 0.3 V and power up to 12.7 W m^{-2} at pH 4.

18.5.2 Improving Performance of Stacked MDCs

Osmotic water transport in stacked MDCs increases the salinity of the dilute and reduces water recovery. To reduce this osmotic water transport, the hydraulic retention time in the stack should be minimized by increasing salt water flow rates. To maintain proper desalination in the dilute effluent, increase in current density is also necessary. At high current densities, ionic separation will be quick enough to reduce hydraulic retention time of saltwater in the stack, thus decreasing osmotic water transport (Zhang and He 2012).

The volume of water transported by osmosis through the IEM is directly proportional to its area of cross section (Fig. 18.7). By reducing the area of cross section of the IEMs, we can decrease osmotic water transport. But the problem is that the internal resistance of the MDC also increases with decrease in the cross sectional area, which would result in increased electrical potential losses. It is

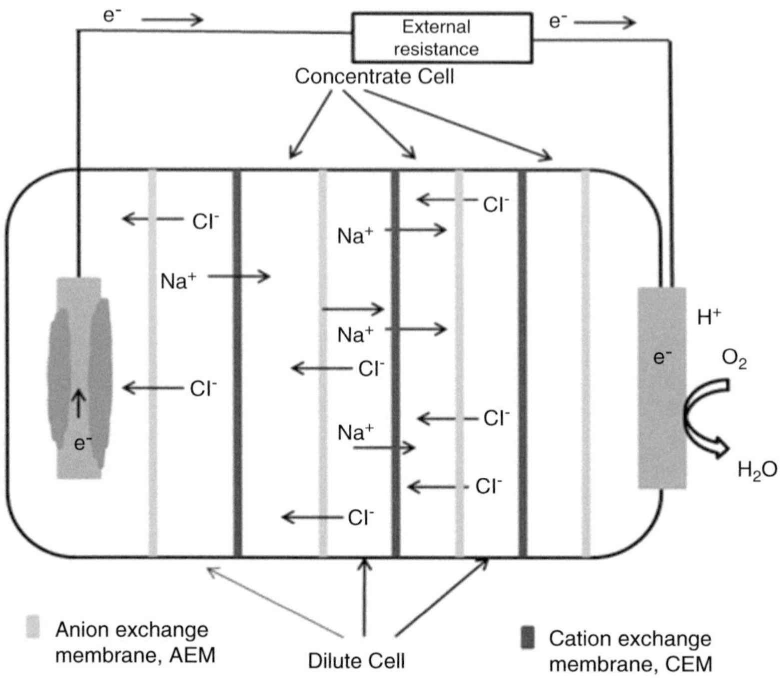

Fig. 18.7 Schematic of a stackable MDC

experimentally observed that decreasing the cross sectional area by an order of magnitude does not affect the current flowing through the stack significantly. Apart from reducing osmotic water transport, reduction in cross sectional area also reduces the capital cost of the MDC because IEMs are very costly. We can also use IEMs with less water content to reduce osmotic water transport through them, although low water content sacrifices high ionic conductivity.

18.5.3 IEM Integrity Under High Microbial Activity

As the AEM lies adjacent to the anode, which is a site of active microbial growth, a biofilm formation in the AEM is inevitable. It has been observed that on using the MDC continuously for long periods of time, bacterial biofilms develop on the AEM surface though it is not clear how they affect the ionic transport through the AEM (Luo et al. 2012)? The growth of biofilm and deposition of organic matter in the wastewater on the IEM may affect the functional groups on the surface of the IEM or interfere with its polymeric structure. However, it has not been studied extensively. Coatings on the AEM, which suppress organic matter transport to the IEM, might enhance the stability and extend the life of the AEM (Logan 2008).

18.5.4 Water Safety

While working with microbes and wastewater, the general question which would always arise is: Will the desalinated water be safe? There is always a possibility for the microbes to pass through the AEM and enter into the middle chamber. It all depends on the properties of the material of the AEM. The pores of the AEM must be small enough to allow only the ions and water molecules to pass through. But in the practical scenario, it is often observed that solutes like acetate and glucose pass through the pores of the AEM quite easily and move into the desalination chamber (Saeed et al. 2015). The osmotic movement of water also drags certain organic molecules in the wastewater along with it into the desalinated water. The problem worsens in the case of MODCs where the pore size of FO membranes is much higher and it may even allow certain microbes to pass into the middle chamber. This allows bacterial growth in the desalination chamber due to availability of organic matter in it. All these phenomena promote the contamination of the desalinated water and proper steps have to be taken to nullify it. It includes using AEMs with very small pore size and proper downstream treatment processes to decontaminate the desalinated water (Mehanna et al. 2010).

18.6 Conclusion

Currently, the MDCs suffer from many drawbacks and generate low power. The integration of MFCs into a wastewater treatment-cum-desalination plant will present greater challenges for bio-fouling of IEMs and in terms of being cost-effective but with newer advances in terms of extra chemical production and more efficient sewage treatment and power production, the expenses may get recovered. For production of potable water MDCs can be used as a pre-treatment step for reverse osmosis. It will reduce the energy requirements in the RO process. It has been proven that for partial desalination of brackish water for irrigational purposes, MDCs are the most suitable treatment method. Further developments in the design would improve the performance of MDCs and diversify their applications.

References

Bard, A. J., & Faulkner, L. R. (2001). *Electrochemical methods: Fundamentals and applications* (2nd ed.). New York: Wiley.

Cao, X., Huang, X., Liang, P., Xiao, K., Zhou, Y., Zhang, X., & Logan, B. E. (2009). A new method for water desalination using microbial desalination cells. *Environmental Science & Technology, 43*, 7148–7152.

Chen, X., Xia, X., Liang, P., Cao, X., Sun, H., & Huang, X. (2011). Stacked microbial desalination cells to enhance water desalination efficiency. *Environmental Science & Technology, 45*, 2465–2470.

Chen, S., Liu, G., Zhang, R., Qin, B., Luo, Y., & Hou, Y. (2012a). Improved performance of the microbial electrolysis desalination and chemical-production cell using the stack structure. *Bioresource Technology, 118*, 507–511.

Chen, X., Liang, P., Wei, Z., Zhang, X., & Huang, X. (2012b). Sustainable water desalination and electricity generation in a separator coupled stacked microbial desalination cell with buffer free electrolyte circulation. *Bioresource Technology, 119*, 88–93.

Forrestal, C., Xu, P., Jenkins, P. E., & Ren, Z. (2012a). Microbial desalination cell with capacitive adsorption for ion migration control. *Bioresource Technology, 120*, 332–336.

Forrestal, C., Xu, P., & Ren, Z. (2012b). Sustainable desalination using a microbial capacitive desalination cell. *Energy & Environmental Science, 5*, 7181–7187.

Greenlee, L. F., Lawler, D. F., Freeman, B. D., Marrot, B., & Moulin, P. (2009). Reverse osmosis desalination: Water sources, technology, and today's challenges. *Water Research, 43*, 2317–2348.

Harnisch, F., & Schröder, U. (2010). From MFC to MXC: Chemical and biological cathodes and their potential for microbial bioelectrochemical systems. *Chemical Society Reviews, 39*, 4433–4448.

Jacobson, K. S., Drew, D. M., & He, Z. (2011a). Efficient salt removal in a continuously operated upflow microbial desalination cell with an air cathode. *Bioresource Technology, 102*, 376–380.

Jacobson, K. S., Drew, D. M., & He, Z. (2011b). Use of a liter-scale microbial desalination cell as a platform to study bioelectrochemical desalination with salt solution or artificial seawater. *Environmental Science & Technology, 45*, 4652–4657.

Kim, Y., & Logan, B. E. (2011). Series assembly of microbial desalination cells containing stacked electrodialysis cells for partial or complete seawater desalination. *Environmental Science & Technology, 45*, 5840–5845.

Kim, Y., & Logan, B. E. (2013). Microbial desalination cells for energy production and desalination. *Desalination, New Directions in Desalination, 308*, 122–130.

Kokabian, B., & Gude, V. G. (2013). Photosynthetic microbial desalination cells (PMDCs) for clean energy, water and biomass production. *Environmental Science: Processes and Impacts, 15*, 2178–2185.

Kokabian, B., & Gude, V. G. (2015). Sustainable photosynthetic biocathode in microbial desalination cells. *Chemical Engineering Journal, 262*, 958–965.

Lakshminarayanaiah, N. (1969). Electroosmosis in ion-exchange membranes. *Journal of the Electrochemical Society, 118*, 338–342.

Lefebvre, O., Tan, Z., Kharkwal, S., & Ng, H. Y. (2012). Effect of increasing anodic NaCl concentration on MFC performance. *Bioresource Technology, 112*, 336–340.

Liu, C., Rainwater, K., & Song, L. (2011). Energy analysis and efficiency assessment of reverse osmosis desalination process. *Desalination, 276*, 352–358.

Logan, B. E. (2008). *MFCs* (1st ed.). Hoboken: Wiley-Interscience.

Luo, H., Xu, P., Jenkins, P. E., & Ren, Z. (2012). Ionic composition and transport mechanisms in microbial desalination cells. *Journal of Membrane Science, 409–410*, 18–23.

McGovern, R. K., Weiner, A. M., Sun, L., Chambers, C. G., Zubair, S. M., & Lienhard, V. J. H. (2014). On the cost of electrodialysis for the desalination of high salinity feeds. *Applied Energy, 136*, 649–661.

Mehanna, M., Saito, T., Yan, J., Hickner, M., Cao, X., Huang, X., & Logan, B. E. (2010). Using microbial desalination cells to reduce water salinity prior to reverse osmosis. *Energy & Environmental Science, 3*, 1114.

Oh, S.-E., & Logan, B. E. (2006). Proton exchange membrane and electrode surface areas as factors that affect power generation in MFCs. *Applied Microbiology and Biotechnology, 70*, 182–189.

Ping, Q., & He, Z. (2013). Improving the flexibility of microbial desalination cells through spatially decoupling anode and cathode. *Bioresource Technology, 144*, 304–310.

Rahimnejad, M., Adhami, A., Darvari, S., Zirepour, A., & Oh, S.-E. (2015). MFC as new technology for bioelectricity generation: A review. *Alexandria Engineering Journal, 54*, 745–756.

Saeed, H. M., Husseini, G. A., Yousef, S., Saif, J., Al-Asheh, S., Abu Fara, A., Azzam, S., Khawaga, R., & Aidan, A. (2015). Microbial desalination cell technology: A review and a case study. *Desalination, 359*, 1–13.

Torres, C. (2012). Improving MFCs. *Membrane Technology, 2012*, 8–9.

Veerman, J., Post, J. W., Saakes, M., Metz, S. J., & Harmsen, G. J. (2008). Reducing power losses caused by ionic shortcut currents in reverse electrodialysis stacks by a validated model. *Journal of Membrane Science, 310*, 418–430.

Watson, V. J., & Logan, B. E. (2011). Analysis of polarization methods for elimination of power overshoot in MFCs. *Electrochemistry Communications, 13*, 54–56.

Wen, Q., Zhang, H., Chen, Z., Li, Y., Nan, J., & Feng, Y. (2012). Using bacterial catalyst in the cathode of microbial desalination cell to improve wastewater treatment and desalination. *Bioresource Technology, 125*, 108–113.

Zhang, B., & He, Z. (2012). Integrated salinity reduction and water recovery in an osmotic microbial desalination cell. *RSC Advances, 2*, 3265–3269.

Zhang, B., & He, Z. (2013). Improving water desalination by hydraulically coupling an osmotic MFC with a microbial desalination cell. *Journal of Membrane Science, 441*, 18–24.

Chapter 19
Biophotovoltaics: Conversion of Light Energy to Bioelectricity Through Photosynthetic Microbial Fuel Cell Technology

Rashmi Chandra, S. Venkata Mohan, Parra-Saldivar Roberto, Bruce E. Ritmann, and Raul Alexis Sanchez Cornejo

19.1 Introduction

Solar radiation led into the planet earth by photosynthesis is the main energy basis for survival. Photosynthesis is a physico-chemical process where photosynthetic organism transform solar energy into chemical energy with simultaneous use of bioenergetic processes (Georgianna and Mayfield 2012; Strik et al. 2011). Photosynthetic microbial fuel cells (PhFCs) are newly established tools that harvest sun energy to yield electricity and has gained major attentiveness in applied and academic research due to its sustainable and renewable nature (Chandra et al. 2012; Venkata Mohan et al. 2014; Venkata Subhash et al. 2013). PhFCs use plants or phototrophic microorganisms to trap sun light and use photo-bioelectrochemical system to produce bioelectricity (El Mekawy et al. 2014). This technology symbolizes a multidisciplinary method to search for renewable energy. It represents the convergence for the life-sciences, chemical and physical science (McCormick et al. 2015; Xiao and He 2014). Microbial fuel cells (MFCs) are mainly focused towards anodic reaction with dark fermentation as microbial metabolic function. Analogous to dark

R. Chandra (✉) • P.-S. Roberto • R.A.S. Cornejo
School of Engineering and Science, Tecnologico de Monterrey, Campus Monterrey, Ave. Eugenio GarzaSada 2501, Monterrey, NL, Mexico, 64849
e-mail: rashmichandra@itesm.mx

S. Venkata Mohan
Swette Center for Environmental Biotechnology, The Biodesign Institute, Arizona State University, Tempe, AZ, USA

B.E. Ritmann
School of Engineering and Science, Tecnologico de Monterrey, Campus Monterrey, Ave. Eugenio GarzaSada 2501, Monterrey, NL, Mexico, 64849

Bioengineering and Environmental Sciences (BEES), CSIR-Indian Institute of Chemical Technology (CSIR-IICT), Hyderabad 500 007, India

D. Das (ed.), *Microbial Fuel Cell*,
https://doi.org/10.1007/978-3-319-66793-5_19

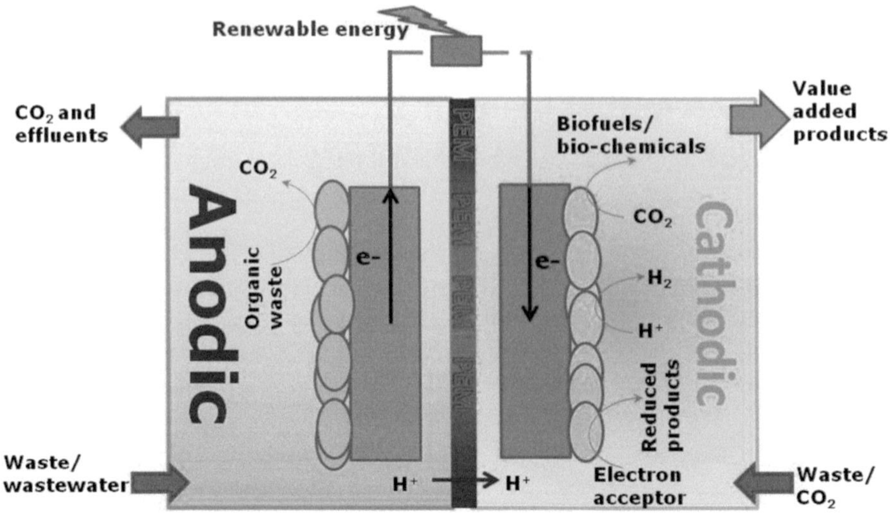

Fig. 19.1 Mechanism of development of potential gradient in biological system

fermentation, photosynthetic fuel cells (PhFC) or biophotovoltaic system (BPV) will also be functioned using photo synthetic bacteria (PSB) or algae/cyanobacteria based on their photosynthetic mechanism (oxygenic or anoxygenic) (Chandra et al. 2012; Rosenbaum et al. 2010a). Oxygenic photosynthesis based autotrophic PhFCs include diverse categories of MFC containing both heterotrophs and autotrophs (Venkata Mohan et al. 2010, 2011) (Fig. 19.1). The current chapter has made an attempt to convey the present photosynthetic mechanism for PhFC application in understanding on the innate potential of photosynthetic mechanism.

19.2 Mechanism of Development of Potential Gradient in Biological System

Light energy and electron in reduce substrates (H_2S, CH_4, organic compounds etc.) begin "electric circuit" which terminates when electron arrives at the electron sink provided by a terminal electron acceptor (Bretschger et al. 2010). An electrical circuit is established between light energy (responsible for excitation of e^-), e^- source (from the substrate metabolism) and the e^- sink (O_2) by inserting an external electrode as transitional acceptor of electron (Venkata Mohan et al. 2008a). PhFC work on the basis of oxidative reaction divided by a PEM (proton exchange membrane) and excitation of e^- by light energy from the anodic oxidation. Anode chamber is a bio-factory that enables the production of e^- and H^+ through organic carbon breakdown in absence of oxygen. Protons goes to the cathode through PEM and helps in development of potential difference and finally to bioelectricity. Light energy captured by pigment (bacteriochlorophyll (*Bchl*) and chlorophyll molecules) molecules is channeled to the reaction centre. Solar energy redirects organic waste / CO_2 as e^- source to activate a series of photo-chemical reactions separating charge across the membrane. Extensively distributed *Bchl a* in anoxygenic photosynthetic bacteria can transfer energy to the reaction centre (RC) which was bounded by a light harvesting complex called RC-LH (Blankenship 2016). This complex helps to develop electromotive force (emf) with help of quinine and membrane-embedded electron transport protein. Electron from the *Bchl-a* enters quinone pool and passes to cytochrome complex before exerting through cell membrane. Solar energy acts as source of energy and electron source was acetate (Eqs. 19.1, 19.2, 19.3 and 19.4) for photo-electrochemical power generation as shown below.

$$C_6H_{12}O_6 + \text{Light} \rightarrow \text{Acetate} + \text{Propionate} + \text{Butyrate} + H^+ + CO_2 \quad (19.1)$$

$$CH_3CH_2COOH + 3H_2O + \text{Light} \rightarrow HCO_3^- + \text{Acetate} + 7H^+ \quad (19.2)$$

$$CH_3CH_2CH_2COOH + 2H_2O + \text{Light} \rightarrow \text{Acetate} + 2H^+ + 2H_2 \quad (19.3)$$

$$CH_3COO^- + 4H_2O \rightarrow 2HCO_3^- + 9H^+ + 8e^- \quad (19.4)$$

In oxygenic photosynthesis, protein complexes like PSI, PSII and cytochrome bf complex are required to operate in series for electron transfer from H_2O to $NADP^+$ and oxygen generation. Electron is transported among these bulky protein complexes by tiny moveable protein-like plastoquinone and plastocyanin. These molecules transport electron to comparatively extended distances showing a distinctive role in solar energy capture and bioelectricity generation.

19.3 Light Harvesting Technologies for Bioelectricity Generation

19.3.1 Chemical Based

The solar radiation is majorly responsible for empowering earth right from its birth. This solar energy can be trapped by various chemical means like photovoltaic and thermal harvesting. Photovoltaic cells (PV) are one of the major interests in solar energy conversion, where sun light is transformed straight to electrical energy or by photoelectrochemical cells (PEC) where light is transformed to energy in chemical. This technology has undergone development and applied in both minor private systems as well as in large-scale installations connected to the national grids (Badawy 2015). A PV cell is a purely solid-state device, where holes or electrons carry the moving charge and there is no accompanying chemical change (Archer 2002). Initially solar cells were built on silicon wafers where solar conversion efficiencies are between 12% and 16%, (Badawy et al. 2015). The next stage of solar cells is based on thin film technology. This device is based on a dye sensitized porous nanocrystalline TiO_2 photoanode with efficiencies greater than 10%, lower than thin film Si solar cells. Uninterrupted research in nanotechnology has directed towards improvement of the present solar cells that are based on nano-porous and nano-crystals materials. These technologies were considerably high cost and sophisticated which led to the very defined application (Badawy 2015). These technologies are confined to few segments and did not introduce energy to the biological system for wastewater treatment, CO_2 sequestration or biofuel generation. In this regards photosynthetic organism plays a major role to trap the solar energy and redirect it to the biosphere for generation of biofuel. This property of PhFC or BPV showed an advantage over PV.

19.3.2 *Biological Based*

19.3.2.1 Anoxygenic Photosynthesis at Anode

Production of molecular hydrogen has been demonstrated and well established as an ubiquitous feature of photosynthetic microbial metabolism (Chandra and Venkata Mohan 2011, 2014). Photosynthetic purple bacteria follow photofermentation process (anoxygenic photosynthesis) to produce H_2 from organic compounds which have been studied in PhFC for immediate hydrogen removal and bioelectricity generation of photosynthetic activity and power production of 120 mW m^{-2} (Chandra et al. 2012). The critical point for PhFC operation is its stability, proton exchange membrane and cost effectiveness (Schröder et al. 2003). Bchl *a* molecules present in anoxygenic photosynthetic organisms are used to trap sunlight as energy source and organic waste or CO_2 as e^- source to direct the reaction centre of the photo-system. Bchl *a* is efficient in energy transfer efficiently (95 to 99%) via reaction centre (RC) bounded to light harvesting complex called LH. This complex converts solar energy via quinon pool and second cytochrome (cyt) *bc1* complex. The exited e^- from the Bchl *a* go in Q pool and then to cytochrome complex and finally excrete through cell membrane (Fig. 19.2). These solar radiations act as source of energy and organic carbon as e^- source. Anoxygenic photosynthesis have benefits in maintaining less DO that help in increasing the power output. It shows remediation of acid-rich effluent generated from the acidogenic hydrogen production process as substrate for power generation.

19.3.2.2 Photosynthetic at Anode with Artificial Mediators Biological Photovoltaics

In these type of photosynthetic microbial fuel cell (PhFC), intracellular carbon storages are metabolized and electrons are collected at the electrode through PhFC. These systems operate under oxygenic photosynthesis which results in release of e^-

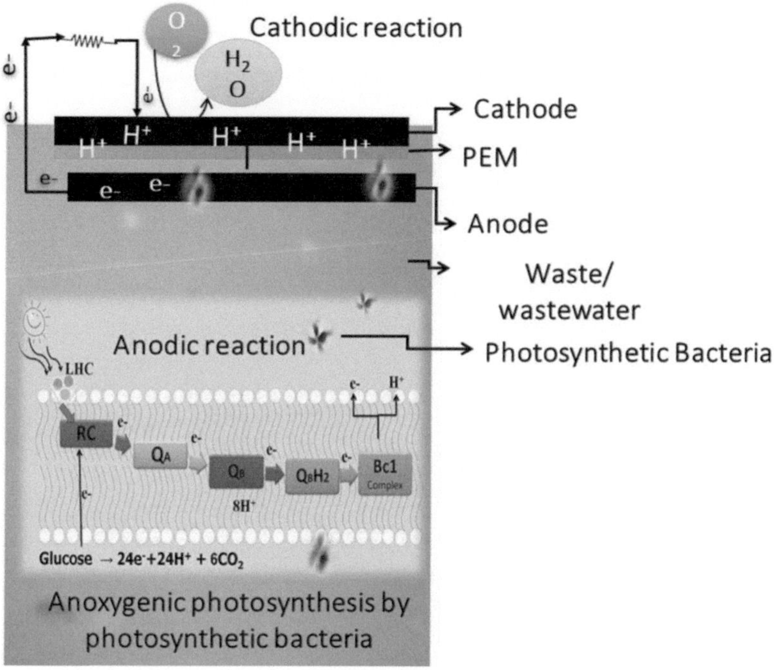

Fig. 19.2 Anoxygenic photosynthesis at anode: *RC* (Reaction centre), Q_A (Quinone A), Q_B (Quinone B), Bc1 Complex (Quinone Complex), Quinone (Q), Plastoquinone (PQ), Cytochrome f (cyt f) and Plastocyanin (PC)

and H^+ along with oxygen by the photophosphorylation of water. Dissolved oxygen will act as a terminal e^- acceptor (TEA) and trap the reducing equivalents generated and lower the electrogenesis activity (Venkata Mohan et al. 2008b; Biffinger et al. 2008). Production of the reducing equivalents has constructive effect on the electrogenic activity but concurrent production of O_2 has negative effect on the complete activity. There are studies reported with 2-hydroxy-1,4-naphtoquinone as an artificial mediator to shuttle the e^- from microorganism to the anode (mediated PhFC). Use of unstainable and environmentally problematic non-natural redox mediators prevent the initial PhFC from practical realization and result in slow down of PhFC research.

19.3.2.3 Oxygenic Photosynthesis at Anode

Oxygenic photosynthesis can easily couple bioelectricity production in a well-defined MFC. In oxygenic photosynthesis and releases e^- and H^+ during photophosphorylation result in discharge of O_2, e^- and H^+. Solar radiation transformation follows a scheme, where H^+ will be captured by the light-harvesting complex

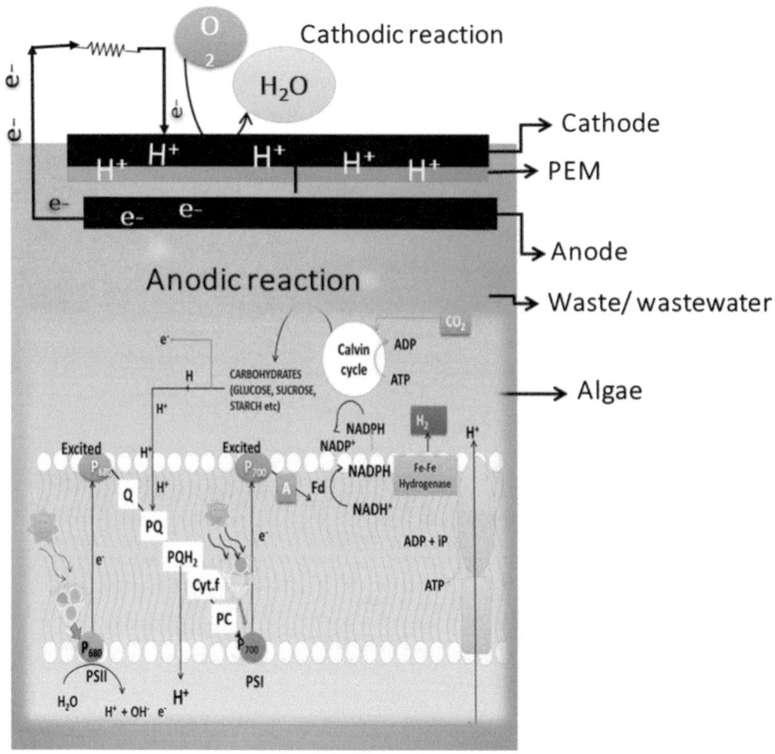

Fig. 19.3 Oxygenic photosynthesis at anode, PSII (Photosystem II), Quinone (Q), Plastoquinone (PQ), Cytochrome f (cyt f) and Plastocyanin (PC), PSI (Photosystem I) and Ferredoxin (Fd)

(LHC) and relocated to the P680 (PSII) which becomes exited followed by splitting one water molecule to H^+, e^- and O_2. At this site light energy gets converted to chemical energy and that can be converted to bioelectricity in presence of electrode-membrane assembly through the produced redox equivalents (Fig. 19.3). These types of PhFC are called oxygenic photo-bioelectrocatalytic fuel cell (PhFCOX); they are capable to divert e^-straight from the electron transport chain (ETC) to external electrode (anode) and neutralizing at cathode (Chandra et al. 2012, Biffinger et al. 2008; Pisciotta et al. 2010). Studies have been performed to evaluate the electrogenic activity of these microorganism made in PhFCOX and showed bioelectrogenic activity (3.55 mW m^{-2}) accompanying with biomass production of 2.87 g L^{-1}. Current generation was higher during the sunlight (46 mV; 0.6 mA) as compared to the dark (6 mV; 0.01 mA). This type of MFCs are self-sustainable and furthermore the biomass can be utilized for other value added products in a biorefinery. These studies offer an economical, renewable and justifiable electricity possibility related with CO_2 mitigation and wastewater treatment.

19.3.2.4 Oxygenic Photosynthesis at Cathode

Mechanical aeration at cathode is the most energy exhaustive process. Microbes play an exceptional role in oxygen reduction reaction at cathode in photosynthetic microbial fuel cell operation (Wetser et al. 2015). Reducing equivalents (H^+ and e^-) reduction at cathodic results in bioelectrogenic activity. Presence of effective terminal e^- acceptor (TEA) at cathode helps in escalating e^- transfer its efficient deployment at cathode and ensuing decrement e^- losses and advanced power output and catalysts like platinum, Fe^{3+}, $K_3[Fe(CN)_6]$ were additionally cast off for the intensification of abiotic cathode process to enhance bioelectrogenic activity (Zhang et al. 2010; Venkata Mohan et al. 2008a; Hamelers et al. 2010). Oxygenic photosynthesis is a probable substitute to electricity powered aeration carried by microalgale and cyanobacteria. O_2 generated during this process via biophotolysis helps in CO_2 sequestration in presence of sunlight. Solar radiation excite the Chl P680 to Chl P680* (manganese centre at the reaction centre) which ruptures H_2O molecules into $4H^+$, $4e^-$ and two oxygen atoms. Q_B a mobile protein accepts two e^- and then takes on two H^+. PQ is the detached plastoquinone molecule; it is moveable and transfers electron through plastoquinone and plastocyanin. They transport comparatively stretched distances and play an exclusive part in photosynthetic energy transformation and finally to NADPH which is used in dark reaction for the CO_2 fixation. The generated oxygen in the whole process acts as terminal e^- acceptor for e^- and H^+ generated at the anode (Fig. 19.4). The

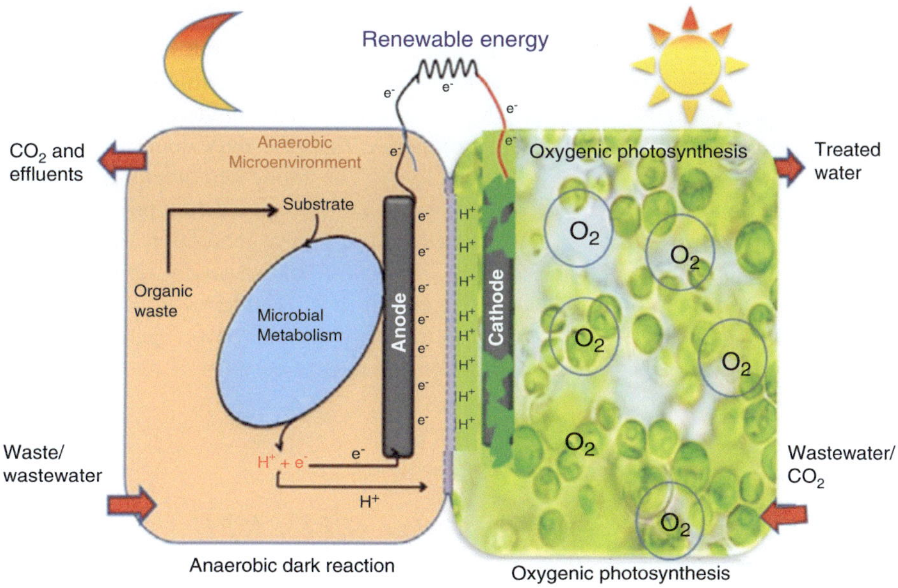

Fig. 19.4 The synergistic association between bacterial fermentation at anode and the oxygenic photosynthesis of microalgae at cathode facilitated good power output as well as treatment efficiency

synergistic connotation among anaerobic bacterial metabolism at anode and the oxygenic photosynthesis of microalgae at cathode assisted notable power output. Its operation throughout spring and cold climate showed higher bioelectrogenic activity (57.0 mW m^{-2}) over summer weather (1.1 mW m^{-2}) because of higher oxygenic photosynthetic activity of microalgae which result in higher dissolved oxygen (DO) levels. However, high temperatures and light intensity in summer result in inactivation of RuBisCO (Venkata Mohan et al. 2014). Algal oxygenic photosynthesis at cathode facilitates to maintain higher DO and removing the necessity of energy exhaustive aeration.

19.3.2.5 Plant MFC (Synergism Between Mixed Heterotrophic Bacteria and Plant)

Plants fix solar energy and CO_2 to carbohydrates by photosynthesis in green tissues of the plants and translocate to rhizosphere via roots of the plants (Helder et al. 2013; Hubenova et al. 2012; Timmers et al. 2012). Root system discharges various organic compounds in to the zone of rhizosphere through rhizo deposition. These machineries of root and rhizosphere not only solitarily serve as a foundation of organic carbon source but also comprise biomolecule that stimulate chemotaxis of microbial consortium to the rhizosphere. Root exudates also act as mediator for e^- transfer and combination of MFC in plant rhizosphere helps in change of sun energy into bio-electricity. Venkata Mohan et al. reported four electrode microbial fuel cell assemblies by changing distances from root in rhizosphere to anode i.e. Anode 1–0; Anode 2–8; Anode 3–12 and Anode 4–16 cm at 2 cm depth from soil-layer (Fig. 19.5). The cell assemblage close to the root revealed maximum electrogenic activity because of the serious part of root-exudates. During light exposure process all MFC showed 10% greater power production than the dark conditions which directly endorsed plant's photosynthetic activity. The anode kept nearer to the root showed maximum power output. Water logging conditions create a positive influence on power generation. A relative assessment of power generation was deliberate with *Arundinella anomala, Spartina anglica* and *Arundo donax* where *S. anglica* documented greater bio-electrogenic activity (Helder et al. 2012; Helder et al. 2010; Strik et al. 2011). In continuation to this power generation by *Pennisetum setaceum* rhizo deposits are formed mostly because of photosynthetic CO_2 capture by the photosynthetic activity.

19.3.3 Ecological Engineered System (EES): MFC to Wetland System

EES is based on environmental ideologies to simplify the recycling of natural resources efficiently (Venkata Mohan et al. 2010). Ecological system has exceptional benefits like narrow or no usage of chemical compounds, easy

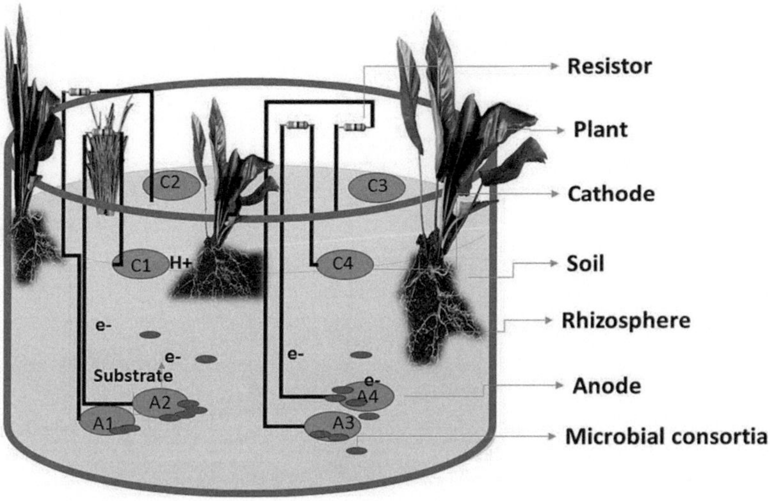

Fig. 19.5 Plant MFC (Synergism between plant and mixed heterotrophic bacteria)

operation, no foul odour and inexpensive. When these types of system with photosynthetic microorganism is integrated with electrode system they result in benthic fuel cell or sediment type fuel cells (SMFC). Incorporating EES with SMFC can facilitate the bioelectricity generation from the active microbial metabolism at sediment beds by metabolizing organic matter through biochemical reactions (Bond et al. 2012; Marsili et al. 2008) or by excretion of redox components (Pandit and Mahadevan 2011; Zhuang et al. 2011). A small floating macrophyte ecosystem with *Eichhornia crassipes* and snails embedded with sediment fuel cells was reported using wastewater (Mohan and Chandrasekhar 2011). MFC arrangements in SMFC revealed feasibility of power generation from domestic wastewater at individual electrode setups which reasonably enhanced with distillery wastewater and reduced further with rise in the load. The electrode assemblages connected in series showed stable power production (780 $\pm$ 22 mV; 4.14 $\pm$ 0.19 mA). A notable treatment efficiency was observed during EES operation. The study computed the synergetic relationship of ecological–electrogenic engineered system on wastewater treatment and bioelectricity production. This process can be easily combined to primary wastewater treatment plants to improve treatment efficiency effluents as secondary or tertiary treatment of wastewater (Fig. 19.6).

19.3.4 Light Harvesting Proteins for Photovoltaic and Photoelectrochemical Devices

Photosynthetic proteins such as cyanobacterial phycocyanin, phycoerathrin, PSI etc. provide an encouraging approach to enhance the photocurrent density. The

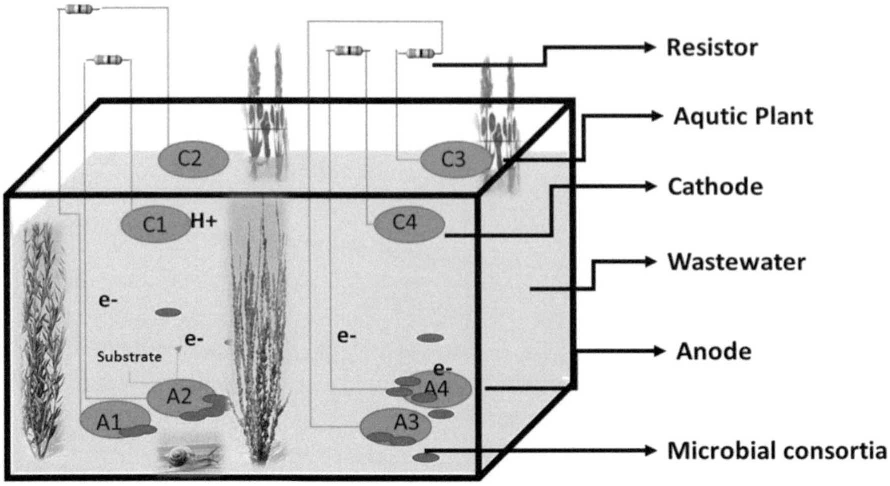

Fig. 19.6 Synergism between phototrophic and mixed heterotrophic bacteria in sediments in ecological system

application of these photosynthetic proteins have very high influence on photoanode assembly in bioinspired or fusion bio-inorganic photovoltaic devices. These light-harvesting proteins alone or entire photosynthetic reaction centres are used to "sensitize" metal and semiconductor surfaces (Ihssen et al. 2014; Lovley 2006). Phycobilisomes are multiprotein-pigment complexes found in photosynthesis apparatus of prokaryotic cyanobacteria, *Rhodophyta* ("red algae") and *Cryptophyta*. The close-fitting association of proteins with the electrode surface is an essential requirement for the appropriate effect of bio-hybrid devices. These photosynthetic proteins can be immobilized on surface by various ways like electrostatic interactions, adsorption, trap in matrixes such as in situ polymerized hydrogels, creation of covalent bonds between the polypeptide chain and chemical groups of the surface, crosslinking or hydrophobic (van der Waals) (Badura et al. 2011; Kathiravan and Renganathan 2009; Suemori et al. 2006). The immobilization depend on the physicochemical behaviours of protein, its interaction with the aqueous phase and covalent linkage (Tran and Balkus 2011). There are various new opportunities of engineering bio-hybrid hematite photoanodes for PEC cells which raise many further questions and leave opportunity for advance developments.

19.4 Applications

Photosynthetic MFCs are encouraging renewable energy sources due to their minimal or no net-CO_2 emission. Despite the PhFCs and MFC advantages their applied aspect is actually restricted to definite areas in a non-industrial way. The goal in the near future is to enhance cost effectiveness and performance efficiency in order to scale-up the PhFC and MFC systems for making them a feasible renewable energy source.

19.4.1 Wastewater Treatment

Microorganisms are able to degrade effluents and generate power at the same time. The PhFC and MFCs technology can be applied to produce electrical power taking as a substrate the effluent of industrial, agricultural and municipal wastewater. Even pollutants like azo dyes, polycyclic aromatic hydrocarbons (PAHs), benzene derivatives and inorganic wastewaters containing sulphide are also reported to act as substrates (Han et al. 2015; Raschitor et al. 2015; Zhang et al. 2016). In recent years, COD removal have reached to about 100% with this technology (Luo et al. 2011) and important advances were made by Wu et al. (2016) for scaling up a reactor of 72 L for wastewater treatment with a COD removal efficiency of 97%.

19.4.2 Powering Underwater Monitoring Devices

Environmental sensors can be used to understand the behaviour and responses of an ecosystem, but these require energy for operation. MFC and PhFCs can perhaps be used to control devices particularly in systems such as creeks, rivers and oceans where it is problematic to regularly access the system to exchange batteries (Bond et al. 2002). Sediment fuel cells are being developed to monitor electrochemical approaches e.g. amperometry, potentiometry, conductometry and water toxicity.

19.4.3 BOD Sensing

Microbial fuel cell machinery is capable to act as a sensing device for pollutant analysis and *in situ* process monitoring. As reported by Kim et al. (2003), an MFC showed long-term stable performance for a period up to 5 years and Di Lorenzo et al. (2014) reported a minimum response time of 2.8 min for this technology. These studies show that the MFC can be a feasible alternative for BOD sensing far extended life span than BOD sensors reported in the literature.

19.4.4 *Biohydrogen Production in PhFC*

In normal conditions of bacterial fermentation in the MFC, hydrogen production is limited to 2–3 mol H_2 (mol glucose)$^{-1}$ due to a biochemical barrier. This process can be adapted to yield more hydrogen instead of bioelectricity resulting in a bioelectrochemically assisted microbial reactor (BEAMR). This was accomplished under anaerobic condition of cathode chamber and supplementing with external voltage of 0.25 V to cathode where H^+ are condensed to H_2. The BEAMR has the potential to produce a total of 8–9 mol H_2 mol^{-1} glucose and would provide a renewable option to fit production to energy needs.

19.5 Conclusion

PhFC is a sustainable biophotovoltanic device which enables numerous bioenergy/byproduct production along with waste reduction. Enactment of PhFC has been studied for diverse applications using an extensive range of substrates from domestic to industrial wastewater to produce energy with simultaneous treatment. This technology has to get several changes in order to make it a sustainable alternate source to the present technologies. For the scaling up feasibility, the manufacturing cost of PhFC should be reduced and its efficiency has to be enhanced. The theoretical value of conceivable potential is only 1.2 V (from single cell); therefore, stacking of fuel cells in series or parallel connection must be improved to get feasible power. To improve the economic viability, studies on the valuable end products like hydrogen or proteins are needed. The efforts of various research groups throughout the world will overcome the limitations of this technology and give to the society an innovative and promising source of alternate energy.

Acknowledgements The authors want to thank Director of Tecnológico de Monterrey Monterrey Mexico; CSIR-IICT, Hyderabad, India; and Arizona State University, Tempe, AZ, USA for their encouragement and support.

References

Archer, M. D. (2002). Photovoltaics and photoelectrochemistry: Similarities and differences. *Physica E: Low-dimensional Systems and Nanostructures, 14*, 61–64.

Badawy, W. A. (2015). A review on solar cells from Si-single crystals to porous materials and quantum dots. *Journal of Advanced Research, 6*, 123–132.

Badawy, W. A., Elmeniawy, S. A., & Hafez, A. N. (2015). Improvement of the photovoltaic characteristics of industrially fabricated solar cells by chemical etching of the Si surface. *Journal of Solar Energy Engineering, 137*, 041007.

Badura, A., Guschin, D., Kothe, T., Kopczak, M. J., Schuhmann, W., & Rogner, M. (2011). Photocurrent generation by photosystem 1 integrated in crosslinked redox hydrogels. *Energy & Environmental Science, 4*, 2435–2440.

Blankenship, R. E. (2016). Molecular evidence for the evolution of photosynthesis. *Trends in Plant Science, 6*, 4–6.

Bond, D. R., Holmes, D. E., Tender, L. M., & Lovley, D. R. (2002). Electrode-reducing microorganisms that harvest energy from marine sediments. *Science, 295*, 483–485.

Bond, D. R., Strycharz-glaven, S. M., Tender, L. M., & Torres, C. I. (2012). On E-transport through *Geobacter* biofilms. *Chem Sus Chem, 5*, 1099–1105.

Bretschger, O., Osterstock, J. B., Pinchak, W. E., Ishii, S., & Nelson, K. E. (2010). Microbial fuel cells and microbial ecology: Applications in ruminant health and production research. *Microbial Ecology, 59*, 415–427.

Chandra, R., Annie Modestra, J., & Venkata Mohan, S. (2015). Biophotovoltaic cell to harness bioelectricity from acidogenic wastewater associated with microbial community profiling. *Fuel, 160*, 502–512.

Chandra, R., Sravan, J. S., Hemalatha, M., Butti, S. K., & Venkata Mohan, S. (2017). Photosynthetic synergism for sustained power production with microalgae and photobacteria in a biophotovoltaic cell. *Energy and Fuels, 31*, 7635–7644. https://doi.org/10.1021/acs.energyfuels.7b00486.

Chandra, R., & Venkata Mohan, S. (2011). Microalgal community and their growth conditions influence biohydrogen production during integration of dark-fermentation and photo-fermentation processes. *International Journal of Hydrogen Energy, 36*, 12211–12219.

Chandra, R., & Venkata Mohan, S. (2014). Enhanced bio-hydrogenesis by co-culturing photosynthetic bacteria with acidogenic process: Augmented dark-photo fermentative hybrid system to regulate volatile fatty acid inhibition. *International Journal of Hydrogen Energy, 39*, 7604–7615.

Chandra, R., Venkata Subhash, G., & Venkata Mohan, S. (2012). Mixotrophic operation of photo-bioelectrocatalytic fuel cell under anoxygenic microenvironment enhances the light dependent bioelectrogenic activity. *Bioresource Technology, 109*, 46–56.

Di Lorenzo, M., Thomson, A. R., Schneider, K., Cameron, P. J., & Ieropoulos, I. (2014). A small-scale air-cathode microbial fuel cell for on-line monitoring of water quality. *Biosensors & Bioelectronics, 62*, 182–188.

El Mekawy, A., Hegab, H. M., Vanbroekhoven, K., & Pant, D. (2014). Techno-productive potential of photosynthetic microbial fuel cells through different configurations. *Renewable and Sustainable Energy Reviews, 39*, 617–627.

Georgianna, D. R., & Mayfield, S. P. (2012). Exploiting diversity and synthetic biology for the production of algal biofuels. *Nature, 488*, 329–335.

Hamelers, H. V. M., Ter Heijne, A., Sleutels, T. H. J. A, Jeremiasse, A. W., Strik, D. P. B. T. B., & Buisman, C. J. N. (2010). New applications and performance of bioelectrochemical systems. *Applied Microbiology and Biotechnology, 85*, 1673–1685.

Han, K., Yueh, P.-L., Qin, L.-J., Hsueh, C.-C., & Chen, B.-Y. (2015). Deciphering synergistic characteristics of microbial fuel cell-assisted dye decolorization. *Bioresource Technology, 196*, 746–751.

Helder, M., Strik, D. P. B. T. B., Hamelers, H. V. M., Kuhn, A. J., Blok, C., & Buisman, C. J. N. (2010). Concurrent bio-electricity and biomass production in three plant-microbial fuel cells using Spartina anglica, Arundinella anomala and Arundo donax. *Bioresource Technology, 101*, 3541–3547.

Helder, M., Strik, D. P. B. T. B., Hamelers, H. V. M., Kuijken, R. C. P., & Buisman, C. J. N. (2012). New plant-growth medium for increased power output of the plant-microbial fuel cell. *Bioresource Technology, 104*, 417–423.

Helder, M., Strik, D. P. B. T. B., Timmers, R. A., Raes, S. M. T., Hamelers, H. V. M., & Buisman, C. J. N. (2013). Resilience of roof-top plant-microbial fuel cells during Dutch winter. *Biomass and Bioenergy, 51*, 1–7.

Hubenova, Y., & Mitov, M. (2012). Conversion of solar energy into electricity by using duckweed in direct photosynthetic plant fuel cell. *Bioelectrochemistry, 87*, 185–191.

Ihssen, J., Braun, A., Faccio, G., Gajda-Schrantz, K., & Thony-Meyer, L. (2014). Light harvesting proteins for solar fuel generation in bioengineered photoelectrochemical cells. *Current Protein & Peptide Science, 15*, 374–384.

Kathiravan, A., & Renganathan, R. (2009). Photosensitization of colloidal TiO_2 nanoparticles with phycocyanin pigment. *Journal of Colloid and Interface Science, 335*, 196–202.

Kim, B. H., Chang, I. S., Cheol Gil, G., Park, H. S., & Kim, H. J. (2007). Challenges in microbial fuel cell development and operation. *Applied Microbiology and Biotechnology, 76*, 485–494.

Lovley, D. R. (2006). Bug juice: Harvesting electricity with microorganisms. *Nature Reviews. Microbiology, 4*, 497–508.

Luo, Y., Zhang, R., Liu, G., Li, J., Qin, B., Li, M., Chen, S. (2011) Simultaneous degradation of refractory contaminants in both the anode and cathode chambers of the microbial fuel cell. Bioresour. Technol. 102, 3827–32.

Marsili, E., Baron, D. B., Shikhare, I. D., Coursolle, D., Gralnick, J. A., & Bond, D. R. (2008). *Shewanella* secretes flavins that mediate extracellular e- transfer. *Proceedings of the National Academy of Sciences of the United States of America, 105*, 3968–3973.

McCormick, A. J., Bombelli, P., Bradley, R. W., Thorne, R., Wenzel, T., & Howe, C. J. (2015). Biophotovoltaics: Oxygenic photosynthetic organisms in the world of bioelectrochemical systems. *Energy & Environmental Science, 8*, 1092–1109.

Mohan, S. V., & Chandrasekhar, K. (2011). Self-induced bio-potential and graphite e-accepting conditions enhances petroleum sludge degradation in bio-electrochemical system with simultaneous power generation. *Bioresource Technology, 102*, 9532–9541.

Pandit, A. V., & Mahadevan, R. (2011). In silico characterization of microbial electrosynthesis for metabolic engineering of biochemicals. *Microbial Cell Factories, 10*, 76.

Raschitor, A., Soreanu, G., Fernandez-Marchante, C. M., Lobato, J., Cañizares, P., Cretescu, I., & Rodrigo, M. A. (2015). Bioelectro-Claus processes using MFC technology: Influence of co-substrate. *Bioresource Technology, 189*, 94–98.

Rosenbaum, M., He, Z., & Angenent, L. T. (2010). Light energy to bioelectricity: Photosynthetic microbial fuel cells. *Current Opinion in Biotechnology, 21*, 259–264.

Schröder, U., Nießen, J., & Scholz, F. (2003). A generation of microbial fuel cells with current outputs boosted by more than one order of magnitude. *Angewandte Chemie, International Edition, 42*, 2880–2883.

Strik, D. P. B. T. B., Timmers, R. A., Helder, M., Steinbusch, K. J. J., Hamelers, H. V. M., & Buisman, C. J. N. (2011). Microbial solar cells: Applying photosynthetic and electrochemically active organisms. *Trends in Biotechnology, 29*, 41–49.

Suemori, Y., Nagata, M., Nakamura, Y., Nakagawa, K., Okuda, A., Inagaki, J., Shinohara, K., Ogawa, M., Iida, K., Dewa, T., Yamashita, K., Gardiner, A., Cogdell, R. J., & Nango, M. (2006). Self-assembled monolayer of light-harvesting core complexes of photosynthetic bacteria on an amino-terminated ITO electrode. *Photosynthesis Research, 90*, 17–21.

Timmers, R. A., Strik, D. P. B. T. B., Hamelers, H. V. M., & Buisman, C. J. N. (2012). Characterization of the internal resistance of a plant microbial fuel cell. *Electrochimica Acta, 72*, 165–171.

Tran, D. N., & Balkus, K. J. (2011). Perspective of recent progress in immobilization of *Enzymes. ACS Catalysis, 1*, 956–968.

Venkata Mohan, S., Mohanakrishna, G., & Sarma, P. N. (2008a). Effect of anodic metabolic function on bioelectricity generation and substrate degradation in single chambered microbial fuel cell. *Environmental Science & Technology, 42*, 8088–8094.

Venkata Mohan, S., Mohanakrishna, G., Reddy, B. P., Saravanan, R., & Sarma, P. N. (2008b). Bioelectricity generation from chemical wastewater treatment in mediatorless (anode) microbial fuel cell (MFC) using selectively enriched hydrogen producing mixed culture under acidophilic microenvironment. *Biochemical Engineering Journal, 39*, 121–130.

Venkata Mohan, S., Mohanakrishna, G., Chiranjeevi, P., Peri, D., & Sarma, P. N. (2010). Ecologically engineered system (EES) designed to integrate floating, emergent and submerged macrophytes for the treatment of domestic sewage and acid rich fermented distillery wastewater: Evaluation of long term performance. *Bioresource Technology, 101*, 3363–3370.

Venkata Mohan, S., Mohanakrishna, G., & Chiranjeevi, P. (2011). Sustainable power generation from floating macrophytes based ecological microenvironment through embedded fuel cells along with simultaneous wastewater treatment. *Bioresource Technology, 102*, 7036–7042.

Venkata Mohan, S., Srikanth, S., Chiranjeevi, P., Arora, S., & Chandra, R. (2014). Algal biocathode for in situ terminal e-acceptor (TEA) production: Synergetic association of bacteria-microalgae metabolism for the functioning of biofuel cell. *Bioresource Technology, 166*, 566–574.

Venkata Subhash, G., Chandra, R., & Venkata Mohan, S. (2013). Microalgae mediated bio-electrocatalytic fuel cell facilitates bioelectricity generation through oxygenic photomixotrophic mechanism. *Bioresource Technology, 136*, 644–653.

Wetser, K., Sudirjo, E., Buisman, C. J. N., & Strik, D. P. B. T. B. (2015). Electricity generation by a plant microbial fuel cell with an integrated oxygen reducing biocathode. *Applied Energy, 137*, 151–157.

Wu, S., Li, H., Zhou, X., Liang, P., Zhang, X., Jiang, Y., & Huang, X. (2016). A novel pilot-scale stacked microbial fuel cell for efficient electricity generation and wastewater treatment. *Water Research, 98*, 396–403.

Xiao, L., & He, Z. (2014). Applications and perspectives of phototrophic microorganisms for electricity generation from organic compounds in microbial fuel cells. *Renewable and Sustainable Energy Reviews, 37*, 550–559.

Zhang, Q., Saito, T., Cheng, S., Hickner, M. A., & Logan, B. E. (2010). Microbial fuel cell cathodes with poly(dimethylsiloxane) diffusion layers constructed around stainless steel mesh current collectors. *Environmental Science & Technology, 44*, 1490–1495.

Zhang, Q., Hu, J., & Lee, D.-J. (2016). Microbial fuel cells as pollutant treatment units: Research updates. *Bioresource Technology, 217*, 121–128. https://doi.org/10.1016/j.biortech.2016.02.006.

Zhang, Q., Ma, J., Qiu, G., Li, L., Geng, S., Hasi, E., Li, C., & Wang, G. (2012). Bioresource technology potential energy production from algae on marginal land in China. *Original Research Article, 109*, 252–260.

Zhuang, K., Vemuri, G. N., & Mahadevan, R. (2011). Respiro-fermentation. *Molecular Systems Biology, 7*, 1–9.

Chapter 20
Application of Microbial Fuel Cell as a Biosensor

Ramya Veerubhotla and Debabrata Das

20.1 Introduction

Microbial fuel cells (MFCs) are being studied as a biosensor for a wide gamut of applications. Currently, since the power generated by an MFC is too low to power any practical devices, it is pertinent to use MFC as a sensor for the accurate and simple measurement of various analytes. The present chapter deals with the various upcoming applications of the MFC technology in the field of sensing with an emphasis on Biochemical Oxygen Demand (BOD) bio-sensing. The application of MFCs as BOD biosensor is one of the most widely studied fields followed by the volatile fatty acid sensing and toxicity sensing. Figure 20.1 depicts various novel applications of MFC as biosensor from the literature.

20.2 Microbial Biosensors

Biosensors are devices which combine a bio-receptor, usually an enzyme, microbe or a biologically active molecule that responds to an analyte or change in its concentration and a transducer that converts the response into a measurable output (D'Souza 2001; Su et al. 2011). Occasionally, an amplifier is also employed which multiplies the signal for implicit applications. The transducer is usually chosen depending upon the analyte as well as the biological sensing element. It is desirable to have biosensors that are portable, reproducible and those which provide rapid and accurate results of the parameters to be measured (Lim et al. 2015). Depending

R. Veerubhotla (✉) • D. Das
Department of Biotechnology, Indian Institute of Technology Kharagpur, Kharagpur 721302, India
e-mail: vramya.90@gmail.com

D. Das (ed.), *Microbial Fuel Cell*,
https://doi.org/10.1007/978-3-319-66793-5_20

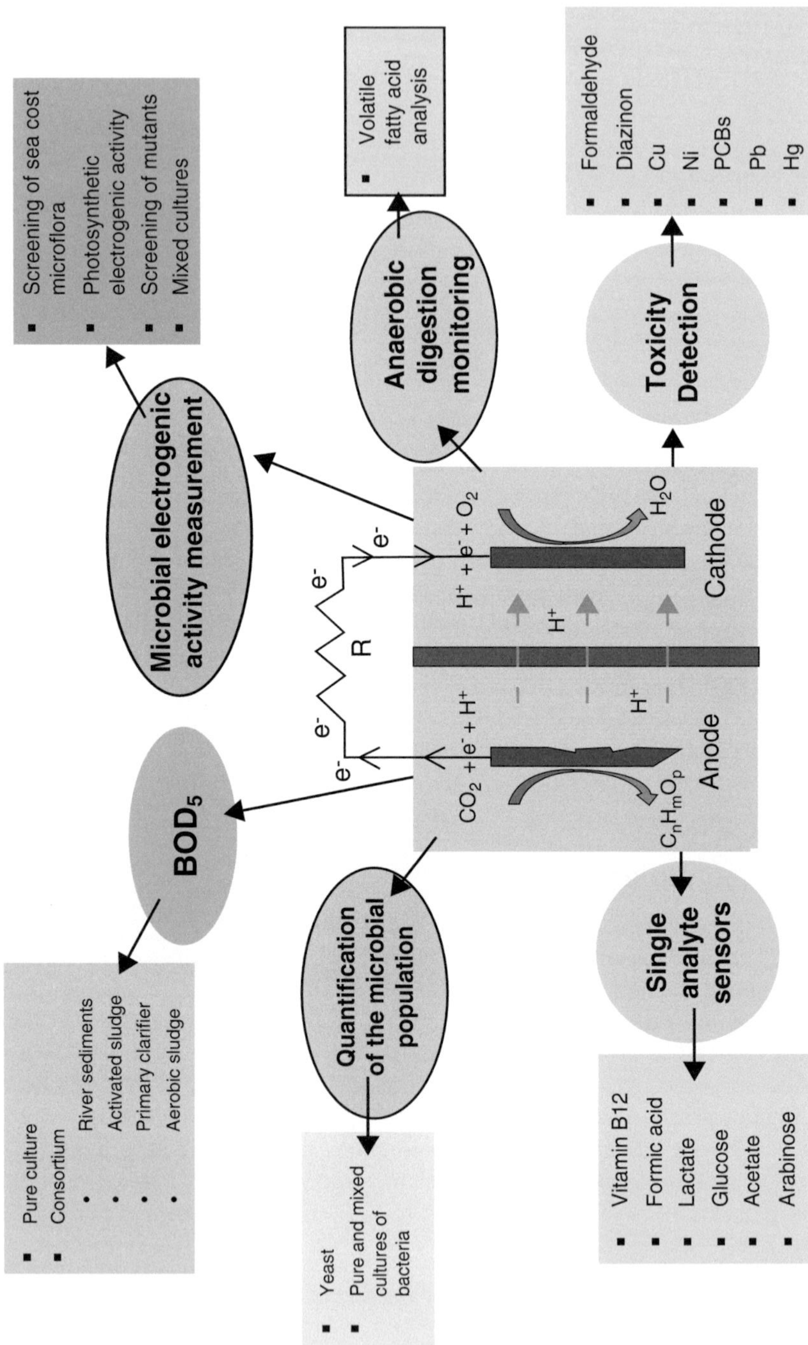

Fig. 20.1 Typical applications of MFC as sensor

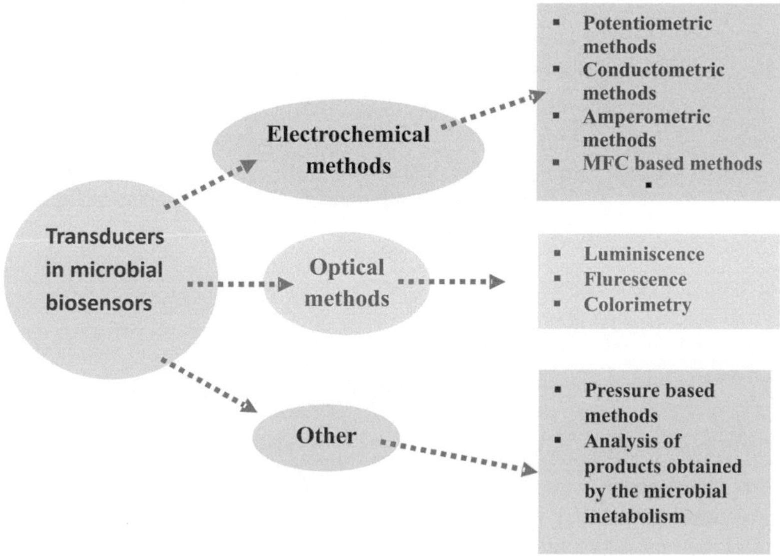

Fig. 20.2 Classification of the transducers of various microbial biosensors

on the type of the transducer, microbial biosensors are classified as depicted in Fig. 20.2.

Microbial biosensors in general pose various advantages as microorganisms are easy to grow and maintain, can work independently, and are easy to adapt and regenerate. Unlike use of the enzyme-based biosensors which require stringent micro-environment to sustain the activity, microbial biosensors can work with the help of a wide variety of substrates and environments depending on the culture conditions. Since purification steps are absent in these sensors, use of microbes is quite inexpensive as compared to enzymatic biosensors. The same microbes can be utilized to monitor various analytes depending on the transducer used. Microbes also offer unique advantages in diverse immobilization methods possible to entrap them while retaining the activity. Further, genetic engineering techniques offer new avenues and applications in the field of microbial biosensors. Among the different transduction methods (Fig. 20.2), electrochemical method of transduction is the most convenient method for the detection of analytes. The output signal is easy to read and process as it does not require sophisticated equipment unlike fluorescence or bioluminescence. MFCs use various techniques of electrochemistry such as amperometry, voltammetry, cyclic or conductometry for the quantification of the BOD.

20.3 Principle of MFC as a Biosensor

An MFC consists of anaerobic anode chamber where electroactive microbes act as biocatalysts utilizing the nutrients added. The electrons generated by the metabolism of the nutrients present in the medium is proportional to the concentration of

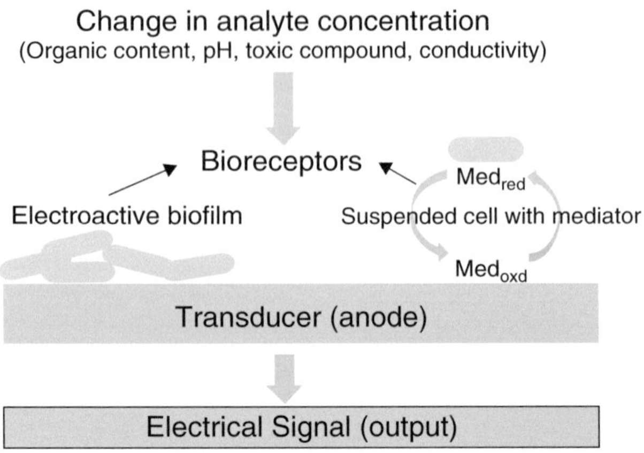

Fig. 20.3 Principle of MFC as a sensor

the nutrient solution in the anode chamber. These electrons travel through the external circuit to the cathode where the oxygen reduction takes place. Both anode and the cathode chambers are separated by a proton exchange membrane that selectively allows the movement of protons to oxygen reduction. The change in the analyte concentration can be precisely determined by the bioreceptors (electroactive bacteria) which is usually immobilized on the transducer (anode) itself. Typically the electrons generated by the metabolism in the anode chamber are measured which acts as the signal for sensing as represented in Fig. 20.3.

The MFC system can be utilized for the principle of biosensing in two possible ways:

1. The concentration of the organic component (analyte) or a specific growth molecule is proportional to the electrons generated by the system which can be measured by the electron flow (current, voltage, coloumbic efficiency). Examples of such sensors include BOD biosensors, single analyte sensors for the detection of glucose, vitamins etc., volatile fatty acid monitoring sensors etc.
2. In other cases, the MFC-based system can also be utilized to monitor the toxic contaminants or pollutants where the concentration or titre of the pollutant added hampers the activity of immobilized electroactive bacteria (Abrevaya et al. 2015). In such cases, the concentration of the toxic compound is inversely proportional to the electrons generated by the bacteria and hence the signal. Typical examples of such toxicity sensors include detection and quantification of formaldehyde, heavy metals etc.

In MFC-based biosensors, the electrons generated by the bacteria can be diverted to anode via two well known mechanisms: (1) Mediator-driven and (2) Mediator-less as demonstrated in Fig. 20.3 (Abrevaya et al. 2015). In mediator-driven systems, a soluble mediator either added externally or produced by microbe itself can shuttle the electrons between the cells and electrodes. Various mediators are known to enhance the electron transfer which are described in the previous chapters. Such systems can operate even with non-electroactive bacteria and the concentration and diffusion of the mediator greatly affects the output signal. On the other hand, mediatorless systems have bacteria immobilized to the electrode. Electrons generated at the anode can result in the generation of the signal either by the outer membrane complex or with the help of bacterial nanowires. In most cases, a biofilm formation is induced on the surface of the anode with time. Such a mechanism does not involve any external addition of the mediator and is hence more suitable for practical applications.

20.4 Advantages of MFC as a Sensor

Use of an MFC as biosensor eliminates the need for a separate transducer by acting as a biofilm anode. Besides, the microbial population expands itself by the reproduction of the microbes in the provided volumes inducing a self-regeneration capability. Thus, an MFC can possess self-start property and work as a stand-alone biosensor. Depending upon the volume of the anode chamber and the bacteria used, the range of current output from the cells is in the range of mA-μA. Thus, the output signal from the MFC (usually in the form of current or voltage) is simple to read and interpret. Further, the fabrication of the devices does not need expensive probes, construction methods and stringent maintenance. The aforementioned unique properties of MFC make it one of the ideal platforms for the sensing as well as quantification of various analytes. The key to accurate sensing with high reproducibility is a stable/consistent biofilm with a defined microbial population at the anode which is resistant to environmental fluctuations. Environmental parameters such as pH, temperature, conductivity, electron acceptors, toxicants, composition of wastewater added will greatly influence the measurement (Abrevaya et al. 2015). Further, inherent system parameters such as ion transport across the membranes used, effective cathode reduction reaction, and electron transport rate, configuration of the device, volume of the device and mode of operation will define the net output signal and its stability with time.

20.5 BOD and Its Importance

BOD is the amount of dissolved oxygen needed by aerobic biological organisms to break down organic material present in a given wastewater sample at certain temperature over a specific time period. Measuring the BOD of water bodies is of immense environmental significance. BOD quantifies how the biodegradable organic content of a water body can affect the oxygen concentration in it. The pollution load of effluents from industries is reflected by means of BOD and is critical as an agreeable amount of dissolved oxygen is necessary for a healthy aquatic life. Hence, there are strict limits as to what level of BOD containing wastewater should be discharged in to the nearby water bodies depending on the

Pollution Control Board of different countries. Furthermore, the wastewater/effluent treatment method adopted and various engineering aspects of wastewater treatment involve an accurate measurement of BOD. Thus, a rapid quantification of BOD is essential for the assessment of water quality (Jouanneau et al. 2014).

20.6 Methods of Assessing BOD

The standard BOD test takes a period of 5 days and it involves the incubation of sludge in specific BOD bottles in dark for a time period of 5 days. Microbes in the wastewater use the dissolved oxygen to oxidize the organic content for deriving energy for metabolism. The dissolved oxygen is measured before and after the stipulated time using a Clark electrode (Jouanneau et al. 2014). Although both COD and BOD measure the organic content of a sample, BOD exclusively measures only the biodegradable content. A standard testing of BOD is essential for any organic body before the disposal of wastewater. However, certain problems associated with the standard method such as need to maintain the probes, variability of the inoculum and a narrow range of detection has encouraged researchers to innovate new methods for a quick estimation of the BOD. The standard method is not a feasible option for the on-site BOD monitoring as it requires a lot of working space and bulky apparatus such as incubators. Since various environmental applications are dependant on the BOD values, it is crucial to have a portable, rapid and inexpensive method for the quantification with minimum deviation from the BOD_5 value. BOD biosensors based on bioluminescence, photometric, manometric methods, modified standard methods and bioreactor-based sensors are already demonstrated in this regard. The following section of this chapter will discuss in detail the merits and demerits of using MFC as an alternate sensor for BOD providing valuable insights on the challenges and the scope of improvement.

20.7 Application of MFC as a BOD Sensor

The key advantage of using an MFC for BOD monitoring is that the time consumed to give a preliminary estimate of the sample BOD is much less as compared to the conventional standard method. Since the reactor volume can be customized by the user, MFC-based sensors enable quick estimation of BOD even from minimal sample volumes. Typically, the response times of the devices are in the range of few minutes to hours. Depending upon the mode of operation and the configuration, it is possible to operate the system with a minimal maintenance. Unlike the standard method of BOD measurement which deploys the dissolved oxygen monitoring using probes, MFCs employ anaerobic chambers enabling an indirect current measurement. The electrons obtained by the degradation or oxidation of the organic content in an anaerobic environment is proportional to the concentration of the organic matter or BOD. The electrochemically active bacteria can metabolize the organic content present in the anaerobic chambers with or without the use of mediators and the generated electrons can be diverted to the anode. This unique ability of the electrochemically active bacteria to divert electrons only takes place in the absence of any readily available electron acceptors in the environment the microbes are placed. The very advantage of the use of an MFC is the fact that this flow of electrons can be detected using a simple amperometer in a precise way eliminating the need for the use of DO probes. However, both current and charge output are monitored and corelated with the organic content depending upon the range of the BOD.

Diverse groups of exoelectrogens can be colonized to the anode depending on the acclimatization approach. Hence, this method can not only have a broad range of measurement but also provides a more reliable BOD measurement depending on the nature, diversity, quantity of microbial population present and the configuration of the system in use. Further, reports suggest that online monitoring of the BOD is possible using a continuously operated MFC system with minimum operational requirements. Such a system is demonstrated by Zhang et al. using a submersible MFC system. Another added advantage of an MFC system is the reduced time spans in obtaining the BOD values. Depending on the maturity of the biofilm, the time requirement for measurement can range from a few minutes to hours (Table 20.1). While some MFCs use continuous systems with an acclimatized biofilm, some systems adopt a batch operation where bacterial colonization on the electrodes need to be induced and further measurement of the organic content is attempted. These characteristics suggest the application of MFC for both field applications as well as lab scale rapid analysis of the BOD. The configuration of the MFC adopted also has a profound role in deciding the response time as well as the accuracy of the system.

20.7.1 MFC as a BOD Biosensor—State of Art

The concept that an MFC can act as an amperometric biosensor for BOD estimation is first introduced by Karube et al. (1977). By using an MFC containing whole cells of *Clostridium butylicum* on the platinum electrode, they could estimate the BOD with a relative error of 10%. As the cells metabolized the organic substrates, hydrogen and formate formed by the cells reacted at the electrodes and resulted in the electricity generation. The steady state current linearly varied with the input strength of organic substrate. This led to the measurement of the BOD within 30–40 min which was much less than the time consumed by conventional methods (Karube et al. 1977). Following this, a continuously operated MFC containing a mixed consortia (obtained from a sewage plant) is proposed (Kim et al. 2003). A practical field application is also demonstrated using the set-up for the online monitoring of the real time samples from a wastewater treatment plant. The results showed encouraging results with a good stability in the current generation. The

Table 20.1 MFC as a BOD biosensor – present state of art

Source of the biocatalyst (Microbes)	Mediator	Measurement time	Range of detection (mg L^{-1})	Mode of operation	Volume (mL)	References
Clostridium butyricum	No	70 min	10–300	Batch	100	Karube et al. (1977)
Consortium (river sediments)	No		5–10	Batch	25	Kang et al. (2003)
Consortium (anaerobic sludge)	No	40 min–2 h	50–400	Continuous	50	Di Lorenzo et al. (2009)
Consortium (primary clarifier)	No	40 min	10–250	Fed batch	500	Zhang and Angelidaki (2011)
Escherichia coli	Poly-neutral red		50–1000			Liu et al. (2012)
Consortium (anaerobic and aerobic sludge)	No	5–20 h	32–1280	Batch	11.8	Modin and Wilén (2012)
Mixed consortia	No	5–1 min.	20–200	Continuous	5	Moon et al. (2004)
Geobacter-dominated biofilms	No	17.5 h	174 to 1200	Batch mode	1	Commault et al. (2016)
Anaerobic sludge	No	3–5 min	0.025 to 25	Batch	100	Kumlanghan et al. (2007)

work also noted the significance of recalibration for long time use of the set-up. Various attempts by the researchers in this regard along with their respective BOD ranges is tabulated in Table 20.1.

Further, Moon et al. (2004) adopted a step change strategy for the quantification of the input fuel stream concentration. They observed that the response time of the devices are directly proportional to the anode chamber volumes. They achieved a drastic reduction in the response time (upto 5 min) using a small chamber volume (5 mL) (Moon et al. 2004). In order to measure the BOD in the lower range, the use of oligotrophic microbes (which are electroactive) in an MFC is first suggested by Kang et al. (2003). Their research suggested that use of a smaller membrane and a highly active cathode can effectively prevent the oxygen diffusion from the cathode to anode. The leakage of the oxygen is thus minimized into the anode chamber in order to maximize the coloumbic efficiency which resulted in obtaining a good reproducibility inspite of using low nutrient concentration. Kumlanghan et al. (2007) used an MFC-based BOD biosensor by integrating the system with an anaerobic digester. The consortium for each batch reaction is taken from the reactor itself. With a response time of 5 min, the system showed good correspondence with the input feed concentration (Kang et al. 2003). Their study also optimized the electrolyte concentration as the ion transport plays an important role in the overall performance (Kumlanghan et al. 2007).

While all the above mentioned preliminary studies used an MFC of double chambered configuration, Di Lorenzo et al. put forth the use of a single chambered MFC system in a continuous mode as a BOD biosensor. A strong correlation between the hydraulic retention time (HRT) and the response time of the MFC is shown successfully (Di Lorenzo et al. 2009). Another interesting approach for the *in situ* monitoring of the BOD content of groundwater is suggested by Zhang and Angelidaki (2011). A submersible MFC is developed by their team with a systematic study of the effect of temperature, pH, conductivity and the input feed concentration on the sensitivity of the system. They suggested that their endeavours can further open doors to monitor the bioremediation of toxic compounds in a non-invasive and simple way (Zhang and Angelidaki 2011).

On the other hand, use of mediator-based BOD biosensor in conjunction with nanoparticle aided pretreatment (TiO_2) is demonstrated by Liu et al. Neutral red, the mediator used in the study, is co-immobilized along with *E. coli* cells on a glassy carbon anode. The authors recommended that photocatalysis of the organic molecules prior to addition into the anodic chamber can substantially cause pronounced signals as compared to the untreated ones (Liu et al. 2012). A novel BOD sensor which works on the principle similar to that of an electrolysis cell was proposed by Modin and Wilén (2012) where external input of 0.2 V is supplied to anode to aid the microbial degradation of the organic compounds. With acetate as the source of carbon, the system could measure a very high range of BOD, probably due to the external voltage supplied. The system also avoided the use of an ion exchange membrane which eliminated the demerits of using membranes such as pH splitting (Modin and Wilén 2012). Upcoming applications demonstrated an MFC for measuring the BOD of cow milk unlike others who measure the organic content of wastewaters (Commault et al. 2016). Selectively adopted *Geobacter* dominated biofilm is utilized in their study by using a new strategy of ethanol acclimatization (Commault et al. 2016).

One such BOD biosensor which works with the help of electrochemically active bacteria is already marketed by KORBI Co. Ltd. as BOD analyzer, HABS Series.

20.7.2 Challenges of MFC-Based BOD Biosensors

One of the potential problems using MFCs as biosensors is the significant diffusion of oxygen from cathode chambers to the anode. Microbes then aerobically respire instead of utilizing the anode causing a drop in the coloumbic efficiency and thus interfere with the measurement (Moon et al. 2005). This can lead to improper results as it becomes difficult to accurately estimate the BOD due to interference from the oxygen, especially when the BOD of low nutrient concentration is being measured. Hence, sparging the influent organic fuel to remove the inherent oxygen content while minimizing the oxygen leakage into anode chamber can improve the coloumbic yield. An endeavour in this direction is attempted by Ayyaru and Dharmalingam (2014) where the use of a sulphonated poly ether ketone membrane to improve the response is suggested. Besides, the use of membrane can itself result in problems such as pH splitting, substrate transfer across the membrane etc. (Christgen et al. 2015). Hence there is a need for the development of membraneless systems or highly efficient membrane cathode assembly units to overcome such effects.

BOD measurement of the real-time wastewater samples using MFCs, however, still remains highly challenging due to the drastic effect of the indigenous properties of the wastewater such as pH, conductivity, presence of external mediators, inhibitors or ready electron acceptors for microbial respiration. This variation in the internal factors also makes the *in situ* monitoring of the organic pollutants impractical. Several studies highlighted the prominence of the pH and conductivity of the anolyte (wastewater) on the performance of an MFC (Zhang and Angelidaki 2011; Kim et al. 2009). Possible approaches to reduce the dependancy of pH on the bioelectric output are the use of membraneless design and to provide an appropriate buffering in the system.

Detecting the presence of inherent electron acceptors in the wastewater samples and quantifying them is tedious and impractical. Furthermore, presence of toxic inhibitors in the wastewaters can alter biofilm structure, microbial community leading to decrease the bacterial viability. Researchers attempted to solve this by the addition of certain inhibitors such as azide and cyanide (Chang et al. 2005). Although such an addition in minute quantities is claimed to not affect the performance of the system in the absence of nitrate or other electron acceptors. It is, however, not encouraged as it can affect the bacterial community of the anodic biofilm. Viability of the biofilm and its sensitivity towards environmental parameters is to be thus considered for stability of the biofilm as well as the overall performance of the sensor. Further, efforts are needed to maintain a uniform electrochemical environment in the cells to eliminate batch to batch variation and ensure accurate measurement. Significant progress is needed in the pretreatment methods to bring down the BOD level to the desirable range and to eliminate unwanted compounds. This can further make the MFC-based BOD biosensors superior as compared to the standard BOD measurement methods in terms of accuracy.

20.8 Upcoming Applications of MFC in the Field of Sensing

In the recent years, there is a growing interest in the novel applications of MFCs as sensors for various applications as depicted in Fig. 20.4.

20.8.1 Screening of Electroactive Microbes

Since electroactive microbes form the basis of various bioelectrochemical systems such as MFCs, electrolysis cells, microbial desalination cells, microbial carbon capture cells, microbial electrosynthesis cells etc., finding appropriate microbes

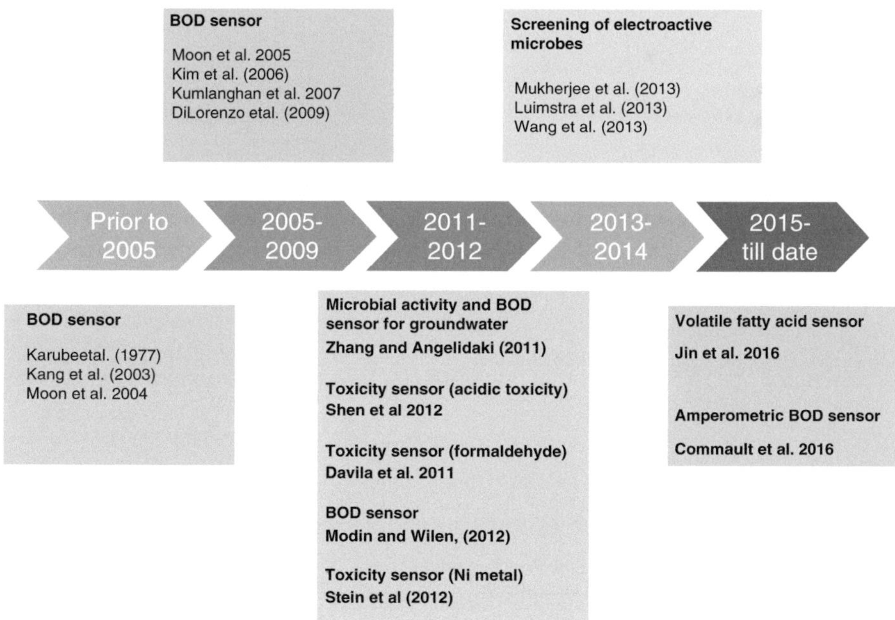

Fig. 20.4 Timeline of the various research works in the field of application of MFC as a sensor

suitable for the system becomes most crucial for the improvement of MFCs. Portable microfluidic platforms are reported for providing a quick assessment of the electrogenic potential of such microbes (Mukherjee et al. 2013). Such platforms are successful in preliminary screening of bioelectric potential of multiple strains simultaneously with tiny sample volumes (Hou et al. 2011). Such techniques are particularly useful for the evaluation of the performance of genetically modified strains. Choi et al. (2015) deviced a paper based disposable and inexpensive 48-well MFC array on a paper platform for practical real world applications.

20.8.2 Toxicity Sensing

MFCs are being applied as potential toxicity sensing elements owing to the change in the bioelectric output in the presence of a toxic compound. Since conventional methods of estimating the toxic elements in the soil, water and sediments are usually time-consuming, and require a lot of equipment, skilled personnel to interpret the results, it is beneficial to use MFC, because the biofilm of the system is highly sensitive to certain toxic compounds. Arsenic biosensor by genetically modified *Shewanella* sp. (Webster et al. 2014); nickel detection by Stein et al. (2012); online biomonitoring of Pb, Hg and PCBs using MFCs (Kim et al. 2007); and microfabricated formaldehyde biosensor (Dávila et al. 2011) are some of the works in the field of toxicity biosensing by MFCs. However, the research is still in a native state and there is a need to precisely detect the pollutants and obtain reproducible results.

20.8.3 VFA Sensing

Instead of measuring the overall BOD or the strength of wastewater, MFCs are now emerging as the biosensors for the volatile fatty acid (VFA) monitoring of the anaerobic digestion process. MFCs have been shown to have a precise correlation with the concentrations of single analyte species and can thus form promising VFA biosensors. This can avoid use of instruments like gas chromatography, tedious sample preparation methods to detect the analytes and enable an accurate and facile detection of the VFAs. Notable attempts in this regard are made by several researchers (Kaur et al. 2013; Jin et al. 2016).

20.9 Conclusion and Future Perspectives

MFC have the potential of being developed into promising biosensor devices that can be marketed for practical applications. To exploit the inherent unique advantages of the system, an interdisciplinary approach with collective efforts from researchers of various fields is needed. Improvement of the sensitivity and reproducibility of the system enhances the practical applicability of the MFC-based biosensors. Miniaturization and multiplexing can also provide rapid and preliminary estimation of various useful analytes in a simple manner.

References

Abrevaya, X. C., Sacco, N. J., Bonetto, M. C., Hilding-Ohlsson, A., & Cortón, E. (2015). Analytical applications of MFCs. Part I: Biochemical oxygen demand. *Biosensors & Bioelectronics, 63*, 580–590.

Ayyaru, S., & Dharmalingam, S. (2014). Enhanced response of MFC using sulfonated poly ether ether ketone membrane as a biochemical oxygen demand sensor. *Analytica Chimica Acta, 818*, 15–22.

Chang, I. S., Moon, H., Jang, J. K., & Kim, B. H. (2005). Improvement of a MFC performance as a BOD sensor using respiratory inhibitors. *Biosensors & Bioelectronics, 20*, 1856–1859.

Choi, G., Hassett, D. J., & Choi, S. (2015). A paper-based MFC array for rapid and high-throughput screening of electricity-producing bacteria. *Analyst, 140*(12), 4277–4283.

Christgen, B., Scott, K., Dolfing, J., Head, I. M., & Curtis, T. P. (2015). An evaluation of the performance and economics of membranes and separators in single chamber MFCs treating domestic wastewater. *PloS One, 10*, 1–13.

Commault, A. S., Lear, G., Bouvier, S., Feiler, L., Karacs, J., & Weld, R. J. (2016). *Geobacter*-dominated biofilms used as amperometric BOD sensors. *Biochemical Engineering Journal, 109*, 88–95.

Dávila, D., Esquivel, J. P., Sabat, N., & Mas, J. (2011). Silicon-based microfabricated MFC toxicity sensor. *Biosensors & Bioelectronics, 26*, 2426–2430.

Di Lorenzo, M., Curtis, T. P., Head, I. M., & Scott, K. (2009). A single-chamber MFC as a biosensor for wastewaters. *Water Research, 43*, 3145–3154.

D'Souza, S. F. (2001). Microbial biosensors. *Biosensors & Bioelectronics, 16*, 337–353.

Hou, H., Li, L., De Figueiredo, P., & Han, A. (2011). Microfluidic MFC Array for multiplexed long-term parallel analysis of microbial activities. In *15th international conference on miniaturized systems for chemistry and life sciences*, Seattle, Washington.

Jin, X., Angelidaki, I., & Zhang, Y. (2016). Microbial electrochemical monitoring of volatile fatty acids during anaerobic digestion. *Environmental Science & Technology, 50*(8), 4422–4429.

Jouanneau, S., Recoules, L., Durand, M. J., Boukabacheb, A., Picotb, V., Primaultc, Y., Lakeld, A., Sengeline, M., Barillonf, B., & Thouand, G. (2014). Methods for assessing biochemical oxygen demand (BOD): A review. *Water Research, 49*, 62–82.

Kang, K. H., Jang, J. K., Pham, T. H., Moon, H., Chang, I. S., & Kim, B. H. (2003). A MFC with improved cathode reaction as a low biochemical oxygen demand sensor. *Biotechnology Letters, 25*, 1357–1361.

Karube, I., Matsunaga, T., Mitsuda, S., & Suzuki, S. (1977). Microbial electrode BOD sensors. *Biotechnology and Bioengineering, 19*, 1535–1547.

Kaur, A., Kim, J. R., Michie, I., Dinsdale, R. M., Guwy, A. J., & Premier, G. C. (2013). MFC type biosensor for specific volatile fatty acids using acclimated bacterial communities. *Biosensors & Bioelectronics, 47*, 50–55.

Kim, M., Youn, S. M., Shin, S. H., Jang, J. G., Han, S. H., Hyun, M. S., Gadd, G. M., & Kim, H. J. (2003). Practical field application of a novel BOD monitoring system. *Journal of Environmental Monitoring, 5*, 640–643.

Kim, M., Hyun, M. S., Gadd, G. M., & Kim, H. J. (2007). A novel biomonitoring system using MFCs. *Journal of Environmental Monitoring, 9*, 1323–1328.

Kim, M., Hyun, M. S., Gadd, G. M., Kim, G. T., Lee, S. J., & Kim, H. J. (2009). Membrane-electrode assembly enhances performance of a MFC type biological oxygen demand sensor. *Environmental Technology, 30*, 329–336.

Kumlanghan, A., Liu, J., Thavarungkul, P., Kanatharana, P., & Mattiasson, B. (2007). MFC-based biosensor for fast analysis of biodegradable organic matter. *Biosensors & Bioelectronics, 22*, 2939–2944.

Lim, J. W., Ha, D., Lee, J., Lee, S. K., & Kim, T. (2015). Review of micro/nanotechnologies for microbial biosensors. *Frontiers in Bioengineering and Biotechnology, 3*, 1–13.

Liu, L., Zhang, S., Xing, L., Zhao, H., & Dong, S. (2012). A co-immobilized mediator and microorganism mediated method combined pretreatment by TiO_2 nanotubes used for BOD measurement. *Talanta, 93*, 314–319.

Modin, O., & Wilén, B. M. (2012). A novel bioelectrochemical BOD sensor operating with voltage input. *Water Research, 46*, 6113–6120.

Moon, H., Chang, I. S., Kang, K. H., Jang, J. K., & Kim, B. H. (2004). Improving the dynamic response of a mediator-less MFC as a biochemical oxygen demand (BOD) sensor. *Biotechnology Letters, 26*, 1717–1721.

Moon, H., Chang, I. S., Jang, J. K., Kim, K. S., Lee, J., Lovitt, R. W., & Kim, B. H. (2005). On-line monitoring of low biochemical oxygen demand through continuous operation of a mediator-less MFC. *Journal of Microbiology and Biotechnology, 5*, 192–196.

Mukherjee, S., Su, S., Panmanee, W., Irvin, R. T., Hassett, D. J., & Choi, S. (2013). A microliter-scale MFC array for bacterial electrogenic screening. *Sensors and Actuators A: Physical, 201*, 532–537.

Stein, N. E., Hamelers, H. M. V., Van Straten, G., & Keesman, K. J. (2012). On-line detection of toxic components using a MFC-based biosensor. *Journal of Process Control, 22*, 1755–1761.

Su, L., Jia, W., Hou, C., & Lei, Y. (2011). Microbial biosensors: A review. *Biosensors & Bioelectronics, 26*(5), 1788–1799.

Webster, D. P., TerAvest, M. A., Doud, D. F. R., Chakravorty, A., Holmes, E. C., Radens, C. M., Sureka, S. I., Gralnick, J. A., & Angenent, L. T. (2014). An arsenic-specific biosensor with genetically engineered *Shewanella oneidensis* in a bioelectrochemical system. *Biosensors & Bioelectronics, 62*, 320–324.

Zhang, Y., & Angelidaki, I. (2011). Submersible MFC sensor for monitoring microbial activity and BOD in groundwater: Focusing on impact of anodic biofilm on sensor applicability. *Biotechnology and Bioengineering, 108*, 2339–2347.

Chapter 21
Microbial Fuel Cell as Alternate Power Tool: Potential and Challenges

Sheela Berchmans

21.1 Introduction

The anxiety of mankind regarding the fast depletion of reserves of oil and natural gas has kindled the invention of viable alternate energy resources. Further the consequence of our heavy dependency on fossil fuels is reflected in the emission of greenhouse gases leading to global warming and ozone layer depletion which is adversely affecting our environment. In order to address these concerns, efforts are being made globally to develop alternate renewable energy technologies which are preferably green. Scientists have learnt, over the centuries, the technologies of energy conversion from one form to another. For example, harvesting of electrical energy is possible from different forms of energies such as tidal, wind, solar, hydro, thermal, chemical and mechanical. Conversion of chemical energy to electrical energy is known from the days of Volta (eighteenth century), the inventor of voltaic pile and who was the contemporary of Luigi Galvani who first observed animal electricity. The existence of electric field in living organisms can be explicitly seen in electric eel and in the electrical activity of human organs like heart (electrocardiography), brain (electroencephalogram), muscle (electromyogram), eye (electroocular) and in the transmission of signals in nerve cells and these phenomena indicate the scope of converting chemical energy available in biological systems to electrical energy.

The chemical energy available in abundant biomass surrounding us can be harnessed in presence of biological catalysts like enzymes (enzymatic fuel cells) and microorganisms (microbial fuel cells, MFCs). Ever since the discovery of Potter (1910) regarding the effect of microorganisms in the decomposition of

S. Berchmans (✉)
Electrodics and Electrocatalysis (Biosensors) Division, CSIR-Central Electrochemical Research Institute, Karaikudi 630003, Tamil Nadu, India
e-mail: sheelab@cecri.res.in

D. Das (ed.), *Microbial Fuel Cell*,
https://doi.org/10.1007/978-3-319-66793-5_21

organic matter, researchers are focussing on the ability of the biological systems to produce electrical energy. In 1911, Potter observed that a maximum voltage of 0.3–0.5 V could be generated in the presence of *Sacchromyces cerevisae* with Pt as electrode and glucose as substrate (Potter 1911). This was followed by the work of Cohen who tried to produce bacterial batteries (Cohen 1931). The interest on MFCs further augmented with the advent of manned space travel where the scope of MFCs was extended for the conversion of biowaste to energy in spacecrafts in 1960s (Sell 2001). The realisation that power production is possible in the absence of chemical mediators spurred further interest in the development of MFCs during last two decades during which the maximum power output derived from MFCs increased by many folds (Kim et al. 1999; Park and Zeikus 2003; Xing et al. 2008). Now for nearly over a century, research on MFCs is continuously being pursued as MFCs provide a great scope for the production of electrical energy directly by the oxidation of organic matter (Rabaey and Rozendal 2010; Pant et al. 2012; Arends and Verstraete 2012).

MFCs make use of microorganisms as catalysts to drive the anodic and/or cathodic reaction to produce electricity. The chemical energy available in biodegradable substances can be converted into electricity by exoelectrogenic microorganisms. At the anode compartment electrons and protons are generated by the oxidation of organics by certain bacteria. The electrons reach the cathode compartment through the external circuit where it reduces the electron acceptor present and at the same time protons generated at the anode are transferred to the cathode through a membrane separator or through the electrolyte in a membrane less cell. The anodic reactions which are based on microbial oxidation are alike in all type of MFC applications. However, different cathodic reactions can be employed in MFCs to produce electrical energy if the overall reaction is thermodynamically favourable. For practical applications, oxygen is generally considered as the most favourable electron acceptor because of its unlimited availability and its positive redox potential. Feasibility of using other electron acceptors with high redox potential such as nitrate, sulphate and any other contaminants in the environment with high redox potential, which are electrochemically or biologically reducible in the cathode chamber, can also be considered. To further increase the cell voltage, permanganate, dichromate, peroxide and ferricyanide are being used in MFCs because of their high redox potential (Yang et al. 2011). Microorganisms that can reoxidise reduced metal oxides and Fe^{2+} species like *Acidithiobacillus ferrooxidans* and *Thiobacillus ferrooxidans* can also be employed as cathodic biocatalysts (Inglesby et al. 2012; Kumar et al. 2015; Prasad et al. 2006).

During the initial period, the focus of the microbial fuel cell research was centred on wastewater treatment, but over a period of one hundred years, the field of MFCs has grown into a much more diverse field of research called bioelectrochemical systems (BES) due to the advent of several related technologies such as microbial electrolysis, microbial desalination, microbial electrosynthesis, photomicrobial cells and so on (Fig. 21.1). Bioelectrochemical systems (BES) are nothing but electrochemical cells that use microorganisms as catalyst at one or both the electrodes where electrons produced at the anode by the microbial

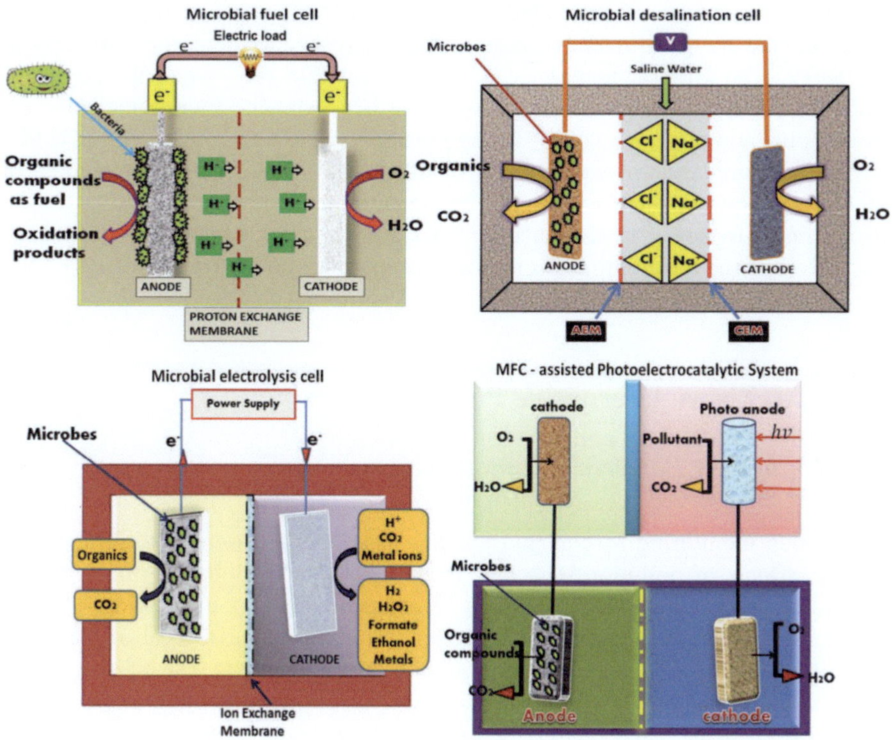

Fig. 21.1 Microbial fuel cell and its analogues

oxidation of organics are transferred to the cathode for a reduction reaction. BESs are believed to be a promising technology for sustainable production of energy or chemicals. There are two types of BESs viz. electron producing MFCs and electron consuming microbial electrolysis cells (MECs). However, according to the terminology of Schroder et al. (2015), BES comprises systems which derive from all subfields of bioelectrochemistry, including enzyme, microbial, protein, DNA or neuroelectrochemistry.

Extra energy is required by MECs to drive the overall reaction to obtain products. During the last decade many applications have been demonstrated with BESs. In one type of application, it has been shown that a wide variety of substrates like acetate, cellulose, starch, domestic wastewater, paper recycling wastewater, food wastes, landfill leachate, complex industrial wastes, sewage sludge, and animal and plant wastes can be treated at the anode compartment of BESs. Another type of application is related to the cathodic reactions in the BESs where products like H_2, H_2O_2, ethanol, formate and metals can be formed by applying an additional voltage over and above MFC voltage. The additional energy required for the cathodic reaction is determined by the thermodynamics of the reactions. The process can become viable only under higher production rate and efficiency.

The voltage and the coulombic efficiencies of the anodic and the cathodic reactions determine the energy efficiency while the internal resistance determines the current efficiency. The BES technology can become feasible only when the paybacks (value of products and efficiency of treatment of wastewater) are higher than the cost (capital and operation costs). Though many reviews on the capability of MFCs for the biodegradation of wide variety of substrates are available, the potential of MFCs as a power source is not well documented. This chapter would cover the efforts made to power devices using MFCs and the power management systems available and the different means to tap the maximum power from MFCs.

21.2 MFCs as Alternative Power Sources

21.2.1 MFCs Powering Remote Sensors

Continuous monitoring of our environment has become essential to observe the adverse effects of pollution such as climate change and global warming. Hence sensors are being deployed in various locations for habitat monitoring, weather monitoring, agricultural applications, fisheries research etc. In most of the situations, it is often necessary to deploy sensors in remote locations. The main difficulty in the deployment of sensors in remote locations lies in providing uninterrupted power for the sensors over a long period. The power requirement becomes significant with the increase in the number of sensors. The reliability, life and the frequency of operation of the sensor depend upon the type of power source employed. It is preferred to make use of renewable power sources which do not require human intervention. While many types of power sources are available, the power requirements of the sensors especially in marine environments are met by batteries which are relatively cheap. However, the need for proper pressure housings and the finite lifetime of batteries which is around three years add further cost. In the case of sensors installed in deep seas, additional expenses are incurred in transportation during replacement of batteries. Further batteries are believed as environmentally hazardous. Therefore, there is a need for finding alternate power sources for remote sensors. As MFCs can operate for a long time using local resources, they would be a good alternative source of power. Sediment microbial fuel cells (SMFCs), also referred to as benthic MFCs, have shown great promise as a novel energy harvesting technology that can provide consistent maintenance-free power over a long period of time, that can cover the lifetime of sensor and communication hardware. Sediment microbial fuel cells are installed in lakes, rivers and oceans where the sediments found at the bottom of lakes, rivers or oceans provide a source of renewable fuel for SMFCs (Fig. 21.2).

SMFCs are made up of two electrodes which are inert and conductive such as graphite, stainless steel, or carbon cloth. The anode is embedded under the sediment and the cathode is positioned in the overlying water. The organics or hydrogen

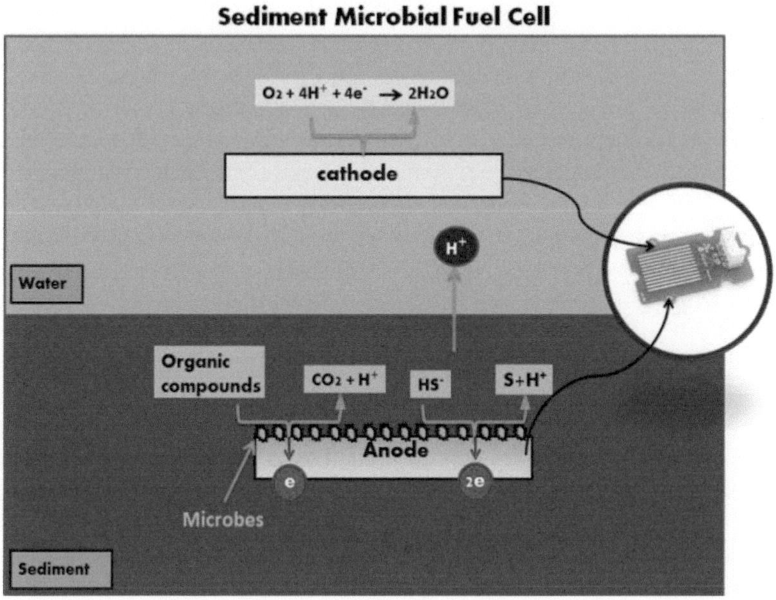

Fig. 21.2 Schematic diagram of a typical sediment microbial cell

sulphide based compounds present in the sediment are oxidised by microorganisms present in the sediment (Aller 1983; Reimers et al. 2006; Ateya and Al-Kharafi 2002) producing protons and electrons according to the following reactions:

$$\text{Oraganic C} + H_2O \longrightarrow CO_2 + H^+ + e^- \quad (21.1)$$

$$HS^- \longrightarrow S + H^+ + e^- \quad (21.2)$$

The electrons formed in the anodic reactions reach the cathode through the external circuit. Oxygen is reduced to form water or hydroxyl ions at the cathode according to the following reactions (Logan and Regan 2006; Logan et al. 2006; Wilcock and Kauffman 1997; Meehan et al. 2011)

$$O_2 + 4H^+ + 4e^- \longrightarrow 2H_2O \quad (21.3)$$

$$O_2 + 2H^+ + 2e^- \longrightarrow 2OH^- \quad (21.4)$$

In one type of SMFC demonstrated for powering wireless sensors, it has been reported that manganese oxide obtained by microbial activity can be reduced on the cathode. (Rhoads et al. 2005). However, microorganisms cannot produce sufficient manganese oxide relentlessly when larger current densities are desirable. Manganese dioxide is exhausted at the cathode owing to continuous discharging at the cathode, after which oxygen reduction controls the cathodic reaction. Recent studies have shown clearly that SMFCs can be practically deployed for powering underwater (Donovan et al. 2013) ground (Donovan et al. 2008) and floating sensors (Donovan et al. 2011; Tender et al. 2008; Nielsen et al. 2007). SMFCs that can deliver power in the range of 3.4–36 mW can be used for powering sensors. It has been demonstrated by Tender et al. that two SMFCs generating a power of 24 mW and 36 mW could supply power for a weather buoy implanted with temperature and humidity sensors (Tender et al. 2008) which were positioned in the Potomac River, at Washington, DC and Tukerton, NJ, USA. Similarly Donovan et al. employed SMFCs delivering power between 11 mW and 2500 mW to operate wireless temperature sensors in a creek at Palouse, WA, USA (Donovan et al. 2008, 2011).

The salient benefits of using SMFCs are that they are sustainable as fuel is continuously being replenished by natural sedimentation processes and their lifetime is not restricted by the electrodes present. The other power sources for underwater sensors, such as seawater batteries (Mijarez et al. 2007) have a limited lifetime. Similarly, the use of hydroelectric power sources is limited by the availability of water flow, which depends upon the location and season. The drawbacks of SMFCs are ascribed to nonlinear scaling, low power generation, and low voltage outputs (Dewan et al. 2008). The power output of a SMFC is not proportional to the size of the electrodes which makes the scaling up a costly matter. This has stimulated the development of new energy harvesting technologies to overcome the problem of low power generation and to avoid scaling up issues. A SMFC continuously generating an average power of 3.4 mW is made to deliver high power in short burst by an energy harvesting scheme and is demonstrated to power a sensor consuming 2.5 W (Donovan et al. 2011).

Besides SMFC, a wearable MFC operated by foot pumped urine has been used to power a wireless transmitter, making it self-reliant and it can send a message every 2 mins to the PC controlled receiver station. The bio-inspiration of the fish circulatory system has enabled Taghavi et al. to design and choose a suitable material for the foot pumping system which indirectly utilises energy from human walking to circulate urine, as the fuel, through the MFCs. The wearable device consisted of 24 individual flexible MFCs located in a pair of socks and the foot pumping part was made of soft tubing and check valves. The soft tubing placed under the heels mimics the involuntary heart muscles which produced the frequent fluid push-pull mechanisms during walking. The 24 MFCs arranged as 12 couples generated an open circuit voltage of 4 V and the normal gaiting of human corresponding to 90 steps min^{-1} provides urine circulation with the flow rate of 45 μL min^{-1} in each leg for each MFC couple (Taghavi et al. 2014, 2015).

21.2.2 MFCs for Robotics

MFC represents an alternative source of energy that can convert a wide variety of organic substrates available globally to electricity and hence it can power devices that need to operate in any type of environment. MFCs would be a good choice to construct autonomous robots which can produce electrical energy from the plethora

Fig. 21.3 Futuristic idea of an autonomous robot acting as a scavenger

of the environment and can function without the aid of human intervention (Fig. 21.3). A single MFC unit can produce absolute power of the order of μW to mW, with a thermodynamic limit attributed to an OCV of 1.1 V. Hence arrays of miniaturised MFCs are required to build robots. The miniaturised cells are advantageous in the sense that they offer shorter migration paths to the substrates and for the exit of proton and electron. Further they occupy less space and mass that can facilitate building efficient robots (Ieropoulos et al. 2012). A gadget similar to a robot called as gastrobot (also known as chew train) was first demonstrated by Wilkinson in 2000. It worked in presence of *E. coli* fed with table sugar for metabolism and the resultant rich fluids were utilised through a stack of fuel cells and the electrical energy produced was used to recharge the onboard batteries that were powering the various actuators present in the device (Wilkinson 2000).

The first illustration of a robot capable of collecting the required fuel in the form of organic matter available in the immediate vicinity was produced by Kelly et al. and called as slugbot as it was designed to collect slug pest from muddy field which would form the real organic feed for the MFC (Kelly et al. 2000). This robot just demonstrated the capability of finding its own fuel from naturally occurring organic sources. It was in 2003, a robot known as Ecobot-1, that was solely powered by MFC, was constructed. This robot could perform phototaxis with a stack of eight MFCs in series and functioned with the help of *E. coli* (biocatalyst), methylene blue (anode) and ferricyanide (cathode). The power output of the MFC stack fed with table sugar was collected by a group of six electrolytic capacitors with a total capacitance of 28.2 mF. The capacitors preset to a desired threshold voltage will require a charging period which depends upon the magnitude of current output and this attributes a discontinuous characteristic to the functioning of robot which would remain idle for 30 s (during charging) and would move towards the light for 3 s. However, ecobot-1 was not tested for its endurance and longevity studies. Ecobot-2, the next version of ecobot series, could function in a similar intermittent behaviour for 12 d. The robot consisted of eight MFCs with mixed sludge culture anode and air cathode without any mediator. This robot could execute, sensing ambient temperature, handling on-board microcontroller actuation, photo taxis and communication (wireless transmission of temperature). Also ecobot-2 was imparted with the capability of consuming rough and complex substrates such as prawn shells, dead flies and rotten fruits in the 25 mL anode compartments of the MFCs present inside the device. This robot took 14 mins to charge the onboard capacitor of 28.2 mF to produce energy for 3 s for its assigned tasks.

Ecobot-3 was additionally bestowed with foraging behaviour and was powered by 48 small scale MFCs (two tiers of 24 MFCs each on top of the other) and was made more independent in the sense that it can perform multiple tasks like collection of food and water from its environment, digestion of the same followed by distribution of the digest through onboard MFCs through mechatronic units and was imparted with the ability to get rid of its own waste. Hence ecobot-3 was able to generate and store sufficient energy to run a total of five motors and four pumps and for the more complex electronic controller. For a multiple number of continuously fed units, the assembly should be able to produce enough energy for the peripheral units and for fuel distribution for the development of a truly self-sustainable system (Melhuish et al. 2006; Ieropoulos et al. 2012).

21.2.3 Paper-Based MFC Devices

The development of paper or paper-like substrates for batteries and other energy storage devices are becoming more popular as paper substrates are multifunctional, versatile, flexible, disposable and are of low cost. Portable, disposable and affordable analytical/diagnostic devices that are suitable for single use especially in remote and resource-poor regions can be fabricated using paper and they can be used as point of care diagnostic tool without the need for time consuming routine clinical analysis (Fig. 21.4). Microwatt level power sources are preferred over conventional batteries to power paper-based diagnostic devices. Paper-based sensing system essentially appended with low power paper microbatteries would make the sensing system complete. Several type of devices with integrated paper-based batteries or energy storage devices have been developed for various applications such as on chip fluorescence assay, for powering biomedical devices, lithium ion paper battery, super capacitor integrated into photoelectrochemical lab on paper device and enzymatic paper-based biofuel cell. Enzymatic and microbial biofuel cells are suitable for implementation on paper-based substrates.

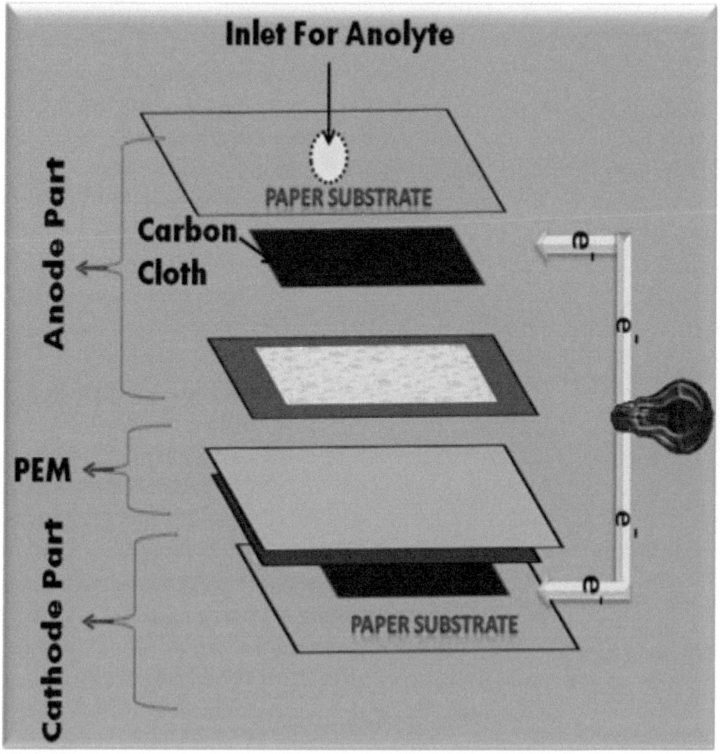

Fig. 21.4 Schematic representation of a paper-based MFC

The first literature report on paper-based MFC (Fraiwan et al. 2013a, b, Fraiwan and Choi 2013) presented (i) a paper-based proton exchange membrane filled with sodium polystyrene sulphonate; (ii) micro-fabricated paper chambers made up of patterned hydrophobic barriers of photoresist; and (iii) reservoirs made up of paper for holding the anolyte and catholyte for longer period of time. The paper-based MFCs produced electricity within a shorter time in comparison with conventional MFCs which require long start-up times ranging from several days to a week due to the slow growth and adaptation of bacteria on the anode of MFCs (Qian et al. 2009; Nguyen et al. 2014). In the case of paper MFC, the hydrophilic paper reservoir quickly absorbs the anolyte and accelerates the attachment of a number of bacteria cells to the anode. A stack of paper-based MFCs can power a red LED for more than 30 min without the use of power management circuits. Such MFCs can be used as affordable, disposable power source for single use diagnostic devices in resource-limited settings as any kind of fuel (waste water/any source of organic content) in the immediate vicinity can act as an excellent source for current generation through microbial activity (Fraiwan and Choi 2014). However, potassium ferricyanide is used as a catholyte, which is not suitable for actual application because it is toxic and expensive. Air breathing cathodes will be preferred due to their low cost, sustainability and lack of waste products. A paper-based, two chambered MFC stack consisting of four cells connected in series, fabricated by applying Origami technique provided an open circuit voltage of 1.6 V and a stable output current of 4 μA. Similar stack of paper-based MFCs would be suitable for powering on chip paper-based biosensors. The bacterial cells flowed through patterned fluidic pathways within the paper matrix and this work represents a blend of origami, the paper folding technology and MFC technology providing a new concept to the fabrication of stacked assembly of paper MFCs (Zhang et al. 2015).

Winfield et al. (2015) have taken pioneering efforts to make use of paper-based MFC technology to power portable emergency locator transmitters (ELTs). The paper-based MFC device, besides being light weight, is also robust and fast in terms of response. Urine is demonstrated as an ideal fuel for MFCs to provide power for ELTs which can also be an indicator of proof of life. Novel origami tetrahedron MFCs (TP-MFCs) using photocopier paper have been fabricated to test different urine-based inoculants. A stack of six abiotic MFCs inoculated with urine could produce a sufficient voltage to energise a power management system after 3 h 15 min. At this stage, the anodes of TP-MFCs can be removed and stored in refrigerator after drying in air for seven days. After four weeks, these MFCs immediately responded to fresh urine and achieved a useful working voltage within 35 mins. Two paper MFCs connected in parallel were able to transmit 85 radio signals and in a series configuration 238 broadcasts over 24 h. These findings provide a proof of concept for the suitability of using urine activated paper MFCs as "proof of life" reporting systems. Besides, such devices are simple, inexpensive, and lightweight (Winfield et al. 2015).

21.2.4 *Pee Power Urinal Field Trials*

Field trials, demonstrating the use of MFCs driven by pee power, were conducted on Frenchay Campus (UWE, Bristol) during February–May 2015 and Glastonbury Music Festival at Worthy Farm, Pilton in June 2015, for internal lighting. The first trial made in the university campus demonstrated the feasibility of modular MFCs for lighting where the MFCs were fed with the urine collected from the toilets with limited users consisting of the staff and students only. The second trial conducted at the music festival with a large number of audiences (~1000 users per day) also revealed the prospective of the MFCs to generate power for internal lighting. The power output recorded for individual MFCs is 1–2 mW and one stack consisted of 36 cells. The electrical output of the pee power urinals was proportional to the number of MFCs used and depended on temperature and flow rate. The campus urinal consisted of 288 MFCs and the Glastonbury urinal consisted of 432 MFCs. It was observed that the COD removal was >95% for the campus urinal and on average 30% for the Glastonbury urinal. This pee power urinal field trials conducted for the first time have demonstrated the feasibility of MFCs for simultaneous electricity generation and direct urine treatment. Urine is a good resource that meets the fuel requirement of MFC where the sustainable microbial community on the anode generates power, water and minerals from urine (Ieropoulos et al. 2016). MFCs show a good scope for maintaining sanitation and community health in public places. Similar urine-fed MFC stacks were also used to charge mobile phone. These studies also demonstrated that ceramic membranes can replace the ion exchange membranes generally being employed and urine can be an alternative organic feedstock to MFC (Ieropoulos et al. 2013).

21.2.5 MFCs Powering Low Power Density Devices

A Barcelona-based company, Arkyne Technologies, use a single plant to charge mobile phones. The company's product, a pot called the Bioo Lite, uses the energy generated by photosynthesis to charge smartphones. Microorganisms in the pot break up the compounds, liberating electrons that pass through nanowires. The resulting electricity powers a typical USB port, where one can plug in their phone or tablet cable (http://www.arkynetechnologies.com).

It has been shown that low power density devices such as digital clocks, display devices and LEDs can be powered by MFCs (Rengasamy and Berchmans 2012). It has also been shown in the author's lab that two chambered MFCs with anodic biofilms formed by the inoculation of food waste could simultaneously treat canteen food wastewater and produce current to power low power density devices (Fig. 21.5).

21.3 Factors Constraining Energy Output of MFCs

The maximum power output of the MFC is limited by the cell voltage which is the difference between the anodic and cathodic potentials. When acetate serves as electron donor at the anode and oxygen as electron acceptor at the cathode, the thermodynamic anode and cathode potentials as calculated by Nernst equation are −0.3 V (vs NHE) and 0.8 V (Vs NHE) respectively which will result in a theoretical OCV of 1.1 V. However, many factors such as activation polarisation, concentration polarisation, ohmic losses, diffusion limitation, membrane resistance and resistance due to the biofilm contribute towards lowering of the cell voltage and hence only a maximum voltage of around 0.7–0.8 V is observed in practical situations. Further the electrode potentials are affected by the nature of electron donors, electron acceptors and microbial inocula used in the system which negatively affects the energy recovery efficiency and the output voltage. Additional energy losses arise when the fuel/substrate loses its electrons to side reactions such as methanogenesis, nitrate and sulphate reductions, H_2 scavenging and aerobic microbial growth. Similar losses are observed at the cathode. When O_2 reaction is employed as the cathodic reduction reaction, the high activation barrier combined

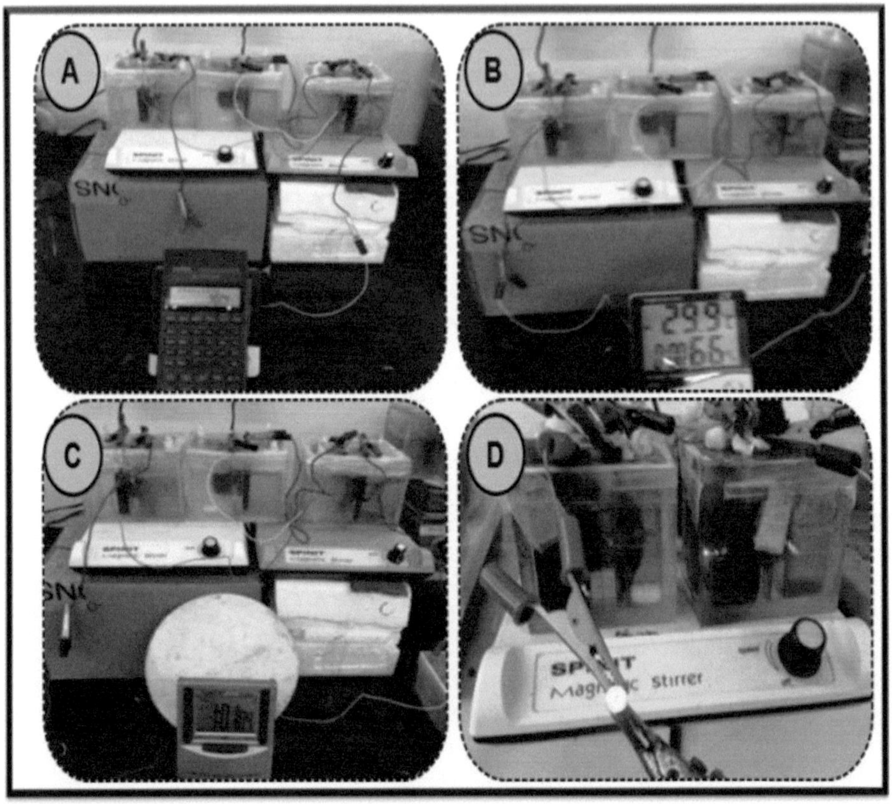

Fig. 21.5 Microbial fuel cells delivering power from canteen waste delivering power for devices such as: (**a**) Calculator; (**b**) Display device; (**c**) Digital time piece; and (**d**) LED lamp

with the complexity of the reduction reaction involving four electrons and several possible mechanistic pathways contribute to lowering of the energy efficiency and the output voltage. Many noble and non-noble metal-based catalysts are being employed as O_2 reduction catalysts. The low solubility of O_2 in water leads to mass transport limitation in the cathode compartment. In case of biocathodes, the growth of biofilm and the mass transfer characteristics within the biofilm also contribute to energy losses. Extra losses are added owing to membrane resistance caused by insufficient ion transport resulting in pH gradients. In the case of anion exchange membranes less energy loss is observed due to pH gradient. But these membranes are susceptible to substrate permeability and deformation. Size selective separators such as micro porous filtration membranes, porous fabrics, glass fibres and nylon meshes can be made to replace ion exchange membranes as they usually display higher ion transport ability and lower internal resistance. However, coulombic efficiency is reduced as a result of increased substrate and O_2 permeation through pores.

Conventionally, the maximum power output that can be derived from an MFC is determined by applying varying external resistances sequentially over fixed intervals of time or by applying linear potential sweep. The resulting semi-circular graph obtained between voltage and current density shows a point of maximum power output when the applied external resistance is equal to the internal resistance. However, depending upon the application, researchers take into consideration either top power point or high current output point. For example if one's goal is to treat wastewater, the priority will lie on maximum removal organic content and the MFC needs to run at the highest current output condition rather than at the top power output point. Similarly, for H_2 producing MECs, the ideal operating condition would be high current region rather than top power output point. In the case of microbial desalination cells (MDCs), if the goal is to achieve high salt removal, the ideal operating condition would be high current region whereas if the goal is to achieve higher energy, the appropriate operating condition would be maximum power output region. Researchers have put in tremendous efforts to harvest energy from MFCs for practical implementation. The direct power output from a single MFC cannot power common electronics. While higher power can be obtained by connecting MFCs in series or parallel, the resulting power output is often non-linear. Hence developing suitable energy harvesting systems and power management systems are essential for MFCs.

21.4 Energy Harvest in MFC

From the foregoing discussion it is clear that the energy obtained from the MFC will not be sufficient to energise any frequently used electric gadgets. New energy harvesting technologies are to be coupled with MFC technology to capture, store and boost the power output of MFC. At present electrochemical capacitors are mainly integrated with MFCs for driving devices that consume low power. The power output characteristics of the MFCs can be increased through alternate charging and discharging of capacitors. The output voltage of four MFCs was found to increase from 0.7 to 2.5 V and the peak power improved by 2.6 times by charging a group of parallel connected capacitors and then discharging them in series, with insignificant energy loss in the circuit. By replacing the MFC anode by a capacitive electrode, the exoelectrogens growing on the capacitive electrode can directly transmit the electrons to the capacitor for storage. However, energy storage remains to be improved compared to the external capacitor.

The use of a capacitor cannot ensure the production of continuous power output, but it can provide a discontinuous supply of high power. Hence, a power management system consisting of charge pump, boost converter and load are used in the circuit. The low current provided by the MFC is taken up by the charge pump to energise the capacitor while the boost converter lifts the output voltage of the capacitor to the voltage level of the load. Such power management systems would be of great use to SMFCs to provide intermittent power for sensors located in remote places. So far reported power management systems mainly focus on improving DC power output in order to energise low power consumption devices. However, alternating current power is required to operate gadgets used for wastewater treatment and hence there is a dire need to develop energy management systems that are able to perform DC-AC power conversion for large scale MFCs. Recent reports describe a DC-AC converter that can generate alternating voltage in any desired frequency at $\geq$95.5% efficiency. However, the question of effectively incorporating such a converter with a capacitor remains to be answered (Wang et al. 2012, 2015; Alaraj et al. 2014).

A new energy harvesting technology based on a super conducting magnetic energy storage system which stores energy in the magnetic field produced by a flow of direct current in a superconducting coil (inductor) is being introduced to this field recently. This technology can store electrical energy as current circulating indefinitely through a coil made up of a super conducting material. Once the superconducting coil is charged, the current will not decay and the magnetic energy can be stored. This new technology can harvest 67% energy from MFC. Though the costs involved in making such a magnetic energy storage device is very high, the minimal energy loss and high quality power output make it a smart option for energy storage. Research efforts in this direction are still in their formative years. At present capacitors are the most appropriate devices for energy management in MFCs (Sun et al. 2016).

21.5 Conclusions

The MFCs research initiated in 1910 by Potter has attracted researchers worldwide and it has evolved into a novel technology of its own right with the inclusion of applications in wide areas. Initially MFCs were considered mainly suitable for wastewater treatment applications and were not seriously considered as renewable alternative source of energy. However, the interesting reports on MFCs powering remote sensors in ocean floors, MFCs powering robots, MFCs producing electricity from urine providing power for lights etc., increase the scope for MFC applications in terms of their suitability as alternative power source. These examples indicate that MFCs can become a practical reality provided they are cost effective. If MFCs are benchmarked against chemical batteries and fuel cells, it will be clear that MFCs are sustainable energy resources. However, batteries and fuel cells outrank MFCs in terms of energy output per unit volume. If MFCs are bench marked against anaerobic digestion, two distinct differences are observed. MFCs cannot handle suspended and particulate organic material whereas anaerobic digestion is capable of handling them. Using MFCs value added products can be obtained in addition to electricity which is not feasible in anaerobic digestion. However, energy conversion rates are lower in MFC technology. Hence to make the MFC technology more viable the challenges are to be understood more clearly. To maximise the power output derived from MFCs proper power management systems should be evolved and integrated with MFCs. Further synthetic biologists can genetically engineer exooelectrogens with ideal electron transfer properties. The future of MFC research looks brighter when we envisage MFC scavenger robots executing the work of cleaning our environment, MFCs powering devices in remote locations, MFCs producing value added products acting like biorefineries and so on.

Acknowledgement The author acknowledges Ms. S. Sundari and Mrs. V. Manju, Research assistants, for the figures and the preparation of the references list respectively.

References

Alaraj, M., Ren, Z. J., & Park, J.-D. (2014). Microbial fuel cell energy harvesting using synchronous flyback converter. *Journal of Power Sources, 247*, 636–642.

Aller, R. C. (1983). The importance of the diffusive permeability of animal burrow linings in determining marine sediment chemistry. *Journal of Marine Research, 41*, 299–322.

Arends, J. B. A., & Verstraete, W. (2012). 100 years of microbial electricity production: Three concepts for the future. *Microbial Biotechnology, 5*, 333–346.

Ateya, B. G., & Al-Kharafi, F. M. (2002). Anodic oxidation of sulfide ions from chloride brines. *Electrochemistry Communications, 4*, 231–238.

Cohen, B. (1931). The bacterial culture as an electrical half-cell. *Journal of Bacteriology, 21*, 18–19.

Dewan, A., Beyenal, H., & Lewandowski, Z. (2008). Scaling up microbial fuel cells. *Environmental Science & Technology, 42*, 7643–7648.

Donovan, C., Dewan, A., Heo, D., & Beyenal, H. (2008). Batteryless, wireless sensor powered by a sediment microbial fuel cell. *Environmental Science & Technology, 42*, 8591–8596.

Donovan, C., Dewan, A., Peng, H., Heo, D., & Beyenal, H. (2011). Power management system for a 2.5 W remote sensor powered by a sediment microbial fuel cell. *Journal of Power Sources, 196*, 1171–1177.

Donovan, C., Dewan, A., Heo, D., Lewandowski, Z., & Beyenal, H. (2013). Sediment microbial fuel cell powering a submersible ultrasonic receiver: New approach to remote monitoring. *Journal of Power Sources, 233*, 79–85.

Fraiwan, A., & Choi, S. (2013). A multi-anode paper-based microbial fuel cell for disposable biosensors. *IEEE SENSORS*, 1908–1911.

Fraiwan, A., & Choi, S. (2014). Bacteria-powered battery on paper. *Physical Chemistry Chemical Physics, 16*, 26288–26293.

Fraiwan, A., Mukherjee, S., Sundermier, S., & Choi, S. (2013a). A microfabricated paper-based microbial fuel cell. In *IEEE 26th international conference on micro electro mechanical systems (MEMS)* (pp. 809–812).

Fraiwan, A., Mukherjee, S., Sundermier, S., Lee, H.-S., & Choi, S. (2013b). A paper-based microbial fuel cell: Instant battery for disposable diagnostic devices. *Biosensors and Bioelectronics, 49*, 410–414.

Ieropoulos, I. A., Greenman, J., Melhuish, C., & Horsfield, I. (2012). Microbial fuel cells for robotics: energy autonomy through artificial symbiosis. *ChemSusChem, 5*, 1020–1026.

Ieropoulos, I. A., Ledezma, P., Stinchcombe, A., Papaharalabos, G., Melhuish, C., & Greenman, J. (2013). Waste to real energy: The first MFC powered mobile phone. *Physical Chemistry Chemical Physics, 15*, 15312–15316.

Ieropoulos, I. A., Stinchcombe, A., Gajda, I., Forbes, S., Merino-Jimenez, I., Pasternak, G., Sanchez-Herranz, D., & Greenman, J. (2016). Pee power urinal – Microbial fuel cell technology field trials in the context of sanitation. *Environmental Science: Water Research & Technology, 2*, 336–343.

Inglesby, A. E., Beatty, D. A., & Fisher, A. C. (2012). Rhodopseudomonas palustris purple bacteria fed Arthrospira maxima cyanobacteria: Demonstration of application in microbial fuel cells. *RSC Advances, 2*, 4829–4838.

Kelly, I., Holland, O., & Melhuish, C. (2000). Slugbot: A robotic predator in the natural world. In M. Sugisaka & H. Tanaka (Eds.), *Proceedings of the international symposium on artificial life and robotics for human welfare and artificial life robotics* (pp. 470–475).

Kim, B. H., Ikeda, T., Park, H. S., Kim, H. J., Hyun, M. S., Kano, K., Takagi, K., & Tatsumi, H. (1999). Electrochemical activity of an Fe(III)-reducing bacterium, Shewanella putrefaciens IR-1, in the presence of alternative electron acceptors. *Biotechnology Techniques, 13*, 475–478.

Kumar, R., Singh, L., Wahid, Z. A., & Din, M. F. M. (2015). Exoelectrogens in microbial fuel cells toward bioelectricity generation: A review. *International Journal of Energy Research, 39*, 1048–1067.

Logan, B. E., & Regan, J. M. (2006). Microbial fuel cells—Challenges and applications. *Environmental Science & Technology, 40*, 5172–5180.

Logan, B. E., Hamelers, B., Rozendal, R., Schröder, U., Keller, J., Freguia, S., Aelterman, P., Verstraete, W., & Rabaey, K. (2006). Microbial fuel cells: Methodology and technology. *Environmental Science & Technology, 40*, 5181–5192.

Meehan, A., Gao, H., & Lewandowski, Z. (2011). Energy harvesting with microbial fuel cell and power management system. *IEEE Transactions on Power Electronics, 26*, 176–181.

Melhuish, C., Ieropoulos, I., Greenman, J., & Horsfield, I. (2006). Energetically autonomous robots: Food for thought. *Autonomous Robots, 21*, 187–198.

Mijarez, R., Gaydecki, P., & Burdekin, M. (2007). Flood member detection for real-time structural health monitoring of sub-sea structures of offshore steel oilrigs. *Smart Materials and Structures, 16*, 1857–1869. IOP Publishing, ISSN: 0957-0233.

Nguyen, T. H., Fraiwan, A., & Choi, S. (2014). Paper-based batteries: A review. *Biosensors and Bioelectronics, 54*, 640–649.

Nielsen, M. E., Reimers, C. E., & Stecher, H. A. (2007). Enhanced power from chambered benthic microbial fuel cells. *Environmental Science & Technology, 41*, 7895–7900.

Pant, D., Singh, A., Van Bogaert, G., Irving Olsen, S., Singh Nigam, P., Diels, L., & Vanbroekhoven, K. (2012). Bioelectrochemical systems (BES) for sustainable energy production and product recovery from organic wastes and industrial wastewaters. *RSC Advances, 2*, 1248–1263.

Park, D. H., & Zeikus, J. G. (2003). Improved fuel cell and electrode designs for producing electricity from microbial degradation. *Biotechnology and Bioengineering, 81*, 348–355.

Potter, M. C. (1910). *On the difference of potential due to the vital activity of microorganisms.* Proceedings of the University of Durham Philosophical Society 1910 (pp. 245–249).

Potter, M. C. (1911). *Electrical effects accompanying the decomposition of organic compounds.* Proceedings of the Royal Society of London B 1911 (pp. 260–276).

Prasad, D., Sivaram, T. K., Berchmans, S., & Yegnaraman, V. (2006). Microbial fuel cell constructed with a micro-organism isolated from sugar industry effluent. *Journal of Power Sources, 160*, 991–996.

Qian, F., Baum, M., Gu, Q., & Morse, D. E. (2009). A 1.5 μL microbial fuel cell for on-chip bioelectricity generation. *Lab on a Chip, 9*, 3076–3081.

Rabaey, K., & Rozendal, R. A. (2010). Microbial electrosynthesis—Revisiting the electrical route for microbial production. *Nature Reviews Microbiology, 8*, 706–716.

Reimers, C. E., Girguis, P., Stecher, H. A., Tender, L. M., Ryckelynck, N., & Whaling, P. (2006). Microbial fuel cell energy from an ocean cold seep. *Geobiology, 4*, 123–136.

Rengasamy, K., & Berchmans, S. (2012). Simultaneous degradation of bad wine and electricity generation with the aid of the coexisting biocatalysts *Acetobacter aceti* and *Gluconobacter roseus*. *Bioresource Technology, 104*, 388–393.

Rhoads, A., Beyenal, H., & Lewandowski, Z. (2005). Microbial fuel cell using anaerobic respiration as an anodic reaction and biomineralized manganese as a cathodic reactant. *Environmental Science & Technology, 39*, 4666–4671.

Schroder, U., Harnisch, F., & Angenent, L. T. (2015). Microbial electrochemistry and technology: Terminology and classification. *Energy & Environmental Science, 8*, 513–519.

Sell, D. (2001). Bioelectrochemical fuel cells. In H.-J. Rehm & G. Reed (Eds.), *Biotechnology. Volume 10: Special processes* (2nd ed., pp. 5–10). Frankfurt am Main: Wiley-VCH.

Sun, M., Zhai, L.-F., Li, W.-W., & Yu, H.-Q. (2016). Harvest and utilization of chemical energy in wastes by microbial fuel cells. *Chemical Society Reviews, 45*, 2847–2870.

Taghavi, M., Greenman, J., Beccai, L., Mattoli, V., Mazzolai, B., Melhuish, C., & Ieropoulos, I. A. (2014). High performance, totally flexible, tubular microbial fuel cell. *ChemElectroChem, 1*, 1994–1999.

Taghavi, M., Stinchcombe, A., Greenman, J., Mattoli, V., Beccai, L., Mazzolai, B., Melhuish, C., & Ieropoulos, I. A. (2015). Self sufficient wireless transmitter powered by foot-pumped urine operating wearable MFC. *Bioinspiration & Biomimetics, 11*, 016001.

Tender, L. M., Gray, S. A., Groveman, E., Lowy, D. A., Kauffman, P., Melhado, J., Tyce, R. C., Flynn, D., Petrecca, R., & Dobarro, J. (2008). The first demonstration of a microbial fuel cell as a viable power supply: Powering a meteorological buoy. *Journal of Power Sources, 179*, 571–575.

Wang, H., Park, J.-D., & Ren, Z. (2012). Active energy harvesting from microbial fuel cells at the maximum power point without using resistors. *Environmental Science & Technology, 46*, 5247–5252.

Wang, H., Park, J.-D., & Ren, Z. J. (2015). Practical energy harvesting for microbial fuel cells: A review. *Environmental Science & Technology, 49*, 3267–3277.

Wilcock, W. S. D., & Kauffman, P. C. (1997). Development of a seawater battery for deep-water applications. *Journal of Power Sources, 66*, 71–75.

Wilkinson, S. (2000). "Gastrobots" – Benefits and challenges of microbial fuel cells in foodpowered robot applications. *Autonomous Robots, 9*, 99–111.

Winfield, J., Chambers, L. D., Rossiter, J., Greenman, J., & Ieropoulos, I. (2015). Urine-activated origami microbial fuel cells to signal proof of life. *Journal of Materials Chemistry A, 3*, 7058–7065.

Xing, D., Zuo, Y., Cheng, S., Regan, J. M., & Logan, B. E. (2008). Electricity generation by rhodopseudomonas palustris DX-1. *Environmental Science & Technology, 42*, 4146–4151.

Yang, Y., Sun, G., & Xu, M. (2011). Microbial fuel cells come of age. *Journal of Chemical Technology & Biotechnology, 86*, 625–632.

Zhang, J., Fraiwan, A., & Choi, S. (2015). Origami paper-based microbial fuel cells for disposable biosensors. *19th international conference on miniaturized systems for chemistry and life sciences, Gyeongju, Korea.*

Chapter 22
Microbially Mediated Electrosynthesis Processes

Mohita Sharma and Priyangshu M. Sarma

22.1 Microbial Electrosynthesis for Bioelectrochemical Processes

As stated by Lovley (2008), an environmental niche of any given microorganism can make it function as electrode reducer or an electrode oxidizer, just as microorganisms can function either as iron reducer or iron acceptor depending on environmental conditions. Electron transfer on microbial-electrode interfaces is a result of evolutionary capabilities of some microbes to perform effective extracellular exchange with insoluble minerals and related natural electron acceptors and donors (Lovley 2012). BES technology is capable of converting chemical energy of organic wastes including low-strength wastewaters and lignocellulosic biomass into electricity or other value added products. All BES consist of an anode where the oxidation reaction occurs and a cathode for the reductions, and at least one of these reactions is microbially catalyzed, hence classified as microbial bioanode and biocathode respectively (Rabaey et al. 2010a). In bioanodes, bacteria called exoelectrogens oxidize organic or inorganic matter anaerobically to discharge electrons, which are transferred through the electron transport chain to the electrode directly or indirectly. In biocathodes, bacteria called electrotrophs receive the electrons from the cathode directly or via some redox mediators to reduce compounds like organics, carbon dioxide, sulphate or nitrate (Gregory et al. 2004;

M. Sharma (✉)
TERI University, Vasant Kunj, New Delhi, India

University of Calgary, Calgary, Canada
e-mail: mohitasharma87@gmail.com

P.M. Sarma
TERI University, Vasant Kunj, New Delhi, India

ONGC Energy Centre, New Delhi, India

D. Das (ed.), *Microbial Fuel Cell*,
https://doi.org/10.1007/978-3-319-66793-5_22

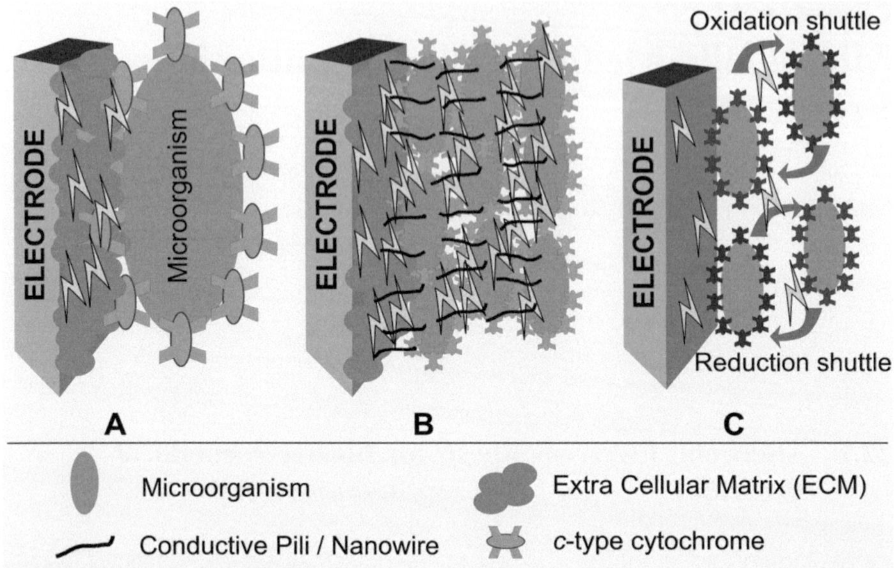

Fig. 22.1 Mechanisms of electron transfer to bioelectrochemical electrode surfaces: (**a**) Direct electron exchange with the electrode surface; (**b**) Formation of putative nanowire like structure consisting of conductive pili and loosely bound *c*-type cytochrome for exchange of electrons; and (**c**) Electron transfer with the help of redox mediators (Adapted from Lovley 2012)

Rabaey et al. 2010a). As compared to conventional fuel cells, the BES operates under relatively mild conditions of temperature and pressure, using a wide variety of organic substrates and mostly without using expensive precious metals as catalysts. BES presents a wide number of advantages when compared to classical routes towards sustainable energy production like generation of electric power, chemical or ecological goods from renewable and non-carbon fuel sources. As a consequence, their applicability is not restricted to their geographic location, thus promoting resource independency, accessibility, decentralization, self-sufficiency, and environmentally safe practices. Electroactive microorganisms have characteristic electron exchange properties with conducting materials (Lovley 2008). For this electron exchange, either they release some redox mediators that facilitate indirect electron transfer by acting as redox shuttles or they directly take part in electron transfer through *c*-type cytochromes, by production of conductive exopolymeric materials or by forming conductive biofilm matrix (Fig. 22.1).

Some microbial catalytic processes also require external electrical input to drive the conversions and overcome cathodic over-potentials, since many of the coupled electrochemical reactions are usually not thermodynamically feasible. Such electricity-driven bioproduction processes are referred to as microbial electrosynthesis (MES), through which microbes reduce inorganic (e.g. CO_2) or organic chemicals (e.g. volatile fatty acids) into extracellular organic compounds (Rabaey and Rozendal 2010). Although the fundamental processes and microbe-cathode interactions are less understood as compared to bioanodes, several proof-of-concept studies have already been reported for MES (Nevin et al. 2011; Sharma et al. 2013a). The possibility of obtaining higher value products from waste streams and integration of these processes into the existing biorefineries make MES very attractive (Desloover et al. 2012; He et al. 2016).

22.2 Factors Affecting the Performance of BES

Modification of architecture, materials, biomass and solution chemistry can significantly influence the performance of BES. For the commercialization of this technology, every aspect has to be collaboratively researched and improvised to bring down the overall production, installation and operational costs. In general, the different factors that affect the performance of BES can be broadly categorized into six parameters (Fig. 22.2) that have been described in the following sections.

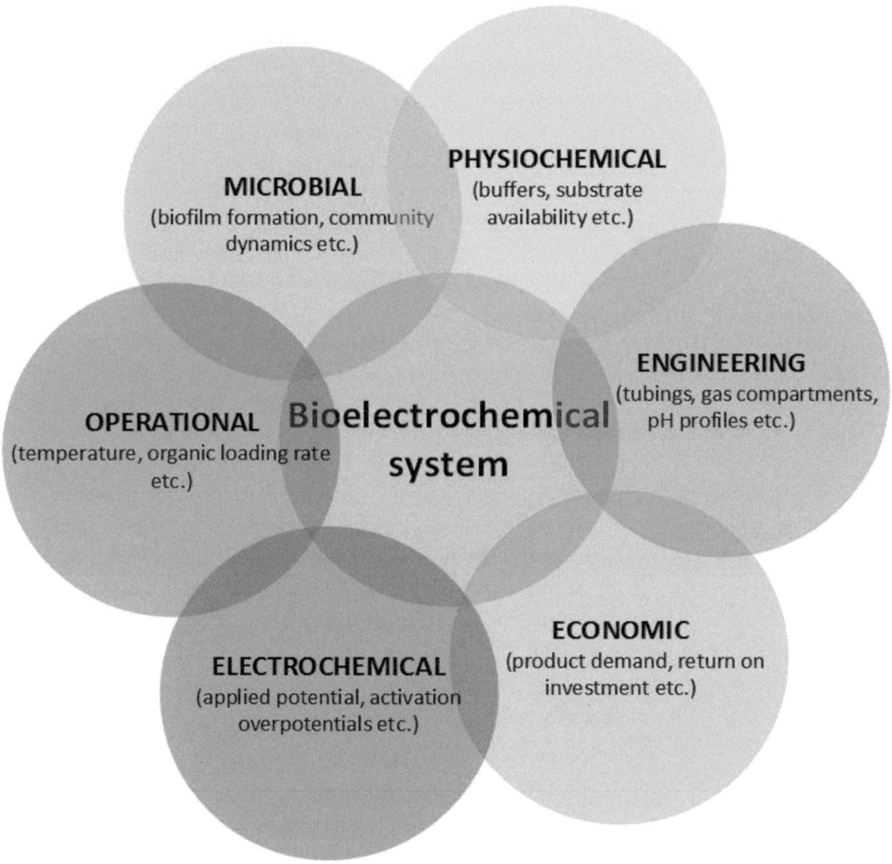

Fig. 22.2 Parameters affecting the performance of a bioelectrochemical system

22.2.1 Electrochemical Parameters

During the monitoring of BES, a number of electrochemical hitches are encountered that might lead to failure of BES operation. Some of the common electrochemical related factors essential to monitor during BES operation include activation overpotential, ohmic polarization, voltage reversal and applied potential at anode and cathode.

22.2.1.1 Activation Polarization

A common problem faced during the operation of BES is losses that occur due to activation overpotential. Overpotential is the difference between thermodynamically determined potential and experimentally observed potential of a half reaction. The energetic losses that occur at electrode contributing to overpotential include energy lost during the activation, ohmic losses and energy lost during microbial growth and maintenance, mainly due to imperfect catalysis of redox reactions (Rabaey et al. 2007).

Different strategies have been adopted to circumvent this problem. These include use of higher oxygen concentration or higher oxygen pressure in the cathodic chamber in case of open-air cathode BES. In studies where thermophiles are used, operation of BES at higher temperatures is feasible. Increase in the biomass load at the surface of the electrode or increase of the roughness and surface area of the electrode material may also help in reduction of activation polarization losses. Other methods include increasing the surface to volume ratio for development of more robust electroactive biofilms (Desloover et al. 2012; Sharma et al. 2014a; Guo et al. 2015).

22.2.1.2 Ohmic Polarization

This occurs due to resistance of flux of electrons through the electrode material and the flux of ions in electrolyte solution and separator membrane. The resistance includes electronic (from electrical circuit), ionic and contact resistance due to losses in the electrode (Rabaey and Rozendal 2010). Ohmic losses can be decreased by use of higher ionic strength electrolyte solution (Rousseau et al. 2013) and use of current collectors (Sharma et al. 2013b). Reduction in the distance between working and reference electrodes (Zhang et al. 2014), reduced electrode spacing (Cheng and Logan 2011), use of optimized electrode to membrane sized ratios and use of membranes with low resistivity (Logan et al. 2006) can also help mitigate these losses.

22.2.1.3 Voltage Reversal

When operated in a serially connected stack, sometimes the individual cells in the stack are unable to provide equal amount of current as the higher power generating cell. In such cases, the weaker cell is charged by the stronger cell and reduces the overall stack performance. This phenomenon is commonly termed as voltage reversal, and is due to substrate starvation in the cells especially in the anode compartment (Gurung et al. 2012) and the difference in the internal resistance of cells of the stacks. Hence, a continuous source of substrate and the internal resistance of the cells of the stacks need to be matched before using them together in the stack.

22.2.1.4 Applied Potential

The potential applied to the working electrode can majorly influence the performance of the BES as non-optimized potential can provide less energy for microbial growth and cell maintenance, but an optimized applied potential can enhance growth rate of bacteria and enhance electron discharge. Too high potential can negatively affect the catalytic activities of enzymes by causing conformational deformities in associated proteins and cell rupture that might eventually lead to reduced output. Hence maintenance of optimal potential becomes essential to balance electron discharge and microbial cell growth. The improvement in performance due to application of potential is generally speculated due to change in the electrode surface topography and chemistry, improved microbial adhesion over the electrode surface, increase in enzymatic activity and establishment of direct electron transfer contacts by formation of electroactive putative nanowire like structures in some cases. Ki et al. (2016) have studied the overpotential losses in BES and observed that cathode related phenomena leads to highest overpotentials. They provided various new innovative strategies to reduce overpotentials like introduction of carbon dioxide to cathode chamber that leads to reduction in pH and subsequently help in reducing the applied voltage. Improvisations in cell design, membrane selection, increased anodic surface area using carbon fibres without increasing distance between the two electrodes and improved electrochemical characterization for better monitoring of BES processes were some of the other strategies also suggested in their study for improving performance of BES.

22.2.2 *Physicochemical Parameters*

The conductivity of the solution used for BES operation can also affect reactor performance and hence it becomes necessary to monitor the buffer and salinities used in electrochemical tests (Liu et al. 2005; Logan 2012; Sharma et al. 2015).

Three important physicochemical parameters that affect the microbial electrode performance namely substrate availability, salinity, buffer concentration etc. have been discussed in the subsequent subsections.

22.2.2.1 Substrate Availability

The amount of power generated varies with the substrate used in the BES. Acetate and glucose have been usually reported to produce higher power densities (Logan 2012). Loading rate also significantly affect the BES performance and becomes a major concern in BES as limited availability of substrates can lower the performance of BES being mostly fed on waste streams. Usage of such substrates leads to depletion in the rate of oxidation of substrate in the anode compartment and the subsequent generation of electrons in the process. A pretreatment like hydrolysis might increase the availability of electron donors required in the anode compartment of the MFC. For other BES operations, operation in continuous mode might ensure ample substrate availability for reactions to efficiently proceed.

22.2.2.2 Salinity

A tremendous increase in internal resistance of the cell occur under highly saline conditions, as it leads to dehydration of anodophyllic cells, which will further lower the electron transfer rate in the fuel cell. Methanogenesis is inhibited at salt concentrations above 5 g L^{-1}, while acidogenesis can be inhibited at concentrations higher than 20 g L^{-1}. Hence during treatment of wastewaters with salinity level above seawater, it will be strategic to pretreat the wastewater before feeding it in BES (Lefebvre et al. 2012). Certain exceptional halophillic bacteria like *Geoalkalibacter subterraneus*, however, have been also reported to produce high current density in hyper saline environments via direct electron transfer (Carmona-Martínez et al. 2013). Rousseau et al. (2013) also reported a bioanode developed from salt marsh inoculum in medium containing upto 45 g L^{-1} of NaCl, which generated current density up to 85 A m^{-2}. Hence a balance needs to be maintained between the amount of salinity desirable for enhanced conductivities in the system and the type of microbe used for BES operation.

22.2.2.3 Concentration Polarization

Slow rate of electron transfer especially in thicker biofilms through exopolysaccharide and the rate of proton movement out of the biofilm lead to concentration polarization. Also when oxygen cathodes are used, limited oxygen solubility in water causes strong mass transfer limitations, which further increases when biocathode are used. In addition, concentration polarization arises due to limitation of fuel and oxidant supply to the electrode. Increasing the buffer concentration and mass transfer rate, lowering the electrolyte pH and using high shear biofilm acclimatization procedure can also help in enhancement of BES performance (Sleutels et al. 2012). BES operation is usually conducted at neutral pH as it is considered most conducive for microbial growth (Liu et al. 2005). Reactor mixing can also reduce concentration polarization effect.

22.2.3 Operational Parameters

Energy recovery from BES depends on essential operational conditions such as temperature, since it controls the bacterial kinetics and rate of mass transfer of protons through the liquid (Liu et al. 2005), organic loading rate, use of buffers and hydraulic retention time. The rate of fuel oxidation also effect energy recovery as it influences the catalytic activity of the anode, fuel diffusion and the diffusion of electrons and protons, rate of substrate degradation and flow mode (Liu et al. 2005). Membraneless fuel cell operation should be avoided and more selective electrocatalysts that have lesser cross–selectivity should be used (Rabaey et al. 2010a). The presence of other electron acceptors in the vicinity of anode might also decrease the coulombic efficiency (CE). Leakage of oxygen into the anodic chamber reduces the CE (Logan et al. 2006).

22.2.3.1 Mediators

Mediators act as shuttles, which can be incorporated in the system to facilitate transfer of electrons between the microbe and the electrode (Rabaey and Rozendal 2010). The mediator chosen should be highly reactive with a reversible electrochemistry, low molecular weight, water solubility, stability in solution for multiple turnovers, should be able to work stably over a wide range of pH and capable of reducing overpotential at the electrode (Rabaey and Verstraete 2005). Soil compounds like humic acid and cysteine serve as redox mediators for dissimilatory iron reducing bacteria. The non-specificity of artificial mediators provides an added advantage specifically for mixed cultures where the diverse microbial community present in the system can use it for the transfer of electrons to the electrode. Other redox mediators reported with the electrodes includes phenazines, thionin, resazurin, 1,4 Naphthaquinone, neutral red and anthraquinone-1,6 disulfonic acid (AQDS) that has been used for various BES operations (Rabaey and Rozendal 2010).

In case of diffusive mediated electron transfer, the mediator that can diffuse across the cell membrane like quinines, ferrocene, nickelocene, and organometallic compounds, organic dyes are specifically used for enzymatic cells. While in case of non-diffusive mediated electron transfer, the mediator does not have to penetrate a cell membrane and are usually immobilized on the electrode surface-like immobilization of methyl viologen, neutral red, redox and conducting polymers, alginate-based hydrogels that can be effectively used for enzyme immobilization when used in combination with metal ions, poly (diallyldimethylammonium chloride) and PPDA (Qin et al. 2012). Mediators add cost, lack long term stability, may be toxic, and are easily washed out in flow through systems, hence cannot be used in open environments, and need to be separated from the products at the end before discharge into the environment (Lovley 2011).

22.2.3.2 pH Splitting

Gradient build up inside the BES can occur due to acidification of the anode side of the compartment as a result of proton generation, while the cathode side turns more alkaline due to utilization of protons (Rabaey et al. 2010b). In order to circumvent these mass transfer losses, a membraneless device can be used or removal of membrane that separates the two compartments. This will not only help in maintaining the pH but also reduces the ohmic resistance of the system. However, the negative impact of removing the membrane will be decreased in CE of the system. A loop configuration can also be used where the effluent of the anode is used as an influent for the cathode. But it may hamper the formation of valuable chemicals at the cathode due to crossover of organics, oxygen consumption and pollution of products. Temporal potential switching or CO_2 addition to the cathode can also be tested to mitigate this problem. A flow through system at an optimal hydraulic residence time can be preferred over batch mode for maintaining the pH of the system under control.

22.2.3.3 Other Operational Consideration

Unsatisfactory mixing and turbulence in each compartment can also become a limiting factor during scaling up. Internal resistance of a BES also increases with electrolyte volume that generates problem of mass flow and diffusion limitations because of the increased distance between the two chambers. Sharma et al. (2015) conducted a study with two electrochemical cells that were run together in a continuous mode to study the impact of altering selected operational parameters on the performance of the MEC consisting of an electroactive SRB based biocathode. The adaptability of this biocathode to different phosphate and ammonium chloride concentrations also highlighted the potential of this biocathode for developing nutrient recovery processes from waste streams (Sharma et al. 2015). Additional factors that contribute to internal resistance and plausible solutions to reduce such losses have been mentioned in Fig. 22.3.

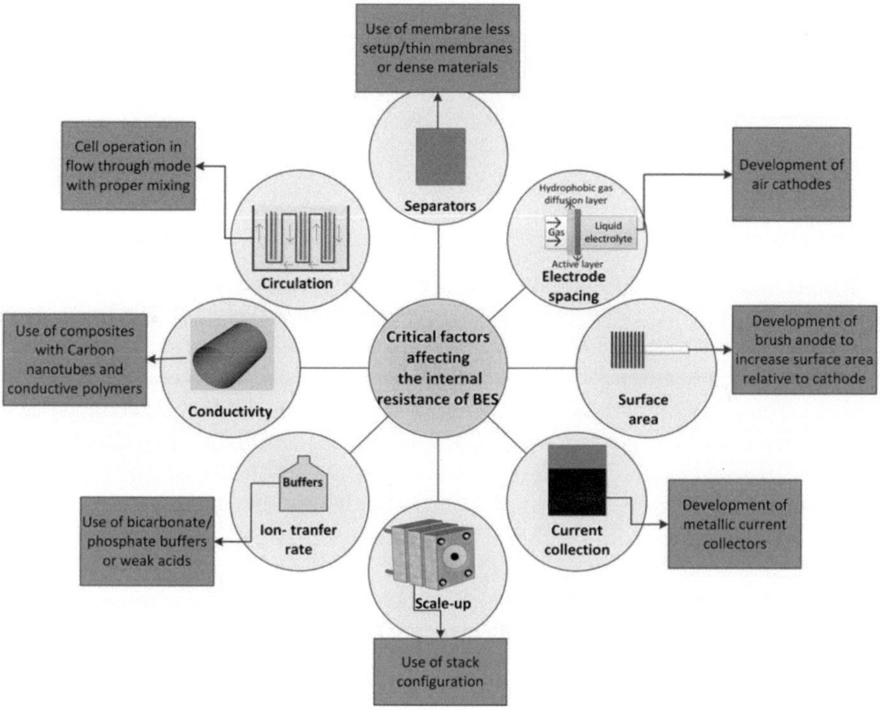

Fig. 22.3 Factors that affect the internal resistance during the operation of BES and plausible solutions to reduce such losses

22.2.4 Engineering Parameters

From an engineering perspective, important considerations include the configuration of the reactor, internal currents, membrane electrode assembly, tubing and compartments which have been discussed in subsequent subsections.

22.2.4.1 Reactor Configuration

The reactor configuration used in the BES also has a major influence on the performance of the reactor (Liu et al. 2005). Tubular (Clauwaert et al. 2007) stacked and baffled stacked are few of the efficient reported designs and their performance have been reported (Pham et al. 2009). Clauwert et al. reported that enlargement of reactors is not an efficient scaling-up approach and it is better to do compartmentalization into smaller units and then combine together and upscale it, because these dynamic smaller units will have lesser internal resistance and will follow shorter migration paths for uptake of substrate and release of protons

(Clauwaert et al. 2007). This prevents the diffusion of air near the anode that will subsequently prevent electrochemical losses and increase not only the CE but power generation as well. However, very close placement can lead to short-circuiting and hence should be avoided.

22.2.4.2 Internal Currents

Internal currents also arise due to detrimental leakage of substrate or oxygen in the cathodic and anodic compartments of the cell respectively, which have been separated by membrane like proton exchange membrane (PEM). It leads to a condition like an internal short circuit, which decreases the overall cell performance. Thus, operating a fully anaerobic BES, where both the bioanode and biocathode chambers are anaerobically maintained, can help in mitigating this problem. However, membraneless system confront even more detrimental cross-over effects and in those cases parasitic internal currents become more pronounced. In order to minimize this problem, membraneless fuel cell operation should be avoided and more selective electrocatalysts that have lesser cross–selectivity should be used (Rabaey et al. 2010a). The presence of other electron acceptors in the vicinity of anode might also decrease CE.

22.2.4.3 Membranes

Membrane electrode assembly influences the performance of the BES. Though it becomes quite practical to recover product from the cathode compartment when a membrane separates the anode chamber, the usage of membrane can contribute to increase in electrolyte resistance and lead to pH gradient formation. Fouling of the membrane because of deposition of biofilm and extracellular polymers development, and replacement of proton binding sites by other cations after a long-term operation also leads to reduction in the ion exchange capacity and conductivity (Xu et al. 2012). Other membranes available include anion exchange membranes (AEM), ultrafilteration membrane, bipolar membranes, nanoporous polymer membrane and interpolymer cation exchange membrane (CEM) (Chen et al. 2008). Bipolar membranes require additional energy for splitting water into protons and hydroxyl ions. Periodic cleaning of cathode assembly might be required as salt crystal formation occurs on the catalyst side of cathode and gradual deformation of the membrane can further degrade the BES performance. Sleutels et al. (2009) demonstrated that MES cells perform better in the presence of AEM than CEM, which was attributed to lower internal resistance of the AEM due to lower transport resistance of ions through the AEM.

22.2.4.4 State-of-the-Art Electrode Materials

Recently lot of research is focused on improved material as it is considered a key to highly efficient MES processes. Guo et al. (2015) have recently elaborately reviewed the effect of electrode material and composition on the performance of BES. Zhang et al. (2013) introduced novel concepts for electrode modification like functionalization of carbon cloth with chitosan for creation of positive charge for better adherence of negatively charged bacteria, or treatment with conductive metals like nickel, gold, palladium nanoparticles and use of cotton-based textile composite cathodes with carbon nanotubes that lead to improved performance. Chen et al. (2012a) reported the fabrication of inexpensive layered corrugated carbon (LCC) material as a high performance electrode material, which could reach a projected current density of 70–390 A m^{-2} depending on the number or layers or stacks of LCC used. Jourdin et al. (2014, 2015) have reported the development of a novel biocompatible highly conductive 3-S electrode manufactured by growth of multiwalled carbon nanotubes on reticulated vitreous carbon (Nano-Web-RVC). These electrodes have been reported to be quite efficient in conversion of CO_2 to acetic acid as a result of enhancement in bacterial attachment, effective mass transfer and enhanced biofilm development on the nanostructured electrode material. Nie et al. (2013) developed an efficient porous nickel nanowires anchored to graphite electrodes, which increased the biocathode's performance by 2.3-fold (282 mM $day^{-1}m^{-2}$) due to increase in interfacial interactions and area. A self-assembled 3D macroporous graphene oxide/bacteria (*Shewanella oneidensis*) hybrid biofilm constructed by Yong et al. (2014) demonstrated better performance due to increased loading of biomass, increased specific surface area and enhanced bidirectional electron transfer as compared to naturally occurring biofilms.

22.2.4.5 Tubings and Compartments

It is also necessary to use proper gas tight reactors particularly in the case of MECs where hydrogen production is an important objective because hydrogen will always leak through long tubing, connectors etc. Continuous gas release methods are also preferable to avoid the inhibition in gas production due to hydrogen gas accumulation. Leakage of oxygen into the anodic chamber reduces the CE (Logan et al. 2006). The gas used for sparging the reactor in the headspace can also direct the products obtained at the end of an MES operation (Sharma et al. 2014b). Mixed culture sulphate reducing bacteria-based biocathode was used in BES at an applied potential of –850 mV vs. Ag/AgCl, succinate, ethanol, hydrogen, glycerol and propionate were observed in a hermetically sealed reactor and this process was considered to be driven by the *in situ* hydrogen production. However, when the reactor with same operational conditions was continuously sparged with nitrogen gas, then the products changed to acetone, propionate, isopropanol, propanol, isobutyrate, isovalerate and heptanoate. This study suggested that electroactive biocathodes serve as efficient biocatalysts and the metabolic routes shifts with alteration of headspace environment.

22.2.5 Microbial Parameters

The biocatalysts present on the electrode surface ultimately govern the performance of BES as they regulate the rate of bioelectrochemical conversions. Whether a pure culture is used or mixed culture, the electroactive component of biomass deposited on the electrode surface directly regulates the performance of BES. Single cultures metabolize only a limited range of organic compounds. Hence generally mixed cultures are preferred which are metabolically more flexible and robust. But it is also necessary to know the community structure of the microbial consortium used, as side reactions like methanogenesis result in loss of electrons, overgrowth of biomass and also reduces the substrate conversion efficacy to current (Pham et al. 2009). The mixing and mass transfer, pH kinetics in the reactor, the bacterial affinity constant K_S for the substrate also influence the performance of BES (Rabaey and Verstraete 2005). Direct electron transfer is commonly observed in iron-rich oligotrophic environments that lead to production of thin and sparse biofilms. This facilitates the movement of electrons between the cell and electrode via direct contact. The capability of *Geobacter sulfurreducens* for production of conductive filaments for long range electron transport highlight the potential prospects of BES in the field of bioenergy and bioelectronics. With the use of molecular tools and the possibility to tweak the genetic makeup of these electroactive bacteria to more efficient biomachines, the use of molecular tools for efficient strain development for MES processes is largely underexplored and has huge potential as reviewed by Sydow et al. (2014).

22.2.6 Economic Parameters

For commercialization of BES technology and scaling up processes, economic factors also need consideration. There is a prerequisite to bring down the investment cost of manufacturing and operation of BES to get significant returns on the investment. Krieg et al. (2014) have provided a very interesting comparative review on advances of reactor design, different commercially available electrode materials, and essential cost consideration for scale up operations. Through life cycle assessment studies, it has been demonstrated that MEC provide significant environmental benefits over MFC due to their additional value added chemical production capabilities because it is unlikely to compete with highly efficient large-scale electricity markets (Rosenbaum and Franks 2014). For crosswise comparison of studies related to BES, correction of current density, impedance, capacitance, and other classical parameters by an appropriate surface indicator is important and was elaborately discussed by Sharma et al. (2014a). The demonstration of this technology indicated several problems in the past. It has been observed that unsatisfactory mixing and turbulence in each compartment can become a limiting factor during scaling up. Giddings et al. (2015) have recently demonstrated an MES operation with biofilms of *Sporomusaovata* with a simplified membraneless reactor design by eliminating potentiostatic control on the cathode and replacing it with direct current power source. Torres (2014) has elaborated on designing large scale reactors and indicated that it is essential to fully understand target and limitations of this technology by research on characterization, cost, production and optimization parameters before commercialization.

22.3 Biocathode Development

Different materials have been previously used as terminal electron acceptor in cathode compartment of BES. Platinum though expensive, was initially used but later other acceptors that were tested include sulphate, nitrate, peroxide, potassium ferricyanide, potassium dichromate, light depending polyiodide besides commonly used oxygen (Zhang et al. 2012). Microbes present at the biocathode lead to catalysis of electrons, protons and oxidant present at the cathode and it has proven to be a cost effective alternative for sustainable BES operation for waste treatment, removal of unwanted compounds, metal recovery and biochemical production. These biocathode can be either aerobic, where oxygen acts as the oxidant and the microbes assist the oxidation of transition metal compounds such as Mn(II) and Fe(II) for the delivery of electron to oxygen. Anaerobic biocathode make use of compounds like nitrate, sulphate, iron, manganese, carbon dioxide, fumarate, urinate, arsenate and selenite as terminal electron acceptors. Simultaneous treatment of two different kinds of substrates is possible in this type of setup. Some of the strains that are reported for good biocathode development like *Geobacter metallireducens* (Marsili and Bond 2007) for nitrate to nitrite or fumarate to succinate reduction and has been reported for cathodic biofilm development. Hydrogen production at cathode has been part of the initial demonstrated cathodic reactions in various MEC setup (Rozendal et al. 2006; Tartakovsky et al. 2009; Call and Logan 2008; Villano et al. 2010a, b). Hence numerous bioelectrochemical conversions are possible in the biocathodic chamber of BES as enlisted in Table 22.1. However, the type of microbe present in the BES either as biocatalyst or a planktonic constituent directs the performance of BES.

Table 22.1 Current densities obtained with various biocathodes in MES (Adapted from Huang et al. 2011 and Sharma et al. 2014a)

Electron acceptor	Source inoculum	Operation mode (potential vs Ag/AgCl)	Current density ($A\ m^{-2}$)	References
Nitrate/ fumarate	Geobacter metallireducens/ sulfurreducens	DC, Nafion membrane, −0.5 V	NM	Gregory et al. (2004)
TCE	Geobacter lovleyi	DC, batch-fed, −0.5 V	0.02	Strycharz et al. (2008)
Sodium fumarate	Pure culture of G. sulfurreducens	SC, batch-fed, SS cathode (2.5 cm^2) half CP at −0.6 V	20.5	Dumas et al. (2008)
CO_2	Enrichment culture	DC; batch-fed, CP at −0.7 to −1 V	8.7 $A\ m^{-3}$	Cheng et al. (2009)
CO_2	Enrichment culture	DC; batch-fed, CP at −0.95 V	7.5	Villano et al. (2010b)
CO_2	Enrichment culture anaerobic sludge	DC with continuous mode, Pt coated Ti mesh anode, graphite felt cathode, CP at −0.95 V	0.78 ± 0.20 0.25 ± 0.04 (water)	van Eerten Jansen et al. (2012)
CO_2	Sporomusa ovata	DC; graphite sticks electrodes, batch, CP at −0.6 V	NM	Nevin et al. (2010)
H_2	Enrichment culture	DC, continuously fed, cathode at −0.7 V	1.9–3.3	Jeremiasse et al. (2010)
Acetate	Enriched sludge from distillery WW	DC batch-fed, graphite felt electrodes, cathode poised at –0.75 V	NM	Steinbusch et al. (2009)
Acetate	Enrichment culture, mainly clostridium Kluyveri	DC; recirculated catholyte, cathode poised at −1.1 V	1.8 ± 0.6	van Eerten-Jansen et al. (2013)
CO_2	Enrichment culture, AS	DC with continuous mode, Pt coated Ti mesh anode, graphite felt cathode, CP at −0.9 V	1.6	van Eerten-Jansen et al. (2013)
CO_2	Mixed culture from AS	Two-chamber, fed-batch, CP at –1.15 V	15	Jiang et al. (2013)
Acetate & butyrate mix	Sulphate reducing mixed culture	SC, batch-fed, CP at −0.85 V	160–210	Sharma et al. (2013a)
CO_2	Clostridium dominated mixed culture from a syngas fermenting LS	H type reactor separated with CEM, Ti rod counter electrode, carbon cloth cathode with area of 9 cm^2	~5	Ganigué et al. (2015)
CO_2	Enriched mixed culture from pond sediments & LS	EPD-3D electrode, −0.85 V vs SHE, fed-batch mode, DC with CEM	102	Jourdin et al. (2015)

(continued)

Table 22.1 (continued)

Electron acceptor	Source inoculum	Operation mode (potential vs Ag/AgCl)	Current density (A m^{-2})	References
CO_2	Actogenic biocathode developed from brewery WW	Two chambers separated with PEM, −590 V vs. SHE	NM	Marshall et al. (2012, 2013b)
CO_2	Pre-enriched acetate consortium from effluent of a bioanode and AD	Three chambered single reactor with AEM and CEM, carbon felt as cathode and titanium coated TiO_2/IrO_2 as anode material	5	Gildemyn et al. (2015)
CO_2	Mixed culture enriched from WW	Carbon felt with SS with CP at −1.1 V	10	Bajracharya et al. (2015)
CO_2	Mixed culture enrichment from LS and algae UASB	Carbon felt electrode, CP of −1.26 to −1.28 V, fed-batch or batch mode	5	Patil et al. (2015)
CO_2 and bicarbonatee	Homoacetogenic consortium	H type, PEM separation, graphite cathode and VITO-core™ as anode CP of −0.6 and −0.8 V	14–42	Mohanakrishna et al. (2016)
Egeriadensa fed with lake WW	Mesophillic mixed culture	SC MEC-AD system Ti/RuO_2 mesh plates electrodes, AV 0.1-1 V	NM	Zhen et al. (2016)
Glucose medium	Clostridium beijerinckii IB4	H type reactor, graphite felt electrodes, CEM, CP of −0.7 V	NM	He et al. (2016)

Note: *WW* wastewater, *AS* anaerobic sludge, *LS* laboratory setup, *AV* applied voltage, *CP* cathode potential, *SC* single chambered, *DC* double chambered, *SS* stainless steel

22.4 Advantages and Application of Bioelectrochemical Conversions

Replacement of precious metal catalysts by biocatalysts in BES have made its operation cost effective as they are sustainable, rejuvenate and adapt themselves to the required conversion activity and can easily decrease the overpotential at the electrode surface to increase BES performance (Lovley 2011; Rabaey and Rozendal 2010). Subsequently, electricity-driven processes have also been applied and widely explored in the context of bioremediation and inorganic/resource recovery (Modin et al. 2012). There is also flexibility in terms of source of electron donors selected as substrate in the BES that makes it operational even at lower temperature and pressure conditions. Amongst the various kinds of fuels used for operation of BES, water is an inexpensive, readily available, plausible, abundant electron source as it can readily split with the release of oxygen and protons at the anode surface (Lovley 2011). There are organic compounds ubiquitously present in wastewater and fermentation effluents. These include acetate, butyrate, lactate etc. in extremely low concentrations that makes their extraction unfeasible (Rabaey and Rozendal 2010). Hence an alternate strategy is to make use of these organics before disposal by recovery in the form of alcohols in BES. This technology is also known to limit the release of odours and perform with high coulombic efficiency conversions. One of the important benefits of BES is the electrical energy recovery that can offset to an extent the operation and energy costs of the treatment process of wastewaters (Sadhukhan et al. 2016).

A wide range of bioelectrochemical processes has been developed from electrical power generation to the production of biofuels and biochemicals (Bajracharya et al. 2016). The most-studied microbial electrocatalysis-based process so far is the wastewater treatment (Logan and Rabaey 2012; Sharma et al. 2013b) and production of electricity in microbial fuel cells (MFCs). The major technological breakthrough, which expanded the application range for microbial electrocatalysis, has been the possibility to apply electricity as a driving force to catalyse the production of high value chemicals, redirect fermentation pathways or as electron acceptors to oxidize wastewater organics (Desloover et al. 2012). The chemical production include hydrogen, caustic and hydrogen peroxide at the cathode, at a lower energy cost compared to more classical electrochemical production processes (Rozendal et al. 2009; Rabaey et al. 2010b). Aulenta et al. (2012) reported the production of H_2 in the presence of *Desulfovibrio paquesii* by direct electron acceptance from polarized graphite electrodes (−900 mV vs. SHE) at the rate of 5–8 mmol/L with coulomic efficiencies of nearly 100%. Methane production in MEC cells have also been considered an important end product (Villano et al. 2010a; Cheng et al. 2009; Lee et al. 2016; Beese-Vasbender et al. 2015).

The production of chemicals especially short chain carboxylic acids from waste streams have drawn attention for MES worldwide (Marshall et al. 2013a). Caproate (precursor for fuel alkanes), caprylate (van Eerten-Jansen et al. 2013), butyrate and combination of one or more of the above mentioned chemicals (Lovley and Nevin 2013; Marshall et al. 2013a, b; Sharma et al. 2013a, b). Sharma et al. (2013a, b) demonstrated the development of an SRB-based biocathode that was capable of reducing organic acids to a number of compounds with commercial value. High cathodic current densities in the range of 160-210 A m^{-2} were successfully achieved for driving these microbially catalysed electrochemical reactions. Marshall et al. (2013a, b) demonstrated the microbial electrosynthesis of acetate from carbon dioxide using an acetogenic biocathode that could produce 17.25 mM d^{-1} acetate. Gildemyn et al. (2015) have recently reported the production of acetate in a single, three-chamber reactor with accumulation of upto 13.5 g L^{-1}. This system also demonstrated improvement in reactor design from conventional designs as it consisted both SEM to separate cathode compartment from a saline extraction compartment and a CEM between the saline extraction compartment and the anode compartment to avoid chlorination of acetic acid anode. This ensured continuous reduction of CO_2 to acetic acid and its easy extraction without product inhibition. *Ganigué* et al. (2015) reported the production of 1.82 mM d^{-1} butyrate from CO_2 as a sole carbon source using microbial electrosynthesis approach.

Microbial electrolysis cells have been recently used for cobalt recovery along with simultaneous methane and acetate production (Huang et al. 2014). A recalcitrant organ fluorine pollutant *p*-Fluronitrobenzene (*p*-FNB) was also reported to be successfully removed from a BES system at an applied voltage of 1.4 V (Feng et al. 2014). Oxalate, an important intermediate produced in the mineralization of various organic pollutants, was also demonstrated to be completely removed from the anode compartment of a BES system with concomitant production of caustic soda and methane in the cathode chamber (Bonmati et al. 2013; Chen et al. 2012b). This combined a microbial desalination cell (MDC) with a microbial electrolysis cell (MEC) to develop new microbial electrolysis desalination and chemical production cell (MEDCC). This cell consisted of four chambers and bipolar membranes and could give a coulombic efficiency of 62–97% at an applied voltage of 0.3–1 V along with acid (HCl) and base (NaOH) production in separate chambers. A recent proof of concept study by Molenaar et al. (2016) on microbial rechargeable battery, as efficient local energy storage device, which combines MFC and MEC in one device, could also be a very promising application of BES. Rosenbaum and Franks (2014) and Zhang and Angelidaki (2014) have provided thought-provoking reviews on the present status of the BES and practical challenges that needs to be addressed for making this technology commercial to meet short-term and long-term BES applications.

22.5 Conclusions

Rising environmental concerns, ever increasing energy needs, search for alternate energy production methods and promotion of carbon neutral processes, have brought innovative technologies like MES in the forefront. New approaches like *in silico* studies to determine the most efficient metabolic pathways promoting energy conservation, electron transport mechanisms and right combination of substrate and product formation can be the next big lead for increased production using microbial electrosynthesis (Kracke and Krömer 2014; Roy et al. 2016). So, the success and commercial acceptance of such new technologies solely depend on *in situ* and *in silico* research advancement and interdisciplinary collaborations for the development of highly efficient production processes with minimum investment on infrastructure and running costs.

References

Aulenta, F., Catapano, L., Snip, L., Villano, M., & Majone, M. (2012). Linking bacterial metabolism to graphite cathodes: Electrochemical insights into the H_2-producing capability of Desulfovibrio sp. *ChemSusChem, 5*(6), 1080–1085.

Bajracharya, S., ter Heijne, A., Benetton, X. D., Vanbroekhoven, K., Buisman, C. J., Strik, D. P., & Pant, D. (2015). Carbon dioxide reduction by mixed and pure cultures in microbial electrosynthesis using an assembly of graphite felt and stainless steel as a cathode. *Bioresource Technology, 195*, 14–24.

Bajracharya, S., Sharma, M., Mohanakrishna, G., Benneton, X. D., Strik, D. P., Sarma, P. M., & Pant, D. (2016). An overview on emerging bioelectrochemical systems (BESs): Technology for sustainable electricity, waste remediation, resource recovery, chemical production and beyond. *Renewable Energy, 98*, 153–170.

Beese-Vasbender, P. F., Grote, J. P., Garrelfs, J., Stratmann, M., & Mayrhofer, K. J. (2015). Selective microbial electrosynthesis of methane by a pure culture of a marine lithoautotrophic archaeon. *Bioelectrochemistry, 102*, 50–55.

Bonmati, A., Sotres, A., Mu, Y., Rozendal, R., & Rabaey, K. (2013). Oxalate degradation in a bioelectrochemical system: Reactor performance and microbial community characterization. *Bioresource Technology, 143*, 147–153.

Call, D., & Logan, B. E. (2008). Hydrogen production in a single chamber microbial electrolysis cell lacking a membrane. *Environmental Science & Technology, 42*(9), 3401–3406.

Carmona-Martínez, A. A., Pierra, M., Trably, E., & Bernet, N. (2013). High current density via direct electron transfer by the halophilic anode respiring bacterium *Geoalkalibacter subterraneus*. *Physical Chemistry Chemical Physics, 15*(45), 19699–19707.

Chen, G. W., Choi, S. J., Lee, T. H., Lee, G. Y., Cha, J. H., & Kim, C. W. (2008). Application of biocathode in microbial fuel cells: Cell performance and microbial community. *Applied Microbiology and Biotechnology, 79*(3), 379–388.

Chen, S., He, G., Liu, Q., Harnisch, F., Zhou, Y., Chen, Y., Hanif, M., Wang, S., Peng, X., Hou, H., & Schröder, U. (2012a). Layered corrugated electrode macrostructures boost microbial bioelectro catalysis. *Energy & Environmental Science, 5*(12), 9769–9772.

Chen, S., Liu, G., Zhang, R., Qin, B., & Luo, Y. (2012b). Development of the microbial electrolysis desalination and chemical-production cell for desalination as well as acid and alkali productions. *Environmental Science & Technology, 46*(4), 2467–2472.

Cheng, S., & Logan, B. E. (2011). High hydrogen production rate of microbial electrolysis cell (MEC) with reduced electrode spacing. *Bioresource Technology, 102*(3), 3571–3574.

Cheng, S., Xing, D., Call, D. F., & Logan, B. E. (2009). Direct biological conversion of electrical current into methane by electromethanogenesis. *Environmental Science & Technology, 43*(10), 3953–3958.

Clauwaert, P., Van der Ha, D., Boon, N., Verbeken, K., Verhaege, M., Rabaey, K., & Verstraete, W. (2007). Open air biocathode enables effective electricity generation with microbial fuel cells. *Environmental Science & Technology, 41*(21), 7564–7569.

Desloover, J., Arends, J., Hennebel, T., & Rabaey, K. (2012). Operational and technical considerations for microbial electrosynthesis. *Biochemical Society Transactions, 40*(6), 1233–1238.

Dumas, C., Basseguy, R., & Bergel, A. (2008). Microbial electrocatalysis with *Geobacter sulfurreducens* biofilm on stainless steel cathodes. *Electrochimica Acta, 53*(5), 2494–2500.

Feng, H., Zhang, X., Liang, Y., Wang, M., Shen, D., Ding, Y., Huang, B., & Shentu, J. (2014). Enhanced removal of *p*-fluoronitrobenzene using bioelectrochemical system. *Water Research, 60*, 54–63.

Ganigué, R., Puig, S., Batlle-Vilanova, P., Balaguer, M. D., & Colprim, J. (2015). Microbial electrosynthesis of butyrate from carbon dioxide. *Chemical Communications, 51*(15), 3235–3238.

Giddings, C. G., Nevin, K. P., Woodward, T., Lovley, D. R., & Butler, C. S. (2015). Simplifying microbial electrosynthesis reactor design. *Frontiers in Microbiology, 6*, 1–6.

Gildemyn, S., Verbeeck, K., Slabbinck, R., Andersen, S. J., Prévoteau, A., & Rabaey, K. (2015). Integrated production, extraction and concentration of acetic acid from CO_2 through microbial electrosynthesis. *Environmental Science & Technology Letters, 2*(11), 325–328.

Gregory, K. B., Bond, D. R., & Lovley, D. R. (2004). Graphite electrodes as electron donors for anaerobic respiration. *Environmental Microbiology, 6*(6), 596–604.

Guo, K., Prévoteau, A., Patil, S. A., & Rabaey, K. (2015). Engineering electrodes for microbial electrocatalysis. *Current Opinion in Biotechnology, 33*, 149–156.

Gurung, A., Kim, J., Jung, S., Jeon, B. H., Yang, J. E., & Oh, S. E. (2012). Effects of substrate concentrations on performance of serially connected microbial fuel cells (MFCs) operated in a continuous mode. *Biotechnology Letters, 34*(10), 1833–1839.

He, A. Y., Yin, C. Y., Xu, H., Kong, X. P., Xue, J. W., Zhu, J., Jiang, M., & Wu, H. (2016). Enhanced butanol production in a microbial electrolysis cell by *Clostridium beijerinckii* IB4. *Bioprocess and Biosystems Engineering, 39*(2), 245–254.

Huang, L., Regan, J. M., & Quan, X. (2011). Electron transfer mechanisms, new applications, and performance of biocathode microbial fuel cells. *Bioresource Technology, 102*(1), 316–323.

Huang, L., Jiang, L., Wang, Q., Quan, X., Yang, J., & Chen, L. (2014). Cobalt recovery with simultaneous methane and acetate production in biocathode microbial electrolysis cells. *Chemical Engineering Journal, 253*, 281–290.

Jeremiasse, A. W., Hamelers, H. V., Saakes, M., & Buisman, C. J. (2010). Ni foam cathode enables high volumetric H_2 production in a microbial electrolysis cell. *International Journal of Hydrogen Energy, 35*(23), 12716–12723.

Jiang, Y., Su, M., Zhang, Y., Zhan, G., Tao, Y., & Li, D. (2013). Bioelectrochemical systems for simultaneous production of methane and acetate from carbon dioxide at relatively high rate. *International Journal of Hydrogen Energy, 38*(8), 3497–3502.

Jourdin, L., Freguia, S., Donose, B. C., Chen, J., Wallace, G. G., Keller, J., & Flexer, V. (2014). A novel carbon nanotube modified scaffold as an efficient biocathode material for improved microbial electrosynthesis. *Journal of Materials Chemistry A, 2*(32), 13093–13102.

Jourdin, L., Grieger, T., Monetti, J., Flexer, V., Freguia, S., Lu, Y., Chen, J., Romano, M., Wallace, G. G., & Keller, J. (2015). High acetic acid production rate obtained by microbial electrosynthesis from carbon dioxide. *Environmental Science & Technology, 49*(22), 13566–13574.

Ki, D., Popat, S. C., & Torres, C. I. (2016). Reduced overpotentials in microbial electrolysis cells through improved design, operation, and electrochemical characterization. *Chemical Engineering Journal, 287*, 181–188.

Kracke, F., & Krömer, J. O. (2014). Identifying target processes for microbial electrosynthesis by elementary mode analysis. *BMC Bioinformatics, 15*(1), 1.

Krieg, T., Sydow, A., Schröder, U., Schrader, J., & Holtmann, D. (2014). Reactor concepts for bioelectrochemical syntheses and energy conversion. *Trends in Biotechnology, 32*(12), 645–655.

Lee, J. Y., Lee, S. H., & Park, H. D. (2016). Enrichment of specific electro-active microorganisms and enhancement of methane production by adding granular activated carbon in anaerobic reactors. *Bioresource Technology, 205*, 205–212.

Lefebvre, O., Tan, Z., Kharkwal, S., & Ng, H. Y. (2012). Effect of increasing anodic NaCl concentration on microbial fuel cell performance. *Bioresource Technology, 112*, 336–340.

Liu, H., Cheng, S., & Logan, B. E. (2005). Power generation in fed-batch microbial fuel cells as a function of ionic strength, temperature, and reactor configuration. *Environmental Science & Technology, 39*(14), 5488–5493.

Logan, B. E. (2012). Essential data and techniques for conducting microbial fuel cell and other types of bioelectrochemical system experiments. *ChemSusChem, 5*(6), 988–994.

Logan, B. E., & Rabaey, K. (2012). Conversion of wastes into bioelectricity and chemicals by using microbial electrochemical technologies. *Science, 337*(6095), 686–690.

Logan, B. E., Hamelers, B., Rozendal, R., Schröder, U., Keller, J., Freguia, S., Aelterman, P., Verstraete, W., & Rabaey, K. (2006). Microbial fuel cells: Methodology and technology. *Environmental Science & Technology, 40*(17), 5181–5192.

Lovley, D. R. (2008). The microbe electric: Conversion of organic matter to electricity. *Current Opinion in Biotechnology, 19*, 564–571.

Lovley, D. R. (2011). Powering microbes with electricity: Direct electron transfer from electrodes to microbes. *Environmental Microbiology Reports, 3*(1), 27–35.

Lovley, D. R. (2012). Electromicrobiology. *Annual Review of Microbiology, 66*, 391–409.

Lovley, D. R., & Nevin, K. P. (2013). Electrobiocommodities: Powering microbial production of fuels and commodity chemicals from carbon dioxide with electricity. *Current Opinion in Biotechnology, 24*(3), 385–390.

Marsili, E., & Bond, D. (2007). Electrochemical methods for observing electron transfer from bacterial biofilms to conductive surfaces. In *Meeting abstracts*. The Electrochemical Society No. 8, Abstract 454.

Marshall, C. W., Ross, D. E., Fichot, E. B., Norman, R. S., & May, H. D. (2012). Electrosynthesis of commodity chemicals by an autotrophic microbial community. *Applied and Environmental Microbiology, 78*(23), 8412–8420.

Marshall, C. W., LaBelle, E. V., & May, H. D. (2013a). Production of fuels and chemicals from waste by microbiomes. *Current Opinion in Biotechnology, 24*(3), 391–397.

Marshall, C. W., Ross, D. E., Fichot, E. B., Norman, R. S., & May, H. D. (2013b). Long-term operation of microbial electrosynthesis systems improves acetate production by autotrophic microbiomes. *Environmental Science & Technology, 47*(11), 6023–6029.

Modin, O., Wang, X., Wu, X., Rauch, S., & Fedje, K. K. (2012). Bioelectrochemical recovery of Cu Pb, Cd, and Zn from dilute solutions. *Journal of Hazardous Materials, 235*, 291–297.

Mohanakrishna, G., Vanbroekhoven, K., & Pant, D. (2016). Imperative role of applied potential and inorganic carbon source on acetate production through microbial electrosynthesis. *Journal of CO_2 Utilization, 15*, 57–64.

Molenaar, S. D., Mol, A. R., Sleutels, T. H., TerHeijne, A., & Buisman, C. J. (2016). Microbial rechargeable battery: Energy storage and recovery through acetate. *Environmental Science & Technology Letters, 3*(4), 144–149.

Nevin, K. P., Woodard, T. L., Franks, A. E., Summers, Z. M., & Lovley, D. R. (2010). Microbial electrosynthesis: Feeding microbes electricity to convert carbon dioxide and water to multicarbon extracellular organic compounds. *MBio, 1*(2), e00103-10.

Nevin, K. P., Hensley, S. A., Franks, A. E., Summers, Z. M., Ou, J., Woodard, T. L., Snoeyenbos-West, O. L., & Lovley, D. R. (2011). Electrosynthesis of organic compounds from carbon dioxide is catalyzed by a diversity of acetogenic microorganisms. *Applied and Environmental Microbiology, 77*(9), 2882–2886.

Nie, H., Zhang, T., Cui, M., Lu, H., Lovley, D. R., & Russell, T. P. (2013). Improved cathode for high efficient microbial-catalyzed reduction in microbial electrosynthesis cells. *Physical Chemistry Chemical Physics, 15*(34), 14290–14294.

Patil, S. A., Arends, J. B., Vanwonterghem, I., Van Meerbergen, J., Guo, K., Tyson, G. W., & Rabaey, K. (2015). Selective enrichment establishes a stable performing community for microbial electrosynthesis of acetate from CO_2. *Environmental Science & Technology, 49* (14), 8833–8843.

Pham, T. H., Aelterman, P., & Verstraete, W. (2009). Bioanode performance in bioelectrochemical systems: Recent improvements and prospects. *Trends in Biotechnology, 27*(3), 168–178.

Qin, C., Chen, C., Xie, Q., Wang, L., He, X., Huang, Y., Zhou, Y., Xie, F., Yang, D., & Yao, S. (2012). Amperometric enzyme electrodes of glucose and lactate based on poly (DADMAC)-alginate-metal ion-enzyme biocomposites. *Analytica Chimica Acta, 720*, 49–56.

Rabaey, K., Rodriguez, J., Blackall, L. L., Keller, J., Gross, P., Batstone, D., Verstraete, W., & Nealson, K. H. (2007). Microbial ecology meets electrochemistry: Electricity-driven and driving communities. *The ISME Journal, 1*(1), 9–18.

Rabaey, K., & Rozendal, R. A. (2010). Microbial electrosynthesis: Revisiting the electrical route for microbial production. *Nature Reviews Microbiology, 8*(10), 706–716.

Rabaey, K., & Verstraete, W. (2005). Microbial fuel cells: Novel biotechnology for energy generation. *Trends in Biotechnology, 23*(6), 291–298.

Rabaey, K., Angenent, L., Schroder, U., & Keller, J. (Eds.). (2010a). *Bioelectro-chemical systems: From extracellular electron transfer to biotechnological application* (1st ed.). London: IWA Publishing.

Rabaey, K., Butzer, S., Brown, S., Keller, J., & Rozendal, R. A. (2010b). High current generation coupled to caustic production using a lamellar bioelectrochemical system. *Environmental Science & Technology, 44*(11), 4315–4321.

Rosenbaum, M. A., & Franks, A. E. (2014). Microbial catalysis in bioelectrochemical technologies: Status quo, challenges and perspectives. *Applied Microbiology and Biotechnology, 98*(2), 509–518.

Rousseau, R., Dominguez-Benetton, X., Délia, M. L., & Bergel, A. (2013). Microbial bioanodes with high salinity tolerance for microbial fuel cells and microbial electrolysis cells. *Electrochemistry Communications, 33*, 1–4.

Roy, S., Schievano, A., & Pant, D. (2016). Electro-stimulated microbial factory for value added product synthesis. *Bioresource Technology, 213*, 129–139.

Rozendal, R. A., Hamelers, H. V., Euverink, G. J., Metz, S. J., & Buisman, C. J. (2006). Principle and perspectives of hydrogen production through biocatalyzed electrolysis. *International Journal of Hydrogen Energy, 31*(12), 1632–1640.

Rozendal, R. A., Leone, E., Keller, J., & Rabaey, K. (2009). Efficient hydrogen peroxide generation from organic matter in a bioelectrochemical system. *Electrochemistry Communications, 11*(9), 1752–1755.

Sadhukhan, J., Lloyd, J. R., Scott, K., Premier, G. C., Eileen, H. Y., Curtis, T., & Head, I. M. (2016). A critical review of integration analysis of microbial electrosynthesis (MES) systems with waste biorefineries for the production of biofuel and chemical from reuse of CO_2. *Renewable and Sustainable Energy Reviews, 56*, 116–132.

Sharma, M., Aryal, N., Sarma, P. M., Vanbroekhoven, K., Lal, B., Benetton, X. D., & Pant, D. (2013a). Bioelectrocatalyzed reduction of acetic and butyric acids via direct electron transfer using a mixed culture of sulfate-reducers drives electrosynthesis of alcohols and acetone. *Chemical Communications, 49*, 6495–6497.

Sharma, M., Jain, P., Varanasi, J. L., Lal, B., Rodríguez, J., Lema, J. M., & Sarma, P. M. (2013b). Enhanced performance of SRB based biocathode using stainless steel mesh on activated carbon fabric electrode. *Bioresource Technology, 150*, 172–180.

Sharma, M., Bajracharya, S., Gildemyn, S., Patil, S. A., Alvarez-Gallego, Y., Pant, D., Rabaey, K., & Dominguez-Benetton, X. (2014a). A critical revisit of the key parameters used to describe microbial electrochemical systems. *Electrochimica Acta, 140*, 191–208.

Sharma, M., Varanasi, J. L., Jain, P., Dureja, P., Lal, B., Dominguez-Benetton, X., Pant, D., & Sarma, P. M. (2014b). Influence of headspace composition on product diversity by sulphate reducing bacteria biocathode. *Bioresource Technology, 165*, 365–371.

Sharma, M., Sarma, P. M., Pant, D., & Dominguez-Benetton, X. (2015). Optimization of electrochemical parameters for SRB based biocathode. *RSC Advances, 5*(49), 39601–39611.

Sleutels, T. H., Hamelers, H. V., Rozendal, R. A., & Buisman, C. J. (2009). Ion transport resistance in microbial electrolysis cells with anion and cation exchange membranes. *International Journal of Hydrogen Energy, 34*(9), 3612–3620.

Sleutels, T. H., ter Heijne, A., Buisman, C. J., & Hamelers, H. V. (2012). Bioelectrochemical systems: An outlook for practical applications. *ChemSusChem, 5*(6), 1012–1019.

Steinbusch, K. J., Hamelers, H. V., Schaap, J. D., Kampman, C., & Buisman, C. J. (2009). Bioelectrochemical ethanol production through mediated acetate reduction by mixed cultures. *Environmental Science & Technology, 44*(1), 513–517.

Strycharz, S. M., Woodard, T. L., Johnson, J. P., Nevin, K. P., Sanford, R. A., Löffler, F. E., & Lovley, D. R. (2008). Graphite electrode as a sole electron donor for reductive dechlorination of tetrachlorethene by *Geobacter lovleyi*. *Applied and Environmental Microbiology, 74*(19), 5943–5947.

Sydow, A., Krieg, T., Mayer, F., Schrader, J., & Holtmann, D. (2014). Electroactive bacteria—Molecular mechanisms and genetic tools. *Applied Microbiology and Biotechnology, 98*(20), 8481–8495.

Tartakovsky, B., Manuel, M. F., Wang, H., & Guiot, S. R. (2009). High rate membraneless microbial electrolysis cell for continuous hydrogen production. *International Journal of Hydrogen Energy, 34*(2), 672–677.

Torres, C. I. (2014). On the importance of identifying, characterizing, and predicting fundamental phenomena towards microbial electrochemistry applications. *Current Opinion in Biotechnology, 27*, 107–114.

van Eerten Jansen, M. C., ter Heijne, A., Buisman, C. J., & Hamelers, H. V. (2012). Microbial electrolysis cells for production of methane from CO_2: Long term performance and perspectives. *International Journal of Energy Research, 36*(6), 809–819.

van Eerten-Jansen, M. C., ter Heijne, A., Grootscholten, T. I., Steinbusch, K. J., Sleutels, T. H., Hamelers, H. V., & Buisman, C. J. (2013). Bioelectrochemical production of caproate and caprylate from acetate by mixed cultures. *ACS Sustainable Chemistry & Engineering, 1*(5), 513–518.

Villano, M., Aulenta, F., Beccari, M., Majone, M., & Roma, I. T. (2010a). Microbial generation of H_2 or CH_4 coupled to wastewater treatment in bioelectrochemical systems. *Chemical. Engineering, 20*, 2–7.

Villano, M., Aulenta, F., Ciucci, T., Giuliano, A., & Majone, M. (2010b). Bioelectrochemical reduction of CO_2 to CH_4 via direct and indirect extracellular electron transfer by a hydrogenophilicmethanogenic culture. *Bioresource Technology, 101*(9), 3085–3090.

Xu, J., Sheng, G. P., Luo, H. W., Li, W. W., Wang, L. F., & Yu, H. Q. (2012). Fouling of proton exchange membrane (PEM) deteriorates the performance of microbial fuel cell. *Water Research, 46*(6), 1817–1824.

Yong, Y. C., Yu, Y. Y., Zhang, X., & Song, H. (2014). Highly active bidirectional electron transfer by a self assembled electroactive reduced graphene oxide hybridized biofilm. *Angewandte Chemie International Edition, 53*(17), 4480–4483.

Zhang, Y., & Angelidaki, I. (2014). Microbial electrolysis cells turning to be versatile technology: Recent advances and future challenges. *Water Research, 56*, 11–25.

Zhang, G., Wang, K., Zhao, Q., Jiao, Y., & Lee, D. J. (2012). Effect of cathode types on long-term performance and anode bacterial communities in microbial fuel cells. *Bioresource Technology, 118*, 249–256.

Zhang, T., Nie, H., Bain, T. S., Lu, H., Cui, M., Snoeyenbos-West, O. L., Franks, A. E., Nevin, K. P., Russell, T. P., & Lovley, D. R. (2013). Improved cathode materials for microbial electrosynthesis. *Energy & Environmental Science, 6*(1), 217–224.

Zhang, F., Liu, J., Ivanov, I., Hatzell, M. C., Yang, W., Ahn, Y., & Logan, B. E. (2014). Reference and counter electrode positions affect electrochemical characterization of bioanodes in different bioelectrochemical systems. *Biotechnology and Bioengineering, 111*(10), 1931–1939.

Zhen, G., Kobayashi, T., Lu, X., Kumar, G., & Xu, K. (2016). Biomethane recovery from Egeriadensa in a microbial electrolysis cell-assisted anaerobic system: Performance and stability assessment. *Chemosphere, 149*, 121–129.

Chapter 23
Recent Progress Towards Scaling Up of MFCs

Dipak A. Jadhav, Makarand M. Ghangrekar, and Narcis Duteanu

23.1 Genesis and Advancement in MFC Research

Microbial fuel cell (MFC) is an advanced bioelectrochemical system for treatment of wastewater which transforms chemical energy available in the organic matter present in wastewater directly into electrical output using electrochemical active bacteria (EAB) as a biocatalyst without causing any harmful effects (Logan 2008). At the anode in anodic chamber, EAB convert biologically oxidizable organic matter into carbon dioxide, protons and electrons (Fig. 23.1a). Electrons (e^-) are travelled to the anode electrode and further passed to the cathode through an electrical circuit. Protons (H^+) are exchanged from anodic chamber to cathodic chamber through a CEM by cation exchange capacity of membrane. In cathodic chamber, the protons and electrons combine with oxygen to form water as an end product during reduction reaction.

Microbial extracellular electron transfer (EET) was first discovered in 1911 by Prof. M. C. Potter (1911), who demonstrated that electrical energy can be produced from cultures of *Saccharomyces* and *E. coli* using platinum (Pt) electrodes. Further Prof. Cohen (1931) developed a series of MFCs with current output of 2 mA. This enigmatic field of research for several years was later revived in 1960s (Davis 1963)

D.A. Jadhav
School of Water Resources, Indian Institute of Technology, Kharagpur,
Kharagpur 721302, India

M.M. Ghangrekar
Department of Civil Engineering, Indian Institute of Technology Kharagpur,
Kharagpur 721302, India

N. Duteanu (✉)
Industrial Chemistry and Environmental Engineering, University Politechnica of Timisoara,
Timisoara, Romania
e-mail: narcis.duteanu@chim.upt.ro

D. Das (ed.), *Microbial Fuel Cell*,
https://doi.org/10.1007/978-3-319-66793-5_23

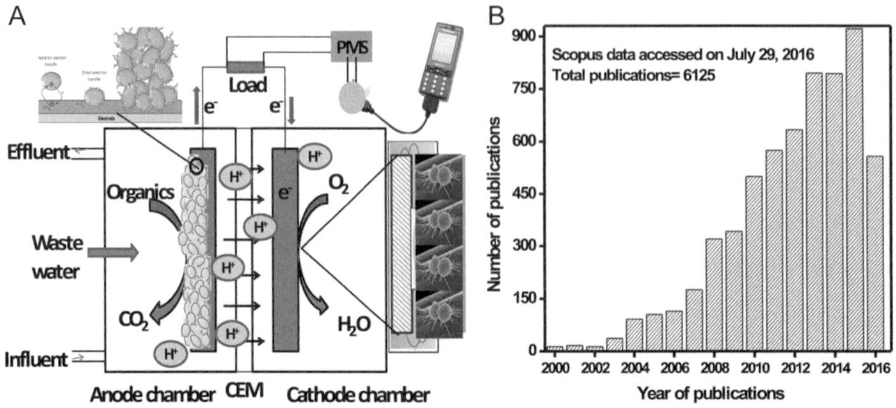

Fig. 23.1 (**a**) Schematic diagram representing working of dual chamber microbial fuel cell; and (**b**) Number of publications on microbial fuel cell over the years (Scopus data 2016)

and 1990s (Allen and Bennetto 1993). This important discovery remained mostly overlooked until researchers in the 1980s found that water soluble mediators (i.e., electron shuttles such as methylene blue, 2-bromoethanesulfonate, humic acid, etc.) artificially added to microbial media results into enhancement in electron transfer from bacterial cells to the anode electrode (Logan 2008). The discovery of pure culture of EAB such as *Geobacter* and *Shewanella* sp. (Kim et al. 2002) found to be capable for direct electron transfer (DET) from their metabolic pathways to the electrode, which has later awakened the interest in multidisciplinary research groups, involving microbiology, electrochemistry, material science and experts from engineering fields. In recent years, researchers are working to find the efficient electrode materials, types and combinations of bacterial cultures used as catalyst in anodic chamber as well as exploring biotic cathode, and electron transfer in MFC. Major breakthrough in MFC research was the development of cloth electrode assembly that leads to power generation of 2.87 kW m^{-3} and Coulombic efficiency (CE) of 83.5% (Fan et al. 2012).

Over 100 years, MFCs varying from μL to few litres volume have been studied to evaluate its performance (Logan 2010; Schroeder 2011). Recently, Feng et al. (2014) developed 230 L pilot scale MFC for treatment of municipal wastewater and achieved about 79% COD removal with power generation of 116 mW. More than 6123 research articles have been published till now addressing various aspects of MFC (Fig. 23.1b) varying from pollutant removal, electrode material modifications, electron transfer, operating parameters, modelling issue, scaling up issues, etc. The advancement in MFC research domain increased power output from nW to kW m^{-3} with simultaneously achieving effective wastewater treatment and other resources recovery (Jadhav and Ghangrekar 2015). Recently, Gude (2016) and Zhang et al. (2016) reviewed the technological development in MFC research for wastewater treatment (Ömeroğlu and Sanin 2016). Even though the idea of harvesting bioenergy generated from wastewater by microbial oxidation process has been around for more than 100 years, researchers have just begun to fully understand the MFC system and how to bring out its true potential for practical applications (Kim et al. 2007). However, considering the status of present research, most of MFC experts are moving towards scaling up of MFC for effective wastewater treatment and developing different approaches to enhance the power output.

23.2 Bottleneck in MFC Research

As compared to other renewable energy sources, MFC is still considered as a low energy producing system because of its thermodynamic limitations and different voltage losses associated with redox reactions. Theoretically, the maximum voltage which can be achieved from a single MFC unit is limited to 1.1 V by using acetate as a carbon source in anodic chamber and oxygen as the cathodic electron acceptor (Logan 2008). However, the actual voltage obtained from an MFC is always lower than this maximum theoretical voltage due to several voltage losses, often called as overpotential (Du et al. 2007; Logan 2008). The major bottlenecks for practical application of MFCs are the lower CE, lower power output and high production cost as compared to other conventional wastewater treatment processes. Although improvement in power and current density was reported from scalable design of MFC, this increase in electrical output is not in proportion with corresponding increase in volume of anodic chamber (Feng et al. 2014).

So far, most of the studies on MFCs have evaluated the performance of laboratory scale MFC (up to few litres in volume). Only a few pilot scale tests have been carried out and the results are not meeting the expectations. The main bottleneck currently perceived in MFC research is the field application of this technology coupled with cathodic electron transfer limitations (Jadhav et al. 2014). The field application of MFC technology is restricted by its design aspects, technological, electrochemical and microbiological limitations. To overcome the cathodic limitations, varieties of catalysts have been investigated for the direct electron transfer from cathode to oxygen in the air, which increases the overall cost of MFC fabrication (Rabaey and Keller 2008). Also, the lower power output from such units restricts the application of this technique from commercialization.

Considering the microbiological limitation, microbes are relatively slow electron transformers even at its fastest growth rate and have to compete to other non-electrogens for food as in case of mixed inoculum. Although CE over 90% has been achieved in few cases, it has little effect on the crucial problem of low reaction rate and voltage loss. Also, other limitation is represented by substrate diffusion towards anode in anodic chamber of larger volume MFC, which is not at sufficient rate to reach acceptable levels of current and cell potential (to minimize the diffusion losses) and it needs proper mixing condition for better proton transfer and to maintain homogenous condition throughout the electrolyte solution (Rabaey et al. 2012). Field scale MFC system demonstrations are required to take a step forward towards commercialization but it may pose new challenges and limitations which need to be addressed in a systematic manner. Even though considerable progress has been achieved to decrease the electrode and separator cost in the last several years, still this technology is far away from its practical applications and more efforts are required from the researchers. Clearly, a major breakthrough is needed to advance the MFC technologies beyond academic research and lab scale.

23.3 Scaling Up of MFC

The development of MFC technology needs selection of suitable and cost effective electrode materials and separator, and engineering design of scalable architecture to find its feasibility for wastewater treatment under optimum operating conditions. Researchers have enhanced the specific power densities by increasing the total surface area of the electrodes per unit volume of anodic or cathodic chamber and by decreasing the total reactor volume. However, proportional volumetric power generation decreased with increase in volume of MFC. The challenges for bringing MFC technologies out of the lab i.e. for practical applications, is to take into account the number of factors influencing performance of the MFC. The main challenge of MFC is to scale the energy output with respect to volume of wastewater being treated. Total power from MFC can be improved either by increasing the capacity (volume) of MFC or by connecting number of MFCs in electrically stacked arrangement (series or parallel). MFCs have been tested in the lab scale with volume ranging from 1.5 μL (Qian et al. 2009) to several litres (Jang et al. 2004; Li et al. 2008; Scott et al. 2007) with most of the MFCs typically using tens to hundreds of millilitres (Logan 2008, 2010; Logan et al. 2006). Out of 6150 research articles published on MFC (Scopus data 2016), less than 1% papers have discussed the issue of scaling up and the challenges of it.

The first pilot scale tubular MFC (1 m^3) consisting of 12 modules was operated at Foster's brewery in Yatala, Queensland (Australia). However, no more later results are available for the same MFC (Web reference 1). The pilot scale stackable horizontal MFC (SHMFC) of 230 L capacity was capable to harvest current of 0.435 A (Table 23.1) and maximum power of 116 mW (Feng et al. 2014). Recently, 90 L capacity stackable pilot scale MFC consisting of five stackable modules produced energy of 0.097 kWh m^{-3} and organic matter removal of 86.3% using brewery wastewater (Dong et al. 2015a). Experimental studies on scaling up of MFCs reported volume ranging from 1 to 1000 L, consisting of short-term lab scale experiments to long-term *in situ* pilot scale investigations (Janicek et al. 2014; Logan 2010; Ghadge et al. 2016). The scaling up studies reported in literature has focused mainly on wastewater treatment issue, long-term operation and economic analysis; however, the power output from such system is still low to operate electronic appliances directly (Table 23.1). To reduce the cost of separators, Ghangrekar et al. have tested the clayware ceramic separator for scaling up MFC studies and found it to be suitable separator to withstand with more hydrostatic pressure (Ghadge et al. 2016; Jadhav and Ghangrekar 2016; Ghadge and Ghangrekar 2015). Such clayware MFC with fabricating cost <1 US$ generated

Table 23.1 Overview of scaling up studies on MFC cited in literature

MFC details	Size (L)	Substrate (Wastewater)	Power (COD removal)	Findings	References
Stackable horizontal MFC	230	Municipal wastewater	116 mW, 435 mA, COD 79; CE 5% at 1 Ω	Cost analysis for scale up study	Feng et al. (2014)
96 module MFC	200	Primary wastewater	50 mA; 76.8%	Field trials of large-scale MFC systems	Ge and He (2016)
Stacked sediment MFC	113	Sediment	2.33 mW	PMS for SMFC & long-term operation	Ewing et al. (2014)
MFC with 36 MEAs	100	Septic tank wastewater	36 mW (CE 5.8%); COD 91%	Effective organic matter removal & disinfection in MFC	Jadhav and Ghangrekar (2016)
Stackable 5 modular MFC	90	Brewery wastewater	0.097 kWh m^{-3}; COD 86.3%	Effective real WWT (no energy input)	Dong et al. (2015a)
FRP MFC with 8 MEAs	45	Synthetic WW	17.63 mW (10 Ω); CE 2.03%	Long-term operation over 1 year	Ghadge et al. (2016)
4-SCMFCs (membraneless)	45	Effluent of the primary clarifier	0.36 kWh kg^{-1} COD, CE 4.8%	Full-scale wastewater treatment plant	Hiegemann et al. (2016)
Dual chamber MFC	40	Municipal wastewater	96 mW m^{-2}; COD 75%	Simultaneous WWT and energy recovery	Tota-Maharaj and Paul (2015)
Air cathode MFC with 3 MEAs	26	Synthetic WW with sucrose carbon source	17.85 mW (CE 5.1%); COD 78%	Developed scalable clayware structure MFC	Ghadge and Ghangrekar (2015)
Oxic-anoxic 2-stage bio-cathode MFC	23	Synthetic WW containing acetate	43.1 W m^{-3}; COD 95%	C and N removal using two-stage process	Liang et al. (2013)
MFC with 12 MEAs	16	Domestic wastewater	3.14 mW; COD 80%	Cathode fouling reduce performance	Jiang et al. (2011)
40 tubular SCMFC	10	Brewery wastewater	38.1 mW (4.1 W m^{-3})	Cathodic limitations in MFC	Zhuang et al. (2012)

*MEA[a]: Multiple electrode assembly; WWT[b]: wastewater treatment; SCMFC[c]: single chambered MFC

a significant power density of 16 W m^{-3}, which is comparable to other MFCs fabricated with expensive polymeric exchange membrane like Nafion. Clayware MFC was scaled up to 100 L volume (Fig. 23.2) with multiple electrode assemblies to treat septic tank sludge and showed capability to charge mobile phone battery with suitability to field level application using bioelectric toilets (Jadhav and

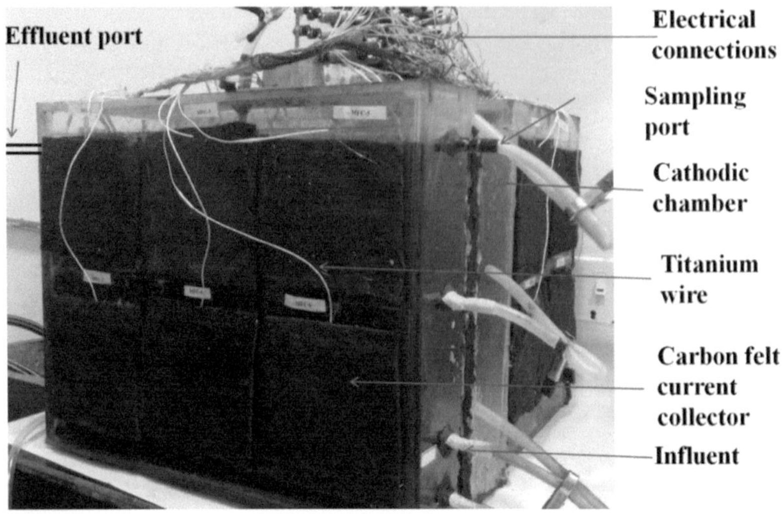

Fig. 23.2 Pictorial view of 100 L bioelectric septic tank MFC

Ghangrekar 2016). To enhance the proportional power output and to study suitability of MFC for field applications, most scale-up concepts suggested to-date needs modularization and replication of the fundamental MFC structure to increase volume and energy output (Table 23.1).

23.4 Hybrid Approach of MFC for Wastewater Treatment

With advancement in MFC research, several studies emerged with integration of one or more existing wastewater treatment units with MFC to increase the effluent quality, to recover the maximum available energy present in the wastewater and to enhance the overall treatment efficiency of combined system. In MFC centred hybrid system, MFC can be used as pre-treatment option (Li et al. 2014) as well as post treatment method (Wang et al. 2015). The products such as biohydrogen, biomethane, biofertilizer, electricity, caustic soda, hydrogen peroxide etc. were formed as an intermediate during the integrated operation of hybrid MFC. A methane producing MFC in combination with anaerobic digestion has been recommended as a polishing post treatment for effluent discharged from classical anaerobic digesters (Gao et al. 2014). The hybrid approach of MFC with anaerobic membrane bioreactor (Malaeb et al. 2013), integrated photo-bioreactor fuel cell (Eomh et al. 2011), anaerobic digester (Higgins et al. 2013), upflow anaerobic sludge blanket (UASB) reactor, microbial electrolysis cell (MEC), rotating biological contactor (Sayess et al. 2013) and other wastewater treatment units were investigated to yield higher quality treated effluent and recover valuable by-products (Gajda et al. 2014) (Table 23.2). In the integration of MFC with

Table 23.2 Review of hybrid approaches used in conjunction with MFC

Hybrid approach	Design aspect	Power (CE, %)	Wastewater treatment	Major findings	References
MFC + MBR	Pt-air cathode vs. biocathode MFC	6.8 W m^{-3}	COD 97%; NH_3-N 97%	Sustainable biological sewage treatment hybrid technology	Malaeb et al. (2013)
MFC + enzymatic air breathing cathode	*Shewanella* bioanode and laccase biocathode	26 W m^{-3}	ND	Enzymatic catalyzed reaction improved treatment	Higgins et al. (2011)
Hybrid photosynthetic fuel cells	*Pseudomonas* in anode and *Chlorella* biocathode	650 mW cm^{-2}; (CE 71.5%)	ND	Glycerol biotransformation in anode and carbon capture in cathode	La RottaHernández et al. (2014)
MFC + EO (electro-oxidation)	Air cathode MFC + EO cells	538 mW m^{-2}; (CE 17%)	COD 82%; TN 68%	Effective coking WWT	Huang et al. (2009)
MFC + AD	Upflow single chambered MFC + AD	330 mW m^{-2} at 10 Ω	COD 74%	Electroanalyses of hybrid system for brewery WW	Katuri and Scott (2010)
MREEC	Reverse electrodialysis + mineral carbonation	377 mw m^{-2} at 493 Ω	35% acid and 86% alkali recovery	Acid, alkali, H_2 recovery along with WWT and power generation	Zhu et al. (2014)
MFC + MEC	Single chambered MFC + MEC	43–82 mA m^{-2}; (CE 33%)	H_2 recovery (2.2 mL L^{-1} d^{-1})	Power from an MFC utilized for H_2 production in MEC	Sun et al. (2008)
MFC + PEC	Air cathode MFC + PEC	(CE 64%)	COD removal 67%	Generate hydrogen gas at zero external bias and effective WWT	Wang (2015)
MFC + Photobioreactor (PBR)	Upflow membraneless MFC + PBR	481 mW m^{-3}	P 99.3%; NH_4^+-N 99%	Enhance pollutants removal and recover energy from WW	Jiang et al. (2013)
MFC + UASB+ BAF	MFC-power; UASB-COD; BAF colour removal	1.4 W m^{-2}; 4.95 A m^{-2}	COD 53%; Sulphur 53%; colour 41.1%	Combine MFC with conventional processes for molasses WWT	Zhang et al. (2009)
MFC + IABF	Dual chamber MFC + IABF	427 mW m^{-2}	COD removal 91.7%	Energy self-sufficient combined system for high-quality effluent	Dong et al. (2015b)
MFC + AFMBR	Air cathode MFC + AFMBR	89 mWm^{-2}	COD removal 92.5%	Energy positive combined system for domestic WWT	Ren et al. (2014)

*[a]MBR- Membrane bioreactor; MRECC[b]: microbial reverse-electrodialysis electrolysis and chemical-production cell; PEC[c]: photoelectrochemical cell; IABF[d]: intermittently aerated biological filter; AFMBR[e]: anaerobic fluidized bed membrane bioreactor

granular activated carbon (GAC) adoption system, high pollutants removal percentage from palm oil mill effluent were achieved within several hours, which is considered very fast as compared to MFC unit alone (Tee et al. 2016). Zhang et al. (2009) designed three-stage combined system of MFC with UASB and biological aerated filter (BAF) for the treatment of high strength molasses wastewater and achieved total COD, sulphate and colour removal of 53.2%, 52.7% and 41.1%, respectively.

The hybrid system is capable to achieve high levels of wastewater treatment and it can provide polishing treatment as well as for the removal of specific pollutants. The BES centred hybrid treatment system is proposed as an example of integrating MFC and relevant wastewater technologies for improving treatment sustainability. However, it demands high energy input and further components to attach, which subsequently leads to increase in overall cost of fabrication. Also, difficulty with reactor scaling-up, primary barrier for practical implementation of hybrid system (containing MFC) and complexity of the combined treatment process strive its practical implementation. Along with several envisioned positive synergies, introducing the other processes may also create several new issues. For example, application of membrane separation process as a polishing step would lead to membrane fouling problems, pH imbalance condition, algae in photosynthesis processes can suffer from self-shading at higher concentration, deposition of hazardous metals during struvite precipitation. The alkaline effluent as well as change in micro-environment may harm the microbial activity in the subsequent treatment steps (Li et al. 2014). Therefore, a balanced assessment of all suitable and unfavourable factors needs to be taken into account to model the process design and operation of hybrid system.

23.5 Life Cycle Assessment of MFC

Life cycle assessment (LCA) is a cradle-to-grave approach in which environmental responsibility coupled with MFC system through all the stages during operation period are taken into account for the systematic and quantitative evaluation of environmental impacts. LCA is the most preferred method of choice in recent years for various kinds of new renewable technologies for bioenergy and carbon sequestration. Already several researchers are planning to upscale this technology as mentioned earlier to convince the end users of its potential use. However, limited studies have been carried out on LCA of MFC (Pant et al. 2011). In the study on LCA of MFC, Foley et al. (2010) concluded that MFC is not favourable substitute in comparison with other conventional treatment options. Similarly, in other LCA study conducted by Pant et al. (2011) interpreted that MFC is much better technique comparing with other wastewater treatment technologies. However, they did not present firm proof to support their remark. Hou (2014) concluded that for MFC centered combined system, large environmental burden is brought by the carbon and graphite materials used for electrode construction and environmental intensive metal (Pt) used for cathode construction when comparing with aeration system during LCA.

Several other studies on LCA showed that electrode and separator materials have the greatest impact on performance of MFC (Helder et al. 2013). Although proof of concept studies have demonstrated the feasibility of bioelectricity production from MFC technology, the choice of a functional unit is an important aspect (i.e. mass of production per kLD) (Patil et al. 2015). To make MFC as a feasible and economical solution, the economic and engineering cost analysis is one of the factors that need to be considered while studying LCA. Fornero et al. (2010) presented an economic justification and evaluation of MFC for revenue in terms of electricity generation and cost for municipal wastewater treatment (based on treatment cost of \$ 0.53 kg^{-1} BOD as per US standards). According to their economic justification, for wastewater flow of 100 m^3 d^{-1} with 2 g BOD L^{-1} (with assuming CE of 20% for real wastewater), the net present wealth equals to US\$ 380,528 for considering life span of 10 years. In comparison, the cost required to treat municipal wastewater in plant having activated sludge process (as a secondary treatment option) is \$ 32,760 per year, which is significantly lower than the revenue from wastewater treatment in MFC. The similar cost justification for MFC turned out to be US\$ 35,731 per year. These engineering economic justifications are governed by both electricity revenue generation and wastewater treatment cost. They also reported that electricity generation from MFC is not an attractive solution considering economics of system. However, organic matter removal can make system more efficient and self-sustained. The byproduct recovery from bioelectrochemical system also makes additional benefits in terms of energy value for the wastewater treatment plants.

From an energy production point of view, industrial wastewater may be a better market for this technology rather than municipal (domestic) wastewater, as the former contains far higher contaminants and COD concentration (and thus greater energy generation potential) and can generate interest for industry person to invest for this technology. From industrial point of view, several start-up companies such as Lebone (USA), Trophos Energy (USA), Plant-e (Netherlands), IntAct Labs LLC (USA), Emefcy (Israel), Indian Oil Corporation Ltd. (India), Hy-SyEnce (USA) and few others based on MFC technology have been already established and are trying to commercialize it (Pant et al. 2011). Therefore, by considering the suggestions and remarks received from the researchers and industrial experts, it is too early to compare the MFC technology with other renewable energy technologies because this technology is still under development and needs experience for field studies and environmental concern.

23.6 Current Challenges and Potential Opportunities

To make MFC technology suitable for real world applications, scaling up of this system is necessary. This requires increasing the size of MFC reactor and treatment capability to a practical level, and also achieving acceptable levels of output energy and scaling up of power output (Logan 2010). As aforementioned, MFC technology is potential alternative to effective wastewater treatment and energy generation. However, to utilize this potential several bottlenecks need to be overcome, to make this technology ready for commercialization. Although a neutral or positive energy balance has been theoretically established in past studies, there has not been an actual operation of energetically self-sustained MFC for wastewater treatment at field level (Zhang et al. 2013). Therefore, the biggest challenge of MFC system designed for wastewater treatment application is how we can simultaneously scale-up the reactor by maintaining same energy output per unit volume as demonstrated by small MFC (<1 L size).

The specific limitations related with MFC scaling up include high internal resistance of system, high overall cost of fabrication, operational stability issues, high material cost, other operational problems over the period of time (Gude 2016), slow pollutant degradation kinetics and lower efficiency of mixed culture biofilm on an electrode (Fornero et al. 2010; Logan 2010). The most important limitation of scaling up lied in increasing the electrochemical voltage losses (overpotential) with enlarged size. Successful demonstrations of large scale system are urgently required but it can come with several new challenges and limitations which need to be systematically addressed in near coming future. Yet, the understanding of EAB is still in its early stages, as the diverse microbial populations have several unnoticed electrochemical capabilities that can be exploited in different applications of MFC (Logan 2009). Continuous development efforts and research focus have also established the applicability of MFC technology towards several specialized and value added applications beyond the electricity generation, such as wastewater treatment and operating electronic appliances. However, to enhance the quality of treated effluent and power generation, the integrated bioprocess approach which combines post- treatment and use of PMS to improve the voltage is coming forward. Overcoming these limitations lead to call for further integrated efforts in reactor design engineering, electrode material development, bioelectrochemistry and biological manipulation.

23.7 MFC: Outlook and Future Perspectives

In about one decade of research and development, the research area of microbial electrochemical technologies has expanded noticeably and the performance has improved exponentially. This MFC technology has received attention in last two decades because they can provide access to a cheap renewable and green energy source. Along with the main products (wastewater treatment, electricity, gases, etc.), MFC shows capability of recovering other valuable products such as heavy metals, nutrients, industrial chemicals and gaseous fuel. The efficiency of MFC can be only comparable with conventional fuel cells if highly efficient system needs to be developed with much more advancement in engineering and microbial aspects. The maximum limit of power that can be produced from MFC and optimum size of

MFC is still unknown fact. Low efficiency of electricity yield has also stymied scaling up of this technology. To overcome these challenges, better understanding of microbes that metabolize the organic matter from wastewater and detailed electron transfer mechanism, need to be modelled. MFCs provide us with a model system and platform to establish the various microbial populations present in the exoelectrogenic biofilm on electrode, and it would be a vital research area in understanding how the microbial ecology and bacterial physiology of electrogens develops over time (Li and Sheng 2011).

In the future, further advancement in research is needed in material science, nanomaterial engineering and bioelectrochemistry to develop the MFCs for continuous improvement in the performance. The field scale application of MFC will also throw light on limitations of scaling up and actual kinetic limitation to make it successful. The rapid pace in development of microbial electrogenic genes and biocompatible electrode material might lead to overcome the limitations towards commercialization. To be a promising candidate for real wastewater treatment, the optimization and modelling efforts are required to make positive energy balance and beneficial life cycle assessment (Pant et al. 2010). Despite the above mentioned limitations, bioelectrochemical systems such as MFCs are a fascinating research subject and they have increased the interest among the research communities over the world.

23.8 Conclusion

Current state of art and rate of improvement in energy output and power production warrants a thorough rethinking of applied value and niches for MFC systems for practical applications (Schroeder 2011). The major breakthrough in MFC research is the development of air cathode MFC, use of pure culture inoculum, application of mediators, development of power management system, scaling up studies and bacterial ecology, which leads to increase the power generation and understanding the kinetic limitations of the system. Moreover, MFC technology is a promising area of research, which can solve some of the energy crisis in future, and reduce the percentage of gases emitted through fossil fuel consumption into the atmosphere. In last decade, power output of MFC boosted in order of several magnitudes. However, it is necessary to increase it further for useful applications to scalable level with innovative design and cost effective materials. The combination of wastewater treatment and resource recovery along with electricity production helps in compensating the cost of wastewater treatment, making hybrid MFC system sustainable.

An understanding of microbiology of the current producing process for limiting growth of non-electrogenic microbes is needed before further advancement in power output are possible. The brief journey from lab scale MFC to few field scale trials shows that MFC can be a potential technology to sit alongside other renewable energy technologies, while simultaneously curbing water pollution. However, field scale implementation of such bioelectrochemical system is not straightforward because certain microbiological, technological, bioelectrochemical and economic challenges need to be resolved that have not occurred previously in other conventional wastewater treatment system which makes life cycle assessment complicated. In order to contribute towards a greener future, the MFC technology will be an integrated part of the waste management processes in future.

Acknowledgement Grants received from Department of Science and Technology (File No. DST/INT/UK/P-101/2014), Government of India to undertake this work is duly acknowledged.

References

Allen, R. M., & Bennetto, H. P. (1993). Microbial fuel cells: Electricity production from carbohydrates. *Applied Biochemistry and Biotechnology, 39*(1), 27–40.

Davis, J. B. (1963). Generation of electricity by microbial action. *Advances in Applied Microbiology, 5*, 51–64.

Dong, Y., Feng, Y., Qu, Y., Du, Y., Zhou, X., & Liu, J. (2015a). A combined system of microbial fuel cell and intermittently aerated biological filter for energy self-sufficient wastewater treatment. *Scientific Reports, 5*, 18070.

Dong, Y., Qu, Y., He, W., Du, Y., Liu, J., Han, X., & Feng, Y. (2015b). A 90-liter stackable baffled microbial fuel cell for brewery wastewater treatment based on energy self-sufficient mode. *Bioresource Technology, 195*, 66–72.

Du, Z., Li, H., & Gu, T. (2007). A state of the art review on microbial fuel cells: A promising technology for wastewater treatment and bioenergy. *Biotechnology Advances, 23*, 464–482.

Eomh, H., Chung, K., Kim, I., & Han, J. I. (2011). Development of a hybrid microbial fuel cell (MFC) and fuel cell (FC) system for improved cathodic efficiency and sustainability: The M2FC reactor. *Chemosphere, 85*(4), 672–676.

Ewing, T., Ha, P. T., Babauta, J. T., Tang, N. T., Heo, D., & Beyenal, H. (2014). Scale-up of sediment microbial fuel cells. *Journal of Power Sources, 272*, 311–319.

Fan, Y., Han, S. K., & Liu, H. (2012). Improved performance of CEA microbial fuel cells with increased reactor size. *Energy & Environmental Science, 5*(8), 8273–8280.

Feng, Y., He, W., Liu, J., Wang, X., Qu, Y., & Ren, N. (2014). A horizontal plug flow & stackable pilot microbial fuel cell for municipal wastewater treatment. *Bioresource Technology, 156*, 132–138.

Foley, J. M., Rozendal, R. A., Hertle, C. K., Lant, P. A., & Rabaey, K. (2010). Life cycle assessment of high-rate anaerobic treatment, microbial fuel cells, and microbial electrolysis cells. *Environmental Science & Technology, 44*(9), 3629–3637.

Fornero, J. J., Rosenbaum, M., & Angenent, L. T. (2010). Electric power generation from municipal, food, and animal wastewaters using microbial fuel cells. *Electroanalysis, 22* (7–8), 832–843.

Gajda, I., Greenman, J., Melhuish, C., Santoro, C., Li, B., Cristiani, P., & Ieropoulos, I. (2014). Water formation at the cathode and sodium recovery using microbial fuel cells (MFCs). *Sustainable Energy Technologies and Assessments, 7*, 187–194.

Gao, H., Scherson, Y. D., & Wells, G. F. (2014). Towards energy neutral wastewater treatment: Methodology and state of the art. *Environmental Science: Processes & Impacts, 16*(6), 1223–1246.

Ge, Z., & He, Z. (2016). Long-term performance of a 200 liter modularized microbial fuel cell system treating municipal wastewater: Treatment, energy, and cost. *Environmental Science: Water Research & Technology, 2*(2), 274–281.

Ghadge, A. N., & Ghangrekar, M. M. (2015). Performance of low cost scalable air-cathode microbial fuel cell made from clayware separator using multiple electrodes. *Bioresource Technology, 182*, 373–377.

Ghadge, A. N., Jadhav, D. A., & Ghangrekar, M. M. (2016). Wastewater treatment in pilot-scale microbial fuel cell using multi-electrode assembly with ceramic separator suitable for field applications. *Environmental Progress & Sustainable Energy*. https://doi.org/10.1002/ep.12403.

Gude, V. G. (2016). Wastewater treatment in microbial fuel cells–an overview. *Journal of Cleaner Production, 122*, 287–307.

Helder, M., Chen, W. S., Harst, E. J., Strik, D. P., Hamelers, H. B. V., Buisman, C. J., & Potting, J. (2013). Electricity production with living plants on a green roof: Environmental performance of the plant-microbial fuel cell. *Biofuels, Bioproducts and Biorefining, 7*(1), 52–64.

Hiegemann, H., Herzer, D., Nettmann, E., Lübken, M., Schulte, P., Schmelz, K. G., Gredigk-Hoffmann, S., & Wichern, M. (2016). An integrated 45 L pilot microbial fuel cell system at a full-scale wastewater treatment plant. *Bioresource Technology, 218*, 115–122.

Higgins, S. R., Lau, C., Atanassov, P., Minteer, S. D., & Cooney, M. J. (2011). Hybrid biofuel cell: Microbial fuel cell with an enzymatic air-breathing cathode. *ACS Catalysis, 1*(9), 994–997.

Higgins, S. R., Lopez, R. J., Pagaling, E., Yan, T., & Cooney, M. J. (2013). Towards a hybrid anaerobic digester-microbial fuel cell integrated energy recovery system: An overview of the development of an electrogenic biofilm. *Enzyme and Microbial Technology, 52*(6–7), 344–351.

Hou, J. (2014). *Life cycle assessment of microbial fuel cell*. M.S. thesis, Philadelphia: University of Georgia.

Huang, L., Chai, X., Chen, G., & Logan, B. E. (2009). Effect of set potential on hexavalent chromium reduction and electricity generation from biocathode microbial fuel cells. *Environmental Science & Technology, 45*(11), 5023–5031.

Jadhav, D. A., & Ghangrekar, M. M. (2015). Effective ammonium removal by anaerobic oxidation in microbial fuel cells. *Environmental Technology, 36*(6), 767–775.

Jadhav, D. A. & Ghangrekar, M. M. (2016). *Scalable stacked bioelectric toilet microbial fuel cell for treatment of human waste and charging the mobile phone battery*. Int. Conf. Emerging Technologies in Agricultural and Food Engineering: IIT Kharagpur.

Jadhav, D. A., Ghadge, A. N., & Ghangrekar, M. M. (2014). Simultaneous organic matter removal and disinfection of wastewater with enhanced power generation in microbial fuel cell. *Bioresource Technology, 163*, 328–334.

Jang, J. K., Pham, T. H., Chang, I. S., Kang, K. H., Moon, H., Cho, K. S., & Kim, B. H. (2004). Construction and operation of a novel mediator- and membrane-less microbial fuel cell. *Process Biochemistry, 39*(8), 1007–1012.

Janicek, A., Fan, Y., & Liu, H. (2014). Design of microbial fuel cells for practical application: A review and analysis of scale-up studies. *Biofuels, 5*(1), 79–92.

Jiang, D., Curtis, M., Troop, E., Scheible, K., McGrath, J., Hu, B., Suib, S., Raymond, D., & Li, B. (2011). A pilot-scale study on utilizing multi-anode/cathode microbial fuel cells to enhance the power production in wastewater treatment. *International Journal of Hydrogen Energy, 36*(1), 876–884.

Jiang, H. M., Luo, S. J., Shi, X. S., Dai, M., & Guo, R. B. (2013). A system combining microbial fuel cell with photobioreactor for continuous domestic wastewater treatment and bioelectricity generation. *Journal of Central South University, 20*(2), 488–494.

Kadier, A., Kalil, M. S., Abdeshahian, P., Chandrasekhar, K., Mohamed, A., Azman, N. F., Logroño, W., Simayi, Y., & Hamid, A. A. (2016). Recent advances and emerging challenges in microbial electrolysis cells (MECs) for microbial production of hydrogen and value added chemicals. *Renewable and Sustainable Energy Reviews, 61*, 501–523.

Katuri, K. P., & Scott, K. (2010). Electricity generation from the treatment of wastewater with a hybrid up-flow microbial fuel cell. *Biotechnology and Bioengineering, 107*(1), 52–58.

Kim, H. J., Park, H. S., Hyun, M. S., Chang, I. S., Kim, M. A., & Kim, B. H. (2002). A mediator-less microbial fuel cell using a metal reducing bacterium, *Shewanella putrefaciens*. *Enzyme and Microbial Technology, 30*, 145–152.

Kim, B. H., Chang, I. S., & Gadd, G. M. (2007). Challenges in microbial fuel cell development and operation. *Applied Microbiology and Biotechnology, 76*(3), 485–494.

La RottaHernández, C. E., Leite, A. L., Dantas, P. V., Ramos, S. P., Perez, M. D. L. A., & Takaki, G. M. D. C. (2014). Hybrid microbial-photosynthetic biofuel cells for simultaneous bacterial glycerol biotransformation and algal carbon dioxide capture. *Journal of the Brazilian Chemical Society, 23*, 560–571.

Lee, J. H., Han, J., Choi, H., & Hur, H. G. (2007). Effects of temperature and dissolved oxygen on Se(IV) removal and Se(0) precipitation by *Shewanella* sp. HN-41. *Chemosphere, 68*, 1898–1905.

Li, W. W., & Sheng, G. P. (2011). *Microbial fuel cells in power generation and extended applications. Biotechnology in China III: Biofuels and bioenergy* (pp. 165–197). Heidelberg: Springer.

Li, D., Hong, Y., Xu, M., Luo, H., & Sun, G. (2008). Progress in construction of microbial fuel cell. *Chinese Journal of Applied & Environmental Biology, 14*(1), 147–152.

Li, W. W., Yu, H. Q., & He, Z. (2014). Towards sustainable wastewater treatment by using microbial fuel cells-centered technologies. *Energy. Environmental Sciences, 7*(3), 911–924.

Liang, P., Wei, J., Li, M., & Huang, X. (2013). Scaling up a novel denitrifying microbial fuel cell with an oxic-anoxic two stage biocathode. *Frontiers of Environmental Science & Engineering, 7*, 913–919.

Logan, B. E. (2008). *Microbial Fuel Cells*. New York: Wiley.

Logan, B. E. (2009). Exoelectrogenic bacteria that power microbial fuel cells. *Nature Reviews. Microbiology, 7*(5), 375–381.

Logan, B. E. (2010). Scaling up microbial fuel cells and other bioelectrochemical systems. *Applied Microbiology and Biotechnology, 85*, 1665–1671.

Logan, B. E., Hamelers, B., Rozendal, R., Schröder, U., Keller, J., Freguia, S., Aelterman, P., Verstraete, W., & Rabaey, K. (2006). Microbial fuel cells: Methodology and technology. *Environmental Science & Technology, 40*(17), 5181–5192.

Lovley, D. R., Phillips, E. J. P., Gorby, Y. A., & Landa, E. R. (1991). Microbial reduction of uranium. *Nature, 350*(6317), 413–416.

Malaeb, L., Katuri, K. P., Logan, B. E., Maab, H., Nunes, S. P., & Saikaly, P. E. (2013). A hybrid microbial fuel cell membrane bioreactor with a conductive ultra-filtration membrane biocathode for wastewater treatment. *Environmental Science & Technology, 47*, 11821–11828.

Ömeroğlu, S., & Sanin, F. D. (2016). Bioelectricity generation from wastewater sludge using microbial fuel cells: A critical review. *CLEAN – Soil, Air, Water*. https://doi.org/10.1002/clen.201500829.

Pant, D., Van Bogaert, G., Diels, L., & Vanbroekhoven, K. (2010). A review of the substrates used in microbial fuel cells (MFCs) for sustainable energy production. *Bioresource Technology, 101*(6), 1533–1543.

Pant, D., Singh, A., Van Bogaert, G., Gallego, Y. A., Diels, L., & Vanbroekhoven, K. (2011). An introduction to the life cycle assessment (LCA) of bioelectrochemical systems (BES) for sustainable energy and product generation: Relevance and key aspects. *Renewable and Sustainable Energy Reviews, 15*(2), 1305–1313.

Patil, S. A., Gildemyn, S., Pant, D., Zengler, K., Logan, B. E., & Rabaey, K. (2015). A logical data representation framework for electricity driven bioproduction processes. *Biotechnology Advances, 33*(6), 736–744.

Potter, M. C. (1911). Electrical effects accompanying the decomposition of organic compounds. *Proceedings of the Royal Society B: Biological Sciences, 84*, 260–276.

Qian, F., Baum, M., Gu, Q., & Morse, D. E. (2009). A 1.5 μL microbial fuel cell for on-chip bioelectricity generation. *Lab on a Chip, 9*(21), 3076–3081.

Rabaey, K., & Keller, J. (2008). Microbial fuel cell cathodes: From bottleneck to prime opportunity? *Water Science and Technology, 57*(5), 655–660.

Rabaey, K., Lissens, G. & Verstraete, W. (2012). Microbial fuel cells: Performances and perspectives. In: *Biofuels for fuel cells: Renewable energy from biomass fermentation* (pp. 377–399).

Reimers, C. E., Tender, L. M., Fertig, S., & Wang, W. (2001). Harvesting energy from the marine sediment – Water interface. *Environmental Science & Technology, 35*(1), 192–195.

Ren, L., Ahn, Y., & Logan, B. E. (2014). A two-stage microbial fuel cell and anaerobic fluidized bed membrane bioreactor (MFC-AFMBR) system for effective domestic wastewater treatment. *Environmental Science & Technology, 48*(7), 4199–4206.

Sayess, R. R., Saikaly, P. E., El-Fadel, M., Li, D., & Semerjian, L. (2013). Reactor performance in terms of COD and nitrogen removal and bacterial community structure of a three-stage rotating bioelectrochemical contactor. *Water Research, 47*(2), 881–894.

Schroeder, U. (2011). Discover the possibilities: Microbial bioelectrochemical systems and the revival of a 100-year–old discovery. *Journal of Solid State Electrochemistry, 15*, 1481–1486.

Scott, K., Murano, C., & Rimbu, G. (2007). A tubular microbial fuel cell. *Journal of Applied Electrochemistry, 37*(9), 1063–1068.

Sun, M., Sheng, G. P., Zhang, L., Xia, C. R., Mu, Z. X., Liu, X. W., & Yang, M. (2008). An MEC-MFC-coupled system for biohydrogen production from acetate. *Environmental Science & Technology, 42*(21), 8095–8100.

Tee, P. F., Abdullah, M. O., Tan, I. A. W., Amin, M. A. M., Nolasco-Hipolito, C., & Bujang, K. (2016). Performance evaluation of a hybrid system for efficient palm oil mill effluent treatment via an air-cathode, tubular upflow microbial fuel cell coupled with a granular activated carbon adsorption. *Bioresource Technology, 216*, 478–485.

Tender, L. M., Gray, S. A., Groveman, E., Lowy, D. A., Kauffman, P., Melhado, J., Tyce, R. C., Flynn, D., Petrecca, R., & Dobarro, J. (2008). The first demonstration of a microbial fuel cell as a viable power supply: Powering a meteorological buoy. *Journal of Power Sources, 179*, 571–575.

Tota-Maharaj, K., & Paul, P. (2015). Performance of pilot scale microbial fuel cells treating wastewater with associated bioenergy production in the Caribbean context. *International Journal of Energy and Environmental Engineering., 6*, 213–220.

Wang, H. (2015). *Solar-driven microbial photoelectrochemical system for energy conversion.* PhD thesis, UC Santa Cruz, California.

Wang, H., Qu, Y., Li, D., Zhou, X., & Feng, Y. (2015). Evaluation of an integrated continuous stirred microbial electrochemical reactor: Wastewater treatment, energy recovery and microbial community. *Bioresource Technology, 195*, 89–95.

Zhang, F., & He, Z. (2015). Scaling up microbial desalination cell system with a post-aerobic process for simultaneous wastewater treatment and seawater desalination. *Desalination, 360*, 28–34.

Zhang, B., Zhao, H., Zhou, S., Shi, C., Wang, C., & Ni, J. (2009). A novel UASB–MFC–BAF integrated system for high strength molasses wastewater treatment and bioelectricity generation. *Bioresource Technology, 100*(23), 5687–5693.

Zhang, F., Ge, Z., Grimaud, J., Hurst, J., & He, Z. (2013). Long term performance of liter-scale microbial fuel cells treating primary effluent installed in a municipal wastewater treatment facility. *Environmental Science & Technology, 47*(9), 4941–4948.

Zhang, Q., Hu, J., & Lee, D. J. (2016). Microbial fuel cells as pollutant treatment units: Research updates. *Bioresource Technology, 217*, 121–128.

Zhu, X., Hatzell, M. C., & Logan, B. E. (2014). Microbial reverse-electrodialysis electrolysis and chemical-production cell for H_2 production and CO_2 sequestration. *Environmental Science & Technology Letters, 1*(4), 231–235.

Zhuang, L., Yuan, Y., Wang, Y., & Zhou, S. (2012). Long term evaluation of 10-liter serpentine type microbial fuel cell stack treating brewery wastewater. *Bioresource Technology, 123*, 406–412.

Chapter 24
Scaling Up of MFCs: Challenges and Case Studies

Jai Sankar Seelam, Christina Theresia Rundel, Hitesh C. Boghani, and G. Mohanakrishna

24.1 Introduction

Rapid commercialization and expansion of biological and biotechnological platforms can contribute significantly towards realizing the concept of global bio-based economy. Bioelectrochemical systems (BESs) is one such emerging bio-based technology developed over the last few decades with multi-faceted utility. They assist in active valorization of resources in the form of bioelectricity (microbial fuel cell, MFC), biohydrogen (microbial electrolysis cell, MEC), value-added bioproducts (microbial electrosynthesis, MES) with concomitant waste management (bioelectro-treatment, BET) (Lovley 2006; Rosenbaum and Franks 2014; Venkata Mohan et al. 2014a, b). Of these, MFCs are heavily studied BES units and scalability is an important indicator in realizing their potential for practical application and global utility (Logan 2010). Scientific investigations and scale-up studies suggested that MFC operation at high reactor volumes (>5 L) are complex and are often challenged by several limitations. In this chapter, the problems associated with critical governing factors have been enlisted into operational, electrochemical and economic limitations. A brief overview of representative

J.S. Seelam (✉)
Wetsus Academy, P.O. Box 1113, 8900 CC Leeuwarden, The Netherlands
e-mail: shankar.jaishi@gmail.com; jai.seelam@wur.nl

C.T. Rundel
Universität Stuttgart, Keplerstraße 7, 70174 Stuttgart, Germany

H.C. Boghani
Sustainable Environment Research Centre, University of South Wales, Pontypridd CF37 1DL, UK

G. Mohanakrishna
Separation & Conversion Technologies, VITO – Flemish Institute for Technological Research, Boeretang 200, 2400 Mol, Belgium

D. Das (ed.), *Microbial Fuel Cell*,
https://doi.org/10.1007/978-3-319-66793-5_24

pilot-scale case studies like Bioelectro MET, Value from Urine, EcoBots and Peepower is presented in subsequent sections. Furthermore, possible technical and technological solutions, and future perspective to overcome the mentioned limitations are also included.

24.2 Limitations in Large Scale Applicationof MFCs

The focal point of expanding MFC technology was aimed at alleviating the energy intensive aerobic wastewater treatment, biofuel production and recovery of harmful metals at low concentrations. Concerted efforts have been laid by several research teams in this direction to boost the practical and long-term usability of MFCs through studies from bench-top and semi-pilot scale to pilot scale (Liu et al. 2008; Logan 2010). These studies have always been attempted in order to achieve optimized operational parameters and design configurations (Table 24.1). A representative power density of about 1 kW m^{-3}, proportionate current density of 5000 A m^{-3} of total anolyte volume or 50 A m^{-2} of projected anode surface area with an average voltage output of 0.2 V has been shown to be attainable and could suffice practical, long-term and commercial utility (Clauwaert et al. 2008). However, most of the MFCs reported maximum current densities between 10 and 25 A m^{-2}, evidencing unacceptable levels of power for operating small electrical appliances (Rabaey et al. 2010). The major bottlenecks include (1) low power density, (2) high capital expense (CAPEX) and (3) high operational expense (OPEX).

The overall functionality of a unit depends on factors like microbial inoculum and concentration, substrate composition and concentration, loading pH, feeding rate and configuration, temperature, electrode material, ion exchange membrane/ separator and reactor configuration (Seelam et al. 2015). Better understanding of cause and effects of these factors is necessary to achieve their better electrical performance. But, these governing parameters are often challenged by electrochemical and operational limitations during scale-up which can further influence the economics. Electrochemical constraints include limitations in performance due to design factors like electrode material, reactor vessel design, electrical configuration (series or parallel), internal resistance, electron transfer and biocatalytic activity. The operational performance on the other hand is bounded by start-up, physico-chemical parameters, substrate composition, feed loading rates, microbial culture and its stability. Lastly, CAPEX and OPEX are still high for the commercial bioelectro-production and renewable technologies such as wind and solar clearly outcompete the MFCs. A short introduction on each limiting factor associated with upscaling of MFCs with potential solutions has been discussed in the upcoming sections.

Table 24.1 An overview of different pilot-scale MFC investigations: Reactor configuration, performance details and limitations

Substrate	MFC configuration (Volume)	Electrodes	Performance details	Limitation(s)	Solution(s)	References
Acetate	Stack cell: 4 MFCs (20 L)	A: Coated Ti plates	PD: 144 W m^{-3}	Cathode limited performance due to oxygen reduction rate and cell reversal	Decreasing pH, purging pure oxygen and increasing flow rate	Dekker et al. (2009)
		C: Coated Ti plates	CD: 2.8 A m^{-2}			
Brewery wastewater	12 MFC modules (1000 L)	A: Carbon fibre brush		Low electrolyte conductivity, excess BOD, biofouling on cathode		Keller and Rabaey[a]
Acetate	MFC (5 L)	A: Pt-Ir coated Ti mesh	PD: 200 W m^{-3}	Cathode limitation and internal resistance		Ter Heijne et al. (2011)
		C: Pt-Ir coated Ti mesh	CD: 4.2 A/m^{-2}			
Brewery wastewater	Tubular air-cathode MFCs: 40 MFCs in series (10 L)	A: Graphite felt	PD: 4.1 W m^{-3}	High cathode alkalization and low humidity level at cathode side	Cathode rising with water	Zhuang et al. (2012a)
		C: Cloth cathode assembly	CD: 0.7 A m^{-3}			
Sediment	Sediment-MFC: 4 MFCs (240 L)	A: Graphite felt	P: 2.33 mW	EABF development		Ewing et al. (2014)
		C: Graphite felt				

A Anode, *C* Cathode, *PD* Power density, *CD* Current density, *BOD* Biochemical oxygen demand, *P* Power output, *EABF* Electroactive biofilm
[a]www.microbialfuelcell.org

24.3 Electrochemical Limitations: Design

24.3.1 Electrodes

Electrodes function as electron source and sink, and provide active sites for bioelectrochemical reactions. In addition, they also provide physical support for exoelectrogenic/electroactive biofilm formation and interface for electron transfer between microbes and electrode surface. Several carbon and metal-based electrodes with different compositions and dimensions have been used as electrode materials in MFCs. Noble metals can deliver higher power production but they are often limited by huge costs associated with material and corresponding chemical treatment (Logan 2010). Use of such materials need to be avoided in MFC constellation for large scale applications. The use of macro-porous electrodes posed a promising potential in terms of power enhancement but they are prone to clogging which hinder their long term usage. Also, carbon and metal-based anodes are susceptible to severe losses due to low mechanical strength and corrosion, respectively (Butti et al. 2016). The use of copper, brass and aluminum as anode displayed toxic nature towards microbial growth and use of such materials need to be avoided (Logan 2010). Investigations aimed at employing robust, durable and composite materials for anode fabrication can display improved performance (Navaneeth et al. 2015). Furthermore, biocompatibility is a critical requirement of good electrode system to promote biofilm growth and efficient electron transfer between microorganisms and electrode surface.

24.3.2 Reactor Vessel Design

Reactor vessel design and construction forms an integral part during the planning of pilot-scale investigation. The choice of material used for construction is determinant in predicting the overall performance of a simple single cell setup or a stacked fuel cell (Oliveira et al. 2013). Reactor volume is another performance driving factor which influences costs and possibility of clogging/biofouling. Large scaled units with/without membrane separators also failed to reach the power delivered by their identical smaller system using air cathode. Other factors like electrode arrangement and their spacing are responsible for their lower performance. Closely spaced electrode pair i.e. anode and cathode can improve the performance in larger setups (Liu et al. 2008). Higher electrolyte resistance can cause poor ionic flow in membraneless reactors and inclusion of membrane can lead to an internal resistance contribution of 38–86% due to their ohmic resistance and thereby decreasing the deliverable power (Fan et al. 2008).

Maximizing the electrode surface area has been a fruitful candidate in improving the MFC output during upscaling as shown by the recent advances in the research. For instance, in the case of sediment MFCs (SMFCs), the enlargement of the electrodes or simply the reactor size was not a performance governing factor as shown by Hsu et al. (2013) and they showed that the power density does not always linearly increase with the mere enlargement of the electrode (anode and cathode) surface area. Investigation by Cheng and Logan (2011) suggests that by increasing the cathode size (twice as big as the anode), about 62% more power could be produced. However, when the same strategy was applied to the anode, the power output increased by merely 12%. This shows that relative increase in anode surface area did not have bigger effect on the output and the relative surface area of the cathode electrode was the limiting factor in higher current production. This was coherent with the study by Liu et al. (2008) who found smaller relative surface area of the cathode to be the limiting factor in their MFC system.

24.3.3 Electrical Connectivity

Electric connectivity is often informed by the arrangement of individual MFC units in scalable systems. These units are principally connected either in series or parallel or combination of both based on the target current and/or voltage requirements (Liu et al. 2008). The electrical arrangement determines the overall power yield and estimates the potential drop due to ionic cross-conduction (Kim et al. 2012; Zhuang and Zhou 2009). In a study using landfill leachate by Gálvez et al. (2009), multiple MFCs having bigger electrodes and connected in series resulted in improved power output and treatment efficiencies. In another study by Ewing et al. (2014), scaled up SMFCs were electrically connected in parallel whose performance was compared with that of a single-equivalent SMFC. Both systems performed similarly for about 5 months but later on, the parallelly connected SMFCs delivered 3.5-fold higher power output (Ewing et al. 2014). An independent case study showed that parallelly connected MFCs in their setup outperformed the system connected in series while treating swine wastewater (Zhuang et al. 2012b). Hence, the research seems to suggest that the MFCs connected in parallel or modularized MFCs yield higher electrical output as compared to the MFCs connected electrically in series.

24.4 Operational Limitations

24.4.1 Start-Up

Start-up acts as a triggering step in the functioning of an MFC. It is the most critical aspect when treating huge volumes of wastewater in an industrial-scale installation. So, it becomes a limiting factor in the initial phase of MFC operation and it can be quantified in terms of start-up or lag time. This value can vary between tens of hours to several months (Feng et al. 2008; Liu et al. 2008). Scalable systems presented start-up values ranging between 60 and 103 days. The nature of substrate, inoculum and reactor configuration significantly governs lag in MFC operation (Aelterman et al. 2006). Commonly used mixed microbial cultures take significant time to adapt and produce acceptable output. Loading pH also influences the biofilm formation and thereby affects the startup (Patil et al. 2011). Bioaugmentation offers a favourable situation for improving the start-up of a bioprocess (Mohan et al. 2007; Wilderer et al. 1991). In this direction, use of acclimatized inoculum or effluent from existing treatment systems fed on same substrate can be candidates for quicker starting up of a reactor (Kim et al. 2007). Though this method looks viable for better start up, larger volumes ($>1\ m^3$) of enriched substrate/culture are required to operate MFCs at industrial scale (~10–20 m^3 reactor volume) which limits their utility. Other approaches aimed at minimizing the start-up problems like supplementing the wastewater with additional substrates and specific electron donors can boost the strength of the anolyte (in terms of chemical oxygen demand, COD) and promote the growth and metabolic activity of electroactive microbes (Liu et al. 2011). Fe (III) or fumarate prior to inoculating an MFC with culture of Geobacter sulfurreducens informed the faster start-up of MFC (Torres et al. 2009; Wang et al. 2010). Anolyte conductivity also dictates the current densities and the composition of the biofilm developed on the anode. So, maintaining desirable conductivity levels can inhibit the growth of non-exoelectrogens and consequently favour the start-up. The anode potential (electrochemical control) can also offer a promising solution (Wang et al. 2009).

24.4.2 *Electrolyte*

The electrolyte constitutes an integral component in the functioning of an MFC because it serves as a carbon/energy/electron donor/acceptors source as well as provides route for counter current ionic flow. To test the versatility of MFCs, diverse array of wastes and wastewaters have been used as fuels for MFC application (Pant et al. 2010; Rozendal et al. 2008). Affluent availability and cheaper costs associated with wastes/wastewaters creates a value-added advantage.

24.4.2.1 Chemical Composition

Greater fraction of wastewater constituents function as electron donors to boost microbial growth, metabolic activity and effectiveness of electroactive biofilm. Every wastewater has its characteristic composition and lends different performances when used as fuel in MFCs. The influent constituents drastically impact the microbial assemblage and biofilm integrity, coulombic efficiency, power density and COD removal efficiency (Seelam et al. 2015). The choice and availability of the waste feedstock is, therefore, critical for long-term and sustainable application. The use of wastewater from households and domestic sector can be challenging for effective operation due to its mild organic/inorganic strength and subsequent poor conductivity (Pant et al. 2010). The use of brewery wastewater displayed positive results when operated in an MFC. This is due to the presence of high biodegradable organic matter (chemical oxygen demand, COD) and sparse concentration of inhibitory agents (Pant et al. 2010). The presence of complex substance in several wastewater feedstocks can adversely affect the coulombic efficiencies (Zhuang et al. 2012a, b). The reason for lower coulombic efficiencies can be attributed to different parameters. For example, super saturated levels of organic matter promotes competition between electroactive and non-electroactive microorganisms which influences the current generation (Oliveira et al. 2013). Side reactions like methane production might occur due to the presence of methanogenic growth promoting compounds in the substrate which limits the substrate availability for the exoelectrogenic biofilm (He et al. 2005). It was found that the MFC operated on real wastewater severely underperformed as compared to the synthetic wastewater (Fornero et al. 2010). Furthermore, few wastewater feedstocks require a pre-treatment step to improve the substrate bioavailability. This treatment actively degrades the complex matter into simple compounds but often incurs additional costs due to additional steps.

24.4.2.2 Substrate Loading

In bulk scale systems, organic loading rate (OLR) and sludge loading rate (SLR) are critical during the startup phase. These parameters dictate the capacity of reactor per unit volume and the amount of microbes to degrade organic substrate (Oliveira et al. 2013). Several studies investigated the influence of OLR and SLR on MFC performance and concluded that these parameters are directly proportional to the power yield and organic matter degradation and inversely to that of coulombic efficiency and internal resistance (Martin et al. 2010; Velvizhi and Venkata Mohan 2012). The use of wastewaters did not prove beneficial when the system was scaled up in terms of cathode surface area and it was found that the energy loss was greater when the loading rate was lower (less COD per unit time) (Cheng and Logan 2011). It has also been shown that power output had decreased with gradual increase in the substrate degradation when OLR was too high due to direct anodic oxidation (Martin et al. 2010; Velvizhi and Venkata Mohan 2012). Lower OLR might enhance the coulombic efficiency due to lower methanogenic activity but internal resistance becomes higher in such systems due to lower OLR (Martin et al. 2010; Velvizhi and Venkata Mohan 2012). A compromise between effective electron recovery and COD removal is necessary during scale up operations. Hence, MFC operation at optimum OLR and SLR values is crucial for maximal power production and organic matter treatment.

24.5 Economic Limitations

MFC technology can promise a self-sustaining bioprocess with concomitant waste remediation but its bulk-scale utility is primarily limited by the economic factors. Considering the costs of lab-scale installation with an estimated life span of 10 years, the capital investment is assumed to be about US\$ 3 kg^{-1} influent COD (Fan et al. 2012). The economic advantage of this cost may vary when large scale application is considered. Comparative analysis with conventional treatment systems can often provide a techno-economic view point. Interestingly, a study presented a net profit for an MFC treating wastewater of about US\$ 0.0005 kg^{-1} COD (considering the electricity price and the energy recovery rate). Relative to conventional treatment with activated sludge where no profit is gained, an MFC shows an advantage (Liu and Cheng 2014). MFC also provides other economic advantages as compared to other treatment technologies. These include (1) lower biomass production and no requirement of aeration or temperature adaption and (2) achievable high energy conversion rates (Liu and Cheng 2014).

Several other research investigations reported promising results towards an economically feasible MFC application but in contrast, long-term commercial adoption is far from being reached. Major obstacle to overcome is the material and fabrication costs for the MFCs (Zhuang et al. 2012a, b). Costs associated with the chemical pretreatment of electrode, use of precious metal electrodes and impregnation of current collector into electrodes also limit the bulk-scale utility (Seelam et al. 2015). Electrode materials, current collectors, catalysts for cathode and ion exchange membrane/separators are expensive and contribute to majority of the costs of the system. While anode materials declined in price, cathodes are still mostly expensive. Use of graphite fibre brush anodes present an economic alternative (Feng et al. 2010). Cathodes are the main costly players which conduce up to 75% of the total CAPEX (Rozendal et al. 2008). For a case study referring to carbon cloth/Pt cathode, the costs are approximately US\$ 1000 m^{-2} (Logan 2010). Reduction in cathodic material costs alone would pull down the capital cost to US\$ 0.1/kg COD (Fan et al. 2012). This forecasts treatment costs close to that of the activated sludge treatment. Power densities of about 23–36 W m^{-3} are observed when cathodes are modified and impregnated with typical iron and nickel metals (Aelterman et al. 2009). Economic fuel cell grade materials like stainless steel and cheap binders offer to prevent cathode damage and improve the current densities. Membrane/separator may also be a large cost contributor. This is often necessary and implanted in bulk-scale MFCs to ensure little electrode distance which in turn brings down reactor volume (Liu and Cheng 2014). Though the use of single chambered MFC may demand less capital relative to dual chambered systems due to lack of separator, bioelectricity output is generally compromised (Butti et al. 2016). The presence of separators can be beneficial as it prevents short circuiting and lends ability to space electrodes (anode and cathode) more closely. This enhances the electric output on volumetric basis. Alternative and inexpensive separators are being investigated, but their long-term operation, stability and performance is yet to be examined (Butti et al. 2016). On an operational front, MFCs are negatively influenced by biofouling and scaling which eventually affects long-term stability and energy output (Liu and Cheng 2014). Membrane cleaning and/or replacement might be necessary in two-chambered installations during continuous operation which drastically increases the costs.

Research studies suggested the use of biocathode in an MFC setup but it can also be a pricy affair. Aeration becomes a necessary step to provide dissolved oxygen in the catholyte for the activity of biocathode (Cheng and Logan 2011). Activated carbon with metal mesh current collector can be a suitable alternative catalyst for oxygen reduction. Another encouraging approach is to employ air-cathodes. This configuration is known as single chamber air-cathode cloth electrode assembly MFC design. It presents relative advantage as it is cheaper, easy to fabricate, operate, and generates high power. Both environmental impact and economic feasibility of MFCs demand reduced operating power requirements. High strength wastewaters offer valuable feedstocks and provide added economic advantage for the MFC application. But Fornero et al. (2010) showed that lower coulombic efficiencies were obtained with high strength wastewaters and concluded that the electric output would be insufficient to recover the investment costs. Another case study by Wang and Ren (2013) highlighted that although the fuel cell grade electrode material costs have lowered, MFC technology remains an expensive wastewater treatment variant. It is worthwhile to mention that MFC is still an emerging technology and these economic limitations suggest that cost challenges are enormous and techno-scientific advancements are further prospected to overcome the economic barrier.

24.6 MFCs Toward Commercial Applications: Case Studies

In the quest to find the apt applicability of MFCs and subsequently commercialize them, several research teams have investigated case studies at different locations worldwide. These case studies generate valuable lessons to boost the successful and practical applications of the MFC technology. They are valuable for planning, innovating, analyzing, interpreting and developing. A brief description on each multi-faceted case study is presented here.

24.6.1 Bioelectro MET

With the initiative to promote energy efficient recovery of metals from metallurgical waste and process streams using microbial fuel cells, a collaborative research project: BioelectroMET (*Bioelectro*chemical systems for *MET*al recovery) is being

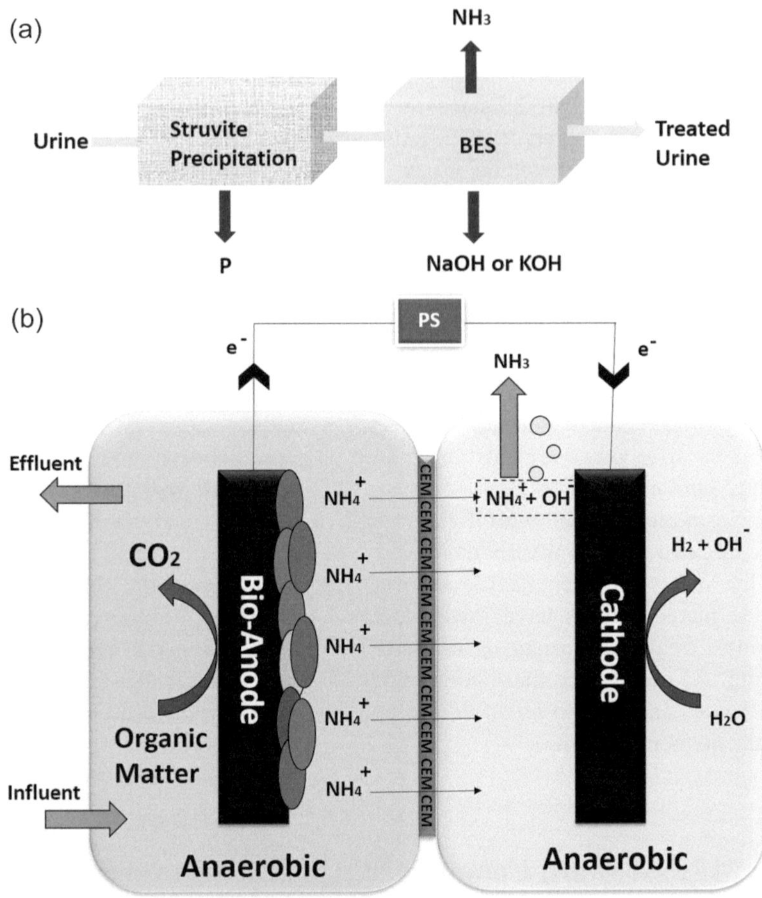

Fig. 24.1 Value from urine – schematic diagram of (**a**) proposed urine treatment concept with simultaneous ammonia recovery with BES; and (**b**) double chambered MEC employed for ammonia recovery fed upon urine as substrate (*PS* Power supply, *CEM* Cation exchange membrane)

investigated at the time of writing this book, within the European Union's Seventh Framework Programme (www.bioelectromet.eu). The principal aim of this case study is to investigate, develop and demonstrate recovery of metals using bioelectrochemical systems at no or fraction of energy input. The metal extraction in this system occurs through electrowinning at the cathode which is assisted by anodic bio-oxidative reactions. For the final phase of the investigation, a prototype with 1 m^2 anode electrode surface area was constructed (Fig. 24.1).

An ideal bioelectrochemical metal recovery would include anodic oxidation of biodegradable organics (electron donor) in wastes/wastewaters coupled with cathodic reduction of metals (electron acceptor) in metallic waste streams. Copper (Cu) recovery using MFCs was the targeted activity in this case study (Ntagia et al.

2015; Rodenas Motos et al. 2015). Theoretically, the use of MFCs for copper recovery at the cathode can deliver a maximum electrical output of 0.49 kWh kg^{-1} Cu when supplemented by wastewater as substrate at the anode. This generates a better cathodic copper extraction, bioelectricity production and waste remediation option where fraction of the produced energy can be utilized for recovery process itself. Conventional electrochemical recovery requires an energy input at a minimum of 2.1 kWh kg^{-1} of Cu (in theory) and in practice, about 2.23 kWh kg^{-1} of Cu using electrolysis cells (Vegliò et al. 2003).

The choice of electron donors and biocatalysts play a vital role in dictating the overall performance of the system. In this direction, the investigations of BioelectroMET were aimed at using extremophiles as biocatalysts and different electron donors in MFCs. Microbial communities surviving harsh conditions (extremophiles) like low pH and concentrated metallic wastewaters were investigated (Sulonen et al. 2015) and later reviewed to display their applicability in BES (Dopson et al. 2016). An MFC catalyzed by mixed acidophilic electroactive microbes was used for anodic oxidation of inorganic compound, tetrathionate, present in mining waste streams (Sulonen et al. 2015). The study reported the presence of extremophiles like *Acidothiobacillus* and *Ferroplasnma* spp. at pH of <2.5. Tetrathionate oxidation at anode was coupled by ferric ion (Fe^{3+}) reduction at cathode under similar pH conditions with simultaneous bioelectricity production at 80 mW m^{-2}. Follow-up investigation studied the tetrathionate disproportionation into sulphate and elemental sulphur, and delivered 2.8-fold higher power density at 225 mW m^{-2} (Sulonen et al. 2016).

Furthermore, MFC operations employed hydrogen and acetate in its anode as electron donors for cathodic copper reduction process. The electroactive biofilms generate electrons for copper recovery by actively extracting electrons from hydrogen and diverting them to the cathode via anode electrode surface and external circuit. This phenomenon was demonstrated by an MFC when acetate-fed biofilm (bioanode) oxidized hydrogen to produce these electrons (Ntagia et al. 2015). The maximum current output with simultaneous copper recovery in this setup amounted up to 250 mW m^{-2}. Also, the utility of acetate as substrate for combined bioelectricity and copper recovery was first demonstrated in a proof-of-principle study by Heijne et al. (2010). In this study, pure copper crystals were electrodeposited at cathode with removal efficiency of >99.88%. A metallurgical MFC with bipolar membrane cum pH separator under aerobic and anaerobic conditions was operated with acetate as fuel. The performance of this system improved by 86% when operated under anaerobic conditions (maximum power density of 0.8 W m^{-2} at a current density of 3.2 A m^{-2}) compared to aerobic conditions (maximum power density of 0.43 W m^{-2} at a current density 1.7 A m^{-2}). Copper recovery efficiency relative to electricity production were reported at 84% (anaerobic) and 43% (aerobic) (Heijne et al. 2010). A novel reactor configuration was adopted as a part of BioelectroMET project to surpass the current and power density demonstrated by this unit. The changes include anion separator instead of bi-polar membrane, reduced internal resistance (electrodes placed closer to each other) and highly conductive copper plate as cathode (Rodenas Motos et al. 2015). These modifications resulted in maximum power density of 5.5 W m^{-2} and current density of 23 A m^{-2}.

24.6.2 Value from Urine

Urine offers to be promising feedstock due to its affluent availability and nutrient-rich composition (about 80% N and 50% P found in wastewaters originate from urine) for bio-based applications (Maurer et al. 2003). Its utility in BESs like MFCs assists in bioelectricity production with concomitant nutrient recovery in the form of fertilizers or value-added chemicals (Kuntke et al. 2012). Research, development and expansion of this concept at commercial scale can present an innovative technology in the field of waste and energy management. In this direction, Value from Urine is a representative case study exhibiting an advanced application of MECs operating on urine as substrate (www.valuefromurine.eu). This project within European Union's Seventh Framework Programme envisions to develop, demonstrate and evaluate a BES platform, which ably treats urine, and also simultaneously recover nitrogen (in the form of ammonia) and phosphorus (in the form of phosphate) before discharging into sewer grid. The adoption of source-separated urine collection system, recovery target of >95% of N and P as ammonium sulphate and struvite, respectively, and waste to energy conversion concepts are the key highlights of the project (Kuntke et al. 2014, 2016; Rodríguez Arredondo et al. 2014). Conventional nitrogen recovery technologies like stripping and electrodialysis are more energy-intensive relative to bioelectro-recovery systems (Maurer et al. 2003; Rodríguez Arredondo et al. 2014). The use of BESs for ammonia treatment is majorly advantageous because they offer a cost-effective platform for both removal and recovery relative to these conventional systems. The case study further prospects efficient urine treatment and production of different chemicals, like e.g. caustic or hydrogen peroxide which would lead to a more advanced system. But, upscaling of BES driven urine treatment for effective ammonia valorization is challenged by several operational and performance parameters (Kuntke et al. 2012).

In pre-project phase, synthetic and real urine was fed as anolyte into a double chambered MFC fitted with a graphite felt anode and gas diffusion cathode (Pt coated Ti felt). Nitrogen was recovered in the form of ammonia through volatilization and acid absorption and overall energy balance was positive which displayed energy self-sufficiency of the MFC. The volatile ammonia was formed from cathodic conversion of diffused ammonium ions due to high pH in the cathode (Kuntke et al. 2012). The inclusion of ammonia recovery neither impacted the performance in terms of current and power densities nor proved toxic to microbes as displayed in earlier investigations (Nam et al. 2010). The exhibited ammonium recovery was 9.57 gN m^{-2} d^{-1} against current density of 2.6 A m^{-2} with real urine which is relatively higher than synthetic urine (Kuntke 2013). However, this system suffered with poor coulombic efficiency (10%), lower ammonia transport to production ratio and scaling due to *in situ* struvite crystallization (Kuntke et al. 2012). In an advanced approach, an attempt was made to valorize ammonia with co-production of hydrogen gas at the cathode in an adapted MFC units often referred as MEC. Such hydrogen production occurs by supplementing about 0.8 V on top of the MFC potential under anaerobic conditions (Logan et al. 2008). The capture of ammonia and current density produced in the MEC (23.07 A m^{-2}) were better than that of the MFC (2.6 A m^{-2}) (Kuntke et al. 2012). The produced hydrogen showcased the advantage of using urine as a fuel in an MEC (Kuntke et al. 2014). Current densities of 14.64–23.07 A m^{-2}, hydrogen production rate of 32–48.6 m^3 H_2 m^{-3} reactor/d, ammonia removal rate of 162.18–173.4 gN/m^2/d and COD removal rate of 130.56–171 g COD/m^2 d^{-1} were reported when operated on five times diluted urine. MEC reactor configuration included a double chambered unit with titanium (Ti) and platinum (Pt) coated flow fields and graphite felt anode. Recent study presented ammonia recovery using hydrophobic and gas permeable membranes. This system generated a stable output with relatively poor current density but enhanced ammonium transport and recovery rate (Kuntke et al. 2016).

24.6.3 EcoBots

The EcoBots concept was developed and investigated by the researchers at the University of West of England, Bristol, UK during the years of 2003 to around 2010 (Ieropoulos et al. 2003, 2005, 2010). The idea was to build an autonomous robot that could be fed on the waste material such as rotten fruits and dead flies for their energy in order to perform locomotive and telemetry tasks. The EcoBots employed MFCs as their powerhouse. The robots exhibited ‘pulsed behaviour’ where the system accumulated energy and waited until the threshold was reached and then moved to a next position and stopped when it was energetically exhausted.

EcoBot-I was constructed using styrene support structure which housed eight MFC units (fed on sugar) in its centre, two photo detecting diodes, two DC geared motors and a control circuit, totaling a 22 cm diameter and 7.5 cm height with 960 g of weight. This platform was mounted onto appropriate caster wheels for the phototactic locomotion. The EcoBot-I was successfully operated by the on-board power generated and moved onto a trajectory to a total of 25 cm in about 400 s (Ieropoulos et al. 2003). EcoBot-II was similar to the EcoBot-I in its construction except that the eight MFCs were now located on the periphery of the circular styrene block. Additionally, this generation robot performed the telemetry task of temperature sensing and data transmission via on-board wireless microprocessor. The EcoBot-II was fed on the refined sugar, rotten peach portions and dead flies. It was successfully demonstrated to move 50 cm in an average of 6 h, in the direction of light source that was kept at 90° angle to the robot (Ieropoulos et al. 2005).

EcoBot-III was significantly improved robot in comparison to the EcoBot-I and EcoBot-II where it could now perform the food and water intake i.e. ingestion and the excretion of the waste i.e. egestion. This robot was 63 cm high and of 29 cm diameter with a total weight of 5.88 kg that housed 48 MFCs arranged in two circular stacks in such a way that the top and bottom stack would result in hydraulic series connection. In addition, it also consisted 120 units of 6800 μF electrolytic capacitors for the charge accumulation. Top part of the robot had a UV light attracter (in the form of hopper) for the flies which were used as substrate. The 300 mL reservoir (stomach of the robot) was kept for the fluid feedstock and a trough was placed at the bottom of the robot for the collection of the egested waste. The robot was mounted on a rail and the ambient temperature was 30 ± 5 °C. The robot was successfully demonstrated to make 110 iterations of movement occurring at an average of 30 h duration in a lab scale experiment which lasted for 7 days (Ieropoulos et al. 2010). These demonstrations of EcoBots present a possibility of powering of scavenging robots that can be fed onto the waste in the wastefield and subsequently cleaning the land. It may be a long way before this kind of dream can be realized but this demonstrates a proof-of-concept for application of MFCs.

24.6.4 Pee Power Urinal

In pursuit of reinventing the toilets for affordable sanitation, Bill and Melinda Gates Foundation launched a challenge in 2011 called, 'Reinvent the toilet challenge' (1999–2016). The team of MFC researchers from University of West of England, Bristol in UK have entered the challenge to tackle the issue by using MFCs. They are set to design the urinal system where urine can be fed directly to the MFCs which produce electrical power (urine-tricity) for charging mobile phones and battery powered consumer electronics. There is an added benefit of the nutrients extracted from the urine which can be used as fertilizers in the agriculture. As reported by Ieropoulos et al. (2016), the Bill and Melinda Gates Foundation and Oxfam funded pee-power urinal was field trialed at the university campus for 3 months and at Glastonbury Music Festival in England in 2015 for 8 days where the urine was voluntarily dispensed by the university male students and staff and the festival goer, respectively. The urinal was a portable toilet from Oxfam which was modified to integrate MFC modules that can be fed directly from the dispensed urine. The MFC stack at university campus contained eight modules each and assembly of 36 modules; totaling 288 individual MFCs electrically connected in combination of series and parallel connections. These modules of MFCs were used to power four LED modules (each of 1.2 W power requirement) fitted inside the urinal via four units of supercapacitors. On an average, 5–10 users visited the urinal in 3-month period of its operation which resulted in hydraulic retention time (HRT) of 2–3 weeks and consequently COD removal of 90–98%. The whole unit generated an average power of 75 mW. The field trial at Glastonbury Music Festival in England was at a larger scale whereby 12 modules (36 MFCs each) containing

432 MFCs with total working volume of 300 L were used. Similar electrical and lighting arrangements as aforementioned were used. As the toilet users were much higher (~1000 per day), the resulting HRT was 0.9 days. The average power reached to ~400 mW in this case. However, the COD removal was only about 25% in an average which they attributed to the lower HRT (Ieropoulos et al. 2016). These trials show the practicality and feasibility of this technology to power safety lights and low powered devices. However, the authors agree that the commercial venture may be followed by the rigorously studied economical and life cycle analysis of this technology.

24.7 Possible Solutions to Overcomethe Limitations

Detailed engineering design of the reactor is essential in reducing the performance related issues of the MFCs. The reactor design should include consideration of operational regime such as fluid flow, temperature, pressure, loading rates and operational maintenance. By the nature of their application, fluid flow seems to be a dominating factor in considering to see how the electrolyte is going to flow through and around the electrode, how much electrode material can you fit into a given volume (to increase the specific electrode area per volume) and also the spacing between the anode and cathode electrodes. For operational maintenance, researchers have shown that controlling the current sourcing of MFCs can be beneficial to their performance. They presented that energy efficiency, start-up time and biocatalytic activity can be improved along with avoiding voltage reversal (Boghani et al. 2013, 2014). There also have been evidences of successful self-sustaining MFCs running on their produced power (including for pumping) (Dong et al. 2015; Ledezma et al. 2013).

24.7.1 Electrode Spacing and Specific Surface Area

One of the significant factors affecting the performance of MFC is the electrode spacing between the anode and the cathode. This is a well-known and established fact. The electrode spacing influences the internal resistance that is created by the electrode and electrolyte material and also the mass transport issues. Decreasing the spacing can decrease the internal resistance, reducing the overpotentials and increasing the power performance (Harnisch and Schroder 2009). One has to take care for the membraneless reactor though, since it may cause short-circuit if the electrodes are placed too close to each other.

It is perhaps plausible to imagine that the number of exoelectrogens residing on the electrode is directly proportional to the electrical current they generate. It is, therefore, sensible to increase the number of these bacteria to increase the electrical power performance. One way to do it may be to increase the surface area to volume ratio. Many researchers have used this approach to increase the volumetric power density using porous carbon material (Cheng et al. 2006; Xie et al. 2012) or geometric arrangement of the electrode (Chen et al. 2011; Michie et al. 2011) or both (Kim et al. 2012). Increase in the electrode material in a given volume comes at a price of increase in the head (or required pumping pressure). There are also chances of blockages in the fluid path due to development of biomass in the pores. Therefore, care must be taken in designing the electrode in order to increase surface area available to the biocatalysts and simultaneously maintain an effective throughput of fluid through the reactor.

24.7.2 Electrolyte Flow Dynamics

The fluid flow plays an important role in transport of reactants and products, to and from the (bio)catalyst sites. It is important to determine the location of dead zones in the reactor and try to minimize them. The electrode activity at dead zone is seriously impaired in the absence or limitation of the substrate due to lack of convective mass transport. This creates unequal distribution of the electrical current density and seriously undermines the overall performance of the electrode as there is gradient of electrolyte potential generated within the reactor. Shear rate on the electrode is also one of the major factor in determining the substrate delivery to the biofilm on (and within the pores of) the electrode (Pham et al. 2008). It is also suggested that mixing enhances the performance of MFCs as it allows movement of ions within the reactor environment. An example of a flat plate and tubular designs of the reactors is considered to demonstrate this. The dimensions of the reactors are arbitrarily chosen but they could represent the lab-scale reactors in reality. Figure 24.2 shows the results of Computational Fluid Dynamics performed on these reactors when considering inflow of water at a rate of 10 mL min^{-1}.

As seen in Fig. 24.2, the dead zones can be clearly visible in the corners of the reactors and in the case of the flat plate reactor, they can occupy larger area of the electrode. The magnitude of velocity of the fluid within the flat plate reactor is much smaller, in comparison to the entry and exit points of the reactor. After entering the reactor, fluid velocity reduces greatly as it finds large volume (expansion of volume in the fluid path) and the converse happens when exiting. So, the shear rate produced by this is unevenly distributed throughout the planar electrode affecting the mass transport as well. If the tubular reactor is considered, the electrode will receive evenly distributed shear rate on its surface as seen in the figure, throughout the electrode length, apart from near the inlet and outlet ends. The dead zones are comparatively small in the tubular design of the reactor in comparison to the flat plate reactor. However, the annular electrode comes with an inherent problem where the mass transport deep within the electrode can be limited if the electrode thickness (annular) is large enough. One way to avoid this problem may be to use a porous electrode but they can also be blocked by the biomass if the pore sizes are reasonably small. Granted that tubular reactor with annular electrode may not be able to give higher surface area to volume ratio to that of the flat plate reactor where

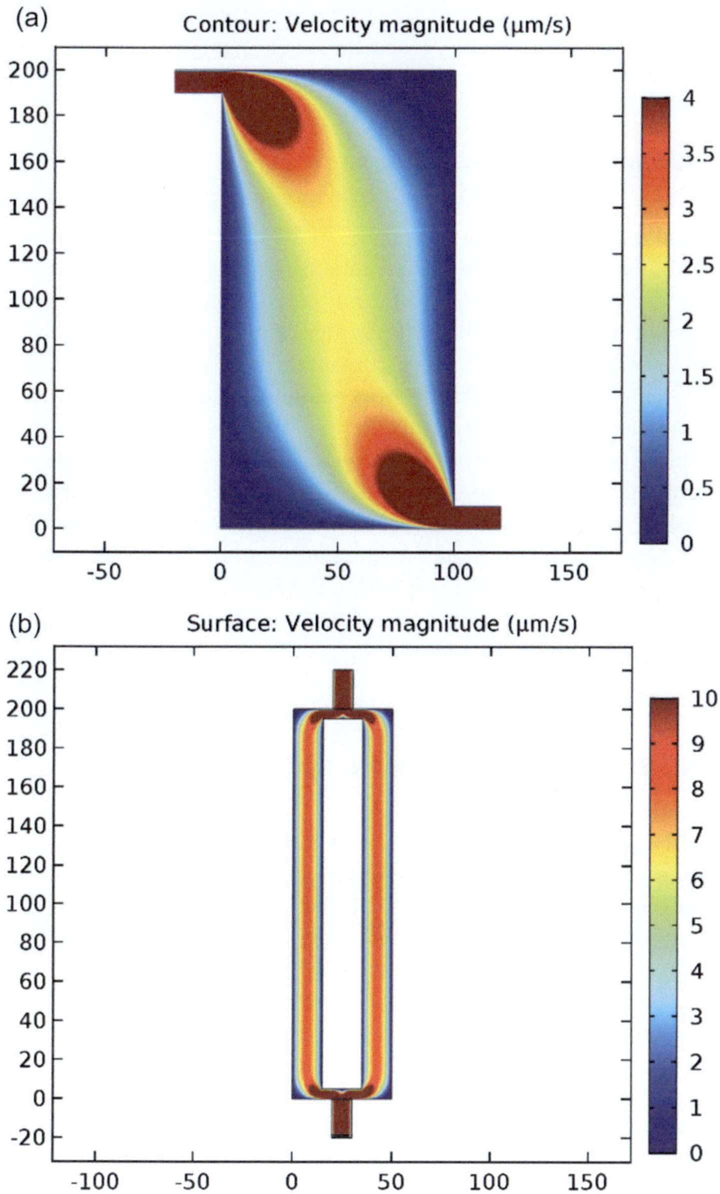

Fig. 24.2 Fluid flow in flat plate and tubular reactors. Axes show geometrical dimensions in mm; Value from Urine – Schematic diagram of (**a**) proposed urine treatment concept with simultaneous ammonia recovery with BES and (**b**) double chambered MEC employed for ammonia recovery fed upon urine as substrate

helical electrode may be an option to increase the surface area (Kim et al. 2012) in a given tubular volume. This also provides a tortuous path to the fluid flow (but at an expense of slight increase in the head) that aids mass transport. A designer would need to consider these issues and at sometimes, one may need to trade off between the choice and design of the reactor, and the performance.

24.7.3 *Minimizing Fabrication Defects*

Designing and manufacturing are correlated in MFC fabrication. When considering planar electrodes, even thickness of the catalyst benefits the performance whereas gaps and patches can develop parasitic current. Leaky chambers can also yield parasitic current if the organic substrate is available to the cathode, some sites can start anodic reactions oxidizing the organic matter and the electrons are taken by the nearby site on the same electrode for the oxygen reduction (Harnisch and Schroder 2009). So, this does not contribute to the overall MFC current. When dissimilar metals are used in the reactors such as copper, tinned copper, stainless steel etc., improper insulation can expose those materials to the electrolyte and there may be possibility of forming a galvanic cell locally. This can impair the power production from MFC. Also, contact resistance is generally very high in the case of loosely connected current collectors. Care should be taken to connect the current collector to the electrode by appropriate means (welding, soldering or compression) to lower the contact resistance as much as possible.

24.8 Conclusions and Future Perspectives

The practical applicability of MFCs to assist the global energy market is imminent but they can still be considered in the development phase. This chapter provided a summary of the fundamental techno-scientific and economic challenges of large-scale MFC investigations. The most concerning issue displayed during upscaling of these systems is the magnitude of electrical energy recovered. Non-traditional and renewable energy producing alternatives powered by wind and solar energy on the other hand, can deliver much higher yield at lower cost. Pilot-scale research demonstrated a multi-faceted alternative through representative case studies with supplementary waste management and can offer an interface to curtail our dependence on depleting fossil fuel reserves. Such studies also exhibit the necessary issues which need attention in order to realize their commercial utility. Expensive electrode materials, complicated reactor architecture, waste feedstock complexity, lower microbial electrocatalytic performance and electrode surface properties are the domains which seek further improvements. Technological advancements aimed at process optimizing variables such as pretreatment of substrate, control of pH, lowering of the internal resistance, integration of biocathode and bioaugmentation, can further be explored in future research. To achieve maximum possible electrical output, MFC operation should be focused in simplified reactor configuration fitted with low cost electrodes and fed upon wastes and wastewaters operated at higher organic loading rates. The use of energy-efficient micro-controller based smart technologies and automation can also bring the practical application of MFC closer to a reality. Research focus directed towards integrating bioprocesses like fermentation, desalination and waste bioelectro-refinery concept can potentially enhance the diversity of MFC utility through simultaneous product recovery, metal recovery, salt separation (desalination) and nutrient valorization. Interestingly, large scale enterprises have taken leap into commercializing MFC technology via energy-efficient wastewater treatment using MFCs (www.emefcy.com) and methane production from waste streams with MECs (Cambrian Innovation Inc., 2013). These attempts and advancements prospect a promising future for BES technology and envision their usage into recovery of nutrients from wastes and wastewaters and/or their self-sufficient treatment.

Acknowledgement J.S. Seelam takes this precious opportunity to thank Anne van Den Ban fonds (ABF) foundation and Wetsus for partially supporting his Masters' study at the Wetsus Academy, The Netherlands. Wetsus Academy is a joint degree program of Water Technology (Master of Science) coordinated by Wetsus, Wageningen University, University of Twente and University of Groningen in Leeuwarden. He also likes to thank his colleagues at Wetsus, Philipp Kuntke and Pau Rodenas Motos, for their comments on Value from Urine and BioelectroMET case studies.

References

Aelterman, P., Rabaey, K., Pham, H. T., Boon, N., & Verstraete, W. (2006). Continuous electricity generation at high voltages and currents using stacked microbial fuel cells. *Environmental Science and Technology, 40*, 3388–3394.

Aelterman, P., Versichele, M., Genettello, E., Verbeken, K., & Verstraete, W. (2009). Microbial fuel cells operated with iron-chelated air cathodes. *Electrochimica Acta, 54*, 5754–5760.

Boghani, H. C., Kim, J. R., Dinsdale, R. M., Guwy, A. J., & Premier, G. C. (2013). Control of power sourced from a microbial fuel cell reduces its start-up time and increases bioelectrochemical activity. *Bioresource Technology, 140*, 277–285.

Boghani, H. C., Papaharalabos, G., Michie, I., Fradler, K. R., Dinsdale, R. M., Guwy, A. J., et al. (2014). Controlling for peak power extraction from microbial fuel cells can increase stack voltage and avoid cell reversal. *Journal of Power Sources, 269*, 363–369.

Butti, S. K., Velvizhi, G., Sulonen, M. L., Haavisto, J. M., Koroglu, E. O., Cetinkaya, A., et al. (2016). Microbial electrochemical technologies with the perspective of harnessing bioenergy: Maneuvering towards upscaling. *Renewable and Sustainable Energy Reviews, 53*, 462–476.

Chen, S., Hou, H., Harnisch, F., Patil, S. A., Carmona-Martinez, A. A., Agarwal, S., et al. (2011). Electrospun and solution blown three-dimensional carbon fiber nonwovens for application as electrodes in microbial fuel cells. *Energy & Environmental Science, 4*, 1417–1421.

Cheng, S., & Logan, B. E. (2011). Increasing power generation for scaling up single-chamber air cathode microbial fuel cells. *Bioresource Technology, 102*, 4468–4473.

Cheng, S., Liu, H., & Logan, B. E. (2006). Increased power generation in a continuous flow MFC with advective flow through the porous anode and reduced electrode spacing. *Environmental Science and Technology, 40*, 2426–2432.

Clauwaert, P., Aelterman, P., Pham, T. H., De Schamphelaire, L., Carballa, M., Rabaey, K., & Verstraete, W. (2008). Minimizing losses in bio-electrochemical systems: The road to applications. *Applied Microbiology and Biotechnology, 79*, 901–913.

Dekker, A., Ter Heijne, A., Saakes, M., Hamelers, H. V. M., & Buisman, C. J. N. (2009). Analysis and improvement of a scaled-up and stacked microbial fuel cell. *Environmental Science and Technology, 43*, 9038–9042.

Dong, Y., Qu, Y., He, W., Du, Y., Liu, J., Han, X., & Feng, Y. (2015). A 90-liter stackable baffled microbial fuel cell for brewery wastewater treatment based on energy self-sufficient mode. *Bioresource Technology, 195*, 66–72.

Dopson, M., Ni, G., Sleutels, T. H. J. A., Abrevaya, X., Sacco, N., Mauas, P., et al. (2016). Possibilities for extremophilic microorganisms in microbial electrochemical systems. *FEMS Microbiology Reviews, 40*, 164–181.

Ewing, T., Ha, P. T., Babauta, J. T., Tang, N. T., Heo, D., & Beyenal, H. (2014). Scale-up of sediment microbial fuel cells. *Journal of Power Sources, 272*, 311–319.

Fan, Y., Sharbrough, E., & Liu, H. (2008). Quantification of the internal resistance distribution of microbial fuel cells quantification of the internal resistance distribution of microbial fuel cells. *Environmental Science & Technology, 42*, 8101–8107.

Fan, Y., Han, S.-K., & Liu, H. (2012). Improved performance of CEA microbial fuel cells with increased reactor size. *Energy & Environmental Science, 5*, 8273–8280.

Feng, Y., Wang, X., Logan, B. E., & Lee, H. (2008). Brewery wastewater treatment using air-cathode microbial fuel cells. *Applied Microbiology and Biotechnology, 78*, 873–880.

Feng, Y., Yang, Q., Wang, X., & Logan, B. E. (2010). Treatment of carbon fiber brush anodes for improving power generation in air-cathode microbial fuel cells. *Journal of Power Sources, 195*, 1841–1844.

Fornero, J. J., Rosenbaum, M., & Angenent, L. T. (2010). Electric power generation from municipal, food, and animal wastewaters using microbial fuel cells. *Electroanalysis, 22*, 832–843.

Gálvez, A., Greenman, J., & Ieropoulos, I. (2009). Landfill leachate treatment with microbial fuel cells: Scale-up through plurality. *Bioresource Technology, 100*, 5085–5091.

Harnisch, F., & Schroder, U. (2009). Selectivity versus mobility: Separation of anode and cathode in microbial bioelectrochemical systems. *ChemSusChem, 2*, 921–926.

He, Z., Minteer, S. D., & Angenent, L. T. (2005). Electricity generation from artificial wastewater using an upflow microbial fuel cell. *Environmental Science and Technology, 39*, 5262–5267.

Heijne, A. T., Liu, F., Van Der Weijden, R., Weijma, J., Buisman, C. J. N., & Hamelers, H. V. M. (2010). Copper recovery combined with electricity production in a microbial fuel cell. *Environmental Science & Technology, 44*, 4376–4381.

Hsu, L., Chadwick, B., Kagan, J., Thacher, R., Wotawa-Bergen, A., & Richter, K. (2013). Scale up considerations for sediment microbial fuel cells. *Royal Society of Chemistry Advances, 3*, 15947–15954.

Ieropoulos, I., Melhuish, C., & Greenman, J. (2003). Artificial metabolism: Towards true energetic autonomy in artificial life. In *Advances in Artificial Life* (pp. 792–799). Berlin/Heidelberg: Springer.

Ieropoulos, I., Melhuish, C., Greenman, J., & Horsfiel, I. (2005). EcoBot-II: An artificial agent with a natural metabolism. *International Journal of Advanced Robotic Systems, 2*, 295–300.

Ieropoulos, I., Greenman, J., Melhuish, C., & Horsfield, I. (2010). EcoBot-III: A robot with guts. In *The 12th international conference on the synthesis and simulation of living systems* (pp. 733–740).

Ieropoulos, I. A., Stinchcombe, A., Gajda, I., Forbes, S., Merino-Jimenez, I., Pasternak, G., et al. (2016). Pee power urinal – Microbial fuel cell technology field trials in the context of sanitation. *Environmental Science: Water Research Technology, 2*, 336–343.

Keller and Rabaey (www.microbialfuelcell.org) (2016). *Urine-tricity++: Electricity from urine.* Bristol Robotics Laboratory.

Kim, J. R., Jung, S. H., Regan, J. M., & Logan, B. E. (2007). Electricity generation and microbial community analysis of alcohol powered microbial fuel cells. *Bioresource Technology, 98*, 2568–2577.

Kim, J. R., Boghani, H. C., Amini, N., Aguey-Zinsou, K.-F., Michie, I., Dinsdale, R. M., et al. (2012). Porous anodes with helical flow pathways in bioelectrochemical systems: The effects of fluid dynamics and operating regimes. *Journal of Power Sources, 213*, 382–390.

Kuntke, P. (2013). *Nutrient and energy recovery from urine*. Wageningen Universiteit (Wageningen University).

Kuntke, P., Śmiech, K. M., Bruning, H., Zeeman, G., Saakes, M., Sleutels, T. H. J. A., Hamelers, H. V. M., & Buisman, C. J. N. (2012). Ammonium recovery and energy production from urine by a microbial fuel cell. *Water Research, 46*, 2627–2636.

Kuntke, P., Sleutels, T. H. J. A., Saakes, M., & Buisman, C. J. N. (2014). Hydrogen production and ammonium recovery from urine by a microbial electrolysis cell. *International Journal of Hydrogen Energy, 39*, 4771–4778.

Kuntke, P., Zamora, P., Saakes, M., Buisman, C. J. N., & Hamelers, H. V. M. (2016). Gas-permeable hydrophobic tubular membranes for ammonia recovery in bio-electrochemical systems. *Environmental Science: Water Research and Technology, 2*, 261–265.

Ledezma, P., Stinchcombe, A., Greenman, J., & Ieropoulos, I. (2013). The first self-sustainable microbial fuel cell stack. *Physical Chemistry Chemical Physics, 15*, 2278–2281.

Liu, W.-F., & Cheng, S.-A. (2014). Microbial fuel cells for energy production from wastewaters: The way toward practical application. *Journal of Zhejiang University: Science A (Appl Phys & Eng), 15*, 841–861.

Liu, H., Cheng, S., Huang, L., & Logan, B. E. (2008). Scale-up of membrane-free single-chamber microbial fuel cells. *Journal of Power Sources, 179*, 274–279.

Liu, G., Yates, M. D., Cheng, S., Call, D. F., Sun, D., & Logan, B. E. (2011). Examination of microbial fuel cell start-up times with domestic wastewater and additional amendments. *Bioresource Technology, 102*, 7301–7306.

Logan, B. E. (2010). Scaling up microbial fuel cells and other bioelectrochemical systems. *Applied Microbiology and Biotechnology, 85*, 1665–1671.

Logan, B. E., Call, D., Cheng, S., Hamelers, H. V. M., Sleutels, T. H. J. A., Jeremiasse, A. W., & Rozendal, R. A. (2008). Microbial electrolysis cells for high yield hydrogen gas production from organic matter. *Environmental Science & Technology, 42*, 8630–8640.

Lovley, D. R. (2006). Microbial fuel cells: Novel microbial physiologies and engineering approaches. *Current Opinion in Biotechnology, 17*, 327–332.

Martin, E., Savadogo, O., Guiot, S. R., & Tartakovsky, B. (2010). The influence of operational conditions on the performance of a microbial fuel cell seeded with mesophilic anaerobic sludge. *Biochemical Engineering Journal, 51*, 132–139.

Maurer, M., Schwegler, P., & Larsen, T. A. (2003). Nutrients in urine: Energetic aspects of removal and recovery. *Water Science and Technology, 48*, 37–46.

Michie, I. S., Kim, J. R., Dinsdale, R. M., Guwy, A. J., & Premier, G. C. (2011). The influence of psychrophilic and mesophilic start-up temperature on microbial fuel cell system performance. *Energy & Environmental Science, 4*, 1011–1019.

Mohan, S. V., Mohanakrishna, G., Veer Raghavulu, S., & Sarma, P. N. (2007). Enhancing biohydrogen production from chemical wastewater treatment in anaerobic sequencing batch biofilm reactor (AnSBBR) by bioaugmenting with selectively enriched kanamycin resistant anaerobic mixed consortia. *International Journal of Hydrogen Energy, 32*, 3284–3292.

Nam, J. Y., Kim, H. W., & Shin, H. S. (2010). Ammonia inhibition of electricity generation in single-chambered microbial fuel cells. *Journal of Power Sources, 195*, 6428–6433.

Navaneeth, B., Hari Prasad, R., Chiranjeevi, P., Chandra, R., Sarkar, O., Verma, A., et al. (2015). Implication of composite electrode on the functioning of photo-bioelectrocatalytic fuel cell operated with heterotrophic-anoxygenic condition. *Bioresource Technology, 185*, 331–340.

Ntagia, E., Rodenas, P., ter Heijne, A., Buisman, C. J. N., & Sleutels, T. H. J. A. (2015). Hydrogen as electron donor for copper removal in bioelectrochemical systems. *International Journal of Hydrogen Energy, 41*, 5758–5764.

Oliveira, V. B., Simoes, M., Melo, L. F., & Pinto, A. M. F. R. (2013). Overview on the developments of microbial fuel cells. *Biochemical Engineering Journal, 73*, 53–64.

Pant, D., Van Bogaert, G., Diels, L., & Vanbroekhoven, K. (2010). A review of the substrates used in microbial fuel cells (MFCs) for sustainable energy production. *Bioresource Technology, 101*, 1533–1543.

Patil, S. A., Harnisch, F., Koch, C., Hbschmann, T., Fetzer, I., Carmona-Martinez, A. A., et al. (2011). Electroactive mixed culture derived biofilms in microbial bioelectrochemical systems: The role of pH on biofilm formation, performance and composition. *Bioresource Technology, 102*, 9683–9690.

Pham, H. T., Boon, N., Aelterman, P., Clauwaert, P., De Schamphelaire, L., van Oostveldt, P., et al. (2008). High shear enrichment improves the performance of the anodophilic microbial consortium in a microbial fuel cell. *Microbial Biotechnology, 1*, 487–496.

Rabaey, K., Butzer, S., Brown, S., Keller, J., & Rozendal, R. A. (2010). High current generation coupled to caustic production using a lamellar bioelectrochemical system. *Environmental Science and Technology, 44*, 4315–4321.

Reinvent the Toilet Challenge Strategy Overview. (1999–2016). Bill and Gates Foundation.

Rodenas Motos, P., ter Heijne, A., van der Weijden, R., Saakes, M., Buisman, C. J. N., & Sleutels, T. H. J. A. (2015). High rate copper and energy recovery in microbial fuel cells. *Frontiers in Microbiology, 6*, 1–8.

Rodríguez Arredondo, M., Kuntke, P., Jeremiasse, A. W., Sleutels, T. H. J. A., Buisman, C. J. N., & ter Heijne, A. (2014). Bioelectrochemical systems for nitrogen removal and recovery from wastewater. *The Royal Society of Chemistry, 1*, 22–33.

Rosenbaum, M. A., & Franks, A. E. (2014). Microbial catalysis in bioelectrochemical technologies: Status quo, challenges and perspectives. *Applied Microbiology and Biotechnology, 98*, 509–518.

Rozendal, R. A., Hamelers, H. V. M., Rabaey, K., Keller, J., & Buisman, C. J. N. (2008). Towards practical implementation of bioelectrochemical wastewater treatment. *Trends in Biotechnology, 26*, 450–459.

Seelam, J. S., Pant, D., Patil, S. A., & Kapadnis, B. P. (2015). Biological electricity production from wastes and wastewaters. In *Microbial factories* (pp. 155–183). New Delhi: Springer.

Sulonen, M. L. K., Kokko, M. E., Lakaniemi, A.-M., & Puhakka, J. A. (2015). Electricity generation from tetrathionate in microbial fuel cells by acidophiles. *Journal of Hazardous Materials, 284*, 182–189.

Sulonen, M. L. K., Lakaniemi, A.-M., Kokko, M. E., & Puhakka, J. A. (2016). Long-term stability of bioelectricity generation coupled with tetrathionate disproportionation. *Bioresource Technology, 216*, 876–882.

Ter Heijne, A., Liu, F., van Rijnsoever, L. S., Saakes, M., Hamelers, H. V. M., & Buisman, C. J. N. (2011). Performance of a scaled-up microbial fuel cell with iron reduction as the cathode reaction. *Journal of Power Sources, 196*, 7572–7577.

Torres, C. I., Krajmalnik-Brown, R., Parameswaran, P., Marcus, A. K., Wanger, G., Gorby, Y. A., & Rittmann, B. E. (2009). Selecting anode-respiring bacteria based on anode potential: Phylogenetic, electrochemical, and microscopic characterization. *Environmental Science and Technology, 43*, 9519–9524.

Vegliò, F., Quaresima, R., Fornari, P., & Ubaldini, S. (2003). Recovery of valuable metals from electronic and galvanic industrial wastes by leaching and electrowinning. *Waste Management, 23*, 245–252.

Velvizhi, G., & Venkata Mohan, S. (2012). Electrogenic activity and electron losses under increasing organic load of recalcitrant pharmaceutical wastewater. *International Journal of Hydrogen Energy, 37*, 5969–5978.

Venkata Mohan, S., Velvizhi, G., Annie Modestra, J., & Srikanth, S. (2014a). Microbial fuel cell: Critical factors regulating bio-catalyzed electrochemical process and recent advancements. *Renewable and Sustainable Energy Reviews, 40*, 779–797.

Venkata Mohan, S., Velvizhi, G., Vamshi Krishna, K., & Lenin Babu, M. (2014b). Microbial catalyzed electrochemical systems: A bio-factory with multi-facet applications. *Bioresource Technology, 165*, 355–364.

Wang, H., & Ren, Z. J. (2013). A comprehensive review of microbial electrochemical systems as a platform technology. *Biotechnology Advances, 31*, 1796–1807.

Wang, X., Feng, Y., Ren, N., Wang, H., Lee, H., Li, N., & Zhao, Q. (2009). Accelerated start-up of two-chambered microbial fuel cells: Effect of anodic positive poised potential. *Electrochimica Acta, 54*, 1109–1114.

Wang, A., Sun, D., Ren, N., Liu, C., Liu, W., Logan, B. E., & Wu, W. M. (2010). A rapid selection strategy for an anodophilic consortium for microbial fuel cells. *Bioresource Technology, 101*, 5733–5735.

Wilderer, P. A., Rubio, M. A., & Davids, L. (1991). Impact of the addition of pure cultures on the performance of mixed culture reactors. *Water Research, 25*, 1307–1313.

Xie, X., Ye, M., Hu, L., Liu, N., McDonough, J. R., Chen, W., et al. (2012). Carbon nanotube-coated macroporous sponge for microbial fuel cell electrodes. *Energy and Environmental Science, 5*, 5265–5270.

Zhuang, L., & Zhou, S. (2009). Substrate cross-conduction effect on the performance of serially connected microbial fuel cell stack. *Electrochemistry Communications, 11*, 937–940.

Zhuang, L., Yuan, Y., Wang, Y., & Zhou, S. (2012a). Long-term evaluation of a 10-liter serpentine-type microbial fuel cell stack treating brewery wastewater. *Bioresource Technology, 123*, 406–412.

Zhuang, L., Zheng, Y., Zhou, S., Yuan, Y., Yuan, H., & Chen, Y. (2012b). Scalable microbial fuel cell (MFC) stack for continuous real wastewater treatment. *Bioresource Technology, 106*, 82–88.

Chapter 25
Challenges in Microbial Fuel Cell and Future Scope

Kuppam Chandrasekhar, Abudukeremu Kadier, Gopalakrishnan Kumar, Rosa Anna Nastro, and Velpuri Jeevitha

25.1 Introduction

The imminent energy crisis due to reserved fossil fuel and global warming due to greenhouse gas emission warrant the necessity for environmentally friendly sources of energy. Fossil fuels adversely affect the nature due to the release of CO_2 into the environment (Venkata Mohan et al. 2011; Venkateswar Reddy et al. 2011a; Kadier et al. 2016a, b). Hence the consumption of fossil fuels based energy sources has harshly threatened human life through its drastic consequences, such as global

K. Chandrasekhar (✉)
School of Applied Bioscience, Agriculture Department, Kyungpook National University, Sangju, South Korea

Bio-Engineering and Environmental Science (BEES), CSIR-IICT, Hyderabad, India
e-mail: chanduibt@gmail.com; chanduibt@knu.ac.kr

A. Kadier
Department of Chemical and Process Engineering, Faculty of Engineering & Built Environment, National University of Malaysia (UKM), 43600 UKM Bangi, Selangor, Malaysia

G. Kumar
Sustainable Management of Natural Resources and Environment Research Group, Faculty of Environmental and Labour Safety, Ton Duc Thang University, Ho Chi Minh City, Vietnam

Center for Materials Cycles and Waste Management Research, National Institute for Environmental Studies, Tsukuba, Japan

R.A. Nastro
Department of Engineering, University of Naples "Parthenope", Centro Direzionale – Isola C4, 80143 Naples, Italy

V. Jeevitha
Department of Environmental Science and Technology, Jawaharlal Nehru Technological University, Hyderabad, Telangana, India

D. Das (ed.), *Microbial Fuel Cell*,
https://doi.org/10.1007/978-3-319-66793-5_25

warming and environmental pollution. As a result, in the present global energy scenario, searching for other energy resources is necessary (Kadier et al. 2015; Chandrasekhar et al. 2015a; Venkata Mohan and Pandey 2013). Last few years, several researchers around the world have made remarkable efforts to find a solution for this energy crisis (Venkata Mohan et al. 2013). In this scenario, fuel cells are a unique addition to the list of alternative energy sources having a negligible CO_2 emission. Production of electricity using microorganisms was firstly reported early in the last century (Potter 1911). Microbial fuel cell (MFC) have been considered as a bio-based reactor that modifies the chemical energy of substrate into electrical energy through biocatalytic action of exoelectrogenic microorganisms under anaerobic circumstances over sequences of metabolic reactions (Du et al. 2007; Kondaveeti and Min 2015; Chandrasekhar et al. 2015b). MFC technology signifies an innovative approach of using microorganisms for bioelectricity generation by the oxidation of organic substrate varied from the synthetic substrate such as acetate, glucose to a complex mixture of the organic substrate including food, dairy, distillery, animal and domestic wastewater (Chandrasekhar and Ahn 2017). In recent years, MFC technology has been developing as one of the popular wastewater treatment based technology to deliver clean water and green energy (Venkata Mohan and Chandrasekhar 2011a, b; Pant et al. 2012; Pandit et al. 2012a, b; Chandrasekhar et al. 2015b). These MFCs overtook other conventional technologies such as an aerated lagoon and anaerobic digester (Logan 2008). Unlike traditional fuel cells, one step conversion of carbon-rich organic waste as a potential substrate to generate bioelectricity in MFCs ensures better conversion capability. MFCs can evade extra gas treatment process due to its CO_2 rich off-gas. Moreover, single chambered or open-air cathode MFCs do not need any external energy input. Hence, it can be advantageous for extensive application in locations lacking electrical amenities (Stams et al. 2006). This chapter stretches an account of the basic principles involved in the working of MFCs and the key applications, challenges and future scope of MFC technology, as it stands today.

25.2 Metabolic Reactions Intricate in Bioelectricity Generation from Exoelectrogens

In the MFC reactor, microbes derive energy from the substrate oxidation for their growth and metabolic activities (Chandrasekhar and Venkata Mohan 2012). However, some amount of the energy generated is reserved by the microorganisms for their growth and metabolic requirements, where rest of the energy possibly will be employed for bioelectricity generation. Therefore, to generate bioelectricity during MFC operation, it is essential for the microbes to compensate for the sum of energy spent for its liveliness and that being transformed into electrochemical energy (Chandrasekhar and Venkata Mohan 2014a, b). Here two major pathways are involved in microorganisms for energy transformation i.e., oxidative pathway (respiration) and fermentation pathway. During the respiration, Gibbs free energy is used by microorganisms for their oxidative pathway mechanism. In this process, electrons flow from one place to another in a respiratory chain and as a final point departures bacterial cell through a series of electron acceptors located in the membrane. In the presence of O_2, substrates get oxidized through four individual steps: Glycolysis, TCA Cycle, Electron transport chain, and Oxidative phosphorylation. In the beginning, during glycolysis pathway, glucose atom (6C) is transformed into two pyruvate molecules (3C) in ten consecutive steps as revealed in Eq. 25.1. The pyruvate further involves in three-phase process and is finally transformed to CO_2 and ATP.

$$\text{Glucose} + 2\text{NAD} + 2\text{ADP} + 2\text{Pi} \rightarrow 2\ \text{NADH} + 2\text{ATP} + 2\text{Pyruvate} + 2\text{H}^+ + 2\text{H}_2\text{O} \tag{25.1}$$

Although aerobic respiration (presence of O_2) is the key practice of bioenergy production in microorganisms, few microbes possibly will produce bioenergy in the absence of O_2 (under anaerobic conditions). In the absence of O_2, glucose molecule (6C) transformed to pyruvate (3C) as in the case of aerobic circumstances, nonetheless, in later stage in its place of the TCA cycle, diverse catabolic pathway will follow precisely to the category of organism viz. eukaryote/prokaryote. For example, few microbes follow alcohol fermentation pathways, while few microorganisms undergo the lactic acid fermentation pathways (Chandrasekhar et al. 2015b). In this fermentation process, substrates get oxidized in the absence of O_2. Hence, inside the bacterial cell, substrates partially get oxidized. The electrons and protons which are generated during this oxidation process are transported out of the bacterial cell with the help of coenzymes. The complete reaction mechanism of MFC is depicted in Eq. 25.2. However, the total mechanism could be separated into half-cell reactions as revealed in Eqs. 25.3 and 25.4. In MFC, the anode chamber can be considered as a bio-factory where reducing equivalents (protons (H^+) and electrons (e^-)) generate through a sequence of reduction and oxidation reactions through substrate oxidation under strict anaerobic circumstances (Kondaveeti et al. 2014). As depicted in Eq. 25.2, e^- released during redox reactions in the anode compartment will move to the anode electrode, while the H^+ will flow to the cathode chamber. Finally, when circuit is closed, the e^- deposited on the anode surface will flow through the circuit to the cathode as shown in Fig. 25.1 and subsequently get reduced with terminal electron acceptor (O_2) in the cathode chamber (this process was almost similar in both single and double chamber MFCs) (Chandrasekhar and Venkata Mohan 2014a, b; Pandit et al. 2014a, b). Overall, substrate oxidation and subsequent redox reactions are responsible for electrical energy generation in MFC.

$$\text{Glucose} + 6\text{O}_2 \rightarrow 6\text{CO}_2 + 6\text{H}_2\text{O} \tag{25.2}$$

$$\text{Glucose} + 6\text{H}_2\text{O} \rightarrow 6\text{CO}_2 + 24\text{H}^+ + 24\text{e}^- \tag{25.3}$$

$$\text{O}_2 + 24\text{H}^+ + 24\text{e}^- \rightarrow 12\text{H}_2\text{O} \tag{25.4}$$

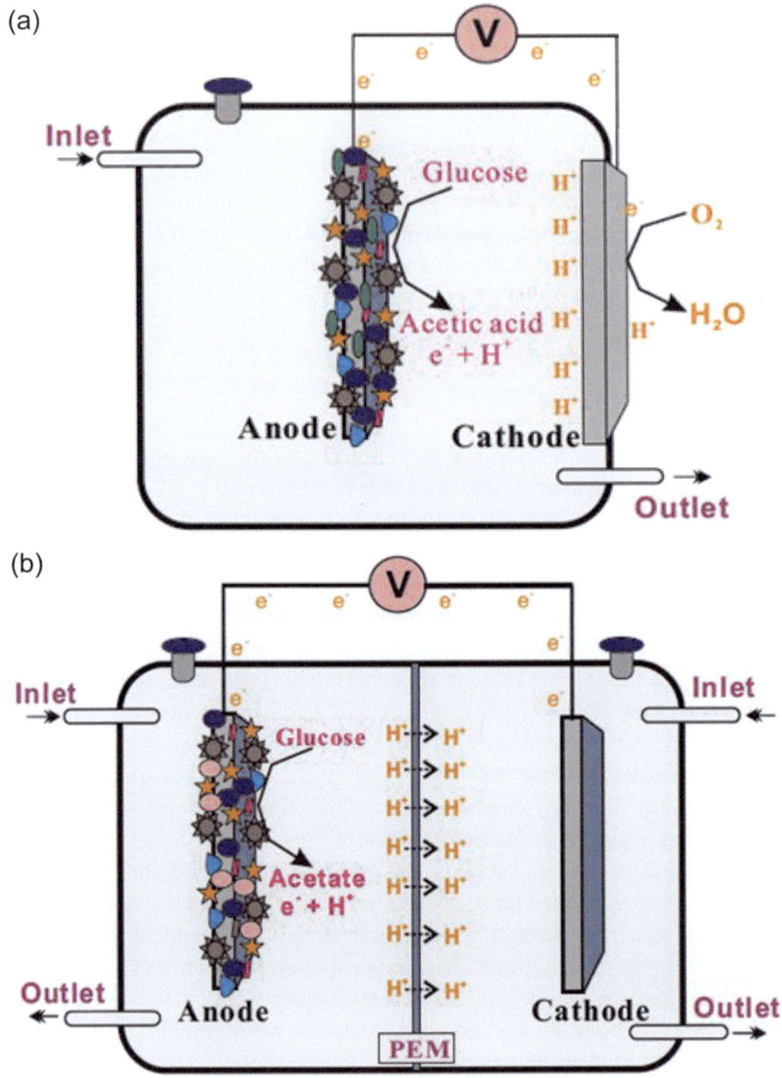

Fig. 25.1 Schematic illustration of (**a**) single chamber microbial fuel cell; and (**b**) double chamber microbial fuel cell

In MFC reactor, the key e^- acceptor which is involved in accepting the e^- generated by exo-electrogenic microorganisms is known as anode. The exoelectrogenic microbial biofilm formation on the anode surface is supposed to be motivated by the utilization of higher quantities of energy by the electrode material in comparison with other e^- acceptors (namely Fe^{3+} or Mn^{4+}oxides).

25.3 MFC Applications

MFC technology showed promise to operate it for bioelectricity generation from renewable biomass and simultaneous wastewater treatment. MFC has the capability to treat different types of waste including agricultural, forestry, industrial and municipal waste (Kiran Kumar et al. 2012; Chandrasekhar et al. 2015b). The current generation has been increased with stackable MFCs. Research has been carried out worldwide to increase the power density level of MFCs and make its operation economical to bring it to large scale (Erable et al. 2010). Along with wastewater treatment, MFC can be used for various applications such as BOD biosensors, bacterial enumeration etc. MFC enriched with electrochemically active bacteria has been used as biochemical oxygen demand (BOD) and toxicity detection biosensor. The presence of toxic substances such as a metal like cadmium, lead, arsenic, chromium (VI), mercury and compound of cyanide, organophosphorus and surfactant induces a change in electric current signal and thereby facilitate in detecting toxicity level in the water samples. At low concentration levels, current generation was found proportional to the concentration of toxic and biodegradable waste substances. The presence of toxic substances in aqueous system can be clubbed with BOD measurement to monitor the quality of wastewater (Mook et al. 2013). Recently sediment MFCs has shown promise for constructed wetland management. Current produced during sediment MFCs operation can be successfully stored in capacitors, subsequently utilize it to drive remote sensors through a power management system. SMFC was built to power to the underwater monitoring devices. MFCs was also found useful to provide long-term, stable power to low power biomedical devices implanted in the human body (Babauta et al. 2012). These *Saccharomyces cerevisiae* bio-catalyzed miniatured MFCs are capable of utilizing glucose in the blood stream to produce electricity. Food spoilage can be traced using MFC. The current generation was found to increase with increasing level of contamination. Henceforth, this kind of technology can be helpful for rapid detection and enumeration of microorganisms in food products (Chandrasekhar et al. 2015b).

25.4 Factors Governing MFC Performance

Operating MFC under optimized anodic condition will allow anodic biocatalyst to grow properly and also to form an electrogenic biofilm on electrode surface during startup which improves the subsequent fuel cell performance in long-term operation. Numerous process parameters such as system pH, alkalinity, substrate nature, substrate concentration and OLR affect the activity of these electrochemically active microorganisms (Gil et al. 2003). The factors influencing the performance of MFC are discussed below.

25.4.1 Biocatalyst

Electrochemically active microorganisms as a biocatalyst play essential role in MFC for power generation. MFCs have been operated with single (pure) and group (mixed) of microorganisms. In the case of pure culture, these are electrochemically active/exoelectrogenic bacteria, which directly transport electrons outside the cell membrane (Pandit et al. 2015). A pure culture such as *Shewanella* sp., *Geobacter* sp., or *Rhodoferax* sp. has been extensively studied in both single and double chamber MFC reactors. These are a group of electrochemically active biocatalyst present in the mixed culture. In general, soil, fresh as well as marine sediments, domestic wastewater, and activated sludge have been widely employed as biocatalyst in MFC, as they are rich in mixed (group of) microorganisms. In MFC, pure exoelectrogenic bacterial culture typically grows very slowly due to strict anaerobic conditions in the anode compartment. However, possibilities of contamination are very high with pure culture as biocatalyst in MFC. Overall performance in terms of the power output of some pure culture is moderately low compared to a mixed culture. On the other hand, mixed culture inoculum takes longer startup time to achieve steady current generation as compared to pure culture.

25.4.2 Substrate

The substrate is one of the factors which influence the performance of electrochemically active bacterial population in the anode compartment. It also determines the existence of dominant bacterial community when mixed culture is used as inoculum. A wide variety of substrates has been used in MFCs from simple carbon source to complex carbon, nitrogen-rich biodegradable wastewater (Chandrasekhar and Venkata Mohan 2012). The pure substrate as acetate, butyrate and glucose is used widely to maintain the homogeneity. Nitrogen-rich substrate like cysteine and proteins is consumed by exoelectrogenic bacteria in the anode compartment for power generation. Different types of wastewater are exploited to treat and concurrently electricity generation in MFC such as domestic wastewater, food waste, dairy waste, real field distillery wastewater, paper industry wastewater, food processing wastewater, and dye wastewater (Pant et al. 2010).

25.4.3 Substrate/COD Concentration

Substrate concentration or chemical oxygen demand (COD) considerably influence the performance of fuel cell operated both in batch feed and continuous mode of operation (Chandrasekhar and Venkata Mohan 2012). In MFC, electricity generation is directly proportional to the substrate concentration which can be measured in terms of COD. However, increasing substrate concentration will increase power generation till it causes feedback inhibition. Hence, it is very essential to determine the optimum substrate concentration to achieve maximum power generation.

25.4.4 Feed pH

Apart from a biotic agent, several abiotic factors also influence the MFC performance. The substrate pH plays a vital role in bioreactor performance (Venkateswar Reddy et al. 2011b). Similarly, the electrolyte pH plays a vital key role in MFC's power output. In MFC operation, power output drastically reduces at the acidic pH range below 6 (Gil et al. 2003). It means, low pH conditions showed an adverse effect on electrochemically active bacterial population, which in turn leads to a drastic drop in power production. It is clear that pH will strongly influence the performance of MFC both in batch feed and continuous mode of operation. An ideal pH range for preferred fuel cell configuration was stated to be in the middle of 7–8 (He et al. 2006). On the other hand, there is a contradiction in the existing literature on the ideal pH conditions for MFC operation possibly due to synergistic or antagonistic interaction influence among these parameters (Jadhav and Ghangrekar 2009).

25.5 Bottlenecks of MFC

25.5.1 Polarization Losses

There are quite a few technical hitches which impedes the function of MFC commonly named as overpotential (Chandrasekhar and Venkata Mohan 2012). Theoretically, MFC can reach an uppermost cell potential of 1.1 V in open circuit circumstances. Nevertheless, practically it experiences various forms of technical hitches and may possibly attain merely 0.8 V (Liu et al. 2004) and about 0.62 V (Rabaey et al. 2005b) in open circuit circumstances and during current generation respectively. So, the difference between measured and theoretical cell voltage together signifies the overpotentials of the electrodes (Chandrasekhar and Venkata Mohan 2012). Nonetheless, during the MFC operation, the flow of e^- from the biocatalyst to the solid electrode material is delayed primarily as a result of the charge transfer resistances commonly named as overpotentials. These overpotentials decrease the potential reached from the fuel cell operation and subsequently drop the process efficacy. These losses can be majorly categorized as activation losses (AL), concentration polarization/losses (CP), and ohmic losses (OL) as shown in Fig. 25.2a (Larminie and Dicks 2000; Chandrasekhar and Venkata Mohan 2012).

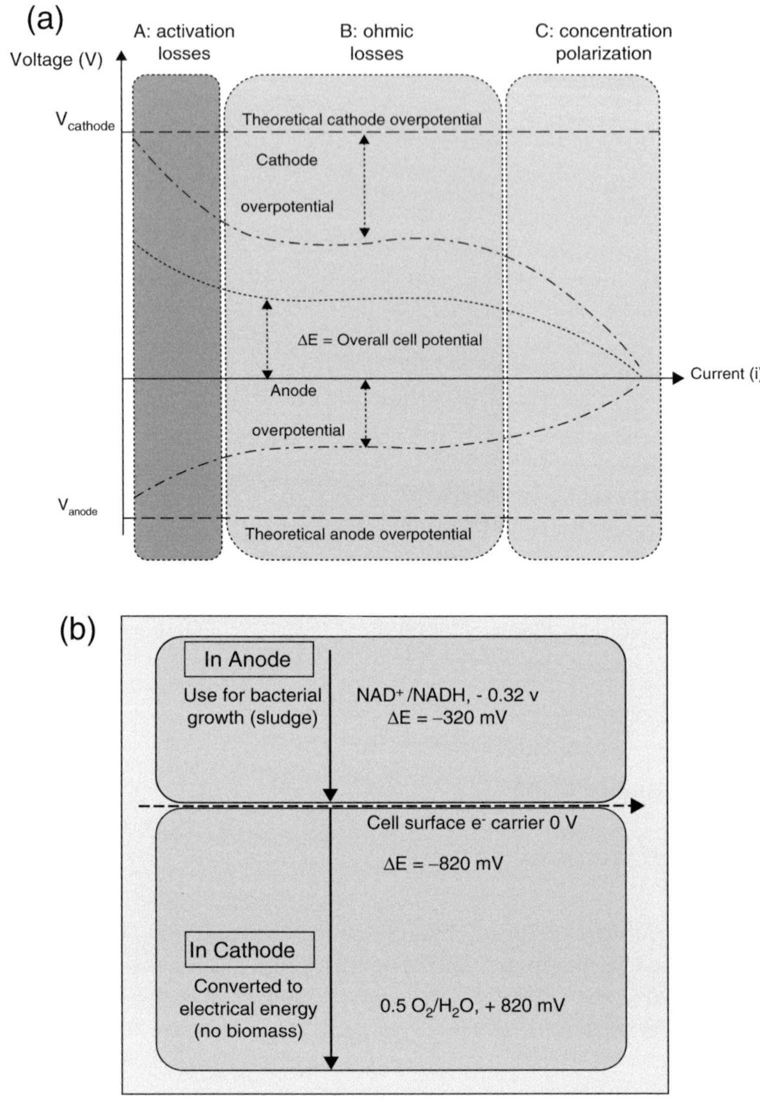

Fig. 25.2 Schematic representation of (**a**) the polarization curve; and (**b**) energy conversion in microbial fuel cell

25.5.2 Activation Losses (AL)

During MFC operation several electrochemical reactions take place at the electrode surface. These electrochemical redox reactions do need certain activation energy. This activation energy essentially is required for electron transfer to either electrolyte/feed oxidation at the anode or to reduce O_2 as terminal e^- acceptor at the cathode (Chandrasekhar and Venkata Mohan 2012). The AL could be diminished by rising the process temperature up to some extents. Likewise, it could be diminished by adding a catalyst to the electrode material which improves electrode catalysis (Schroder et al. 2003; Chandrasekhar and Venkata Mohan 2012). On the other hand, low AL can be achieved by the formation of biofilm on the electrode surface and also by increasing the electrode surface area.

25.5.3 Concentration Polarization (CP)

CP happens when the substrate is being oxidized more rapidly by the side of the electrode than they might be transferred toward the electrode (anode) surface. Generally, this may possibly happen as a result of higher oxidative strength of the anode in the MFCs. However, this phenomenon would be problematic, where flow is extremely slowed down, for instance, a dense non-conductive anodophilic bacterial biofilm, hydrodynamics and geometrical features of MFC reactor design (Chandrasekhar and Venkata Mohan 2012).

25.5.4 Ohmic Losses (OL)

OL in MFC occurred mainly due to resistance to the transfer of e^- through electrode material, interconnections, and resistance to the transfer of ions through the membranes, and electrolyte. Nonetheless, the resistance over the MFC can rise promptly due to improper connections or reduced conductivity of the electrolyte. The arrangement of the anode material must maintain the free stream of the electrolyte, bacterial biofilm formation, good conductivity, and enough turbulence aimed at sufficient H^+ flow in the direction of the separator and cathode electrode materials. Usage of active exoelectrogenic microorganisms or materials as a biocatalyst/catalyst are possible ways to decrease cathode resistance (CR). Increment in the specific surface area and the mass transport rate at the cathode could also decrease the CR (Chandrasekhar and Venkata Mohan 2012). Moreover, in recent years, researchers revealed that the increment in the cathode size/surface area over anode size/surface area will considerably reduce the CR (Fan et al. 2008).

The choice of a separator (e.g. proton exchange membrane) among anode and cathode compartments signifies a choice among two different benefits: one is for great discernment for H^+ and the second one is great stability in microbial colloidal and high nutrient circumstances. Nafion 117 is a proton exchange membrane (PEM) which has been extensively employed for MFCs and has the huge benefit of being very selective for H^+ (Bond and Lovley 2003). The constraint for using separator lies in the improvement of a pH gradient among electrodes (cathode and anode) that implicates additional energy loss. Otherwise, maintaining alkaline pH in the cathode compartment allows a poorer CR, wherein few approaches were examined to lower this pH ramp that is CO_2 supplementation. The use of a more general cation exchange membrane is another option (second approach). Generally, this kind of separator has a higher resistance and low selectivity but usually exhibit higher stability, which has been stated to execute passable for more than 90 days (Rabaey et al. 2005a). Several investigations reported that MFC challenges affected by separator, pH ramp inside the anode compartment, O_2 diffusion, H_2 loss, and concentration polarization crossways the separator (Torres et al. 2008). Conversely, losses happen in the cathode chamber as a result of charge transfer resistance. To reduce the AL, biocatalyst must be supplemented to the electrode or an appropriate mediator is anticipated to deliver e^- from the cathode surface to O_2 which acts as terminal e^- acceptor. Commonly, platinum is used as a potential catalyst in the electrode material entailing a considerable cost (Schroder et al. 2003). Nevertheless, usage of activated carbon (AC) open air cathodes in MFCs are economically best choice as compared to platinum catalyzed electrodes (Elmekawy et al. 2013).

25.5.5 *Microbial Interaction with the Electrode Surface*

The microbial interaction with the electrode surface plays a key role in MFC performance. Electrochemical behaviour of MFC is mainly reliant on how well the exoelectrogenic microbes interact with the electrode/anode surface. The bacterial biofilm adhered properly onto the anode surface is a key prerequisite here. This bacterial adhesion on the electrode can be assumed by the concept of surface charges. The greater parts of the microbes in nature are negatively charged and, therefore, move towards positively charged surfaces. Consequently, numerous surface alteration methods have been used to assist this charge attraction procedure. For instance, researchers made an attempt to treat the electrode surface with NH_3 successful, which enables the negatively charged microorganisms to freely attach on the electrode (anode) surface which is positively charged. Moreover, the power output of the MFC by using NH_3 pretreated electrode material documented superior performance over untreated electrode material. Nonetheless, requirements of high temperature for this NH_3 pretreatment have made this process more expensive, hence it became economically less feasible technology.

25.5.6 *Choice of Anode Biocatalyst*

MFCs can also be functioned by means of pure cultures. Nonetheless, mixed microbial populations are very much appropriate when we fed MFCs with composite substrates such as carbon-rich wastewater (Chandrasekhar and Ahn 2017). But in the case of MFCs operated with pure culture as anodic biocatalyst usually metabolize relatively a narrow range of substrates. 16S rDNA studies exposed that the microbial diversity in the anode compartment was altered from that of the seed culture (Jong et al. 2006). In the course of this culture enrichment practice, an electrochemically active bacterial biofilm progresses on the anode as in MFCs that operated with single cultures. Further, e^- flow from biofilm in the direction of the anode surface is enabled from side to side contact between electrode and microbe as in the case of insoluble Fe^{3+} reduction used as an e^- acceptor or from end to end special pili branded as 'nanowires' (Xiong et al. 2006). It is value saying that the nature and type of the electrode material impacts the formation of biofilm on the surface of the electrode, and that the diversity of anodophilic microorganisms varies among parts of the anode near and away from the membrane (Lowy et al. 2006). Nonetheless, mixed exoelectrogenic bacterial culture for MFCs may possibly be enriched at psychrophilic and thermophilic circumstances (Jong et al. 2006).

25.5.7 *Proton (H^+) Mass Transfer*

In MFCs, H^+ mass transfer limitations hamper the overall process efficiency. The H $^+$ mass transfer would not be restricted as the similar cation exchange membranes (CEM) are employed in MFCs as in the case of chemical fuel cells (CFCs). Mostly, as in comparison with CFCs, MFCs generate much lesser current, because the number of H^+ intake per minute at the electrode (cathode) surface was higher than the number of H^+ transfer per minute through the membrane. The characteristic variances among CFCs and MFCs might be the probable reason for this phenomenon. In MFCs, to maintain the microbial growth, a wide variety of inorganic substances are needed, whereas in the case of CFCs, only pure fuel is sufficient. The chemical substances supplemented to the anode compartment produce cations (for example K^+, Na^+, NH_4^+, Ca^{2+} and, Mg^{2+}), which are typically several times greater than that of H^+ (at neutral pH), which can slow down H^+ transfer through the CEM. Usage of an aqueous solution in the anode/cathode compartment as an electrolyte can also be considered as one of the major limitation aimed at H^+ mass transfer in MFCs.

In recent years, several researchers investigated the consequence of cation transportation through a PEM (Nafion 117) on cathode electrolyte pH and also overall accomplishment of MFC. In the case of "H" type double chambered MFC, the sum of cations other than H^+ transferred from the anode chamber to the cathode chamber was equal to the sum of e^- transported through the circuit (Rozendal et al. 2006). In the case of open-air cathode MFCs, cation transfer through membrane reduce the cathodic redox reactions by the establishment of thick layer of carbonate salts on the surface of the electrode (Pham et al. 2005). In the case of single chamber MFCs, a membraneless operation documented higher efficiency when both the anode and cathode electrodes were placed very closely to evade short-circuiting among two electrodes (Venkata Mohan and Chandrasekhar 2011a; Ghangrekar and Shinde 2007). In membraneless MFCs, when the cathode chamber was fed with a salt or acid solution, the current generation was greater than before which indicated that H^+ diffusion in an aqueous phase is a slow process and is improved in the electrolyte by providing more salt (Jang et al. 2004).

25.5.8 O_2 *Reduction by the Cathode*

In the cathode compartment of the MFCs, O_2 in employment as the terminal e^- acceptor is usual practice. However, the deprived O_2 reduction kinetics has been well recognized as the major impeding step in MFCs performance (Zhao et al. 2006). Non-catalyzed graphite materials are widely used in MFCs as electrode due to its low cost and non-corrosive nature. However, it is a deprived catalyst for O_2 reduction reactions. To enhance the catalytic property, replacement of graphite electrode with platinum electrode material or platinum catalyst coating on the surface of graphite is suggested to achieve 3–4 times higher process efficiency over non-catalyzed graphite electrode (Pham et al. 2004). Nevertheless, application of the platinum as potential cathode electrode material may increase the overall cost of the process, particularly when MFC is operated with wastewater for waste remediation with simultaneous energy generation. At ambient temperature, while H_2O is saturated with atmospheric air, O_2 is barely dissolving in H_2O (8 mg L^{-1}). In double chamber MFCs operated with H_2O as catholyte, DO levels will drop drastically to the level where it hampers the O_2-utilizing reactions while the utilization of O_2 is greater than its solubilization.

25.5.9 *Electron Acceptors Other Than* O_2

The permanganate and hexacyanoferrate are widely used as terminal e^- acceptors other than O_2 in MFC operations (You et al. 2006). Employing permanganate as the terminal e^- acceptor in the cathode compartment documented 4.5 and 11.3 times higher power density (115.6 mW m^{-2}) than that generated by employing hexacyanoferrate (25.6 mW m^{-2}) and O_2 (10.2 mW m^{-2}), respectively. Moreover, the application of permanganate as terminal e^- acceptor in the cathode compartment may possibly document greater open circuit potential (OCP). However, the usage of these oxidizing agents as e^- acceptors in the cathode reaction cannot be recommended due to secondary reactions and ecological side effects when they release into the environment. The role of terminal e^- acceptors in the cathodic reactions can be considered as serious challenge in MFC performance.

25.6 MFC as a Wastewater Treatment System

MFCs, also known as hybrid bio-electrochemical treatment (BET) system, was recently familiar for wastewater treatment apart from power generation (Chandrasekhar and Venkata Mohan 2012). In the presence of O_2, microbes exploit all the chemical energy present in the substrate/wastewater, although only a minor quantity of the chemical energy is accessible to anodic microbial population in MFCs for their growth and metabolic activities, as a huge quantity is altered to electrical energy. Exoelectrogenic microorganisms have a redox potential of about 0 V against SHE as specified above (Park et al. 2001); exoelectrogenic biocatalyst in MFC utilize free energy accessible from the substrate/wastewater oxidation to reduce the anode through the membrane surface (Fig. 25.2b). In this direction, minor quantity of the energy derived from the substrate oxidation is utilized by the microorganisms, and the left over energy is altered to electrical energy means that the yield in MFCs is almost 1/5th that of an aerobic microorganisms. This make evident that MFCs can significantly decrease sludge disposal expenses in wastewater treatment process.

25.7 Future Scope

There is a necessity to talk about low power densities in MFC operation by existing optimization of the design to reduce the losses affected by activation, ohmic, and concentration overpotentials (Nastro et al. 2015). Further, losses caused by unnecessary reactions, for example, the direct oxidation of fuel by O_2 diffusion into the anodic chamber or microbial metabolic reactions, which do not benefit the process, must also be targeted. On the other hand, increasing the system volumetric capacity must be attained short of internal energy losses. In this direction, stacking MFCs is a common choice to evade catastrophic losses. Furthermore, tubular and other stacked methodologies continue to be explored. Also, the action is essential to raise the exoelectrogenic microbial population density, which seems to be restricted for causes excluding the accessibility of attachment positions on the electrode surface, and possibly will control by bioaugmentation and possible field effects existent in the electrode by the advantage of its morphology and conductivity. Ongoing efforts are being made to establish better e^- transfer mechanisms among the electrode and the biocatalyst by the modification of electrode surfaces and also by applying active catalyst coating on the electrode surface.

25.8 Conclusion

MFCs have great future as a bio-based reactor that modifies the chemical energy of substrate into electrical energy through a series of bio-catalytic reactions under anaerobic circumstances principally for stationary power generation usages. MFC technology signifies an innovative approach of using microorganisms for bioelectricity generation by the oxidation of organic substrate varied from the synthetic substrate such as acetate or glucose to a complex mixture of the organic substrate including food, dairy, distillery, animal and domestic waste water. Operating MFC under optimized anodic condition will allow anodic biocatalyst to grow properly and also to form an electrogenic biofilm on electrode surface during startup which improves the subsequent fuel cell performance in long-term operation. Moreover, single chambered or open-air cathode MFCs do not need any external energy input. Hence, it can be advantageous for extensive application in locations lacking electrical amenities. Nevertheless, bottlenecks linked with active mass and charge transfer together with the difficulties associated with making the microorganisms act together with the permanent e^- acceptors have hampered this MFC technology approaching to technological realization as yet. Recent developments in material science together with nanotechnology may provide unique tools to efficiently produce, transportation and utilization of the electrical energy harvested from MFCs for beneficial applications in the upcoming future.

References

Babauta, J., Renslow, R., Lewandowski, Z., & Beyenal, H. (2012). Electrochemically active biofilms: Facts and fiction. A review. *Biofouling, 28*, 789–812.

Bond, D. R., & Lovley, D. R. (2003). Electricity production by Geobacter sulfurreducens attached to electrodes. *Applied and Environmental Microbiology, 69*, 1548–1555.

Chandrasekhar, K., & Ahn, Y.-H. (2017). Effectiveness of piggery waste treatment using microbial fuel cells coupled with elutriated-phased acid fermentation. *Bioresource Technology, 244*, 650–657.

Chandrasekhar, K., & Venkata Mohan, S. (2012). Bio-electrochemical remediation of real field petroleum sludge as an electron donor with simultaneous power generation facilitates biotransformation of PAH: Effect of substrate concentration. *Bioresource Technology, 110*, 517–525.

Chandrasekhar, K., & Venkata Mohan, S. (2014a). Bio-electrohydrolysis as a pretreatment strategy to catabolize complex food waste in closed circuitry: Function of electron flux to enhance acidogenic biohydrogen production. *International Journal of Hydrogen Energy, 39*, 11411–11422.

Chandrasekhar, K., & Venkata Mohan, S. (2014b). Induced catabolic bio-electrohydrolysis of complex food waste by regulating external resistance for enhancing acidogenic biohydrogen production. *Bioresource Technology, 165*, 372–382.

Chandrasekhar, K., Lee, Y. J., & Lee, D. W. (2015a). Biohydrogen production: Strategies to improve process efficiency through microbial routes. *International Journal of Molecular Sciences, 16*, 8266–8293.

Chandrasekhar, K., Amulya, K., & Venkata Mohan, S. (2015b). Solid phase bio-electrofermentation of food waste to harvest value-added products associated with waste remediation. *Waste Management, 45*, 57–65.

Du, Z., Li, H., & Gu, T. (2007). A state of the art review on microbial fuel cells: A promising technology for wastewater treatment and bioenergy. *Biotechnology Advances, 25*, 464–482.

Elmekawy, A., Hegab, H. M., Dominguez-Benetton, X., & Pant, D. (2013). Internal resistance of microfluidic microbial fuel cell: Challenges and potential opportunities. *Bioresource Technology, 142*, 672–682.

Erable, B., Duţeanu, N. M., Ghangrekar, M. M., Dumas, C., & Scott, K. (2010). Application of electro-active biofilms. *Biofouling, 26*, 57–71.

Fan, Y., Sharbrough, E., & Liu, H. (2008). Quantification of the internal resistance distribution of microbial fuel cells. *Environmental Science & Technology, 42*, 8101–8107.

Ghangrekar, M. M., & Shinde, V. B. (2007). Performance of membrane-less microbial fuel cell treating wastewater and effect of electrode distance and area on electricity production. *Bioresource Technology, 98*, 2879–2885.

Gil, G.-C., Chang, I.-S., Kim, B. H., Kim, M., Jang, J.-K., Park, H. S., & Kim, H. J. (2003). Operational parameters affecting the performance of a mediator-less microbial fuel cell. *Biosensors & Bioelectronics, 18*, 327–334.

He, Z., Wagner, N., Minteer, S. D., & Angenent, L. T. (2006). An upflow microbial fuel cell with an interior cathode: Assessment of the internal resistance by impedance spectroscopy. *Environmental Science & Technology, 40*, 5212–5217.

Holmes, D. E., Nicoll, J. S., Bond, D. R., & Lovley, D. R. (2004). Potential role of a novel psychrotolerant member of the family Geobacteraceae, Geopsychrobacter electrodiphilus gen. nov., sp. nov., in electricity production by a marine sediment fuel cell. *Applied and Environmental Microbiology, 70*, 6023–6030.

Jadhav, G. S., & Ghangrekar, M. M. (2009). Performance of microbial fuel cell subjected to variation in pH, temperature, external load and substrate concentration. *Bioresource Technology, 100*, 717–723.

Jang, J. K., Pham, T. H., Chang, I. S., Kang, K. H., Moon, H., Cho, K. S., & Kim, B. H. (2004). Construction and operation of a novel mediator- and membrane-less microbial fuel cell. *Process Biochemistry, 39*, 1007–1012.

Jong, B. C., Kim, B. H., Chang, I. S., Liew, P. W. Y., Choo, Y. F., & Kang, G. S. (2006). Enrichment, performance, and microbial diversity of a thermophilic mediatorless microbial fuel cell. *Environmental Science & Technology, 40*, 6449–6454.

Kadier, A., Simayi, Y., Chandrasekhar, K., Ismail, M., & Kalil, M. S. (2015). Hydrogen gas production with an electroformed Ni mesh cathode catalysts in a single-chamber microbial electrolysis cell (MEC). *International Journal of Hydrogen Energy, 40*, 14095–14103.

Kadier, A., Simayi, Y., Abdeshahian, P., Azman, N. F., Chandrasekhar, K., & Kalil, M. S. (2016a). A comprehensive review of microbial electrolysis cells (MEC) reactor designs and configurations for sustainable hydrogen gas production. *Alexandria Engineering Journal, 55*, 427–443.

Kadier, A., Kalil, M. S., Abdeshahian, P., Chandrasekharc, K., Mohamed, A., Azman, N. F., Logroño, W., Simayi, Y., & Hamid, A. A. (2016b). Recent advances and emerging challenges in microbial electrolysis cells (MECs) for microbial production of hydrogen and value-added chemicals. *Renewable and Sustainable Energy Reviews, 61*, 501–525.

Kim, B. H., Park, H. S., Kim, H. J., Kim, G. T., Chang, I. S., Lee, J., & Phung, N. T. (2004). Enrichment of microbial community generating electricity using a fuel-cell-type electrochemical cell. *Applied Microbiology and Biotechnology, 63*, 672–681.

Kim, G. T., Webster, G., Wimpenny, J. W., Kim, B. H., Kim, H. J., & Weightman, A. J. (2006). Bacterial community structure, compartmentalization and activity in a microbial fuel cell. *Journal of Applied Microbiology, 101*, 698–710.

Kiran Kumar, A., Venkateswar Reddy, M., Chandrasekhar, K., Srikanth, S., & Venkata Mohan, S. (2012). Endocrine disruptive estrogens role in electron transfer: Bio-electrochemical remediation with microbial mediated electrogenesis. *Bioresource Technology, 104*, 547–556.

Kondaveeti, S., & Min, B. (2015). Bioelectrochemical reduction of volatile fatty acids in anaerobic digestion effluent for the production of biofuels. *Water Research, 87*, 137–144.

Kondaveeti, S., Lee, J., Kakarla, R., Kim, H. S., & Min, B. (2014). Low-cost separators for enhanced power production and field application of microbial fuel cells (MFCs). *Electrochimica Acta, 132*, 434–440.

Larminie, J., & Dicks, A. (2000). *Fuel cell systems explained* (p. 308). Chichester: Wiley.

Liu, H., Cheng, S., & Logan, B. E. (2004). Production of electricity from acetate or butyrate using a single-chamber microbial fuel cell. *Environmental Science & Technology, 39*(2), 658–662.

Logan, B. E. (2008). *Microbial fuel cells*. Hoboken: Wiley-Interscience.

Lowy, D. A., Tender, L. M., Zeikus, J. G., Park, D. H., & Lovley, D. R. (2006). Harvesting energy from the marine sediment–water interface II: Kinetic activity of anode materials. *Biosensors & Bioelectronics, 21*, 2058–2063.

Mook, W. T., Aroua, M. K. T., Chakrabarti, M. H., Noor, I. M., Irfan, M. F., & Low, C. T. J. (2013). A review on the effect of bio-electrodes on denitrification and organic matter removal processes in bio-electrochemical systems. *Journal of Industrial and Engineering Chemistry, 19*, 1–13.

Nastro, R., Falcucci, G., Toscanesi, M., Minutillo, M., Pasquale, V., Trifuoggi, M., Dumontet, S. & Jannelli, E. (2015). Performances and microbiology of a microbial fuel cell (mfc) fed with the organic fraction of municipal solid waste (ofmsw). In: *Proceedings of EFC15 conference.*

Pandit, S., Ghosh, S., Ghangrekar, M. M., & Das, D. (2012a). Performance of an anion exchange membrane in association with cathodic parameters in a dual chamber microbial fuel cell. *International Journal of Hydrogen Energy, 37*, 9383–9392.

Pandit, S., Nayak, B. K., & Das, D. (2012b). Microbial carbon capture cell using cyanobacteria for simultaneous power generation, carbon dioxide sequestration and wastewater treatment. *Bioresource Technology, 107*, 97–102.

Pandit, S., Khilari, S., Bera, K., Pradhan, D., & Das, D. (2014a). Application of PVA–PDDA polymer electrolyte composite anion exchange membrane separator for improved bioelectricity production in a single chambered microbial fuel cell. *Chemical Engineering Journal, 257*, 138–147.

Pandit, S., Khilari, S., Roy, S., Pradhan, D., & Das, D. (2014b). Improvement of power generation using Shewanella putrefaciens mediated bioanode in a single chambered microbial fuel cell: Effect of different anodic operating conditions. *Bioresource Technology, 166*, 451–457.

Pandit, S., Khilari, S., Roy, S., Ghangrekar, M. M., Pradhan, D., & Das, D. (2015). Reduction of start-up time through bioaugmentation process in microbial fuel cells using an isolate from dark fermentative spent media fed anode. *Water Science and Technology: a Journal of the International Association on Water Pollution Research, 72*, 106–115.

Pant, D., Van Bogaert, G., Diels, L., & Vanbroekhoven, K. (2010). A review of the substrates used in microbial fuel cells (MFCs) for sustainable energy production. *Bioresource Technology, 101*, 1533–1543.

Pant, D., Singh, A., Van Bogaert, G., Irving Olsen, S., Singh Nigam, P., Diels, L., & Vanbroekhoven, K. (2012). Bioelectrochemical systems (BES) for sustainable energy production and product recovery from organic wastes and industrial wastewaters. *RSC Advances, 2*, 1248–1263.

Park, H. S., Kim, B. H., Kim, H. S., Kim, H. J., Kim, G. T., Kim, M., Chang, I. S., Park, Y. K., & Chang, H. I. (2001). A novel electrochemically active and Fe (III)-reducing bacterium phylogenetically related to *Clostridium butyricum* isolated from a microbial fuel cell. *Anaerobe, 7*, 297–306.

Pham, T. H., Jang, J. K., Chang, I. S., & Kim, B. H. (2004). Improvement of cathode reaction of a mediator-less microbial fuel cell. *Journal of Microbiology and Biotechnology, 14*, 324–329.

Pham, T. H., Jang, J. K., Moon, H. S., Chang, I. S., & Kim, B. H. (2005). Improved performance of microbial fuel cell using membrane-electrode assembly. *Journal of Microbiology and Biotechnology, 15*, 438–441.

Potter, M. (1911). Electrical effects accompanying the decomposition of organic compounds. *Proceedings of the Royal Society of London. Series B, Containing Papers of a Biological Character (1905–1934), 84*(571), 260–276.

Rabaey, I., Ossieur, W., Verhaege, M., & Verstraete, W. (2005a). Continuous microbial fuel cells convert carbohydrates to electricity. *Water Science and Technology, 52*, 515–523.

Rabaey, K., Lissens, G., & Verstraete, W. (2005b). Microbial fuel cells: Performances and perspectives. In P. Lens, P. Westermann, M. Haberbauer, & A. Morino (Eds.), *Biofuels for fuel cells: Biomass fermentation towards usage in fuel*. London: IWA Publishing.

Rozendal, R. A., Hamelers, H. V. M., & Buisman, C. J. N. (2006). Effects of membrane cation transport on pH and microbial fuel cell performance. *Environmental Science & Technology, 40*, 5206–5211.

Schroder, U., Niessen, J., & Scholz, F. (2003). A generation of microbial fuel cells with current outputs boosted by more than one order of magnitude. *Angewandte Chemie (International Ed. in English), 42*, 2880–2883.

Sleutels, T. H., ter Heijne, A., Buisman, C. J., & Hamelers, H. V. (2012). Bioelectrochemical systems: An outlook for practical applications. *ChemSusChem, 5*, 1012–1019.

Stams, A. J. M., de Bok, F. A. M., Plugge, C. M., van Eekert, M. H. A., Dolfing, J., & Schraa, G. (2006). Exocellular electron transfer in anaerobic microbial communities. *Environmental Microbiology, 8*, 371–382.

Torres, C. I., Kato Marcus, A., & Rittmann, B. E. (2008). Proton transport inside the biofilm limits electrical current generation by anode-respiring bacteria. *Biotechnology and Bioengineering, 100*, 872–881.

Venkata Mohan, S., & Chandrasekhar, K. (2011a). Self-induced bio-potential and graphite electron accepting conditions enhances petroleum sludge degradation in bio-electrochemical system with simultaneous power generation. *Bioresource Technology, 102*, 9532–9541.

Venkata Mohan, S., & Chandrasekhar, K. (2011b). Solid phase microbial fuel cell (smfc) for harnessing bioelectricity from composite food waste fermentation: Influence of electrode assembly and buffering capacity. *Bioresource Technology, 102*, 7077–7085.

Venkata Mohan, S., & Pandey, A. (2013). Biohydrogen production: An introduction. In A. Pandey, J.-S. Chang, P. Hallenbeck, & C. Larroche (Eds.), *Biohydrogen* (pp. 1–24). Amsterdam: Elsevier. ISBN: 978-0-444-59555-3.

Venkata Mohan, S., Devi, M. P., Venkateswar Reddy, M., Chandrasekhar, K., Juwarkar, A., & Sarma, P. N. (2011). Bioremediation of petroleum sludge under anaerobic microenvironment: Influence of biostimulation and bioaugmentation. *Environmental Engineering and Management Journal, 10*, 1609–1616.

Venkata Mohan, S., Chandrasekhar, K., Chiranjeevi, P., & Babu, P. S. (2013). Biohydrogen production from wastewater. In A. Pandey, J.-S. Chang, P. Hallenbeck, & C. Larroche (Eds.), *Biohydrogen* (pp. 223–257). Amsterdam: Elsevier. ISBN: 978-0-444-59555-3.

Venkateswar Reddy, M., Chandrasekhar, K., & Venkata Mohan, S. V. (2011a). Influence of carbohydrates and proteins concentration on fermentative hydrogen production using canteen based waste under acidophilic microenvironment. *Journal of Biotechnology, 155*, 387–395.

Venkateswar Reddy, M., Devi, M. P., Chandrasekhar, K., Goud, R. K., & Venkata Mohan, S. V. (2011b). Aerobic remediation of petroleum sludge through soil supplementation: Microbial community analysis. *Journal of Hazardous Materials, 197*, 80–87.

Xiong, Y. J., Shi, L., Chen, B. W., Mayer, M. U., Lower, B. H., Londer, Y., Bose, S., Hochella, M. F., Fredrickson, J. K., & Squier, T. C. (2006). High-affinity binding and direct electron transfer to solid metals by the Shewanella oneidensis MR-1 outer membrane c-type cytochrome OmcA. *Journal of the American Chemical Society, 128*, 13978–13979.

You, S. J., Zhao, Q. L., Zhang, J. N., Jiang, J. Q., & Zhao, S. Q. (2006). A microbial fuel cell using permanganate as the cathodic electron acceptor. *Journal of Power Sources, 162*, 1409–1415.

Zhao, F., Harnisch, F., Schroder, U., Scholz, F., Bogdanoff, P., & Herrmann, I. (2006). Challenges and constraints of using oxygen cathodes in microbial fuel cells. *Environmental Science & Technology, 40*, 5193–5199.

Index

D. Das (ed.), *Microbial Fuel Cell*,
https://doi.org/10.1007/978-3-319-66793-5

MIX
Papier aus verantwortungsvollen Quellen
Paper from responsible sources
FSC® C105338

If you have any concerns about our products,
you can contact us on
ProductSafety@springernature.com

In case Publisher is established outside the EU,
the EU authorized representative is:
Springer Nature Customer Service Center GmbH
Europaplatz 3, 69115 Heidelberg, Germany

Printed by Libri Plureos GmbH
in Hamburg, Germany